Embedded DSP Processor Design

Application Specific Instruction Set Processors

The Morgan Kaufmann Series in Systems on Silicon

Series Editor: Wayne Wolf, Georgia Institute of Technology

Embedded DSP Processor Design

Application Specific Instruction Set Processors

Dake Liu

AMSTERDAM • BOSTON • HEIDELBERG • LONDON
NEW YORK • OXFORD • PARIS • SAN DIEGO
SAN FRANCISCO • SINGAPORE • SYDNEY • TOKYO

Morgan Kaufmann Publishers is an imprint of Elsevier

Morgan Kaufmann Publishers is an imprint of Elsevier.
30 Corporate Drive, Suite 400, Burlington, MA 01803, USA

This book is printed on acid-free paper.

Library of Congress Cataloging-in-Publication Data
Liu, Dake, 1957-
    Embedded DSP processor design: application specific instruction set processors / Dake Liu.
        p. cm. – (The Morgan Kaufmann series in systems on silicon)
    Includes index.
    ISBN 978-0-12-374123-3
1. Embedded computer systems. 2. Signal processing–Digital techniques. 3. Digital integrated circuits.
4. Application-specific integrated circuits. I. Title.
    TK7895.E42D35 2008
    621.39'16–dc22

                                                                        2008012910

ISBN: 978-0-12-374123-3

For information on all Morgan Kaufmann publications,
visit our website at *www.mkp.com* or *www.books.elsevier.com*

Printed and bound by CPI Group (UK) Ltd, Croydon, CR0 4YY
Transferred to Digital Print 2011

To Meiying and Angie

Contents

Preface

In the late 1990s, when I was preparing a course called "Design of Embedded DSP (Digital Signal Processing) Processors" at Linköping University, Sweden, I could not find a textbook describing the fundamentals of embedded processor design. It became my first and prime motivation for writing this book. During my work time in industry, I could not find any suitable and comprehensive reference book either, which led to my second motivation for writing such a book. It has been my belief that this book will be a valuable textbook or reference book for anyone interested in the design of embedded systems in all its aspects, from hardware design to firmware design. Although this book was written mainly for ASIP (application-specific instruction set processor) or ASIC (application-specific integrated circuit) designers, it will also benefit software programmers who want more hardware knowledge, such as DSP application engineers.

While reading this book, you will get opportunities to go through the design of a programmable device for a class of applications, step by step. The material in this book is suitable for teaching senior undergraduate students and graduate students in Electrical Engineering and Computer Engineering. This book can also be used as a reference book for engineers who are designing or want to design application-specific DSP processors, general processors, accelerators, peripheral modules, and even microcontrollers. Embedded system designers (e.g., DSP firmware designers) will also benefit from the knowledge of real-time system design elaborated in this book. Classical CPU designers will benefit from the exposed difference between CPU and ASIP. This book is also a fundamental reference book for researchers.

Fundamental DSP theory and basic digital logic design are addressed in this book as background knowledge. Very basic concepts and methods were used without redundant introduction. Readers without the fundamental knowledge should read relevant books about DSP theory, logic design, and computer architecture before reading this book.

DSP, as opposed to general-purpose computing systems, has been a major technology driven by embedded applications and the semiconductor technologies. This is evident from the growing market of DSP-based products. The increasing need for DSP and DSP processors can be found everywhere in today's society in areas such as multimedia, wireless communications, Internet terminals, car electronics, robot, healthcare, environment monitoring and control, education, scientific computing, industrial control, transportation, and defense.

DSP is used widely for various applications such as data enhancement, data compression, pattern recognition, simulation, emulation, and optimization. Signal recovery in advanced digital communications is a good example of data enhancement. Other data enhancement applications include error correction, echo cancellation, and noise suppression. Data compression is another important area in today's daily used facilities: for example, voice, music, image, video, and Internet data needs to be compressed to fit into a limited bandwidth for transmission and storage. If voice was not compressed by a DSP processor in a mobile phone, the cost of a mobile phone call would be 10 to 15 times higher. If video signals were not compressed, DVD players and digital video broadcasting would not be possible.

Pattern recognition techniques are used for voice recognition, language recognition, and image target recognition for healthcare, car driving, and defense. Physical simulation has been used for gaming, training (education), scientific computing, defense, and experiments that are expensive or even impossible to realize in the real world.

The global market share of DSP processors and microcontrollers is more than 95% of the total volume of processors sold in 2006. DSP processors for embedded applications have led to a major shift in the semiconductor industry. The sales of DSP processors have reached 20% of the global semiconductor market since 2002. Taking only the DSP processors in mobile phones as an example, the total sales in 2006 was more than $10 billion US.

General-purpose DSP processors (commercial off-the-shelf DSP processors) usually have a high degree of flexibility, a friendly design environment, and sufficient design references. General-purpose DSP processors are preferred when requirements on power, performance, and silicon area are not very critical. When these requirements are strict, embedded DSP processors as ASIP will become a necessity. Figure P.1 shows the trend of the different DSP market shares. The figure clearly shows that the future of ASIP DSP is obviously exciting—this is my third motivation for writing this book.

Most DSP applications can be categorized as streaming signal processing, in which the processing speed is higher than the speed of incoming signals. Classic DSP hardware for streaming signal processing was usually implemented on non-programmable ASICs to minimize the silicon cost. Recently programmability has become a vital issue because the complexity as well as design costs keep going higher. Programmability has been required by industries in order to support multimodes or multistandard applications. Thanks to the ongoing progress of modern

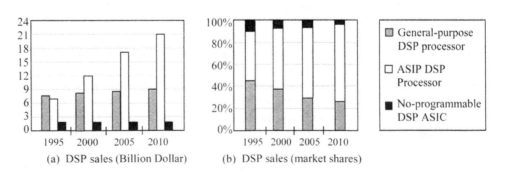

FIGURE P.1

Trend of different DSP market shares (FreehandDSP, Sweden).

VLSI (very large scale integrated circuits) technologies, programmable features have been realistic since the 1990s.

An ongoing trend is that the architecture of MPU (microprocessors as the central processor in personal computers) is converging. The architecture of DSP processors is diverging. One reason is that applications running on DSPs are diverging. However, the functionalities will be relatively fixed when a DSP processor is embedded in a system. Another reason is that the requirements are very critical on silicon efficiency, power consumption, and design cost of embedded processors or ASIP.

General-purpose processor designers think of ultimate performance and ultimate flexibility. The instruction set must be general because the application is unknown, and the programmer's behavior is unknown. ASIP designers have to think about the application and cost first. Usually the biggest challenge for ASIP designers is the efficiency issue. Based on the carefully specified function coverage, the goal of an ASIP design is to reach the highest performance over silicon and the highest performance over power consumption as well as the highest performance over the design cost. The requirement on flexibility should be sufficient instead of ultimate. The performance is application-specific instead of the highest. This book will expose and analyze the differences between general-purpose processor designers and ASIP designers.

An ASIP is often a SIP (Silicon Intellectual Property or Silicon IP, or IP). More SoC (system on a chip) solutions use ASIP IP. Therefore, the focus of this book will be the design of IP cores of ASIP for embedded systems on a chip. Silicon IP has been used as components in silicon chip designs since the mid-1990s. The requirements for quality design of silicon IP became higher after 2000 because silicon IP has been well accepted and widely used. In Figure P.2, the system design complexity is divided into the system design complexity and the component design complexity. Around the middle to late 1980s, RTL components (for example, multipliers and adders) were optimized as the lowest level components of system designs. RTL components took a certain degree of design complexity from the system design so that the system could be relatively more advanced comparing the system designed on a transistor level. During the mid-1990s, the system design became so advanced and complicated that programmable IP has to be used as the lowest level component to relax the system design complexity.

Because an IP usually is designed by the third party or another design team, the system complexity can therefore be shared. Also, because an IP usually is designed for multiusers, the design cost usually is shared by multiusers; relatively high IP design cost is therefore acceptable. This is the fourth motivation of writing the book—to show ways to design high quality programmable IP as components for multiusers.

The fourth motivation became even more important when the platform-based design concept was introduced recently. A platform is a partly designed application-specific system that can be used to adapt to a custom design with minimum cost. The platform-based system design requires the minimum design cost while plugging a programmable IP on the platform and running firmware on it. It means

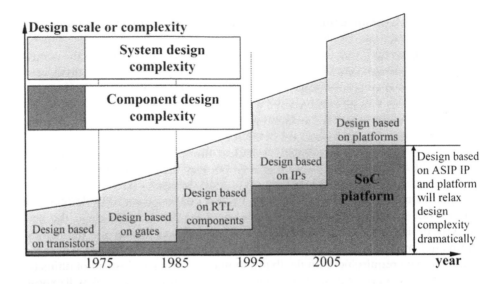

FIGURE P.2

Handling complexity of the design using ASIP IP and platform.

that the design of ASIP must be both silicon-efficient and platform-oriented. The platform-adaptive ASIP design skills offered by this book will thus be even more interesting.

I deeply acknowledge the research and teaching contributions from my PhD students in my Division of Computer Engineering, Department of Electrical Engineering at Linköping University, Sweden. The labs of the course and part of the contents of this book are based on their research work. Di Wu and Johan Eilert read through the book and provided enormous suggestions. Per Karlström managed all MS Word problems and figure formatting using his fantastic Word-VBA programming skills. Andreas Ehliar went through all code examples in the book. Per Karlström, Johan Eilert, Andreas Ehliar, Di Wu, and Master students Vinodh Ravinath and Bobo Svangård implemented the **Senior** DSP processor core. Research engineer Anders (S) Nilsson made the assembler and instruction set simulator of **Senior**, the processor used as the example of the book. Acknowledgment also goes to other PhD students: Rizwan Asghar, Dr. Anders Nilsson, Dr. Eric Tell, Dr. Tomas Henriksson, Dr. Daniel Wiklund, Dr. Ulf Nordqvist, and Lic. Mikael Olausson. I thank all Master students who participated in the course "Design of Embedded DSP Processors" from 1999 to 2007.

My sincere thanks go to Freehand DSP AB (Ltd.), Sweden (or VIA Tech Sweden after 2002), a leading company developing DSP processors for communications and home electronic applications. Special thanks to my friend and boss, CEO Harald Bergh, who went through several chapters and provided very professional and valuable suggestions. I was the cofounder, CTO, and vice president of Freehand

DSP AB (Ltd.) Stockholm, Sweden, during 1999 and 2002, which was later acquired by VIA Technologies in 2002.

I thank Coresonic AB (Ltd.), Linköping, Sweden, a leading DSP core SIP company for programmable radio baseband solutions. I am a cofounder and currently the CTO of this company. I sincerely thank my best friend, Professor Christer Svensson, at the Department of Electrical Engineering (ISY), Linköping University, Sweden, who had been my supervisor (1990–1994) during my research toward my technology doctor degree. Christer is the cofounder and the Chairman of the Board of Coresonic. I also thank cofounders Dr. Eric Tell, Dr. Anders Nilsson, and Daniel Svensson for many useful discussions and encouragements. All staff of Coresonic AB are greatly acknowledged.

I greatly acknowledge the following experts for their insightful discussions: Vodafone chair Professor Gerhard Fettweis, TU Dresden; Professor Christoph Kessler of Linköping University; Professor Lars Svensson of Chalmers University; Professor Viktor Öwall of Lund University; Professor Petru Eles of Linköping University; Dr. Carl-Fredrik Lenderson of Sony Ericsson; Professor Dr. Xiaoning Nie of Infineon Munich; Dr. Franz Dielacher, CSO, Infineon connections Villach; and Infineon fellow Professor Dr. Lajos Gazsi, Düsseldorf.

Finally, the most acknowledgment and gratitude goes to my dear wife Meiying and my daughter Angie. Without their love, understanding, and support, this book would never have been possible.

Dake Liu
December 2007
Linköping, Sweden

REFERENCES

[1] Strauss, W. (2000). Digital signal processing, the new semiconductor industry technology driver. *IEEE Signal Processing Magazine*, March, 52–56.

[2] http://www.fwdconcepts.com.

[3] BDTI, DSP selection guide http://www.bdti.com.

[4] Claasen, T. (2006). An industry perspective on current and future state-of-the-art in system-on-chip (SoC) technology. *Proceedings of the IEEE* 94(6).

[5] http://www.da.isy.liu.se/~dake.

[6] http://www.viatech.se.

[7] http://www.coresonic.com.

List of Trademarks and Product Names

ADI processors

ARM processors

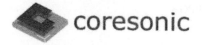

CEVA DSP processors

Coresonic LeoCore processors

FreehandDSP

Freescale

Infineon Camel processor

Intel Pentium and 8x86 processors

NXP EVP16 baseband processor

Openrisc of Opencores

SPI CELL (Sony Panasonic IBM) processors

TI (Texas Instrument) DSP processors

Xilinx FPGA

Tools and programs include MATLAB and
Simulink of Mathworks

Design compiler of Synopsys

LISA and Processor Designer of CoWare

ZSP of LSI

GCC from GNU

Introduction

DSP theory, DSP applications, and implementation of DSP subsystem hardware and algorithms will be reviewed in this chapter. Some real-time DSP applications will be covered briefly in order to provide an introduction to implementation costs (execution time and hardware complexity). Also, DSP architecture, embedded systems, and other essential background knowledge will be collected and reviewed.

This book is intended for readers who already have some fundamental knowledge in logic design, computer architecture, programming using C and HDL languages, DSP theory, and digital VLSI design. Application knowledge in digital communication or data compression is useful, but not necessary.

To make reading easier, a reading guide follows.

1.1 HOW TO READ THE BOOK

This book is organized into four parts. The first four chapters constitute Part 1, which provides some fundamental knowledge. The chapters of Part 1 are organized as follows:

1. Knowledge review, including DSP fundamentals, processor architectures, real-time systems, and design of embedded systems.

2. Numerical representation and precision control of fixed-point number systems in fixed-point processors.

3. Review of ASIP and DSP architectures for different requirements and applications.

4. Introduction to design methodologies for ASIP and DSP firmware.

The second part of this book covers DSP assembly instruction set design and design methodologies. The chapters of Part 2 cover the following topics:

5. Identification of drawbacks and problems of **Junior**, a DSP processor with an intuitive assembly instruction set following C language primitives.

1

6. Profiling for ASIP instruction set design with focus on the difference between profiling for quality software design and profiling for ASIP design.

7. Instruction set design technique is discussed in detail based on identified drawbacks of **Junior** and profiling knowledge. Instruction set design includes trade-off decisions, design of RISC subset, CISC subset, addressing, coding, and design for HW acceleration.

8. Introduction to the toolchain for assembly programming including code analyzer, C compiler, assembler, linker, simulator, and debugger.

9. Benchmarking of assembly instruction set based on DSP kernel algorithms and applications.

The third part of this book covers DSP microarchitecture design and the design of DSP core in detail. Part 3 discusses the following topics:

10. Microarchitecture design in general and ASIP microarchitecture design.

11. RF (Register File) design and implementation.

12. ALU (Arithmetic and Logic Unit) design and implementation.

13. MAC (Multiplication and Accumulation) unit design and implementation.

14. Control path design and implementation.

15. Design and implementation of memory subsystems and address generators.

Part 4 covers advanced issues of application-specific DSP design:

16. Design and implementation of DSP peripherals including core peripheral modules (Interrupt Handler, Timer, and DMA module), interprocessor buses, and chip peripherals (I/O ports).

17. Design for functional acceleration and DSP accelerators.

18. Firmware design for both DSP processors and ASIP.

19. Integration and verification of ASIP.

20. Design of parallel ASIP for streaming signal processing. This chapter is not written for undergraduate students.

The appendix contains the senior instruction set manual, its benchmarking, and RTL implementation.

This book is written for both university students and engineers. The targeted readers are categorized in Table 1.1.

For university teachers and research supervisors, this book can be organized for either a half-year course or a quarter-year course using Table 1.2 and related material available from the web page [1]. We suggest that teachers apply and modify the lab work according to their own curriculum plans.

Table 1.1 Targeted Readers.

Chapters	1	2	3–4	6	5,7,9	8	10	11–15	16,17	18	19	20
Contents	Introduction	Finite-length DSP	HW/SW Architecture	Profiling for ASIP design	Design of instruction set	Design of toolchain	Microarchitecture	Core module design	Advanced ASIP design	Firmware design skills	Processor integration	Design of streaming DSP
CS & EE undergraduates	A	A	A	A	A	B	A	A	B	B	B	B
CS & EE graduate students	A	A	A	A	A	A	A	A	A	A	A	A
ASIP engineers	A	A	A	A	A	A	A	A	A	A	A	A
ASIC and HW engineers	A	A	B	C	C	C	A	A	A	C	A	B
Embedded system designers	A	A	A	A	B	A	A	A	A	A	A	A
Firmware designers	A	A	C	C	–	A	–	–	–	A	C	A
Project, product managers	A	C	A	A	A	–	–	–	–	B	B	B

*CS: students of Computer Science;
EE: students of Electrical Engineering;
A: mandatory or very important;
B: very useful but not mandatory;
C: good to know;
–: not necessary.

As a half-year course (**H**), detailed understanding and comprehensive design skills are mandatory. The time budget for students will be around 240 hours including course participation (48h); book reading (96h); homework (40h); lab work (20h + 20h) and exam, including exam preparation (16h). The quarter-year course (**Q**) can be organized as an introduction to ASIP design. The total time cost for students will be around 120 to 160 hours including time for the exam. A light version of the lab work is prepared for the quarter-year course. A homework solution manual is available on the web for teachers with password control. For postgraduate (PhD) students, added materials will be available from the author by request.

The focus of this book is to teach ASIP hardware implementation skills, especially embedded DSP processors. The following areas will not be covered in the book: fundamental knowledge, such as DSP theory; fundamentals of DSP applications, such as theory of digital communication; compiler theory; and task scheduling theory.

Table 1.2 Reference for Teachers.[1]

Chapter	For half-year course	For quarter-year course	Homework tutorial	Lab work
1	2×45 minutes see H01	2×45 minutes Q01		
2	2×45 minutes see H02	2×45 minutes Q02	45 minutes see T1	
3	3×45 minutes see H03	1×45 minutes Q03		
4	2×45 minutes see H04	1×45 minutes Q03		
5	1×45 minutes see H05	1×45 minutes Q04		
6	2×45 minutes see H06	1×45 minutes Q04	45 minutes see T2	
7	3×45 minutes see H07	1×45 minutes Q05		
8	2×45 minutes see H08	1×45 minutes Q05		
9	1×45 minutes see H09		45 minutes see T3	Lab 1 simulator, profiler
10	2×45 minutes see H10	2×45 minutes Q06		
11	1×45 minutes see H11	1×45 minutes Q07		
12	2×45 minutes see H12	1×45 minutes Q07	45 minutes see T4	
13	2×45 minutes see H13	2×45 minutes Q08	45 minutes see T5	Lab 2 design RF/ALU
14	2×45 minutes see H14	2×45 minutes Q09		Lab 3 design MAC
15	2×45 minutes see H15	2×45 minutes Q10	45 minutes see T6	Lab 4 control path
16	2×45 minutes see H16			
17	1×45 minutes see H17			
18	3×45 minutes see H18	1×45 minutes Q11	45 minutes see T7	
19	2×45 minutes see H19	1×45 minutes Q11		Lab 5 Integration
20	3×45 minutes see H20		Review the book	

Except for Chapters 8 and 18, the focus is on hardware design, including instruction set design and microarchitecture design of a processor. Algorithm design and optimization is not the focus of this book as there are many other books available

---

[1]Power point files are available at the web page of the book.

in this area. However, it is not easy to find a book that guides engineers, senior students, and researchers step by step through an ASIP design. Meanwhile, since DSP architectures may vary a lot to adapt a different class of applications, the demand on the ASIP designer is surging, which further motivates the need of such a book.

C-like pseudocode will be used to describe the behavior of DSP functions. The hardware description language will be a Verilog-like pseudocode. The terminologies and abbreviations will be explained when they show up the first time. If this book is not read page by page, the reader can find terminologies and abbreviations in the glossary.

1.2 DSP THEORY FOR HARDWARE DESIGNERS

Many theoretical books about DSP algorithms are available [2, 3]. It is not this book's intent to review DSP algorithms; instead, DSP algorithms will be discussed from a profiling point of view in order to expose the computing cost.

The remaining part of this chapter is a collection of background knowledge. If you feel that it is insufficient, we recommend that you read a basic DSP book. Experienced engineers and senior PhD students can skip this part and start with Chapter 2. However, all readers are recommended to carefully read through Section 1.3.

1.2.1 Review of DSP Theory and Fundamentals

A signal is generated by physical phenomena. It has detectable energy and it carries information; for example, variation in air pressure (sound), or variation in electromagnetic radiation (radio). In order to process these signals, they first must be converted into electrical analog signals. These signals must be further converted into digital electrical signals before any digital signal processing can take place. A digital signal is a sequence of (amplitude) quantized values. In the time-domain, this sequence is in fixed order with fixed time intervals. In the frequency-domain, the spectrum sequence is in fixed order with fixed frequency intervals. A continuous signal can be converted into a digital signal by sampling. A sampling circuit is called an analog-to-digital converter (ADC). Inversely, a digital signal can be converted into an analog signal by interpolation. A circuit for doing this conversion is called a digital-to-analog converter (DAC).

A system is an entity that manipulates signals to accomplish certain functions, yielding new signals. The process of digital signal manipulation therefore is called digital signal processing (DSP). A system that handles digital signal processing is called a DSP system. A DSP system handles signal processing either in the time-domain or in the frequency-domain. Transformation algorithms translate signals between time- and frequency-domains.

DSP is a common name for the science and technology of processing digital signals using computers or digital electronic circuits (including DSP processors). Processing here stands for running algorithms based on a set of arithmetic kernels.

An arithmetic operation is specified as an atomic element of computing operations, for instance, addition, subtraction, multiplication, and division. Special DSP

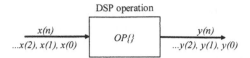

FIGURE 1.1

A simple DSP operation or a DSP system.

arithmetic operations such as guarding, saturation, truncation, and rounding will be discussed in more detail later. If it is not specially mentioned, two's complement is the default data representation in this book.

An algorithm is the mathematical representation of a task. An algorithm specifies a group of arithmetic operations to be performed in a certain order. However, it does not specify how the arithmetic operations involved are implemented. An algorithm can be implemented in software (SW) using a general-purpose computer or a DSP processor, or in hardware (HW) as an ASIC.

A simplified digital signal processing (DSP) system is shown in Figure 1.1. It has at least one input sequence $x(n)$ and one output sequence $y(n)$ that is generated by applying operation OP{} to $x(n)$.

Signals $x(n)$ and $y(n)$ are actually $x(nT)$ and $y(nT)$, where T is a time interval representing the sampling period. This means that the time interval between $x(n)$ and $x(n-1)$ is T. OP{} in Figure 1.1 can be a single operation or a group of operations. A DSP system can be as simple as pure combinational logic, or as complicated as a complete application system including several processors such as video encoders.

In Figure 1.2, fundamental knowledge is classified and depicted in the top part. Applications using fundamental knowledge are listed in the middle, and basic knowledge related to implementations is listed in the bottom part of the figure.

1.2.2 ADC and Finite-length Modeling

As mentioned earlier, a continuous electrical signal can be converted into a digital signal by an ADC. A digital signal can be converted into an analog signal by a DAC. Figure 1.3 shows a simplified and general DSP system based on a digital signal processor (DSP in this figure).

During the conversion from analog to digital and the subsequent processing, two types of errors are introduced. The first type is aliasing, which occurs if the sampling speed is close to the Nyquist rate. The second type is quantization error due to the finite word-length of the system. The ADC performs amplitude quantization of the analog input signal into binary output with finite-length precision. The maximum signal-to-quantization-noise ratio, in dB, of an ideal N-bit ADC is described in [4].

$$\text{SNR}_{Q-\max} = 6.02N + 4.77 - 3 = 6.02N + 1.77 \text{ (dB)} \tag{1.1}$$

1.77 dB is based on a sinusoidal waveform statistic and varies for other waveforms. N represents the data word length of the ADC. The $\text{SNR}_{Q-\max}$ expression

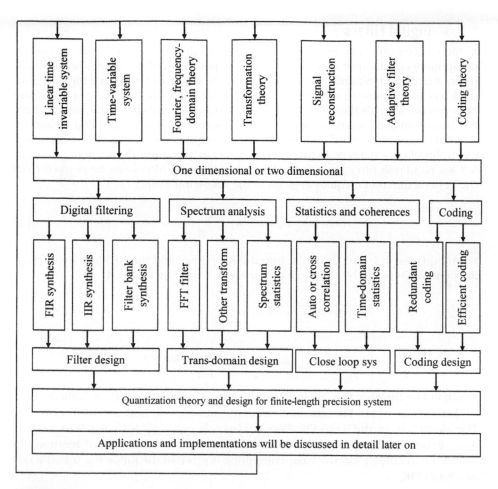

FIGURE 1.2

Review of DSP theory and related activities for ICT (Information and Communication Technology).

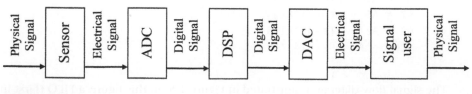

FIGURE 1.3

Overview of a DSP system.

gives a commonly used rule of thumb of 6 dB/bit for the SNR of an ADC. The maximum signal-to-noise ratio of a 12 bits ADC is 74 dB. The principle of deriving the SNR_{Q-max} can be found in fundamental ADC books.

1.2.3 Digital Filters

A filter attenuates certain frequency components of a signal. All frequency components of an input signal falling into the pass-band of the filter will have little or no attenuation, whereas those falling into the stop-band will have high attenuation. It is not possible to achieve an abrupt change from pass-band to stop-band, and between these bands there will be always be a transition-band. The frequency where the transition-band starts is called transition start frequency f_{pass}. The frequency where the transition-band ends is called stop frequency f_{stop}. Figure 1.4 illustrates examples of typical filters. In this figure, (a) is a low-pass filter, (b) is a high-pass filter, (c) is a band-pass filter, and (d) is a band-stop filter. All filters can be derived from the general difference equation given in the following form.[2]

$$y(n) = \sum_{k=0}^{K-1} a_k x(n-k) - \sum_{l=1}^{L-1} b_l y(n-l) \tag{1.2}$$

From Equation 1.2, the FIR (Finite Impulse Response) and the IIR (Infinite Impulse Response) filters can be defined by Equation 1.3 and Equation 1.4:

$$y(n) = \sum_{k=0}^{K-1} a_k x(n-k) \tag{1.3}$$

$$y(n) = \sum_{k=0}^{K-1} a_k x(n-k) - \sum_{l=1}^{L-1} b_l y(n-l) \tag{1.4}$$

When FIR or IIR later appears in this book, it will implicitly stand for "a filter."

Let us take a 5-tap FIR filter as an example and explore how an FIR filter works. The algorithms of the 5-tap FIR filter is $y(n) = \Sigma x(n-k)*b(k)$, $k = 0$ to 4. Here, $x(n-k)$ is the input signal, $b(k)$ is the coefficient, and k is the number of iterations. To unroll the iteration loop, the basic computing is given in the following pseudocode of a 5-tap FIR.

```
//5-tap FIR behavior code
{
    A0 = x(n)*h(0);
    A1 = A0 + x(n-1)*h(1);
    A2 = A1 + x(n-2)*h(2);
    A3 = A2 + x(n-3)*h(3);
    Y(n) = A3 + x(n-4)*h(4);
}
```

The signal flow diagram is illustrated in Figure 1.5. In this figure, a FIFO (First In First Out) buffer consists of $n-1$ memory positions keeping $n-1$ input values. Z^{-1}

---

[2]Equations representing fundamental DSP knowledge can be found in any DSP book. We will not explain and derive equations of fundamental DSP in this book. You can find details in [2, 3].

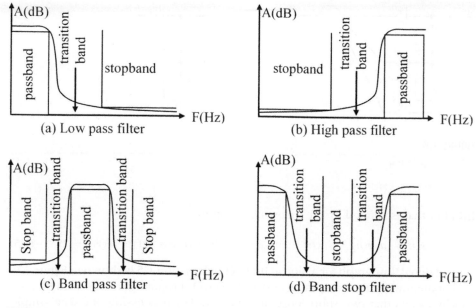

FIGURE 1.4

Filter specifications.

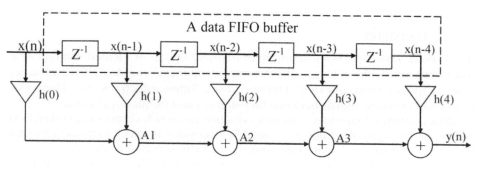

FIGURE 1.5

A 5-tap FIR filter with a FIFO data buffer.

denotes the delay of the input signal or the next position of the FIFO. The triangle symbol denotes the multiplication operation. If the FIFO data and the coefficients are stored in memories, FIR computing of one data sample usually consists of K multiplications, K accumulations, and $2K$ memory accesses. Here K denotes the number of taps of the filter.

Equation 1.4 illustrates an IIR filter. Because the filter output is fed back as part of the inputs for regressive computing, the filter may be unstable if it is not properly designed. The most used IIR filter is the so-called Biquad IIR. A Biquad IIR is a two-tap

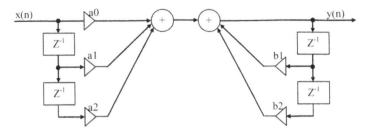

FIGURE 1.6

A Biquad IIR filter.

IIR filter that is defined by the following equation:

$$y(n) = a_0 x(n) + a_1 x(n-1) + a_2 x(n-2) + b_1 y(n-1) + b_2 y(n-2) \qquad (1.5)$$

The signal flow graph for a Biquad IIR filter of Equation 1.5 is shown in Figure 1.6.

For minimizing the computing cost, b_0 is set to 1. Programming a Biquad IIR filter usually means that two short loops are unrolled. For processing one data sample, up to 19 operations are required including five multiplications, four additions, and ten memory accesses.

1.2.4 **Transform**

In this book, transform is an algorithm to translate signals or elements in one style to another style without losing information. Typical transforms are FFT (Fast Fourier Transform) transform, and DCT (Discrete Cosine Transform) transform. These two transforms translate signals between time-domain and frequency-domain.

Time-domain signal processing may consume too much computing power, thus signal processing in other domains might be necessary. In order to process a signal in another domain, a domain transformation of the signal is required before and after the signal processing. Digital signal processing in the frequency-domain is popular because of its explicit physical meaning and lower computational cost in comparison with time-domain processing. For example, if a FIR filter is implemented in the time-domain, the filtering operation is done by convolution. The computing cost of a convolution is $K*N$, where K is the number of taps of the filter and N is the number of samples. If the FIR filter is implemented in the frequency-domain, the computing cost will be reduced to N because the filtering operation in the frequency-domain is $Y(f) = H(f)^*X(f)$. Here, Y is the output in frequency-domain, H is the system transfer function in frequency-domain, and X is the input signal in the frequency-domain. The total cost of frequency-domain signal processing includes the computing cost in the frequency-domain, and the cost of the transform and inverse transform. In general we can summarize that the condition to select frequency-domain algorithms is:

$$TD_{cost} > FFT_{cost} + FD_{cost} + IFFT_{cost} \qquad (1.6)$$

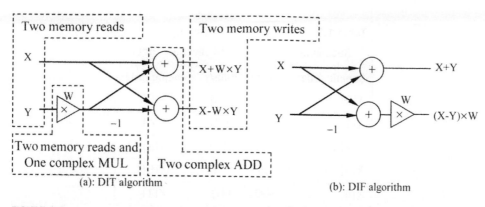

(a): DIT algorithm

(b): DIF algorithm

FIGURE 1.7

DIT butterfly and DIF butterfly of Radix-2 FFT.

TD_{cost} is the execution time cost in time-domain, the FD_{cost} is the execution time cost in frequency-domain, the FFT_{cost} and $IFFT_{cost}$ are the execution time cost of FFT and inverse FFT.

The computational complexity of a direct Fourier transform (DFT) involves at least N^2 complex multiplications or at least $2N(N-1)$ arithmetic operations, without including extra addressing and memory accesses. The computing cost can be reduced if FFT (Fast Fourier Transform) is introduced. FFT is not a new transform from the theoretical perspective; instead it is an efficient way of computing the DFT (Discrete Fourier Transform). If the DFT is decomposed into multiple 2-point DFTs, it is called a radix-2 algorithm. If the smallest DFT in the decomposition is a 4-point DFT, then it is called a radix-4 algorithm. Detailed discussions on FFT and DFT can be found in [2].

There are two main radix-2 based approaches, decimation in time (DIT) and decimation in frequency (DIF). The complexities of both algorithms are the same considering the number of basic operations, the so-called butterfly operations. The butterfly operations are slightly different between the DIT and DIF implementations. Both butterfly schematics (DIT and DIF) are illustrated in Figure 1.7.

X and Y in Figure 1.7 are two complex data inputs of the butterfly algorithm. W is the coefficient of the butterfly algorithm in complex data format. Computing a DIT or DIF butterfly consists of 10 operations including a complex data multiplication (equivalent to six integer operations), two complex data additions (equivalent to four integer operations), and five memory accesses of complex data (two data load, one coefficient load, and two data store, equivalent to 10 integer data accesses). For the DIT algorithm, the input data should be in bit-reversed order. Bit-reversal addressing is given in Figure 1.8 for an 8-point FFT. The bit-reversal computing cost is at least N.

An 8-point FFT signal flow is given as an example in Figure 1.8.

It can be seen in Figure 1.8 that the computation consists of $\log_2 N$ computing layers. Each layer consists of $N/2$ butterfly subroutines. The total computing cost

Table 1.3 Bit-Reversal Addressing.

Sequential Order		Bit-reversed Order	
Sample order	Binary	Binary	Sample order
$X[0]$	000	000	$X[0]$
$X[1]$	001	100	$X[4]$
$X[2]$	010	010	$X[2]$
$X[3]$	011	110	$X[6]$
$X[4]$	100	001	$X[1]$
$X[5]$	101	101	$X[5]$
$X[6]$	110	011	$X[3]$
$X[7]$	111	111	$X[7]$

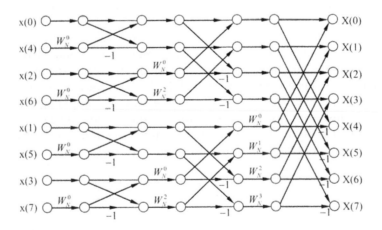

FIGURE 1.8

Signal flow of an 8-point Radix-2 DIF FFT.

is at least $N + 0.5N \times$ (the cost of a butterfly) $\times \log_2 N$. Therefore, the computing cost of a 256-point FFT is at least $0.5 \times 256 \times 10 \times \log_2 256 = 10240$ arithmetic operations and 10240 basic memory accesses (not including bit-reversal addressing illustrated in Table 1.3, which will be discussed in Chapter 7).

1.2.5 Adaptive Filter and Signal Enhancement

In order to adapt to the dynamic environment, such as dynamic noise, dynamic echo, or dynamic radio channel of a mobile radio transceiver, some filter behaviors should be updated according to the change of the environment.

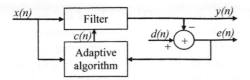

FIGURE 1.9

General form of an adaptive filter.

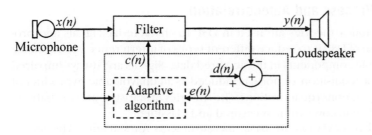

FIGURE 1.10

A room acoustic echo canceller.

An adaptive filter is a kind of filter where the coefficients can be updated by an adaptive algorithm in order to improve or optimize the filter's response to a desired performance criterion. An adaptive filter consists of two basic parts: the filter, which applies the required processing to the incoming signal, and the adaptive algorithm, which adjusts the coefficients of the filter to improve its performance.

The structure of an adaptive filter is depicted in Figure 1.9. The input signal, $x(n)$, is filtered (or weighted) in a digital filter, which provides the output $y(n)$. The adaptive algorithm will continuously adjust the coefficients $c(n)$ in the filter in order to minimize the error $e(n)$. The error is the difference between the filtered output $y(n)$, and the desired response of the filter $d(n)$.

A room acoustic echo canceller, shown in Figure 1.10, is a typical application of an adaptive filter. The purpose of the filter is to cancel the echo sampled by the microphone.

In this acoustic echo canceller, the new coefficient is calculated by the following adaptive algorithm:

$$c(n)_{\text{new}} = Kx(n) - c(n)_{\text{old}} \tag{1.7}$$

Here $K = f(e(n))$ is the convergence factor. The filter is usually a FIR filter operating in the time-domain using convolution. A convolution of a data sample consists of N multiplications, N accumulations, and $2N$ memory read operations. Similar to convolution, coefficient adaptation during one data sample consists of N multiplication and N accumulator operations. There is no extra memory load because the access of the old coefficient and the data were counted for by the computing of the filter. However, the new coefficient should be updated and written back to the coefficient memory. Thus there will be N memory write operations. To conclude, the processing cost of an adaptive filter during one data sampling period could be up to $4N$ computations and $3N$ memory accesses if the coefficients are updated continuously. The computing cost and the memory accesses are at least twice that of a fixed FIR filter.

1.2.6 Random Process and Autocorrelation

Statistics and probability theories are used in DSP to characterize signals and processes. One purpose of DSP is to reduce different types of interference such as noise and other undesirable components in the acquired data. Signals are always impaired by noise during data acquisition, or noise is induced as an unavoidable byproduct of finite-length DSP operations (quantization). The theories of statistics and probability allow these disruptive features to be measured and classified.

Mean and standard deviation are basic measurements for statistics. The *mean*, μ, is the average value of a signal. The mean is simply calculated by dividing the sum of N samples by N, the number of samples. The standard variance is achieved by squaring each of the deviations before taking the average, and the standard deviation σ is the square root of the standard variation.

$$\sigma^2 = \frac{1}{N-1} \sum_{i=0}^{N-1} (x - \mu)^2 \tag{1.8}$$

The mean indicates the statistics behavior of a signal and is an estimation of the expected signal value. The standard deviation gives a creditable measurement of the estimation. The computing cost of mean is $2N + 1$ (N additions, N memory accesses, and the final step of $1/N$). Similarly, the computing cost of standard deviation is $4N + 2$ (three arithmetic and one memory access operations in each tap).

Two other algorithms, autocorrelation and cross-correlation, are useful for signal detection. Autocorrelation is used for finding regularities or periodical features of a signal. Autocorrelation is defined as:

$$a(k) = \sum_{i=0}^{N-1} x(i)\, x(i - k) \tag{1.9}$$

The computing cost of autocorrelation is similar to the cost of FIR. The main difference is that autocorrelation uses two variables from the same data array.

Cross-correlation is used for measuring the similarity of a signal with a known signal pattern. Cross-correlation is defined as:

$$y(k) = \sum_{i=0}^{N-1} c(i)\, x(i - k) \tag{1.10}$$

The computing cost is exactly the same as the cost of a FIR.

1.3 THEORY, APPLICATIONS, AND IMPLEMENTATIONS

The scope of DSP is huge, and DSP as an abbreviation has been used by different people with different meanings. In this book, DSP is divided into three categories: DSP theory, DSP applications, and DSP implementations.

DSP theory is the mathematical description of signals and systems using discrete-time or discrete-frequency methods. Most DSP books have DSP theory as the main subject, and this area is today a mature science. On the other hand, applications and implementations based on DSP theory have become major challenges in today's DSP academia and industry. The three DSP categories are summarized in Figure 1.11.

The theory of digital signal processing (DSP) is the mathematical study of signals in a digital representation and the processing methods of these signals.

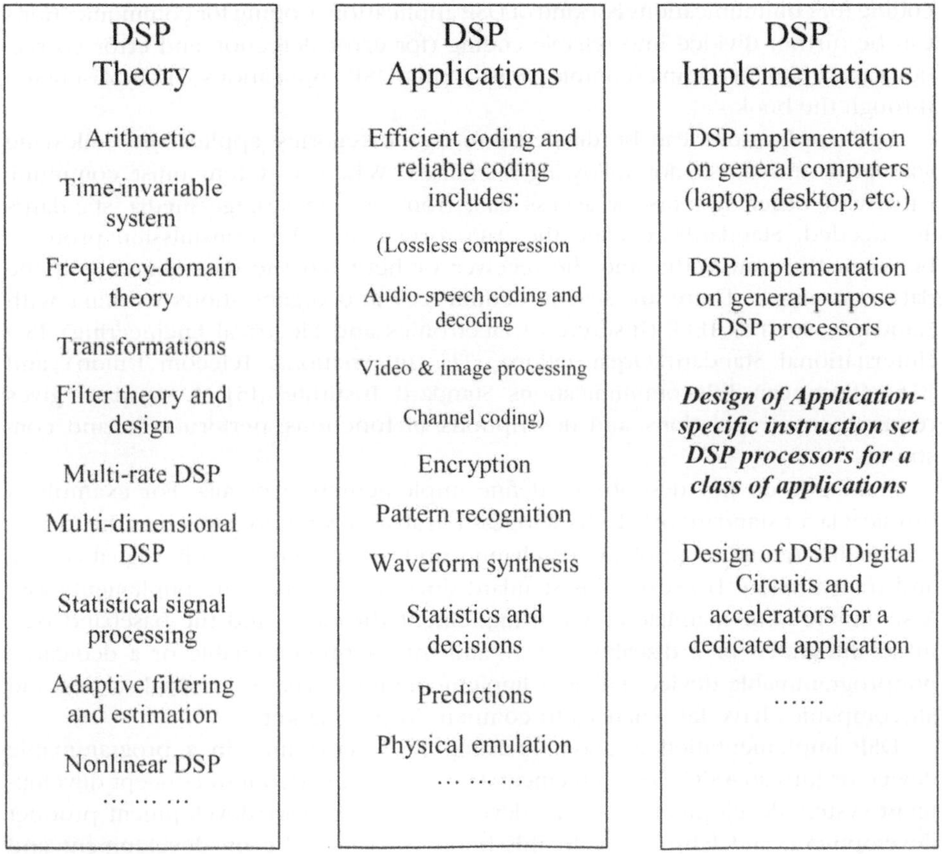

DSP Theory	DSP Applications	DSP Implementations
Arithmetic	Efficient coding and reliable coding includes:	DSP implementation on general computers (laptop, desktop, etc.)
Time-invariable system	(Lossless compression	
Frequency-domain theory	Audio-speech coding and decoding	DSP implementation on general-purpose DSP processors
Transformations	Video & image processing	
Filter theory and design	Channel coding)	*Design of Application-specific instruction set DSP processors for a class of applications*
Multi-rate DSP	Encryption	
Multi-dimensional DSP	Pattern recognition	Design of DSP Digital Circuits and accelerators for a dedicated application
Statistical signal processing	Waveform synthesis	
Adaptive filtering and estimation	Statistics and decisions	
Nonlinear DSP	Predictions	......
... ...	Physical emulation	
	... ...	

FIGURE 1.11

DSP categories for ICT (Information Communication Technology).

By using DSP theory, a system behavior is modeled and represented mathematically in a discrete-time-domain or discrete-frequency-domain. The scope of DSP theory is huge, and the main activities (related to this book) are selectively listed in Figure 1.11.

DSP has turned out to be one of the most important technologies in the development of communications systems and other electronics. After Marconi and Armstrong's invention of basic radio communication and transceivers, users were soon unsatisfied with the poor communication quality. The noise was high and the bandwidth utility was inefficient. Parameters changed due to variations in temperature, and the signals could not be reproduced exactly. In order to improve the quality, analog radio communication had to give way to digital radio communication.

DSP applications in this book are processes through which systems with specific purposes can be modeled using the knowledge of DSP theory. DSP applications based on established DSP theory can be found everywhere in daily life. For example, coding for communications is a kind of DSP application. Coding for communications can be further divided into reliable coding (for error detection and error correction) and efficient coding (compression). Other DSP applications will be discussed through the book.

DSP applications can be divided into two categories: applications following standards and those not following standards. When a system must communicate with other systems, or access data from certain storage media, standards are needed. Standards regulate the data format and the transmission protocol between the transmitter and the receiver or between the data source and the data consumer. There are several committees and organizations working with standards, such as IEEE (Institute of Electronics and Electrical Engineering), ISO (International Standard Organization), ITU (International Telecom Union), and ETSI (European Telecommunications Standard Institute) [5]. A standard gives requirements, regulations, and descriptions of functions, performance, and constraints.

Standards do not describe or define implementation details. For example, a physical layer standard of WLAN (Wireless Local Area Network) specifies the format and quality requirement of a radio channel and regulations for radio signal coding and transmission. However, this standard does not regulate the implementation. A semiconductor manufacturer can implement the radio and the baseband part in an integrated or a distributed way, and into a programmable or a dedicated nonprogrammable device. Efficient implementations make successful stories, and all companies have fair chances to compete in the market.

DSP implementation is about realizing DSP algorithms in a programmable device or into an ASIC. DSP implementation can be divided into concept development, system development, software development, hardware development, product development, and fabrication. In this book, focus will be on development and implementation of hardware systems.

1.4 **DSP APPLICATIONS**

In this section, DSP applications are introduced briefly through examples. The motivation is to provide examples of real-time systems where digital signal processing is intensively used, because real-time applications form the implicit scope throughout this book. For entry-level readers and students, it might be a bit hard to fully understand every application case presented here. However, this will not prevent you from understanding other parts of the book. From an engineering point of view, the examples selected in this section are simpler than those real-world designs in industrial products. Simplification is necessary because understanding a real-world design might take a very long time and may even be confusing sometimes.

As ASIP designers, you should know more about the methods to analyze the cost of applications instead of a deep understanding of the theory behind them. You should understand where the cost is from, and how to analyze and estimate it. To obtain more details, visit the related company websites [6].

1.4.1 **Real-Time Concept**

There are two types of systems: real-time and non-real-time [7, 8]. A real-time system is the simplification of a real-time computing system. The system processes data in a time-predictable fashion so that the results are available in time to influence the process being monitored, played, or controlled. The definition of a real-time system is not absolutely unambiguous; however, it is generally accepted that a real-time DSP subsystem is a system that processes periodical signals. A real-time system in this book, therefore, creates the output signal samples at the same rate by which the input signals arrive. It means that the processing capacity must be enough for processing one sample within the time interval of two consecutively arriving samples. For example, digital signal processing for voice communication in a mobile phone must be real-time. If the decoding of a voice data packet cannot be finished before the arrival of the next data packet, the computing resource has to abort the current computation and information will be lost. To formalize, a real-time system processes tasks within a time interval and finishes the processing before a certain deadline.

Non-real-time digital signal processing has no strict requirements on the execution time. For instance, image encoding in a digital camera does not necessarily have to be real-time because the slow execution will not introduce any system error (though it may annoy the camera user).

1.4.2 **Communication Systems**

A communication system is usually a real-time system, which transmits a message stream from the source to the destination through a noisy channel. The message

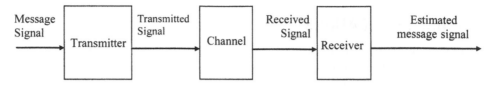

FIGURE 1.12

A communication system.

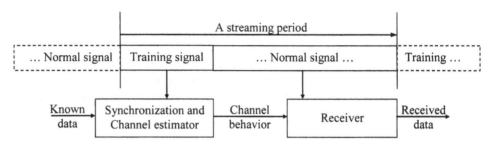

FIGURE 1.13

Streaming signal between a transceiver pair.

source is called transmitter, and the destination is called receiver. A simplified communication system is given in Figure 1.12 [9].

The transmitter and the receiver are usually at different physical locations. The channel is the physical medium between them, which can be either the air interface of a radio system (e.g., a mobile phone and a radio base station) or a telephone wire.

The received signal that has propagated through the channel is interfered with the external environment and distorted internally by the receiver itself due to the physical characteristics of the channel. Thus, the signal received is actually different from the signal sent by the transmitter. For example, a receiver may get continuous interference from other neighboring radio systems and interference such as burst glitches from surrounding equipment. The receiver may receive both the wave in the direct path with normal delay and waves from paths experiencing multiple reflections with longer delays. The signal attenuation may change quickly according to the change of the relative position between the transmitter and the receiver.

The function of the receiver in Figure 1.13 is to recover the transmitted data by estimating the received signal and dynamically modeling the channel behavior. The function of the transmitter is not only to send data, but also to code it in order to enable and facilitate the estimation of the channel impulse response at the receiver side. This enables the receiver to compare the received signal with a known training data and to find the difference between the expected signal and the signal actually received, in order to finally calculate the approximation of the channel impulse response. As soon as the channel is correctly modeled, data can be received and recovered by symbol detection with sufficient accuracy. All these heavy computations in the transmitter and the receiver are called radio baseband signal processing, and include coding, channel model estimation, signal reception, error detection, and error correction. All radio baseband signal processing is a class of DSP applications based on well-known DSP theory.

The computing cost of baseband signal processing varies significantly according to the channel condition and the coding algorithms. An advanced radio receiver requires several hundred to several thousand operations (including memory accesses) for recovering one data bit received from a radio channel. Thus, receiving one megabit per second might require several Giga operations per second. The principle of typical radio baseband signal processing is illustrated in Figure 1.13. The receiver gets a training signal packet first, which is the known signal used for detecting the current channel model as the transfer function $H(f) = Y(f)/X(f)$ by comparing the known training signal stored in the receiver (X) to the received signal (Y). Note that this computation must be conducted in a very short time in order to process the normal signal (payload) in time [10]. The estimated channel model H is used for recovering the normal signal during data reception.

1.4.3 Multimedia Signal Processing Systems

Multimedia signal processing is an important class of DSP applications. Here, the concept of multimedia covers various information formats such as voice, audio, image, and video. Data can be stored in different ways—for example, on CD, hard disk, or memory card, and transmitted via fixed-line or radio channel. Both transmission bandwidth and storage volume will directly affect the end cost. Therefore, data must be compressed for transmission and storage. The compressed data needs to be decompressed before being presented [11].

There are two efficient coding techniques for compression and decompression of data, lossy compression and lossless compression. Lossy compression can be used for voice-, audio-, and image-based applications because the imperfect human hearing and visual capability allows certain types of information to be removed. Lossless compression is required when all information in the data has to be fully recovered.

Lossless Compression

Lossless compression is a method of reducing the size of a data file without losing any information, which means that less storage space or transmission bandwidth is needed after the data compression. However, the file after compression and decompression must exactly match the original information. The principle of lossless compression is to find and remove any redundant information in the data. For example, when encoding the characters in a computer system, the length of the code assigned to each character is the same. However, some characters may appear more often than the others, thus making it more efficient to use shorter codes for representing these characters.

A common lossless compression scheme is Huffman coding, which has the following properties:

- Codes for more probable symbols are shorter than those for less probable symbols.

- Each code can be uniquely decoded.

A Huffman tree is used in Huffman coding. This tree is built based on statistical measurements of the data to be encoded. As an example, the frequencies of the different symbols in the sequence ABAACDAAAB are calculated and listed in Table 1.4.

The Huffman tree is illustrated in Figure 1.14.

In this case, there are four different symbols (A, B, C, and D), and at least two bits per symbol are needed. Thus $10 \times 2 = 20$ bits are required for encoding the string ABAACDAAAB. If the Huffman codes in Table 1.4 are used, only $6 \times 1 + 2 \times 2 + 3 + 3 = 16$ bits are needed. Thus four bits are saved. Once the Huffman codes have been decided, the code of each symbol can be found from a simple lookup table.

Decoding can be illustrated by the same example. Assume that the bit stream 01000110111 was generated from the Huffman codes in Table 1.4. This binary code will be translated in the following way: $0 \rightarrow A, 10 \rightarrow B, 0 \rightarrow A, 0 \rightarrow A, 110 \rightarrow C$, and $111 \rightarrow D$. Obviously the Huffman tree must be known to the decoder.

Table 1.4 Symbol Frequencies and Huffman Codes.			
Symbol	**Frequency**	**Normal code**	**Huffman code**
A	6	00	0
B	2	01	10
C	1	10	110
D	1	11	111

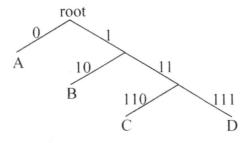

FIGURE 1.14

Huffman tree.

Without special hardware acceleration, the performance of Huffman coding and decoding is usually much lower than one bit per instruction. With special hardware acceleration, two bits per instruction can be achieved. In reference [12], the throughput of Huffman encoding was enhanced from 0.083 bit per clock cycle to more than 2 bits per clock cycle by using a special hardware architecture. Huffman coding and decoding are typical applications for a FSM (Finite State Machine).

Voice Compression

In order to reach a high compression ratio, lossy compression can be used. In this case, data cannot be completely recovered. However, if the lost information is not important to the user or is negligible for other reasons, lossy compression can be very useful for some applications [13]. For example, modern voice compression techniques can compress 104 kbits/s voice data (8 KHz sampling rate with 13 bits resolution) into a new stream with a data rate of 1.2 kbits/s with reasonable (limited) distortion using an eMELP voice codec. The compression ratio here is 86:1 [14].

Voice (speech) compression has been thoroughly investigated, and advanced voice compression today is based on voice synthesis techniques. A simplified example of a voice encoder can be found in Figure 1.15.

The voice encoder depicted in Figure 1.15 [13] synthesizes both the vocal model and the voice features within a period of time. A typical period is between 5 and 20 milliseconds. The vocal mode is modeled using a 10-tap IIR filter emulating the shape of the mouth. The voice features are described by four parameters: gain (volume of the voice), pitch (fundamental frequency of the voice providing the vocal feature), attenuation (describes the change in voice volume), and noise pattern (describes the consonants of the voice).

Instead of transferring or storing the original voice, the vocal model and voice patterns will be transferred or stored. Taking the voice codec in Figure 1.15 as an example, the data size of the original voice during 20 milliseconds is $104 \, kb \times 0.02 = 2.08 \, kb$. After compression, the vocal and voice patterns can be coded using $10 \times 8 + 4 \times 8 = 112$ bits. This corresponds to a compression ratio of 18:1. For implementing the encoder and the decoder, many DSP algorithms will be used such

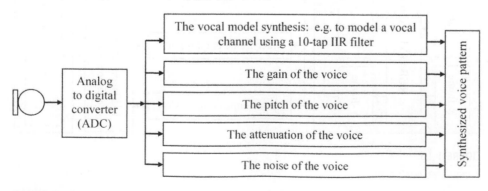

FIGURE 1.15

A voice encoder based on voice synthesis technique.

as autocorrelation, Fast Fourier Transform (FFT), adaptive filtering, quantization, and waveform generation. The computing cost of a complete voice encoder is around 10 to 50 million operations per second for compressing 104 kbits per second using the voice synthesis technique. This is equivalent to about 100 to 500 operations per voice sample.

A voice codec (coder and decoder) is a real-time system, which means that the complete coding and decoding must be finished before the next data arrives (in other words, the data processing speed must be faster than the data rate).

Image and Video Compression

Generally, compression techniques for image and video [15] are based on two-dimensional signal processing. Both lossy and lossless compression techniques are used for these applications. Image and video compression techniques are illustrated in Figure 1.16.

The complete algorithm flow in Figure 1.16 is a simplified video compression flow, and the shaded subset is a simplified image compression flow. The first step in image and video compression is color transformation. The three original color planes R, G, B (R = red, G = green, B = blue) are translated to the Y, U, V color planes (Y = luminance, U and V = chrominance). Because the human sensitivity to chrominance (color) is lower than the sensitivity to luminance (brightness), the U and V planes can be down-sampled to a quarter of their original size. Therefore, a frame with a size factor of 3 (3 color planes) is down-sampled to a frame with the size factor of $1 + 1/4 + 1/4 = 1.5$. The compression ratio here is 2.

Frequency-domain compression is executed after the RGB to YUV transformation. The Discrete Cosine Transform (DCT) is used for transferring the image from

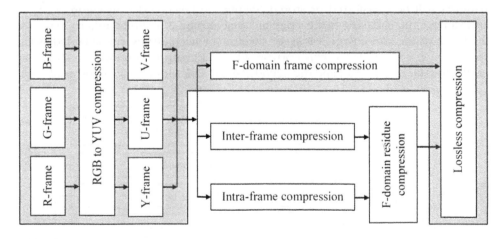

FIGURE 1.16

Image and video compression.

the time-domain to the frequency-domain. Because the human sensitivity to spatially high-frequency details is low, the information located in the higher frequency parts can be reduced by quantization (the information is represented with lower resolution).

After the frequency-domain compression, a lossless compression such as Huffman coding will finally be carried out.

The classical JPEG (Joint Picture Expert Group) compression for still images can reach 20:1 compression on average. Including the memory access cost, JPEG encoding consumes roughly 200 operations and decoding roughly 150 operations, for one RGB pixel (including three color components). The processing of Huffman coding and decoding can vary a lot between different images, so the JPEG computing cost of a complete image cannot be accurately estimated.

Video compression is an extension of image compression. Video compression is utilizing two types of redundancies: spatial and temporal. Compression of the first video frame, the reference frame, is similar to compression of a still image frame. The reference frame is used for compressing later frames, the inter-frames.

If the difference between a reference frame and a neighbor frame is calculated, the result (motion vectors) can be used for representing this frame. The size of this data is usually very small compared to the size of the original frame. Often there is little difference between the corresponding pixels in consecutive frames except for the movement of some objects. The data transferred will consist of a reference frame and a number of consecutive frames represented by motion vectors. The video stream can therefore be significantly compressed.

The classical MPEG2 (Moving Picture Expert Group) video compression standard can reach a 50:1 compression ratio on average. The advanced video codec H.264/AVC standard can increase this ratio to more than 100:1. The computing cost of a video encoder is very dependent on the complexity of the video stream and the motion estimation algorithm used. Including the cost of memory accesses, a H.264 encoder may consume as much as 4000 operations and its decoder about 500 operations for a pixel on average. As an example, for encoding a video stream with QCIF size (176 $\times$ 144) and 30 frames per second, the encoder requires about 3×10^9 operations per second. The corresponding decoder requires about 4×10^8 operations per second.

1.4.4 Review on Applications

You may have realized that the application examples discussed in this section are applicable to a high-end mobile phone with a video camera. The total digital signal processing cost, including 3G (third-generation) baseband, voice codec, digital camera, and video camera, is actually much more than the capacity of a Pentium running at 2 GHz!

How can a DSP subsystem in a mobile phone supply so much computing capacity while keeping the power consumption low? The answer is the use of application-specific processors! Both the radio baseband and media (voice, audio, and video) processors are application-specific instruction set processors or ASIC modules. They are designed for a class of applications and optimized for low power and low silicon cost.

For designing an ASIP, the computing cost of the target applications is an essential design input. In this section, the cost is measured as the unit cost, the cost per sample (a voice sample, a pixel in a picture, or a bit of recovered data from radio channel). We strongly recommend that you use this way of counting the computing cost. In the initial stages of reading this book, you might not be used to this way of thinking. However, through reading this book and finishing the exercises, this way of analysis can be gradually developed.

1.5 DSP IMPLEMENTATIONS

A DSP application can be implemented in a variety of ways. One way is to implement the application algorithms using a general-purpose computer, like a personal computer or a workstation. There are two reasons for implementing a DSP application on a general-purpose computer:

1. To quickly supply the application to the final user within the shortest possible time.

2. To use this implementation as a reference model for the design of an embedded system.

The discussion of the DSP implementation using a general computer in this section follows the first reason. The implementation according to the second reason will be discussed in the chapter on firmware design.

Many DSP applications are implemented using a general-purpose DSP (off-the-shelf processor). Here, general-purpose DSP stands for a DSP available from a semiconductor supplier and not targeted for a specific class of DSP applications. For example, the TMS320C55X processor from Texas Instruments [6] is available on the market and can be used for many DSP applications requiring a computing performance of less than 500 million arithmetic operations per second.

DSP applications can also be implemented using an ASIP DSP. An ASIP is designed for a class of applications such as radio baseband processing in a multimode mobile phone, audio processing in an MP3 player, or image processing in a digital camera.

Another alternative for implementing DSP applications is the nonprogrammable ASIC (Application Specific Integrated Circuit). Many DSP applications were implemented this way in the 1980s due to limitations in silicon area and performance. Recently the nonprogrammable DSP ASIC has been taken over gradually by the DSP ASIP in accordance with increasing requirements on flexibility.

An FPGA (Field-Programmable Gate Array) is also an alternative as an intermediate or the final implementation of DSP applications.

1.5.1 **DSP Implementation on GPP**

Many DSP applications, with or without real-time requirements, can be implemented on a general-purpose processor (GPP). Recently, applications for media entertainment have become popular in personal computers, for example, audio and video players. A video player is a video decoder that is decompressing data according to international standards (e.g., ISO/IEC MPEG-4) or proprietary standards (e.g., Windows Media Video from Microsoft).

The video player must be able to decode the video stream in real-time using the operating system (OS). However, most operating systems for desktops are not originally designed for real-time applications. When designing a high quality media player using such an OS, attention must be paid to the real-time features by correctly setting the priorities of the running tasks.

Another type of DSP application, without real-time requirements, is high performance computing. Typical examples are analysis of the stock market, weather forecast, or earthquakes. This type of analysis often is implemented in software running on a personal or supercomputer. The execution time of such a software program is not strictly regulated and this software therefore is defined as general software instead of real-time DSP software.

1.5.2 **DSP Implementation on GP DSP Processors**

A general-purpose DSP can be bought off-the-shelf from a number of different DSP suppliers (e.g., Texas Instruments, Analog Devices, or Freescale) [6]. A general-purpose DSP has a general assembly instruction set that provides good flexibility for many applications. However, high flexibility usually means fewer application-specific features or less acceleration of both arithmetic and control operations. Therefore, a general-purpose DSP is not suitable for applications with very high performance requirements. High flexibility also means that the chip area will be large. A general-purpose DSP processor can be used for initializing a product because the system design time will be short. When the volume has gone up, a DSP ASIP could replace the general-purpose processor in order to reduce the component cost.

No general-purpose DSP is 100% general. Most general-purpose DSP processors are actually designed with different target applications in mind. General-purpose DSP processors can be divided into processors targeted for either low power or high performance. For example, TMS320C2X of Texas Instruments is designed for low-cost applications, whereas TMS320C55 of Texas Instruments is designed for applications with medium performance and medium power. TMS320C6X of Texas Instruments is designed for applications requiring high performance. General-purpose DSP processors can also be divided into floating-point processors and fixed-point processors.

"General-purpose" implies only that the DSP is designed neither for a specific task nor for a class of specific tasks. It is available on the device market for all possible applications. The instruction set and the architecture must be general, meaning that the instruction set covers all basic arithmetic functions and sufficient control functions. The instruction set is not designed for accelerating a specific group of algorithms. At the same time, peripherals and interfaces must be comprehensive in order to be able to connect to various microcontrollers, memories, and peripheral devices.

Since the processor is off-the-shelf, the main development of a DSP product will be software design. The hardware development is limited to peripheral design, which means connecting the DSP to the surrounding components, including MCU (microcontroller), main memory, and other input–output components. The best way of designing peripherals for an available DSP is to find a reference design from the DSP supplier such as a debug board [6].

1.5.3 DSP Implementation on ASIP

A DSP ASIP has an instruction set optimized for a single application or a class of applications. On one hand, a DSP ASIP is a programmable machine with a certain level of flexibility, which allows it to run different software programs. On the other hand, its instruction set is designed based on specific application requirements making the processor very suitable for these applications. Low power consumption, high performance, and low cost by manufacturing in high volume can be achieved. In case the processor is used for applications for which it was not intended, poor performance can be expected. For example, using a video or image DSP for radio baseband applications will result in catastrophically poor performance. DSP ASIPs are suitable in volume products such as mobile phones, digital cameras, video camcorders, and audio players (MP3 player).

An ASIP DSP has a dedicated instruction set and dedicated data types. These two features will be discussed and implemented throughout this book. You will find later in this book that one instruction of an ASIP DSP could be equivalent to a kernel subroutine or part of a kernel subroutine running on a general-purpose DSP. For supporting algorithm-level acceleration, special functions will be implemented using specific hardware. This yields better performance, lower power consumption, and higher performance/silicon ratio compared to a general-purpose DSP. At the same time, the range of application is limited due to simplification of the instruction set.

An ASIP can be an in-house product or a commercial off-the-shelf (COTS) component available on the market. In-house means a dedicated design for a specific product within a company. COTS means that the processor is designed as an ASSP (application-specific standard product).

Designing an ASIP DSP is the focus of the remaining part of this book. The following chapters will demonstrate methods of designing ASIP from the instruction set specification down to the microarchitecture and the firmware implementation.

1.5.4 DSP Implementation on ASIC

There are two cases when an ASIC is needed for digital signal processing. The first is to meet extreme performance requirements. In this case, a programmable device would not be able to handle the processing load. The second case is to meet ultra-low power or ultra-low silicon area, when the algorithm is stable and simple. In this case, there is no requirement on flexibility, and a programmable solution is not needed.

A typical ASIC application example is decimation for synthetic aperture radar baseband processing. The requirement for this application is up to 10^{11} MAC (multiplication and accumulation) operations per second with relative low power consumption. Nonprogrammable devices can give such performance (as of 2006). Another typical application is a hearing aid device that includes a band-pass filter bank (synthesizer) and an echo canceller. About 40 MIPS (million instructions per second), high data dynamic range, and the power consumption below 1 mW are required. A programmable device will not likely meet these requirements (as of 2006).

ASIC implementation is to map algorithms directly to an integrated circuit [16]. Comparing a programmable device supplying the flexibility at every clock cycle, an ASIC has very limited flexibility. It can be configurable to some extent in order to accommodate very similar algorithms, but typically it cannot be updated in every clock cycle.

When designing an ASIP DSP, functions are mapped to subroutines consisting of assembly instructions. When designing an ASIC, the algorithms are directly mapped to circuits. However, most DSP applications are so complicated that mapping functions to circuits is becoming increasingly difficult. On the other hand, mapping DSP functions to an instruction set is becoming more popular because the challenge of complexity is handled in both software and hardware, and conquered separately.

Example 1.1 exposes the way to map functions directly to a circuit.

Example 1.1

Mapping a 5-tap FIR filter to a hardware circuit: The algorithm of the 5-tap FIR filter is $y(n) = \Sigma x(n - k)*h(k)$. Here, k is the number of iterations from $k = 0$ to $k = 4$.

The algorithm can be mapped to hardware after unrolling the iteration. The pseudocode of a 5-tap FIR becomes:

```
//5-tap FIR behavior
{
     A1 = x(n)*h0 + x(n-1)*h1;
     A2 = A1 + x(n-2)*h2;
     A3 = A2 + x(n-3)*h3;
     Y(n) = A3 + x(n-4)*h4;
}
```

The pseudo HDL code of a 5-tap FIR becomes:

```
//5-tap FIR HDL
{
     A0[32:0] <= x(n)[15:0]*h0[15:0]
     A1[32:0] <= A0[32:0] + x(n-1)[15:0]*h1[15:0];
     A2[33:0] <= A1[32:0] + x(n-2)[15:0]*h2[15:0];
```

```
        A3[34:0] <= A2[33:0] + x(n-3)[15:0]*h3[15:0];
        A4[35:0] <= A3[34:0] + x(n-2)[15:0]*h4[15:0];
        Y(n)[15:0] <= saturation (round (A4[35:0]));
    }
    clk = '1' and clk'event
    {
        X(n-4)<= x(n-3);  x(n-3)<= x(n-2);  X(n-2)<= x(n-1);
        x(n-1)<= x(n);
    }
```

Finally the circuit is shown in Figure 1.17.

In order to avoid accumulation-induced overflow in this example, the data width should be increased after each accumulation. Because the result $y(n)$ shall have the same data type as the input $x(n)$, a rounding operation is necessary before truncating the lower 16 bits. Overflow might happen, so saturation is necessary before using the result. Rounding and saturation will be discussed in the next chapter.

We can see in Example 1.1 that mapping a FIR to hardware is simple. However, when algorithms or applications are complicated, especially when the algorithm details cannot be decided during the system design, this method cannot be used. In this case mapping applications to an instruction set is the only solution.

1.5.5 Trade-off and Decision of Implementations

In Figure 1.18, the power consumption (based on a 90 nm digital silicon process) is shown as a function of the MOPS (million operations per second) figure for different DSP implementations. The power consumption of the memory is very dependent on the application-specific hardware configuration, so the memory cost is not included in this figure.

The general DSP processor typically is used for applications with not more than a thousand MOPS. The DSP ASIC can support applications with very low

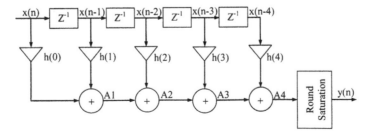

FIGURE 1.17

A 5-tap FIR filter.

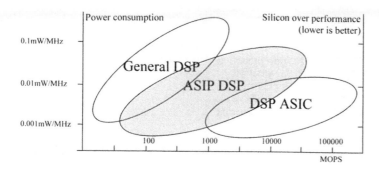

FIGURE 1.18

Comparing three types of DSP implementations.

power consumption (less than 0.01 mW/MHz) and high performance but without requirements on flexibility. The DSP ASIP, however, is used mostly when a trade-off between silicon/power cost, performance, and development effort is required.

1.6 REVIEW OF PROCESSORS AND SYSTEMS

Most DSP applications are implemented using either general-purpose or ASIP DSP processors. A digital signal processor is a programmable integrated circuit for data manipulation. A DSP processor is designed for performing arithmetic functions like add, subtract, multiply, and shift as well as logic functions.

1.6.1 DSP Processor Architecture

Learning processor design is an iterative process, which means that the organization of this book is not an easy task. In this chapter, you will get a bird's-eye view of the methodology of DSP processor design and the processor architecture. In later chapters, you may encounter the same concept several times. In this way, deeper and more complete knowledge will be achieved after each step of iteration.

What Is Inside a DSP?

Similar to other types of processors, a DSP contains five key components:

- Program memory (PM): PM is used for storing programs (in binary machine code). PM is part of the control path.

- Programmable FSM: It is a programmable finite state machine consisting of a program counter (PC) and an instruction decoder (ID). It supplies addresses to the program memory for fetching instructions. Meanwhile, it also performs instruction decoding and supplies control signals to the data processing unit and data addressing unit.

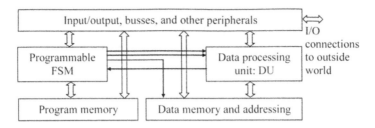

FIGURE 1.19

DSP processor architecture.

- Data memory and data memory addressing: DM stores information to be processed. Three types of data are stored in DM: input/output data, intermediate data in a computing buffer (a part of the data memory), and parameters or coefficients. The data memory addressing unit is controlled by programmable FSM and supplies addresses to data memories.

- Data processing unit (DU): The data processing unit, or datapath, performs arithmetic and logic computing. A DU includes at least a register file (RF), a multiplication and accumulation unit (MAC), and an arithmetic logic unit (ALU). A data processing unit may also include some special or accelerated functions.

- Input/output unit (I/O): I/O serves as an interface for functional units connected to the outside world. I/O also handles the synchronization of external signals. Memory buses and peripherals are also included.

A simplified block diagram is given in Figure 1.19.

1.6.2 DSP Firmware

Before introducing firmware in detail, you should simply accept that firmware is fixed software running in an electronic product. Only real-time firmware will be discussed in this book. Real-time DSP firmware usually consists of an infinite loop that processes real-time signals continuously and periodically. Figure 1.20 depicts a typical top-level infinite loop in a real-time system. One data unit (one data packet or one data sample) is processed in the figure through one complete execution of the infinite loop.

The executable binary code is developed in four steps:

1. **Design the behavior source code.** The behavior source code is the original description that models an application or an algorithm.

2. **Design the hardware dependent source code.** The behavior source code must be modified or rewritten in order to adapt it to the hardware. Hardware adaptation includes modifying hardware constraints and utilizing hardware features.

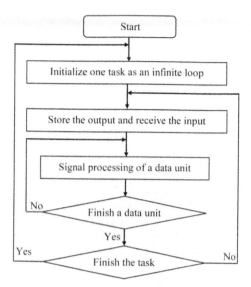

FIGURE 1.20

An infinite loop in a DSP processor.

3. **Design the assembly code.** The hardware-dependent source code is translated to assembly code, or the assembly code is written using the hardware-dependent source code as a reference.

4. **Generate and debug the binary machine code.** Finally, the assembly code is assembled and linked to executable binary machine code. The binary machine code is verified by executing it on an assembly instruction set simulator.

Most DSP applications do not require ultra-high data precision. Therefore, fixed-point processors can be used in order to reduce the silicon cost. For a fixed-point DSP the silicon cost can be less than half compared to a standard floating-point DSP. As fixed-point DSP processors are dominant on the DSP market, the discussions in this book will be focused on this type of processor.

Quantization (noise) error and offset of frequency response are unwanted behaviors when using a fixed-point processor. These errors are due to the finite data length, and handling them will increase firmware complexity. The firmware must maintain the quality of input and intermediate data in the computing buffers. Data quality here stands for a measure of the quantization error (which should be minimized) and the dynamic range (which should be maximized). The final firmware for a fixed-point DSP will contain both functional firmware and firmware for data quality control. The firmware for data quality control includes data quality measurement and scaling (to be discussed with Figure 1.22 and in detail in Chapter 18). Data quality can be optimized by data scaling. Normally, the execution of the data quality control firmware does not occur very often, and it is only needed when the amplitude of the input data is changed or the algorithm in the firmware is changed.

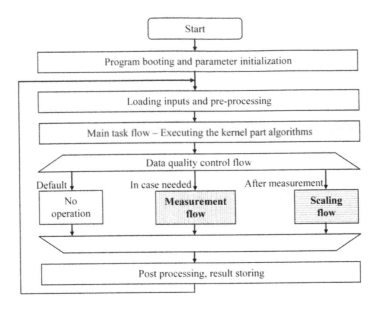

FIGURE 1.21

A typical fixed-point DSP firmware flow.

The program flow is shown in Figure 1.21 and further illustrated in Figure 1.22. The complete firmware running on a fixed-point processor can be divided into three flows. The main flow is shown with gray background, the measurement flow is shown with a background of vertical filling lines, and the scaling flow is shown with a background of horizontal filling lines.

In Figure 1.21 and Figure 1.22, the main flow is executed once for each input streaming data. The data quality control flows are not executed very often. In most cases, the quality control part is skipped (via the default path in Figure 1.21) until it is needed. The measurement flow monitors inputs and intermediate results. In the measurement flow, parameters such as signal-to-noise ratio, signal level on average, and maximum/minimum values will be measured. Other parameters may also be observed, for example, event counting of overflow and saturation. Counters are used for collecting statistics of certain events. The results from measurements will be used in the scaling flow. Finally, scaling factors for input, output, and internal data are modified.

Firmware design is not the focus of this book. Instead, firmware knowledge introduced in this chapter is intended only as background knowledge of source code profiling for instruction set design.

1.6.3 Embedded System Overview

An embedded system is a system that is inside a product system or a product [18]. An embedded system is a special-purpose computer system designed to perform one or a class of dedicated functions. In contrast, a general-purpose computer, such

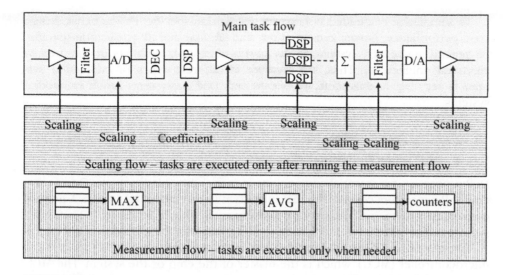

FIGURE 1.22

Relation between function firmware and quality control firmware.

as a personal computer, can do many different tasks, depending on programming. The product user may not even be aware about the existence of the embedded system, although it may play an important role for the function of the product. An embedded system could be a component of a personal computer such as a keyboard controller, mouse controller, or a wireless modem. An embedded system could also be a digital subsystem inside a mobile phone, a digital camera, a digital TV, or in medical equipment. Except for general computers, most microelectronic systems are embedded systems.

A general computer system is not designed for any specific purpose. A desktop computer can be a general-purpose computing engine, a home electronics system, a documentation editing system, a media terminal, or a network terminal. An embedded system is an application-specific system, different from a general computer system. Within the specific application domain, the embedded system may have much higher performance or much lower power consumption compared to a general computer system. An embedded system, such as a radar signal processing system or a computer tomography (CT) processing system, can have Teraflops performance (~1000 times more than the performance of a Pentium). Ultra low power consumption in the range 30 to 50 MIPS/mW is possible for hearing aid embedded systems.

However, it will be impossible to use an embedded processor for applications outside the specific domain. For example, a processor designed for controlling a washing machine can never be used as a DSP for video applications. On the other hand, a desktop computer can handle both these tasks. But flexibility has a price tag. A processor for a washing machine can cost less than one dollar, and a Pentium 4 in a desktop costs much more than $100 (in 2005).

To summarize: embedded systems are application-specific. Product cost, design cost, performance, power consumption, and lifetime are all application-specific. In general, embedded system design covers almost all activities in the area of electrical engineering. Thus, the number of different types of embedded systems is very big. In this book, we discuss only DSP subsystems inside embedded systems.

1.6.4 DSP in an Embedded System

DSP processors are essential components in many embedded systems. One or several DSP processors consist of a DSP subsystem in an embedded system. A general embedded system, including a DSP subsystem, is shown in Figure 1.23. Such a system is also called a system on a chip (SoC) platform for embedded applications.

The system in Figure 1.23 can be divided into four parts. The first part is the microcontroller (MCU), which is the master of the chip or the system. The MCU is responsible for handling miscellaneous tasks, except computing for real-time algorithms. Typical miscellaneous tasks are operating system, connection protocols, Java programs, human-machine-interface, and hardware management.

The second part is the ASIP DSP subsystem including accelerators, which is the main computing engine of the system. All heavy computing tasks should be allocated to this subsystem. The DSP subsystem could include a single processor with accelerators or a multicore processor cluster. For example, in a 3G mobile phone, the DSP subsystem usually has two processors: a baseband processor and an application processor.

The third part is the memory subsystem, which supports data and program storage for the DSP subsystem and the MCU. A SoC usually has multiple levels of memories. Within the MCU core and the DSP core, local memories or level-1 caches

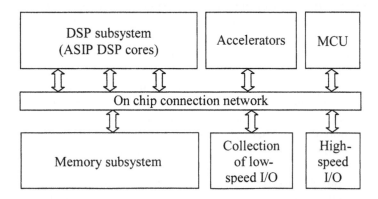

FIGURE 1.23

DSP processor in an embedded system.

can be found. At SoC level, a level-2 cache or an on-chip main memory can be found. Level three in the memory hierarchy is the off-chip main memory.

The fourth part consists of peripherals including high-speed and low-speed I/Os. Analog circuits could be part of the low-speed I/O.

1.6.5 Fundamentals of Embedded Computing

Within a DSP subsystem, embedded computing can be divided into three parts, computing using ASIC, computing using HW/SW (processors + accelerators), and computing using SW. In this book, the focus is to discuss the computing using HW/SW and SW instead of using ASIC. Embedded SW can be divided further into three categories: operating system, SW for real-time computing, and SW for best-effort embedded computing. In this book, the focus is the SW for real-time computing.

SW for real-time computing must be executed on a specific DSP platform based on real-time scheduling. The platform must supply enough computational capacity, and the real-time scheduling guarantees that the execution will consume less time than the time interval between arriving input data.

In an embedded system, the complexity or dependency is relatively known before runtime. Static scheduling can therefore be used to enhance the utilization of the time and hardware resources. In a general computing system, on the contrary, the management of running applications is dynamic, and the issue of a task is based on a dynamic priority table.

Another special feature of embedded computing is the application-specific precision. Both the precision of operands and the precision of computing are specified according to the application. The precision of audio processing can be rather high (around 24 bits), whereas the precision of video can be as low as 8 to 12 bits.

Example 1.2

Classify the following systems. Which is a real-time system, and which is not a real-time system?

- A mobile phone receiving and sending voice data.

- A mobile phone receiving and sending short message packets.

- A person calculating the statistics on the quality of stored data.

- A computer that analyzes stock information.

Answers

- *Mobile phone sending and receiving voice data.* It is a real-time system because the decoding of the received voice packet must be finished before the arrival of the next voice data packet.

- *Mobile phone receiving and sending short messages.* The baseband part is a real-time system because a symbol must be processed before the arrival of the next symbol. The application part for message display is not a real-time system, because the display of the short message can be delayed for a while without affecting the functionality of the system.

■ *Statistics on stored data. It is not a real-time system.* Your boss might ask you to speed up the process. Nevertheless, the arriving new data will not be lost if your processing speed is low.

■ *Stock analysis.* Even though the processing result must be available in time, it is not a real-time system because the new arriving data will not be lost when the processing is slow.

1.7 DESIGN FLOW

The system development process from conceptualization to manufacturing includes modeling, implementation, and verification. This is known as the design flow. The way to conduct designs using the design flow is called design methodology. In this section, methodologies for designing hardware, including processors, will be briefly reviewed. The ASIP DSP design methodology will be discussed in detail in Chapter 4. The purpose of this early introduction to design methodology is to make the following chapters easier to understand.

1.7.1 Hardware Design Flow in General

The design of an embedded system includes implementation of complete and correct functions with specified performance (not the highest), affordable cost, reasonable reliability, and within a limited amount of design time. In most cases, the product lifetime is also a design parameter. In order to design a system, with complex functions optimized and allocated to both hardware and software, an efficient and reliable design flow or methodology is required.

A design consists of several transformation steps from a high-level to a low-level description. More hardware, control information, and constraints are inserted during each transformation to a lower level. A transformation from one level to the next lower level can be executed via description or synthesis. Two basic types of transformations, described as two design methodologies, are introduced in most methodology books: the capture-and-simulate methodology and the describe-and-synthesize methodology [19].

Capture-and-simulate has been the dominant methodology since the 1960s. According to this method, the system is described at every level and each description is proved by simulation. This method is not efficient, because it consumes a lot of design time and designs cannot be sufficiently optimized. However, this is still the only proven way (as of 2007) of high-level design from system specification down to RTL coding.

The describe-and-synthesize methodology was introduced during the late 1980s by the success of logic synthesis. This methodology can be further divided into two levels of abstraction. At the higher level, behavioral synthesis translates the behavioral description into a structural RTL description, including the dataflow at bit level and the control details at cycle and bit level. The main tasks for behavioral synthesis are allocation, scheduling, and binding. At the lower level, RTL synthesis translates the RTL code to a gate level description, the netlist.

Behavioral synthesis is currently (in 2007) far from mature. In most cases, behavioral synthesis of an ASIP is impossible today. There is hardly any automatic method for translating an assembly instruction set to a good HW architecture. One reason for this is the extremely large design space that makes it difficult to optimize the design toward HW multiplexing, high performance, and power efficiency.

A well-known methodology for embedded system design is to divide the design activities using the famous Y-chart, shown in Figure 1.24, which was proposed by Professor Daniel Gajski [19].

By using the Y-chart, an assumption is made that each design can be modeled in three basic ways, emphasizing different properties no matter how complex the design is. The Y-chart has three axes representing design behavior (function, specification), design structure (net list, block diagram), and physical design (layout, boards, packages).

Behavior design of a system means that the design is represented as a black box with specified input and output data relations. The black box behavior does not indicate in any way how it is implemented or how its internal structure looks.

Structure design gives a description of the hardware partition and the relation (interconnection) between the black boxes. Each black box should be further refined by a more detailed structure description including functionality, timing cost, and interconnects until the design reaches the register transfer level.

Physical design implements functions based on logic gates or transistors and interconnects. Physical features include the size and the position of each

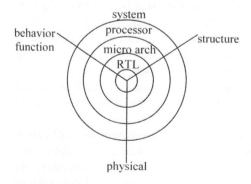

FIGURE 1.24

Y-chart.

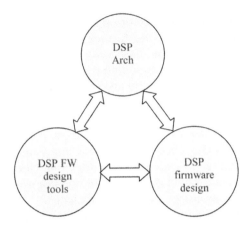

FIGURE 1.25

Knowledge required in ASIP design.

component, the size, the wire routing between components, the placement, physical delay, the power consumption, and the thermo behavior. Physical parameters are taken into account, such as parasitic capacitance and resistance of interconnects.

1.7.2 ASIP Hardware Design Flow

Processor design is a complicated process. Without an advanced design flow, a processor cannot be designed in time and the quality of the design will not be high. The design flow is therefore essential for complicated systems such as ASIP.

The ASIP design flow is introduced briefly here; the detailed discussion will be found in Chapter 4. The ASIP design flow is divided into three parts: architecture design, design of programming tools, and firmware design, as depicted in Figure 1.25. In this book, focus will be on architecture design, and in particular on system architecture and hardware development flow. Other topics such as design of programming tools and design of application firmware will be addressed briefly.

The first and most important step in the design of a processor is the instruction set design. This design step is complicated, and no one can really claim that a certain instruction set is the best. The instruction set design is a trade-off among a multitude of parameters including performance, functional coverage, flexibility, power consumption, silicon cost, and design time. In Figure 1.26, a simplified design flow is described including the basic flow for the design of an instruction set architecture.

The starting point of the design of an ASIP is the application analysis. Application coverage should be specified first and then translated to functional (algorithm) coverage. Application coverage is the process of reading and understanding specifications and standards of the relevant applications. Functional coverage of an ASIP is decided based on both the current standard specifications and carefully collected knowledge (e.g., books and research publications) in order to add extra

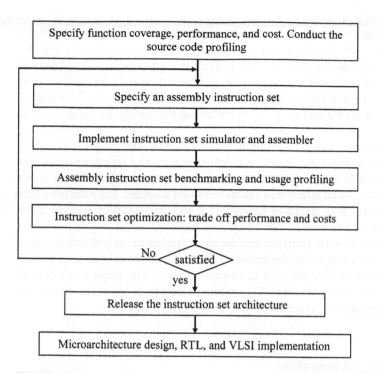

Specify function coverage, performance, and cost. Conduct the source code profiling

Specify an assembly instruction set

Implement instruction set simulator and assembler

Assembly instruction set benchmarking and usage profiling

Instruction set optimization: trade off performance and costs

satisfied — No / yes

Release the instruction set architecture

Microarchitecture design, RTL, and VLSI implementation

FIGURE 1.26

ASIP design flow.

features for future usage. Performance and cost should also be specified as design constraints.

After the functional coverage is determined, the partitioning of hardware and software should be decided through profiling of the source code. Hardware/software partitioning for an ASIP is to meet the performance constraint by defining what functions should be accelerated by application-specific instructions and what functions should be implemented as software routines using conventional instructions. This is an important design step of an instruction set, which is called the 10%–90% code locality. The locality rule means that 10% of the instructions run 90% of the time and 90% of the instructions appear only 10% of the time during execution. In other words, ASIP design is to find the best instruction set architecture optimized for the 10% most frequently used instructions and to select those among the 90% of the not often used instructions in order to guarantee the functional coverage. Application analysis and HW/SW codesign will be discussed in detail in Chapters 4 and 6.

During the process of hardware and software partitioning, the instruction set of the ASIP is gradually specified. The next design step is to implement the instruction set, which includes instruction coding, design of the instruction set simulator, and benchmarking. The coding of the instruction set includes the design of the assembly syntax and the design of the binary machine codes. The instruction set simulator must be implemented after the instruction set has been coded. Finally, the instruction set must be evaluated by benchmarking. The performance of the instruction set and the usage of each instruction will be exposed as inputs for further optimization. Instruction set design will be discussed in detail in Chapter 7. Design of assembly instruction set simulator and benchmarking will be discussed in detail in Chapters 8 and 9.

The ASIP architecture can be specified when the assembly instruction set is released. The microarchitecture design is a refinement of the architecture design including fine-grained function allocation and hardware pipeline scheduling, specifying hardware modules, and interconnections between modules. Microarchitecture design will be discussed in detail in Chapter 10, and case studies will be given throughout Chapters 11 to 15.

The ASIP design flow starts from the requirement specification and finishes after the microarchitecture design. The design of an ASIP is based mostly on experience, and it is essential to minimize the cost of design iteration. The implementation of the microarchitecture, which involves RTL coding, is not the focus of this book. You can find more information in references [17] and [22].

1.7.3 ASIP Design Automation

This sub-section is written for researchers and project managers. Custom design of an ASIP DSP Processor is based on experience and is error prone. Design automation has been investigated and can be used to replace the custom ASIP design. ASIP design automation tools is summarized in Figure 1.27.

Figure 1.27 presents the research on ASIP design automation. The ASIP design automation can be divided into three steps. The first step is the architecture exploration (selecting or generating an architecture and assembly instruction set according to the application analyses). Different profilers were designed by researchers, but the constraint specification tool has not been investigated. The tool to merge multiple CFG has not been extensively investigated. CFG in this step stands for control flow graph, and it will be discussed in detail in Chapters 6 and 8.

The second step is to specify an ADL (Architecture Description Language) to model the instruction set and architecture. It is very difficult. The language must be easy so that ASIP designers can use it in modeling the design. However, if an ADL is easy, it cannot carry sufficient information to generate all tools and architectures. For example, tool generation requires sufficient modeling of the instruction set; and the hardware (datapath, control path) generation requires sufficient modeling of microarchitecture (for example, structure of hardware multiplexing) and its function (to be discussed in Chapter 10). If the ADL carries sufficient information for generating tools and architectures, the ADL will not be readable and cannot be used by ASIP designers. Details are beyond the scope of this book and can be found in our NoGAP research [38, 39].

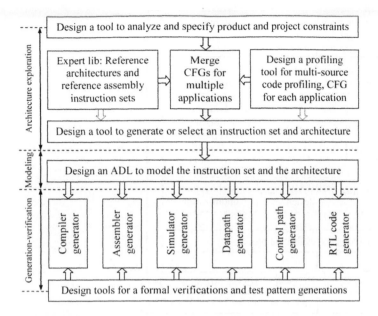

FIGURE 1.27

Automatic ASIP design flow (tool researcher's view).

The third step involves generations and verifications. Enormous research on generation of tools and architectures can be found in Tensilica [32, 33] and the famous LISA project [34,35]. As of 2007, there have been few research contributions to the ASIP formal verification.

It is good for designers to know the basic concepts behind design automation tools. However, this is not the purpose of this book. This book will give more information on how to select and use ASIP design automation tools. Designers' interests actually focus on how to use the tool to generate instruction set, architecture, and assembly programming tools, as well as support for design verifications. It is summarized in Figure 1.28.

Architecture and assembly instruction set exploration according to application profiling is the first step of processor design automation. This is actually the most difficult part because the distance between CFGs (control flow graphs) of multiple applications and an ASM (assembly instruction set) is very large, and there are too many choices to select different instruction set architectures. To manage the large gap, another design step (constraint specification) might be needed.

Application profiling will be discussed in Chapter 6, and instruction decision will be discussed in Chapter 7. However, ways to make decisions by tools are not much investigated. Tools to generate accelerator instructions are available, but the tool to generate a complete processor instruction set does not exist.

Instruction set architecture of a processor is proposed and decided by designers, and the instruction set and the architecture selected will be inputs of

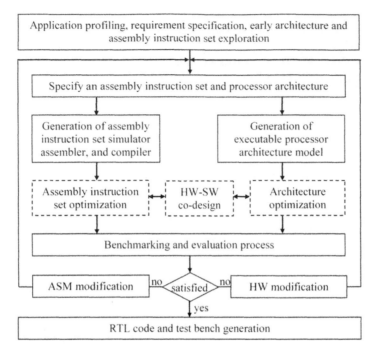

FIGURE 1.28

Automatic ASIP design flow (designer's view).

processor modeling. The processor model will be used for generating the instruction set simulator, the compiler, assembler, and the architecture behavior model. After benchmarking of the instruction set and architecture, RTL code will be finally generated by the ASIP automation design tools. In Table 1.5, selected ASIP design tools are briefly discussed [25–39].

The last column in the table shows a feature of ASIP design automation tools, the instruction set and architecture selection may or may not be limited by a built-in architecture or not. Limited by an architecture means that the automation tool offers instruction extension based on a built-in processor instruction set architecture. The drawback is obviously that the built-in processor may not be suitable for all applications. Therefore, "no" in that column is superior to "limited".

MIMOLA is possibly the first processor design automation tool. It was proposed by Professor Zimmermann in 1976 and was the research product of Professor Zimmermann and Professor Marwedel of Kiel University and TU Dortmund in Germany. Cathedral-I and II proposed by Professor Rabaey and Professor De Man IMEC in Belgium was possibly the first tool successfully used by industry (Mentor DSP station). Target was a successful research spin-off of Dr. Gert Goossens of IMEC, Belgium. ARC and Tensilica are successful companies that supply programmable hardware acceleration based on their processor core as the master of the platform.

Table 1.5 Review of ASIP Design Automation Tools.

Tool or solution	Profiling and architecture exploring	Compiler generator	Assembler generator	Simulator generator	Cycle accurate architecture behavior model generator	Instruction set and architecture Optimizer	Datapath RTL code generator	Control path RTL code generator	Limited by an architecture
MIMOLA				yes			yes	yes	no
Cathedral-II	yes			yes	yes		yes	yes	no
Target	yes	yes	yes	yes	yes		yes	yes	no
ARC		yes	yes	yes	yes		yes	yes	limited
Tensilica	yes	yes	yes	yes	yes	yes	yes	yes	limited
LISA	yes	yes	yes	yes	yes		yes	yes	limited
MESCAL	yes		yes	yes	yes		yes	yes	no
PEAS-III			yes	yes	yes		yes	yes	limited
NoGAP	yes		yes	yes	yes		yes	yes	no
...									

LISA of CoWare was the research spin-off of Professor Meyr and Professor Leupers of ISS Aachen University, Germany. MESCAL is the research project of GigaScale of UC, Berkeley, California. NoGAP is the research project of Linköping University, Sweden.

1.8 CONCLUSIONS

The scope of DSP is huge, and includes DSP theory, DSP-related standards and applications, and DSP implementations. The intention of this book is to make a good partitioning of DSP concepts into these three categories, and to focus on the hardware implementation of application-specific instruction set DSP processors.

In this chapter, knowledge required by the ASIP designer was briefly reviewed. DSP implementation was discussed based on four platforms—the general computer, the general-purpose DSP, the ASIP DSP, and the ASIC. DSP architectures were briefly introduced. Furthermore, the concepts of embedded system and embedded computing also have been introduced because the DSP subsystem is an essential part of most embedded systems. ASIP design flow and methodologies of ASIP design automation were briefly introduced. All topics covered in this chapter are only introductions; they will be discussed in more detail in the following chapters.

We strongly recommend that you study the listed reference books [2, 17, 21, 22] if you are not acquainted with the background knowledge discussed in this chapter.

EXERCISES

1.1 Read a CPU hardware book (for example, [21]) and understand the following concepts (not necessary if you already understand CPU principles):

   a. Computer architecture basics.

   b. Datapath and control path.

   c. Pipeline and parallelism.

   d. Bus and memory.

   e. Processor, interface, and peripherals.

1.2 Read a DSP theory book (for example, [2, 3]) and understand the following concepts (not necessary if you already have basic knowledge of DSP theory):

   a. Time- and frequency-domain signal representation.

   b. DFT, FFT, Z-transform, convolution.

   c. FIR, IIR, decimation, interpolation.

   d. The finite word-length effect.

   e. Adaptive filter.

1.3 (For PhD students) Read a design methodology book (for example, [22]).

1.4 Give two examples of real-time systems. Give two examples of non-real-time systems.

1.5 Explain the difference between DSP theory and DSP Applications.

1.6 What is the relation between program memory and PC?

1.7 In which kind of (general-purpose or ASIP) DSP processor can you find many instructions for algorithm accelerations?

1.8 Discuss and recognize the main differences between general computer systems and embedded systems.

1.9 What are the inputs and what are the outputs of an ASIP design?

1.10 Which of following statement(s) is (are) true?

    a. Runtime of a real-time system cannot be predictable.

    b. A general-purpose DSP processor is the best solution for volume product.

    c. DSP implementation on ASIP can decrease the design cost regardless of the volume.

    d. Infinite loop mentioned in this chapter is used for performance demanding applications.

REFERENCES

[1] http://www.da.isy.liu.se/.

[2] Sanjit, V., Mitra, K. (1998). *Digital Signal Processing, A Computer-Based Approach.* McGraw-Hill.

[3] Madisetti, V. K., Williams, D. B. (1997). *The Digital Signal Processing Handbook.* CRC Press, IEEE Press.

[4] Norsworthy, S. R., Schreier, R., Gabor, Temes, C. (eds.) (1997). *Delta-Sigma Data Converters, Theory, Design, and Simulation.* IEEE Press.

[5] For example, www.ieee.org; www.iso.org; www.itu.org; and www.etsi.org.

[6] For example, www.ti.com, www.adi.com, www.ceva.com, and www.freescale.com.

[7] Krishna, C. M., Shin, K. G. (1997). *Real-Time Systems.* McGraw-Hill.

[8] Laplante, P. A. (2004). *Real-Time Systems Design and Analysis,* 3rd ed. IEEE Press and Wiley.

[9] Cibson, J. D. (ed.). (1999). *The Mobile Communications Handbook,* 2nd ed. CRC Press and IEEE Press.

[10] Liu, D., Tell, E. (2005). *Low Power Baseband Processors for Communications, Low Power Electronics Design,* Chapter 23, C. Piguet, ed. CRC Press.

[11] Salomon, D. (2006). *Data Compression, the Complete Reference,* 3rd ed. Springer.

[12] Kumaki, T. et al. (2005). Multi-port CAM based VLSI architecture for Huffman coding with real-time optimized code word table. *48th Midwest Symposium on Circuits and Systems* **1**, 55–58.

[13] Goldberg, R. Riek, L. (2000). *A Practical Handbook of Speech Coders.* CRC Press.

[14] Collura, J. S., Brandt, D. F., Rahikka, D. J. (1999). The 1.2Kbps/2.4Kbps MELP Speech Coding Suite with Integrated Noise Pre-Processing. *IEEE MILCOM* **2**, 1449–1453.

[15] www.mpeg.org and www.jpeg.org.

[16] Wanhammar, L. (1999). *DSP Integrated Circuits.* Academic Press.

[17] Smith, M. J. S. (1997). *Application-Specific Integrated Circuits (ASIC).* Addison-Wesley VLSI Systems Series.

[18] Wolf, W. (2001). *Computers as Components, Principles of Embedded Computing System Design.* Morgan Kaufmann.

[19] Gajski, D. D., Vahid, F., Narayan, S., Gong, J. (1995). *Specification and Design of Embedded Systems, Technique or Embedded Systems.* Prentice Hall.

[20] Lapsley, P., Bier, J., Shoham, A., Lee, E. A. (1997). *DSP Processor Fundamentals, Architectures and Features.* IEEE Press.

[21] Patterson, D. A., Hennessy, J. L. (2004). *Computer Organization & Design, The Hardware/Software Interface.* Morgan Kaufmann.

[22] Keiting, M., Bricaud, P. (2007). *Reuse Methodology Manual for System On Chip Designs*, 3rd ed. KAP.

[23] Kuo, S. M., Gan, W-S. (2005). *Digital Signal Processors, Architectures, Implementations, and Applications.* Prentice Hall.

[24] Fettweis, G. P. (1997). DSP Cores for Mobile Communications: Where are we going?. *IEEE International Conference on Acoustics, Speech, and Signal Processing (ICASSP'97), page(s)* 279–282, 21.-24.04.97

[25] Zimmermann, G. (1979). The MIMOLA design system, a computer aided digital processor design method, *16th Design Automation Conference.* 1979.

[26] Marwedel, P. (1984). The MIMOLA design system: Tools for the design of digital processors, *21st Design Automation Conference.* 1984.

[27] Krüger, G. (1980). Entwurf einer Rechnerzentraleinheit für den Maschinenbefehlssatz des SIEMENS Systems 7.000 mit dem MIMOLA-Rechnerentwurfssystem, *Diploma Thesis, University of Kiel, Kiel.*

[28] De Man, H., Rabaey, J., Six, P., and Clesen, L. (1986). Cathedral-II, A Silicon Compiler for Digital Signal Processing, *IEEE J. of Design and Test of Computers, December.* 1986.

[29] http://www.retarget.com/

[30] Goossens, G., Lanneer, D., Geurts, W., Van Praet, J. (2006). Design of ASIPs in multi-processor SoCs using the Chess/Checker retargetable tool suit, *IEEE International Symposium on System-on-Chip.* 2006.

[31] http://www.arc.com/

[32] http://www.tensilica.com/

[33] Rowen, C. (2004). *Engineering the Complex SOC.* PTR.

[34] Hoffmann, A., Meyr, H., Leupers, R., (2002). Architecture Exploration for Embedded Processors with LISA, *Kluwer Academic Publishers, ISBN 1-4020-7338-0, Dec. 2002.*

[35] http://www.coware.com/PDF/products/ProcessorDesigner.pdf

[36] Mihal, A., Kulkarni, C., Moskewicz, M., Tsai, M., Shah, N., Weber, S., Yujia Jin, Keutzer, K., Vissers, K., Sauer, C., Malik, S. (2002). Developing architectural platforms: a disciplined approach, *Design & Test of Computers, IEEE Volume 19, Issue 6, Nov.-Dec. 2002 Page(s):6-16.*

[37] Itoh, M., Higaki, S., Sato, J., Shiomi, A., Takeuchi, Y., Kitajima, A., and Imai. M. (2000): Peas-iii: an asip design environment. In *Computer Design, 2000. Proceedings. 2000 International Conference on,* pages 430–436, 2000.

[38] Karlström, P., Liu, D. (2008). NoGAP, a Micro Architecture Construction Framework, *19th IEEE International Conference Application-specific Systems, Architectures and Processors, July 2008 Belgium.*

[39] http://www.da.isy.liu.se/research/nogap/

Numerical Representation and Finite-Length DSP

In this chapter, numerical representations will be introduced. To reach low silicon costs and low power consumption, fixed-point DSP processors will be introduced. Because the design for fixed-point DSP processors will be the focus of the book, the finite-length data behavior will be discussed. Skills will be introduced to keep the highest data precision and dynamic range using low-cost hardware with limited procession. The usage of integer and fractional data for multiplications will be discussed. Design for finite-length datapath with minimized bias and quantization errors are discussed using examples. Handling overflow in finite-length datapath will also be discussed.

2.1 FIXED-POINT NUMERICAL REPRESENTATION

To limit the silicon cost and keep acceptable performance (processor speed) as well the low power consumption, fixed-point DSP processors and other embedded processors are superior to general-purpose floating-point processors for volume embedded products. To get relative high data precision and dynamic range is one of the main design challenges when designing embedded products using fixed-point numerical presentations. Precision and dynamic range will be discussed in detail in later sections. Simple definition of precision (resolution) of a fixed-point data is its smallest value of the numerical representation system. For example, precision of a 16-bit signed data is 2^{-15}. Dynamic range is the largest value over the smallest value of a numerical representation system. For example, the dynamic range of a 16-bit signed data is $(1 - 2^{-15})/2^{-15} = 2^{15} - 1$.

Limited by fixed-point hardware, the precision of the computing results easily could become unacceptable if finite data length problems were not managed carefully. Specific knowledge and design skills to be discussed in this chapter are essential to keep the best data and processing quality using low-cost hardware with limited precision. To gain necessary hardware and firmware design experiences, a deep understanding of fixed-point numerical presentations and finite-length data behaviors are needed. The purpose of this chapter is to introduce these concepts.

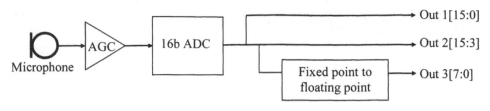

FIGURE 2.1

Checking the voice quality of three voice data formats.

2.1.1 An Intuitive Example

Before discussing the trade-off between data quality and silicon costs, let us intuitively check the quality of phone voice based on three data formats. Let us start data sampling from a high quality microphone followed by a high quality audio amplifier with AGC, automatic gain control, and a high quality ADC (analog to digital converter) in Figure 2.1.

The precision of ADC is 16 bits. The voice from out 1 is represented by the original 16 bits of data; the voice from out 2 has 13 bits of out 1—three of the lower out 1 bits are truncated; the voice from out 3 is the 8 bits floating-point representation of out 2. The function of the fixed-point to floating-point module is specified in Table 2.1.

Example 2.1

Using C-language to mask a source voice code to 16 bits format out1.wav and to 13 bits format out2.wav; Translating out2 to out3.wav based on the specification in Table 2.1.

```
/* EXAMPLE 1: 16-bit data */
out1_16b = in_16b;
/* END EXAMPLE 1 */

/* EXAMPLE 2: 13-bit data */
out2_13b = (in_16b >> 3) & 0x1fff;
/* END EXAMPLE 2 */

/* EXAMPLE 3: 8-bit floating-point data */
/* Get the 13 bit sample. */
sample = out2_13b;
sign = (sample >> 12) & 1;
/* Shift left until the two most significant
 * bits of the 13-bit number are different,
 * but not more than 7 steps. */
exponent = 7;
temp = sample ^ (sample<<1);
```

```
while((((temp & 0x1000) == 0) && (exponent > 0)) {
  exponent--;
  temp <<= 1;
  sample <<= 1;
}
if(exponent == 0) {
  /* Special case for exponent==0: */
  mantissa = (sample >> 8) & 0xf;
} else {
  /* For all other cases: */
  mantissa = (sample >> 7) & 0xf;
}
out3_8b = (sign << 7) | (exponent << 4) | mantissa;
/* END EXAMPLE 3 */
```

Table 2.1 Translation between 8 Bits Floating-Point and 13 Bits Linear Code.

13 bits linear voice samples	8 bits floating-point		
	Sign	Exponent	Mantissa
$01Q_3Q_2Q_1Q_0$XXXXXXX	0	111	$Q_3Q_2Q_1Q_0$
$10Q_3Q_2Q_1Q_0$XXXXXXX	1	111	$Q_3Q_2Q_1Q_0$
$001Q_3Q_2Q_1Q_0$XXXXXX	0	110	$Q_3Q_2Q_1Q_0$
$110Q_3Q_2Q_1Q_0$XXXXXX	1	110	$Q_3Q_2Q_1Q_0$
$0001Q_3Q_2Q_1Q_0$XXXXX	0	101	$Q_3Q_2Q_1Q_0$
$1110Q_3Q_2Q_1Q_0$XXXXX	1	101	$Q_3Q_2Q_1Q_0$
$00001Q_3Q_2Q_1Q_0$XXXX	0	100	$Q_3Q_2Q_1Q_0$
$11110Q_3Q_2Q_1Q_0$XXXX	1	100	$Q_3Q_2Q_1Q_0$
$000001Q_3Q_2Q_1Q_0$XXX	0	011	$Q_3Q_2Q_1Q_0$
$111110Q_3Q_2Q_1Q_0$XXX	1	011	$Q_3Q_2Q_1Q_0$
$0000001Q_3Q_2Q_1Q_0$XX	0	010	$Q_3Q_2Q_1Q_0$
$1111110Q_3Q_2Q_1Q_0$XX	1	010	$Q_3Q_2Q_1Q_0$
$00000001Q_3Q_2Q_1Q_0$X	0	001	$Q_3Q_2Q_1Q_0$
$11111110Q_3Q_2Q_1Q_0$X	1	001	$Q_3Q_2Q_1Q_0$
$00000000Q_3Q_2Q_1Q_0$X	0	000	$Q_3Q_2Q_1Q_0$
$11111111Q_3Q_2Q_1Q_0$X	1	000	$Q_3Q_2Q_1Q_0$

Original voice waveform samples can be found from the book home page or from the MPEG home page. By listening and comparing the results out1, out2, and out3, almost no differences can be identified if the source code is "only human voice without music background." There are two reasons:

- The human ear has the logarithm sensitivity on sound volume. (Voice can be compressed by logarithm data representation.)

- The weak sound is masked by strong sound and cannot be recognized by human being. (High mantissa resolution is not necessary.)

Normal people therefore cannot identify the difference among voices represented using 16-bit, 13-bit, and logarithm 8-bit formats, the opportunities can be utilized to minimize the cost of hardware and data bandwidth.

The conclusion is that custom applications require custom data types. The principle of representing 13 bits linear voice data using 8 bits floating-point voice data in Table 2.1 has been used in telephone transmission to reduce the cost of bandwidth [1]. If the ADC sampling rate is 8 kHz, the data rate of out1 is 128 kb/s, the data rate of out2 is 104 kb/s, and the data rate of out3 is only 64 kb/s. Both transmission bandwidth and silicon costs of using data format out 3 is only half the cost of using the data format of out 1 and about 62% of the cost using the data format of out 2. The conclusion from the intuitive example in Figure 2.1 is that acceptable data precision gives cost efficiency opportunities for a class of applications.

To further explore opportunities of cost minimization by using fixed-point data and hardware, fixed-point numerical representations should be reviewed briefly. Detailed knowledge of fixed-point representation and binary computing arithmetic can be found in reference [2].

2.1.2 Fixed-Point Numerical Representation

Numerical representation is a discrete representation of data by numerals. It consists of rules and principles to specify numerical values with finite precision and dynamic range for DSP implementations. These rules can support the designer when designing DSP instruction set and hardware arithmetic functions with finite precision. It also gives rules to design DSP software using DSP behavior language (C or MATLAB) and rules to design for the data storage in a DSP processor. Thus, understanding numerical representations is the essential step in learning DSP implementations.

Detailed specifications of numerical presentations can be found from the language specification of ISO-C and DSP-C. DSP-C is a C language extension to ISO-C (ISO/IEC IS 9899:1990) for DSP firmware and hardware development. The specification is called ISO/IEC/JTC1/SC22 WG14/N854, which defines different data types, memory space, arrays, and circulation buffers.

By definition, a fixed-point number representation defines a real data type for a number that has a fixed number of digits before and after the radix point, "." (It is also called decimal notation. For decimal systems, it is also called binary point or simply "point" in this book). Fixed-point number representation stands in contrast to floating-point number representation, which is discussed in the next section of this chapter. Fixed-point numbers are useful when floating-point unit (FPU) is not used. Actually, most low-cost embedded processors, including DSP processors, do not have an FPU. In this book, we design mainly fixed-point DSP processors without hardware FPU.

A fixed-point number may be written as $M.F$, where M represents the magnitude (the integer part), "." is the radix point, and F represents the fractional part. A fixed-point number system is defined according to Equation 2.1:

$$V = \sum_{k=\alpha}^{N_d+\alpha-1} \varphi_k d_k b^k \qquad (2.1)$$

where N_d is the number of digits, $\alpha = -F$ is the position of the least significant digit, and b is the current base (e.g., 2 for binary, 10 for decimal, or 16 for hexadecimal). In this book, only decimal will be discussed, and b is implicitly equal to 2. The d is the digit weight (e.g., $0-1$ in binary, $0-9$ in decimal, or $0-F$ in hexadecimal), and φ_k is positive or negative weight (1 or -1). For decimal and hexadecimal systems, φ_k is 1 for positive data and φ_k is -1 for negative data.

Example 2.2

Represent the decimal numbers 99.09, -99.09 and the hexadecimal numbers FA.2B, $-FA.2B$ following Equation 2.4.

V for decimal 99.09 is $V = 1 \times 9 \times 10^1 + 1 \times 9 \times 10^0. + 1 \times 0 \times 10^{-1} + 1 \times 9 \times 10^{-2}$

V for decimal -99.09 is $V = -(1 \times 9 \times 10^1 + 1 \times 9 \times 10^0. + 1 \times 0 \times 10^{-1} + 1 \times 9 \times 10^{-2})$

V for hexadecimal $FA.2B$ is $V = 1 \times F \times 16^1 + 1 \times A \times 16^0. + 1 \times 2 \times 16^{-1} + 1 \times B \times 16^{-2}$

V for hexadecimal $-FA.2B$ is $V = -(1 \times F \times 16^1 + 1 \times A \times 16^0. + 1 \times 2 \times 16^{-1} + 1 \times B \times 16^{-2})$

2.1.3 Fixed-Point Binary Representation

Fixed-point binary representation is a special case of Equation 2.1 when $b = 2$. There are different kinds of binary number representations such as sign magnitude, one's complement, and two's complement. Among these representations, two's complement number representation is especially suitable for arithmetic computing hardware. To adapt the two's complement number representation, Equation 2.1 should be slightly modified to become Equation 2.2:

$$V = \sum_{k=\alpha}^{N_d+\alpha-1} \varphi_k d_k 2^k \qquad (2.2)$$

Here, the differences comparing to Equation 2.4 are:

■ d is the digit weight 1 or 0 following two's complement coding.

■ φ_k is 1 for all $k \neq N_d - 1$, and φ_k is -1 when $k = N_d - 1$.

A two's complement fixed-point data is represented as:

$$-d_{m-1}2^{m-1} + d_{m-2}2^{m-2} + \cdots + d_1 2^1 + d_0 2^0(.) + d_{-1}2^{-1} + d_{-2}2^{-2} + \cdots + d_{-f}2^{-f}$$

The radix point is on the right side of $d_0 2^0$. There are m magnitude bits on the right size of the radix point and f fractional bits on the right side of the radix point.

In terms of binary numbers, each magnitude bit represents a power of two. Each fractional bit represents an inverse power of two. Thus the magnitude of the first fractional bit is $1/2$, the magnitude of the second fractional bit is $1/4$, the magnitude of the third fractional bit is $1/8$, and so on. For signed fixed-point numbers in two's complement format, the upper bound is given by $2^{m-1} - 2^{-f}$, and the lower bound is given by -2^{m-1}, where m and f are the number of bits in M (magnitude) and F (Fractional), respectively.

An example is given in Figure 2.2, exactly following the definition of Equation 2.2.

Example 2.3

Represent 99.5 and −99.5 (decimal) using binary formats based on Equation 2.4.

$$(99.5)D = -0 \times 2^7 + 1 \times 2^6 + 1 \times 2^5 + 0 \times 2^4 + 0 \times 2^3 + 0 \times 2^2 + 1 \times 2^1 + 1 \times 2^0. + 1 \times 2^{-1}$$

$$= 0 + 64 + 32 + 2 + 1 + 0.5 = 99.5$$

$$(-99.5)D = -1 \times 2^7 + 0 \times 2^6 + 0 \times 2^5 + 1 \times 2^4 + 1 \times 2^3 + 1 \times 2^2 + 0 \times 2^1 + 0 \times 2^0. + 1 \times 2^{-1}$$

$$= -128 + 16 + 8 + 4 + 0.5 = -99.5$$

2.1.4 Integer Binary Representation

Integer binary representation is a special case of fixed-point binary representation when $F = 0$, meaning no fractional part. Equation 2.3 as the simplification

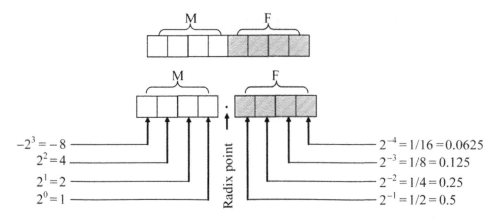

FIGURE 2.2

An example of 8 bits fixed-point data.

of Equation 2.2 represents integer two's complement numbers. The weight factor d_k and φ_k has the same specification as that in Equation 2.2.

$$V = \sum_{k=0}^{N_d-1} \varphi_k d_k 2^k \qquad (2.3)$$

The upper bound is given by $2^{m-1} - 1$, and the lower bound is given by -2^{m-1}, where m is the number of bits in M (magnitude). The bound of fractional data can also be represented as $[-2^{m-1}, 2^{m-1})$. Here, "(" or ")" represents "not including the boundary value" and "[" or "]" represents "including the boundary value."

Example 2.4

Represent 99 (decimal) and -99 (decimal) using two's complement binary formats based on Equation 2.6.

$$(99)D = -0 \times 2^7 + 1 \times 2^6 + 1 \times 2^5 + 0 \times 2^4 + 0 \times 2^3 + 0 \times 2^2 + 1 \times 2^1 + 1 \times 2^0$$

$$= 0 + 64 + 32 + 2 + 1 = 99$$

$$(-99)D = -1 \times 2^7 + 0 \times 2^6 + 0 \times 2^5 + 1 \times 2^4 + 1 \times 2^3 + 1 \times 2^2 + 0 \times 2^1 + 1 \times 2^0$$

$$= -128 + 16 + 8 + 4 + 1 = -99$$

2.1.5 Fractional Binary Representation

Fractional binary representation is a special case of fixed-point binary representation when $M = 0$, meaning no integer (magnitude) part. Equation 2.4 as the simplification of Equation 2.2 represents integer two's complement numbers. The weight factor d_k and φ_k has the same specification as that in Equation 2.2.

$$V = \sum_{k=\alpha}^{0} \varphi_k d_k 2^k \qquad (2.4)$$

The upper bound is given by $2^0 - 2^{-f}$, and the lower bound is given by -2^0, where f is the number of bits in F (Fractional). The bound of fractional data can also be represented as $[-1, 1)$.

Example 2.5

Represent 0.8125 (decimal) and -0.8125 (decimal) using two's complement binary formats based on Equation 2.7.

$$0.8125_D = -0 \times 2^0 + 1 \times 2^{-1} + 1 \times 2^{-2} + 0 \times 2^{-3} + 1 \times 2^{-4}$$

$$= 0 + 1/2 + 1/4 + 0 + 1/16 = 0.5 + 0.25 + 0.0625 = 0.8125_D = 0.1101_B$$

$$-0.8125_D = -1 \times 2^0 + 0 \times 2^{-1} + 0 \times 2^{-2} + 1 \times 2^{-3} + 1 \times 2^{-4}$$

$$= -1 + 0 + 0 + 1/8 + 1/16 = -1 + 0.125 + 0.0625 = -0.8125_D = 1.0011_B$$

Fixed-point DSP processors use integer and fractional data formats. Advantages of fixed-point processors include high performance (the simple hardware can run under higher clock speed), low hardware design cost (datapath and pipeline design is easier), low power consumption, and less silicon area costs (less gate counts and smaller memory size). The native data width (the width of the memory bus) in a fixed-point DSP hardware can be 16 bits, 24 bits, and others. 16 bits fixed-point DSP processors are used for most voice-, image-, video-, and control-based algorithms. 24 bits is used for high quality audio and other applications requiring high data resolution.

2.1.6 Fixed-Point Operands

In most cases, operands in fixed-point processors use either integer or fractional representations, but there are exceptions. For example, after guarding operation (adding the sign bit copy to the left side of the sign bit to enlarge data range), an original fractional number becomes a general fixed-point number including both magnitude and fractional. After scaling, an integer might become a general fixed-point number including both magnitude and fractional. During arithmetic computing in a datapath, micro operations such as guarding, scaling, rounding, truncation, and saturation need to be specified. To formalize specifications for micro operations, a simple nomenclature of denoting operands and results of micro operations should be necessary.

According to the definition of fixed-point numerical representation, the nomenclature should be based on $M.F$ format, where M is the magnitude and F is the fractional. In this book, we use $Q_D(m.f)$ or simply $m.f$ to denote the fixed-point operand; D stands for base, for binary system, $D = 2$ [13]. An example of $Q_2(4.4)$ was given in Figure 2.2. In another example, $Q_2(1.15)$ represents a 16-bit fractional data and $Q_2(16.0)$ represents a 16-bit integer data. Here the data width $= M + F$. A complete set of two's complement 16 bits fixed-point numerical presentations is depicted in Table 2.2.

For example, to prevent the result from getting overflow, a guard bit above the sign bit of the 16 bits fractional data (s.xxxxxxxxxxxxxxx$_2$) should be adopted in ALU operations, which means that the temporal data format inside a 16-bit ALU with one guard bit is 17 bits. The 17 bits fixed-point data is $Q_2(2.15)$ (the input of ALU is fractional data $Q_2(1.15)$), and the binary code of the result is xx.xxxxxxxxxxxxxxx$_2$. The point is at the right side of the second MSB and the binary value sites in the range of $[-2, 2)$ or between -2 to 1.999969482421875 or $-2 \le x < (2 - 2^{-15})$. If the operand is a 16-bit integer, the new 17 bits data format will be $Q_2(17.0)$ after guarding. The data range, precision, and maximum SNR are given in Table 2.3.

By looking at Table 2.2, we can recognize that the dynamic range depends only on N_d, the number of digits, no matter where the radix point. In a signed 16-bit numerical system with 15 significant bits (not including sign bit), the dynamic range is 90 dB and $90/15 = 6$. It is about 6 dB/bit. Remember that 1.76 dB was added to

Table 2.2 Fixed-Point Data Representation.

$Q_D(M. F)$	Range of binary	Minimum negative value	Largest positive value	Binary presentation
$Q_2(1.15)$	$[-2^0, 2^0)$	-1.00000	0.999969482421875	S.XXXXXXXXXXXXXXX
$Q_2(2.14)$	$[-2^1, 2^1)$	-2.00000	1.99993896484375	SX.XXXXXXXXXXXXXX
$Q_2(3.13)$	$[-2^2, 2^2)$	-4.00000	3.9998779296875	SXX.XXXXXXXXXXXXX
$Q_2(4.12)$	$[-2^3, 2^3)$	-8.00000	7.999755859375	SXXX.XXXXXXXXXXXX
$Q_2(5.11)$	$[-2^4, 2^4)$	-16.0000	15.99951171875	SXXXX.XXXXXXXXXXX
$Q_2(6.10)$	$[-2^5, 2^5)$	-32.0000	31.9990234375	SXXXXX.XXXXXXXXXX
$Q_2(7.9)$	$[-2^6, 2^6)$	-64.0000	63.998046875	SXXXXXX.XXXXXXXXX
$Q_2(8.8)$	$[-2^7, 2^7)$	-128.000	127.99609375	SXXXXXXX.XXXXXXXX
$Q_2(9.7)$	$[-2^8, 2^8)$	-256.000	255.9921875	SXXXXXXXX.XXXXXXX
$Q_2(10.6)$	$[-2^9, 2^9)$	-512.000	511.984375	SXXXXXXXXX.XXXXXX
$Q_2(11.5)$	$[-2^{10}, 2^{10})$	-1024.00	1023.96875	SXXXXXXXXXX.XXXXX
$Q_2(12.4)$	$[-2^{11}, 2^{11})$	-2048.00	2047.9375	SXXXXXXXXXXX.XXXX
$Q_2(13.3)$	$[-2^{12}, 2^{12})$	-4096.00	4095.875	SXXXXXXXXXXXX.XXX
$Q_2(14.2)$	$[-2^{13}, 2^{13})$	-8192.00	8911.75	SXXXXXXXXXXXXX.XX
$Q_2(15.1)$	$[-2^{14}, 2^{14})$	-16384.0	16383.5	SXXXXXXXXXXXXXX.X
$Q_2(16.0)$	$[-2^{15}, 2^{15})$	-32768.0	32767.0	SXXXXXXXXXXXXXXX.

Equation 1.1 in Chapter 1. It is added for only ADC marketing and shall not be used in a DSP subsystem.

Pay attention to the fact that the position of the radix point appears only in the behavior C code. It is neither specified in arithmetic computing hardware nor specified in the assembly language. It is up to the specification from the firmware designer. The exact configuration usually is decided by analyzing high-level models of the system.

2.1.7 Integer or Fractional

There are two kinds of two's complement fixed-point representations mostly used in DSP hardware and firmware: the integer and the fractional numerical representation. An example gives the integer and the fractional representations depicted in Figure 2.3.

Table 2.3   Fixed-Point Data Representation and Precision.

$Q_D(M. F)$	Range of data	Precision	Dynamic range	Fractional SNR_{max}	Magnitude SNR_{max}
1.15	$[-2^0, 2^0)$	0.00003	32767 or 90 dB	90 dB	0 dB
2.14	$[-2^1, 2^1)$	0.00006	32767 or 90 dB	84 dB	6 dB
3.13	$[-2^2, 2^2)$	0.00012	32767 or 90 dB	78 dB	12 dB
4.12	$[-2^3, 2^3)$	0.00024	32767 or 90 dB	72 dB	19 dB
5.11	$[-2^4, 2^4)$	0.00048	32767 or 90 dB	66 dB	26 dB
6.10	$[-2^5, 2^5)$	0.00097	32767 or 90 dB	60 dB	32 dB
7.9	$[-2^6, 2^6)$	0.00195	32767 or 90 dB	54 dB	38 dB
8.8	$[-2^7, 2^7)$	0.00390	32767 or 90 dB	50 dB	44 dB
9.7	$[-2^8, 2^8)$	0.00781	32767 or 90 dB	44 dB	50 dB
10.6	$[-2^9, 2^9)$	0.01562	32767 or 90 dB	38 dB	54 dB
11.5	$[-2^{10}, 2^{10})$	0.03125	32767 or 90 dB	32 dB	60 dB
12.4	$[-2^{11}, 2^{11})$	0.06250	32767 or 90 dB	26 dB	66 dB
13.3	$[-2^{12}, 2^{12})$	0.12500	32767 or 90 dB	19 dB	72 dB
14.2	$[-2^{13}, 2^{13})$	0.25000	32767 or 90 dB	12 dB	78 dB
15.1	$[-2^{14}, 2^{14})$	0.50000	32767 or 90 dB	6 dB	84 dB
16.0	$[-2^{15}, 2^{15})$	1.00000	32767 or 90 dB	0 dB	90 dB

The integer representation is, as stated before, derived from Figure 2.3 using the parameters defined in Equation 2.3. The fractional representation in Figure 2.3 uses parameters defined in Equation 2.4. It is important to notice that the arithmetic operation based on integer representation is right-aligned because the point is on the right side of the LSB, and the arithmetic operation based on the fractional representation is left-aligned because the point is on the right side of MSB. Equation 2.5 gives the definition of integer-to-fractional conversion and the fractional-to-integer conversion. Here N is the data width, for example, $N = 16$ for a 16-bit data.

$$Fractional\ data = Integer/2^{(N-1)}$$
$$Integer\ data = Fractional*2^{(N-1)} \tag{2.5}$$

Example 2.6

Convert the bit string $7E02_{16}$ to 16 bits fractional format data.

Solution

The integer result is trivially $7E02_{16} = 323358_{10}$

$$trunc_{16}\left(\frac{7E02_{16}}{8000_{16}}\right) = 0.9844$$

In this case, since the native width is 16 ($N = 16$) and $2^{N-1} = 8000_{16} = 32768_{10}$, the fractional value is then $0.9844_{10} = 0.111\ 1110\ 0000\ 0010_2$.

Example 2.7

Convert 0.9844_{10} to 16 bits integer format and give the bit string in hexadecimal notation.

Solution

In this case, $N = 16$, $2^{N-1} = 8000_{16} = 32768_{10}$. The bit string is found as $0.9844 \times 2^{15} = 0.9844 \times 32768 = 32257_{10} = 7E01_{16}$.

Discussion

After the truncation operation, the bit string $7E02_{16}$ from Example 2.6 becomes $7E01_{16}$ in this example.

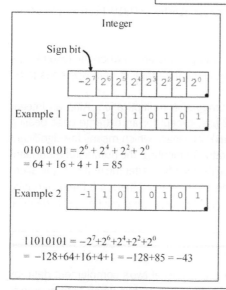 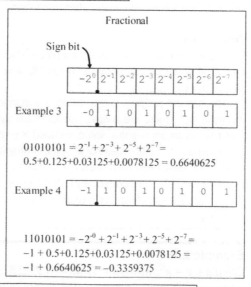

Integer and fixed (fractional) data format

Integer

Sign bit

$$\boxed{-2^7 \mid 2^6 \mid 2^5 \mid 2^4 \mid 2^3 \mid 2^2 \mid 2^1 \mid 2^0}$$

Example 1 $\boxed{-0 \mid 1 \mid 0 \mid 1 \mid 0 \mid 1 \mid 0 \mid 1}$

$01010101 = 2^6 + 2^4 + 2^2 + 2^0$
$= 64 + 16 + 4 + 1 = 85$

Example 2 $\boxed{-1 \mid 1 \mid 0 \mid 1 \mid 0 \mid 1 \mid 0 \mid 1}$

$11010101 = -2^7 + 2^6 + 2^4 + 2^2 + 2^0$
$= -128 + 64 + 16 + 4 + 1 = -128 + 85 = -43$

Fractional

Sign bit

$$\boxed{-2^0 \mid 2^{-1} \mid 2^{-2} \mid 2^{-3} \mid 2^{-4} \mid 2^{-5} \mid 2^{-6} \mid 2^{-7}}$$

Example 3 $\boxed{-0 \mid 1 \mid 0 \mid 1 \mid 0 \mid 1 \mid 0 \mid 1}$

$01010101 = 2^{-1} + 2^{-3} + 2^{-5} + 2^{-7} =$
$0.5 + 0.125 + 0.03125 + 0.0078125 = 0.6640625$

Example 4 $\boxed{-1 \mid 1 \mid 0 \mid 1 \mid 0 \mid 1 \mid 0 \mid 1}$

$11010101 = -2^{-0} + 2^{-1} + 2^{-3} + 2^{-5} + 2^{-7} =$
$-1 + 0.5 + 0.125 + 0.03125 + 0.0078125 =$
$-1 + 0.6640625 = -0.3359375$

Conversion between Integer and fractional data formats:

Integer to fractional: $85/128 = 0.6640625$ // Fractional = Integer / $2^{(N-1)}$
Fractional to integer: $-0.3359375 * 128 = -43$ // Integer = Fractional * $2^{(N-1)}$

FIGURE 2.3

Conversion between 8-bit integer and fractional data formats.

Integer and fractional representations coexist in C code. It is easier to use integer representation for addition, subtraction, logic, shift, and bit manipulation operations. It is also easier to use hexadecimal and binary mixed representation based on an integer computing system. Integer and fractional data are treated as the same operand in addition arithmetic hardware. However, integer and fractional data are treated differently in multiplication arithmetic hardware.

Because the data width of operands is different from the width of the result after integer multiplication, the integer representation is not so convenient for multiplication arithmetic. A problem is highlighted by Example 2.8: the representation of an integer result could be a big problem.

Example 2.8

Multiply 16-bits integer $02FF_{16}$ by 16-bits integer 0110_{16} using a 16-bit fixed-point datapath and save the result in a 16-bit register. Discuss the result.

Solution

A 16×16 bit multiplier is used since it is in a 16-bit datapath; therefore the result will have 32 bits. The result of the multiplication is $00032EF0_{16}$. To save the result into a 16-bit register, we have to make a truncation while translating the 32-bit result into a 16-bit format. The integer is right-aligned with the point at the right side of LSB. Therefore, the result after truncation must be aligned at the right side of the result, and the truncation must be performed from the left side. The value of the result after truncation becomes $2EF0_{16}$ or $0010\ 1110\ 1111\ 0000_2$.

Discussion

We find that the result of an integer multiplication is wrong and overflow when the result keeps the same precision as operands. A reliable way is to limit the range of both operands to be $|x| < 007F_{16}$.

The second problem is the reduced dynamic range after multiplication. After an integer multiplication, there are always two identical sign bits, except for an extreme case dealing with the maximum negative value, $(-\max) \times (-\max) = (+\max)$, which means the significant precision is 2^{N-1} bits instead of 2^N bits. During iterative computing, the number of redundant bits will increase because more and more bits will be the sign bits. After the first multiplication, an extra sign bit is generated, and after the second multiplication, there are three sign bits. If there are L steps of integer multiplications in an iterative loop, there will be $L + 1$ identical sign bits.

Example 2.9

Calculate $y = x^8$, where $x = 7F_{16}(= 0.9921875)$ is a fractional two's complement data, by using an 8-bit integer datapath. The output of the multiplier is 16 bits when both operands are 8-bit. Discuss the results.

The iteration performs the operation $y = x^8$, where $x = 7F_{16}$. Because the hardware multiplier is an 8×8 bit device, the intermediate 8-bit operand for multiplication has to be $y_n[7:0] = 2^{-8} \times \text{Round}\ (ff00 \times Y_n)$ to match the data type of an operand. (It is not a good way, yet the only way to manage the iteration.)

Results of the iteration are given in the following table, step by step.

Steps	Iterations	Results
Step 1	$Y_1 = x^2 = 7F \times 7F$	$Y_1 = 3F01_{16} = 0011\ 1111\ 0000\ 0001_2$
Step 2	$Y_2 = x^3 = y1 \times 7F$	$Y_2 = 1F41_{16} = 0001\ 1111\ 0100\ 0001_2$
Step 3	$Y_3 = x^4 = y2 \times 7F$	$Y_3 = 0F61_{16} = 0000\ 1111\ 0110\ 0001_2$
Step 4	$Y_4 = x^5 = y3 \times 7F$	$Y_4 = 0771_{16} = 0000\ 0111\ 0111\ 0001_2$
Step 5	$Y_5 = x^6 = y4 \times 7F$	$Y_5 = 0379_{16} = 0000\ 0011\ 0111\ 1001_2$
Step 6	$Y_6 = x^7 = y5 \times 7F$	$Y_6 = 017D_{16} = 0000\ 0001\ 0111\ 1101_2$
Step 7	$Y_7 = x^8 = y6 \times 7F$	$Y_7 = 007F_{16} = 0000\ 0000\ 0111\ 1111_2$

Discussion

The resolution of the result is lower after every step of the iteration. Integer-based multiplication is not good for iterative computing.

Let us introduce the fractional representation to DSP hardware for multiplication arithmetic. Remember the definition of the fractional data, that is, $Q_2(1.F)$; there is only one sign bit, and the range of the value is $[-1, 1)$. Using the fractional data type, both the operands and the result are left-aligned.

Normally, an available physical multiplier from a synthesis library conducts integer two's complement multiplication. From a hardware point of view, all multipliers as RTL components are integer multipliers. The fractional multiplication will be carried out by an integer multiplier. To understand from mathematical perspectives how a fractional multiplication is performed using an integer multiplier, the following procedure is taken. Principally, Equation 2.6 can be used, and the integer multiplier can be used for fractional multiplication.

There is an $m \times n$ bits integer multiplier producing an $m + n$ bits result. First an integer operand is converted to a fractional operand by multiplying the fractional operand with the largest absolute integer value following Equation 2.5. Then the multiplication is performed using the integer multiplier. Finally the value is divided back to the correct range according to the fractional representation.

To formalize the procedure described, Equation 2.6 shows the resulting formula, where A is an m-bit fractional data. With $m - 1$ bits left shift, $(2^{m-1}A)$ becomes an equivalent integer operand. B is an n-bit fractional data. With $n - 1$ bits left shift, $(2^{n-1}B)$ becomes an equivalent integer operand. The integer multiplication $(2^{m-1}A) \times (2^{n-1}B)$ therefore is conducted. By right-shifting $2^{m+n-1-1}$ bits, the result of the multiplication $(2^{m-1}A) \times (2^{n-1}B)$ becomes the result of $A \times B$. In fact, there is no hardware shift operation before and after multiplication. Equation 2.6 proves that an integer multiplier can be used for fractional multiplication.

$$\frac{(2^{m-1}A) \times (2^{n-1}B)}{2^{m+n-1-1}} = A \times B \qquad (2.6)$$

By right-shifting $2^{m+n-1-1}$ bits following Equation 2.9, two sign bits exist in the result, and it does not match the definition of fractional data $Q_2(1.F)$. For example, if $m = n = 8$, the format of a fractional operand is s.xxx-xxxx, and the format of the multiplication result will be ss.xx-xxxx-xxxx-xxxx (s stands for "sign bit"). According to the definition of fractional number representation, the result should be s.xxx-xxxx-xxxx-xxxx instead. It means that the real physical fractional multiplication using integer multiplier is intuitively $2 \times A \times B$, or left-shifting one bit after the multiplication.

Right-shifting one bit after the multiplication is to remove the higher sign bit. Normally two most-left bits are both sign bits, and they have the identical value after a multiplication; thus the sign bit in effect is duplicated. There is an exception when running fractional multiplication of $(-1) \times (-1) = 1$. Remember that the range of fractional data is $[-1, 1)$, or the upper limit is $1 - 2^{-m+1}$ instead of 1. The result of the physical multiplication $(-1) \times (-1)$ must be $1 - 2^{-m+1}$ or 0.111 1111 1111 1111.

In the following example, an extreme case gives an exceptional result. If $m = n = 8$, the result of $(-1) \times (-1)$ or $(10000000) \times (10000000) = 0100\ 0000\ 0000\ 0000$. By left-shifting one bit, it becomes $1000\ 0000\ 0000\ 0000$. However, as discussed earlier, the result should be 0.111 1111 1111 1111 instead. Because this is the only exception of the fractional multiplication, it can be compensated in a special way. To compensate the extreme case, the result must be forced to $1 - 2^{-m+1}$ or 0.111 1111 1111 1111 when both operands are (-1) or 1000-0000-0000-0000. The physical operation for the case of $(-1) \times (-1)$ is to saturate the result after one bit left shift, which will be discussed later in this chapter. In this example, the saturation operation is to check if two sign bits of the result are not identical. When they are not identical and the left sign bit is 0, the result will be forced to $1 - 2^{-m+1}$, or 0.111 1111 1111 1111 in this example.

An implementation issue should be discussed. After one bit left shift, a bit should be filled in to the LSB of the result. Following the history of voice processor design, a zero bit is filled in LSB after one bit left shift. This way has been accepted by industry since the 1980s.

To summarize, fractional numerical representation is introduced for multiplication, see Figure 2.4. The integer two's complement multiplier can be used for fractional multiplication. After the multiplication, one of the two sign bits should be removed. A saturation operation is required before removing one of the sign bits to adapt the case when $(-1) \times (-1)$.

Example 2.10
Perform 4×4 bit integer and fractional multiplications of 0111×0111.

Solution
The solutions are:

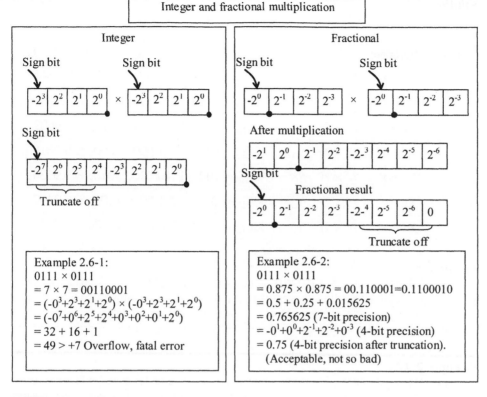

FIGURE 2.4

Integer and fractional multiplications.

Discussion

As discussed, it is not suitable to conduct integer multiplication if the ranges of operands are unknown.

The fractional numerical representation is left aligned with one sign bit. The truncation operation is required while changing the result of a longer data to data with normal width. To discard the lower part of the result, only the truncation errors will be introduced instead of fatal computing errors from integer multiplication. Fractional arithmetic and numerical representations support better finite length fixed-point computing because truncation will never induce fatal errors. Equation 2.5 is for the conversion between fractional and integer data. Furthermore, there is only one sign bit after multiplication; thus the dynamic range will be maximally utilized.

Example 2.11

Repeat Example 2.9 using fractional arithmetic, keeping only one sign bit.

Solution

The result of the iteration is given in the following table step by step.

Steps	Iterations	Results
1	$Y_1 = x^2 = 7F \times 7F$	$Y_1 = 7E02_{16} = 0111\ 1110\ 0000\ 0010_2 = 0.9844$
2	$Y_2 = x^3 = Y1 \times 7F$	$Y_2 = 7D04_{16} = 0111\ 1101\ 0000\ 0100_2 = 0.9769$
3	$Y_3 = x^4 = Y2 \times 7F$	$Y_3 = 7C06_{16} = 0111\ 1100\ 0000\ 0110_2 = 0.9689$
4	$Y_4 = x^5 = Y3 \times 7F$	$Y_4 = 7B08_{16} = 0111\ 1011\ 0000\ 1000_2 = 0.9612$
5	$Y_5 = x^6 = Y4 \times 7F$	$Y_5 = 7A0A_{16} = 0111\ 1010\ 0000\ 1010_2 = 0.9534$
6	$Y_6 = x^7 = Y5 \times 7F$	$Y_6 = 790C_{16} = 0111\ 1001\ 0000\ 1100_2 = 0.9457$
7	$Y_7 = x^8 = Y6 \times 7F$	$Y_7 = 780E_{16} = 0111\ 1000\ 0000\ 1110_2 = 0.9379$

Discussion

The real result of $y = (7F_{16})^8 = 0.93918_{10}$; the radix point is right after the sign bit. The result with finite precision hardware in Example 2.11 is 0.9379. The error introduced by truncation is about 0.00128. The result of the fractional multiplication is acceptable.

Example 2.12

Compute $7FFE_{16} \times 7FFF_{16}$ based on integer arithmetic and fractional arithmetic using 16 bits fixed-point datapath, and save the result in an accumulation register with custom data width of 30 bits.

Solution

The result of the multiplication is $3FFE8002_{16}$ or $0011\ 1111\ 1111\ 1110\ 1000\ 0000\ 0000\ 0010_2$. Converted to an integer value, the result is 1073643522_{10}. To save the result into a 30-bit register, we have to make a truncation first and transfer the 32-bit result into a 30-bit format. When performing the truncation for an integer, right-aligned, we will get the result of $11\ 1111\ 1111\ 1110\ 1000\ 0000\ 0000\ 0010_2$, which is -277763—an overflow occurred!

When using fractional representation, $7FFF_{16} = 0.99996948242187_{10}$ in fractional format. The result is $7FFE_{16} \times 7FFF_{16} = 3FFE8002_{16} \times 2 = 7FFD0004_{16}$ or $0111\ 1111\ 1111\ 1101\ 0000\ 0000\ 0000\ 0100_2$. The result $7FFD0004_{16}$ can be converted to a fractional value, and it is $7FFD0004_{16}/80000000_{16} = 0.99990844 9_{10}$. After truncating the 32-bit result to 30 bits, the result is $01\ 1111\ 1111\ 1111\ 0100\ 0000\ 0000\ 0001_2$, or $1FFF4001_{16}$, which is interpreted as a fractional value, 0.99990845_{10}.

Discussion

The finite-length result using the fractional data type is almost the same as the result before truncation. An overflow is from integer multiplication.

2.1.8 **Other Binary Data Formats**

The two's complement representation is used widely since the addition and subtraction (added by a positive or a negative number) operations can be executed using the same full adder hardware. There is only one representation for zero, namely, "00...0". The range of an n bit integer number in a two's complement representation system is with the unsymmetrical range $-2^{n-1} \le x \le 2^{n-1} - 1$. Two's complement binary data format is the default data format used in all digital arithmetic computing hardware. Seldom can other data formats be found in hardware.

The signed-magnitude representation is seldom used due to complications when performing addition and subtraction. However, it can be suitable when the number of additions and subtractions is low. Researchers have tried to use this system to reduce power consumption in a multiplier in case the white noise is the dominant input. Let us take a short sound sample of a silent room. The results are shown in Table 2.4.

The noise is very small and close to zero. When a system receives only the white noise instead of useful signals, the power consumption of the system should be the minimum. Recall the very basic knowledge of how to measure the power consumption intuitively from digital CMOS circuits. The power consumption depends on the difference of the "next value" of the sampling sequence to the circuit, and it is independent of the value itself. Diff in the table is the measure of the number of different bits between the current sample and the next sample. The measure of the difference from the two's complement data is much larger than that of the

Table 2.4   White Noise from a Silent Room.

Sample	Decimal	Two's complement binary	Diff	Signed-magnitude	Diff
1	−0.00003	1111 1111 1111 1111	x	1000 0000 0000 0001	x
2	−0.00009	1111 1111 1111 1101	1	1000 0000 0000 0011	1
3	+0.00003	0000 0000 0000 0001	14	0000 0000 0000 0001	1
4	−0.00003	1111 1111 1111 1111	15	1000 0000 0000 0001	1
5	−0.00003	1111 1111 1111 1111	0	1000 0000 0000 0001	0
6	+0.00003	0000 0000 0000 0001	14	0000 0000 0000 0001	1
7	−0.00003	1111 1111 1111 1111	14	1000 0000 0000 0001	1
8	+0.00009	0000 0000 0000 0011	13	0000 0000 0000 0011	2
9	+0.00000	0000 0000 0000 0000	2	1000 0000 0000 0000	3
10	+0.00003	0000 0000 0000 0001	1	0000 0000 0000 0001	2

signed-magnitude data. The power consumption by a circuit using signed-magnitude data could be much lower when signal is small. In reality, small signal appears most of the time as inputs of most systems. The implementation signed-magnitude multiplier is easy (according to basic digital CMOS books). However, the ADD/SUB operation based on signed-magnitude data is complicated. That is why the signed-magnitude system is not popular even though the power consumption of the system is low.

One major problem with this system is the representation of number zero, which can be done in two ways, by either $10\ldots0$ or $00\ldots0$. The range of an n bit number of the signed-magnitude representation is the symmetrical range $-2^{n-1} + 1 \leq x \leq 2^{n-1} - 1$.

The one's complement representation is, for example, used for the IP header checksum algorithm of the internet protocol (IP). The number zero, in one's complement format, like in the sign-magnitude case, can be represented in two ways by

Table 2.5 Summary of Numerical Representations.

Decimal data	Two's complement	Sign magnitude	One's complement	Unsigned
7	0111	0111	0111	0111
6	0110	0110	0110	0110
5	0101	0101	0101	0101
4	0100	0100	0100	0100
3	0011	0011	0011	0011
2	0010	0010	0010	0010
1	0001	0001	0001	0001
0	0000	0000	0000	0000
−0	—	1000	1111	—
−1	1111	1001	1110	—
−2	1110	1010	1101	—
−3	1101	1011	1100	—
−4	1100	1100	1011	—
−5	1011	1101	1010	—
−6	1010	1110	1001	—
−7	1001	1111	1000	—
−8	1000	—	—	—

Table 2.6   The Values of an 8-bit Integer.

Binary value	Two's complement interpretation	Unsigned interpretation
00000000	0	0
00000001	1	1
00000010	2	2
...	...	...
01111110	126	126
01111111	127	127
10000000	−128	128
10000001	−127	129
10000010	−126	130
...	...	...
11111110	−2	254
11111111	−1	255

either $11\ldots1$ or $00\ldots0$. The range of an n bit number in the one's complement representation system is the symmetrical range $-2^{n-1} + 1 \leq x \leq 2^{n-1} - 1$.

Unsigned data is used for logic shift and logic data manipulations, see Table 2.5. The comparison of two's complement and unsigned values is given in Table 2.6.

2.2 DATA QUALITY MEASURE

2.2.1 Noise, Distortion, Dynamic Range, and Precision

Relative accurate or acceptable quality is enough for many embedded systems, especially for those with analog inputs or outputs. Voice and audio systems with analog input and output are typical systems evaluated by "relative acceptable quality." A video system is also a typical system requiring relative quality because the final display is analog visual pictures from a human point of view. Many other systems, such as vehicle speed control systems and robot control systems are also with analog inputs and outputs.

Now the question is how should we define "relative accurate or acceptable quality" in this book and in our design? Accuracy and quality are actually the accuracy and quality of parameters. From an engineering point of view, the system parameters could be divided into frequency-domain parameters (bandwidth, frequency range), time-domain parameters (delay, response time, and amplitude parameters), and in other domains. In this chapter, the focus of the discussion is on the amplitude parameters. The definition of "relative accurate or acceptable quality" will therefore be equivalent to the signal-to-noise ratio, SNR, and distortion in this book. The SNR describes the impact of the LSB (least significant bit) of the data and hardware of a finite precision system. The distortion will usually describe the impact of the MSB (most significant bit). An intuitive illustration is given in Figure 2.5. The following Equation 2.7 will further help the understanding of the figure. Floating-point data representations will be discussed later in this chapter.

Distortion is the alteration of the original shape of an object, such as waveform or other form of information or representation. Distortion is usually unwanted. In this chapter, we are talking about amplitude distortion. It occurs in a system when the output amplitude is not a linear function of the input amplitude under specified conditions. In a fixed-point digital system in Figure 2.5, the distortion is specially defined as the case where an input, output, or intermediate data is larger than the MSB. In these cases, data cannot be represented by the fixed-point system of the hardware.

Noise is an intrinsic problem in our daily life. Noise in this book stands for noise signal, which is the unwanted signal interfering with the useful signal. Noise can refer to the acoustic noise in an audio system or to the visual noise commonly seen as snow on the screen. In signal processing it can be considered data without meaning. Noise can be analog noise (dominated by thermal noise induced by the movement of electrons) and digital noise (quantization noise from limited precision of digital presentation and hardware; see Figure 2.5).

When using fixed-point hardware, the quantization error occurs due to the finite word length of the system. Equation 2.7 describes truncation of a value from N_c bit data to N_b bit (fractional) data ($N_c > N_b$).

$$V = \sum_{k=0}^{N_c-1} a_k b^{-k} = \sum_{k=0}^{N_b-1} a_k b^{-k} + \sum_{k=N_b}^{N_c-1} a_k b^{-k} = Q_T [V] + E \qquad (2.7)$$

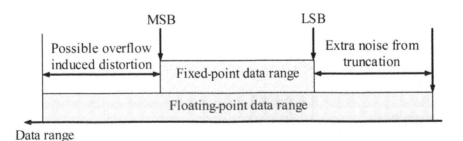

Data range

FIGURE 2.5

Noise and distortion of a fixed-point digital system.

Precision (resolution) is used to specify the quality of digital representation of data. Precision and resolution can be mixed in this book. However, resolution is used more to specify the quality of digital device, and precision is used more to specify the quality of computing result.

The data precision before truncation is represented by the left part in Equation 2.7. It can also be represented using two parts, as the middle part in Equation 2.7. The lower part will be truncated as E. The higher part $Q_T[V]$ is the remaining value with the truncation error. Therefore, $Q_T[V]$ is always smaller than V. When $N_c - N_b$ is large enough, for example, 32 bits in general computers, the precision problem can be negligible. When $N_c - N_b$ is not large enough, for example, 16 bits in a low-power DSP ASIP, the precision problem could be significant.

The maximum SNR_{max}, of a finite-length digital system using two's complement data representation can be represented using Equation 2.8:

$$SNR_{max} = 20 \log \left(positive_{MAX} \, / \, positive_{MIN} \right) \tag{2.8}$$

For example, the MAX positive two's complement data of a 16-bit digital system is $2^{15} - 1 = 32767$ and the MIN positive two's complement data of a 16-bit digital system is 1. In this case, the maximum signal-to-noise ratio of a 16-bit digital system using two's complement data representation is $20 \log(32767/1) = 90.31dB \approx 90dB$.

Pay attention to the SNR definition of the book, which is different from the definition of the SNR of AD converters. The definition of SNR for digital systems is not unique. Sometimes, resolution of data (source or result) is defined by the uncertainty, which is the $1/2$ value of the LSB. For example, the resolution of a 16-bit fractional data is $1/2$ of 2^{-15}. Therefore, the maximum signal-to-noise ratio could even be SNR $= 20 \times \log$ (positive maximum/$(0.5 \times$ positive minimum)$) = 6 + 20 \times \log(Pmax/Pmin)$. After reading this chapter, you will realize that this definition is valid only when rounding algorithms is used to generate the 16 bits two's complement data from a longer data. The problem is that we actually might not know the way of acquiring input data with 16 bits precision. In this book, we should conservatively use the definition given by Equation 2.8.

When discussing the SNR_{max} of a finite length digital system using two's complement data representation, the ratio of MAX/MIN was introduced. This ratio has an actual definition as the hardware dynamic range. The hardware dynamic range normally is defined as the range from the smallest representable value to the largest representable value by the hardware. The SNR_{max} of a finite-length hardware digital system using two's complement data is:

$$SNR_{max} = 20 \log(Dynamic\_range) \tag{2.9}$$

To further continue, the precision of a finite-length digital system is the measurement of the ability to distinguish between the closest values; in other words, it is the smallest representable value of the finite-length digital system. For example, using a 16-bit fixed-point processor, the precision of integer data is 0000 0000 0000 0001.

The precision is determined by the word length in the fixed-point format and by the number of bits in the mantissa in accordance with the floating-point format.

2.2.2 Quantitative Concept of Dynamic Range and Precision

Dynamic range and precision are required by an application as its quality measurement. Quality here is specified between "acceptable" and "good enough", and quality shall not be overspecified. The question, therefore, is popping up again— what will be the lower threshold representing the "acceptable value" and the higher threshold representing the quality of "good enough"?

Quantization means losing message; when the losing message is negligible by users, the quality is OK. The definition of acceptable dynamic range and precision of system outputs can therefore be "no uncomfortable feeling" or "no influence on reliability" or something similar. The definition of good enough can be "much higher than the lower threshold, yet no significant cost impact."

It is much easier if the dynamic range and precision/resolution have been specified by standards. For example, many standards for voice codec give bit accurate specifications for the compliance test. However, in most cases, the dynamic range and precision must be specified by designers. Table 2.7 gives dynamic range and SNR_{max}.

The precision of the examples in Table 2.7 is "1" because the examples are based on integer data representation and the dynamic range is the same as the maximum positive value in the table.

The thermal noise at room temperature can be estimated as $-174\,dBm/Hz$. It means that there is no reason to have the SNR of output signals up to 186 dB using 32-bit resolution. For example, 8-bit resolution is actually enough for normal

Table 2.7 Dynamic Ranges and Precisions.

Data length	Maximum positive	The maximum SNR	Dynamic range
32b	2147483647	186.64 dB	2147483647
24b	8388607	138.47 dB	8388607
20b	524287	114.39 dB	524287
18b	131071	102.35 dB	131071
16b	32767	90.31 dB	32767
14b	8191	78.27 dB	8191
12b	2047	66.23 dB	2047
10b	511	54.17 dB	511
8b	255	42.08 dB	255

video terminals. If the volume of a voice output signal can be controlled, dynamic range of 60 dB and precision of 40 dB can be enough for a phone. Different systems require different resolutions or dynamic ranges. Excessive high quality will induce unnecessary costs.

There are two kinds of dynamic ranges and precisions: dynamic range and precision of the hardware processor (hardware dynamic range), and dynamic range and precision of the data (inputs dynamic range and result dynamic range) to and from the hardware. The dynamic range and precision of input data or output data is always less than that of hardware. To avoid overflow, the data range of the hardware cannot be fully utilized.

The result of the dynamic range and SNR of an application (the design of a firmware system) implemented on a fixed-point processor is thus much less than the dynamic range and SNR of the fixed-point processor. In Chapter 18, the real dynamic range and SNR of an application will be discussed further. The skill in designing a high-quality signal processing system using limited precision hardware will be discussed throughout the book.

2.3 FLOATING-POINT NUMERICAL REPRESENTATION

To improve the dynamic range in a fixed-point system the data width has to be increased. However, increasing the data width will bring increased silicon costs and decreased machine speed of the processor. Since normally high computing speed and low silicon cost are desired, a trade-off is important.

Requirements on dynamic range and resolution can vary for different applications. Most applications have higher requirements on the dynamic range and relative lower requirements on resolution of results. For example, the requirement on the dynamic range for baseband DSP and audio decoding DSP could be more than 100 dB, and requirement on the relative resolution or the SNR (the signal over the noise at the same sampling time) can be just 40 to 60 dB. Some special applications with very long iterative computing require both high dynamic range and high resolution, for example, the audio echo canceller for a large music hall.

Floating-point numerical representation increases the dynamic range by using an exponent. The dynamic range and the resolution can be optimized separately, because the dynamic range is controlled by the exponent other than the mantissa. This is the main reason to introduce the floating-point number system to DSP.

Floating-point has one further advantage over fixed-point. The computing hardware automatically scales (normalize) each data to utilize the full word length of the mantissa after each arithmetic operation; the best resolution is maintained even for small numbers.

The second motivation for using floating-point numerical representation is to reduce the firmware development time. When using fixed-point numerical representation, data quality control such as fine tune of the scaling is essential for inputs, step results, and final results. The firmware therefore becomes complicated, and the development time is longer. Special algorithms are required to relax the impact of finite-length. By using floating-point numerical representation, however, the dynamic range of an application will not be more than the dynamic range of the floating-point data. Overflow or underflow can therefore be naturally eliminated, and complicated scaling and data quality measurement-control processes are therefore not necessary. Since it reduces the firmware development time dramatically, the floating-point numerical representation is useful when the TTM (Time to Market) is short.

At the same time, there are drawbacks from introducing the floating-point numerical representation to hardware. The main drawbacks are the extra hardware cost and relatively low machine speed.

In a floating-point system, the dynamic range is decided mostly by the length of the exponent, and the resolution is decided by the length of the mantissa. The binary floating-point representation is defined according to Equation 2.10, where MS is the mantissa significant and E the exponent.

$$V = MS \cdot b^E \tag{2.10}$$

The binary mantissa is usually a signed value in a range of $[-1, -0.5]$ and $[0.5, 1)$. The binary exponent is an integer that represents the place of the radix point of the mantissa. When mentioning the mantissa of a floating-point binary code, we imply that the mantissa is a normalized mantissa, meaning that the two most significant bits including sign bit are either 01 as a positive data or 10 as a negative data. In DSP hardware with floating-point data format, the normalization is executed implicitly. How to use the exponent is decided by firmware designers. For example, an 8-bit exponent can cover the range of right-shift mantissa 128 bits and left-shift mantissa 127 bits, or $-128 < E < 127$.

There are different definitions for floating-point arithmetic. IEEE standard 754.1985 is one of these standardized definitions. At the same time, the format definition of DSP floating-point arithmetic can be custom;—for example, the floating-point definition of TI processors, which is not standardized but is widely accepted by industries. IEEE standard 754.1985 [3] defines four floating-points in two groups, basic and extended, each having two widths, single and double. Table 2.8 lists the sizes used.

In Table 2.8, p is the length of the mantissa (resolution) or the number of significant bits, E_{max} is the maximum exponent, and E_{min} is the minimum exponent. Exponent width in bits is the length of the exponent in binary format, and the format width in bits is the total binary coding length for a floating-point numeric representation, which is the length of exponent plus the length of mantissa. The basic single format of IEEE 754 floating-point number representation is given in Figure 2.6 (note that the sign bit of the mantissa is separated from the mantissa field).

S is the sign bit of the mantissa (0 is positive and 1 is negative); EXPONENT, following IEEE standard 754.1985 basic single precision format, is an unsigned 8-bit field that determines the location of the radix point of the number being encoded.

Table 2.8   IEEE Standard 754.1985 Format Parameters.

Parameter	Basic single	Extended single	Basic double	Extended double
Format width in bits	32	43	64	79
Width of Mantissa p	24	32	53	64
Exponent width in bits	8	11	11	15
MAX Exponent E_{max}	+127	+1023	+1023	+16383
MIN Exponent E_{min}	−126	−1022	−1022	−16382

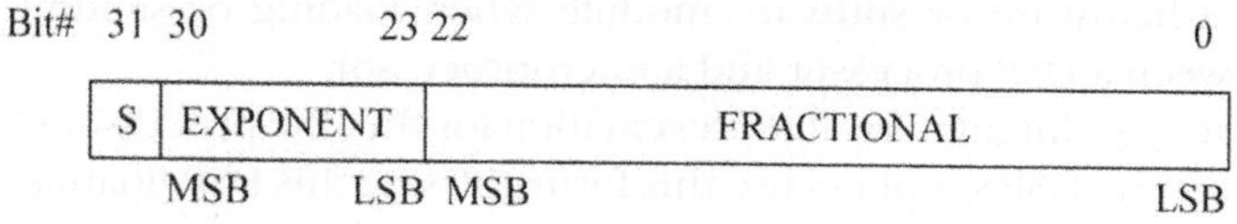

Bit# 31 30   23 22   0

S	EXPONENT	FRACTIONAL

  MSB   LSB MSB   LSB

FIGURE 2.6

IEEE 754 floating-point data format.

Table 2.9   Floating-point Corner Values.

Type	Exponent	Mantissa	Value
Zero	0000 0000	000 0000 0000 0000 0000 0000	0.0
One	0111 1111	000 0000 0000 0000 0000 0000	1.0
Denormalized number	0000 0000	100 0000 0000 0000 0000 0000	Less than 5.9×10^{-39}
Largest normalized number	1111 1110	111 1111 1111 1111 1111 1111	3.4×10^{38}
Smallest normalized number	0000 0001	000 0000 0000 0000 0000 0000	1.18×10^{-38}
Infinity	1111 1111	000 0000 0000 0000 0000 0000	Infinity
NaN	1111 1111	Not zero	NaN

FRACTIONAL is a 23-bit field containing the fractional part of the mantissa without sign bit. LSB is the least significant bit, and MSB is the most significant bit.

Table 2.9 lists corner values of 32-bit single-precision floating-point representation by IEEE. In the table, NaN stands for "not a number." Denormalized numbers (subnormal numbers) is defined as filling the gap around zero in floating-point

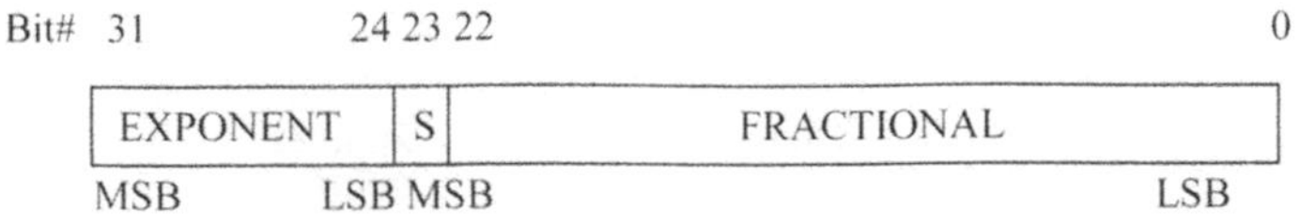

Bit# 31   24 23 22   0

EXPONENT	S	FRACTIONAL

MSB   LSB MSB   LSB

FIGURE 2.7

An example of a DSP floating-point data format.

arithmetic. Any nonzero number that is smaller than the smallest normal number is a denormalized number.

Most DSP processors, however, use a special format of floating-point representation for arithmetic computing and data storage. DSP hardware may not be directly compatible with the IEEE 754 standard. Therefore, we need to insert numerical format conversion as a hardware or software module when loading or sending floating-point data between a DSP processor and a microprocessor.

Figure 2.7 is the floating-point numerical representation for the DSP processor of Texas Instruments [4]; several DSP suppliers use this format. Using this DSP floating-point data format, the sign bit is together with the mantissa. Therefore, the hardware design and verification of the floating-point datapath are easier.

Implementation of a floating-point datapath is complicated. However, silicon cost could be minimized significantly by using some special floating-point arithmetic computing for specific applications in ASIP. Since the acoustic perceptual capability of humans is not so advanced, the noise 40 dB lower than the current audio wave will not be noticed. At the same time, high dynamic range is needed since the perception of sound levels follows a square relation. The dynamic range and precision of computing hardware and results are shown in Table 2.10 from both floating-point arithmetic computing and fixed-point computing. It was proven that 16-bit floating-point data gives enough audio decoding quality, enough precision, and ultra high dynamic range, and keeps low storage costs [5]. In the paper by Eilert et al. [5], the processor uses floating-point data consisting of five unsigned exponent bits and 11 mantissa bits. The cost of data memory is at least 30% lower because most other MP3 players use 24-bit data memory.

Programmable radio baseband processors can also use floating-point data types, mainly targeting SDR (software-defined radio). The use of floating-point data is driven by the large different requirements on dynamic range and precision of results. The dynamic range of inputs is high, more than 80 dB; the result precision can be low, about 20 to 40 dB [6–8].

The floating-point hardware gives much higher SNR and dynamic range. The reason is simple: the dynamic range and precision are represented separately by exponent and mantissa, see Figure 2.8.

Because the floating-point arithmetic unit and the fixed-point arithmetic unit are not used at the same time, the multiplier in hardware can be used for both the mantissa multiplication of the floating-point data and for fixed-point multiplication of the fixed-point data. The hardware overhead in datapath can be negligible.

Table 2.10 Trade-off Reference by Selecting Data Formats.

	Fixed-point	General floating-point	Custom floating-point	Block floating-point
Dynamic range	Low	Very high	High	High
Precision	Acceptable	Very high	Acceptable	Acceptable
Worst SNR	Poor	Very high	High	High
HW design costs	Low	Very high	High	Low
HW speed	Very High	Low	High	High
Memory costs	Low	High	Low	Low
Total HW silicon costs	Low	High	Very low	Low
FW design costs	High	Low	Very low	High

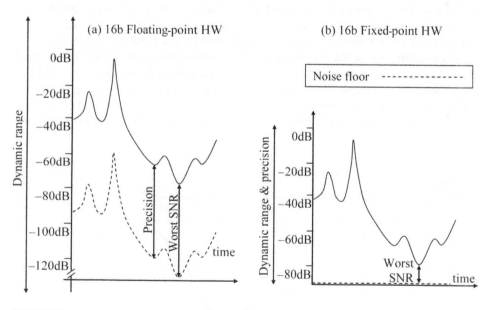

FIGURE 2.8

Comparing the dynamic range of floating-point and fixed-point.

2.4 BLOCK FLOATING-POINT

Terminologies in this chapter shall be used carefully. A fixed-point processor stands for the hardware of a DSP processor that does not have a floating-point unit. Fixed-point means that the datapath of the processor is based on fixed-point hardware for fixed-point arithmetic computing instead of floating-point hardware for floating-point arithmetic computing. However, it does not mean that a fixed-point processor cannot manipulate floating-point data. A fixed-point processor can emulate computing of floating-point data.

Floating-point computing can be emulated by fixed-point DSP processors when a larger dynamic range is required. Block floating-point (also called dynamic signal scaling technique) is an emulation method using a fixed-point DSP processor hardware. Block floating-point data is not floating-point data while it uses an exponent register to denote the exponent or the scaling factor of a block of adjacent data. Block floating-point data is stored and processed in a fixed-point DSP processor in the way shown in Figure 2.9. The strong solid line represents the real data, and the dashed line is the data after scaling. The scaling factor of the data block is stored as the exponent of the data block.

The dynamic range is larger than that of a fixed-point processor. The block floating-point data consists of the scaled data as the mantissa and the exponent of the data block. The exponent is the scaling factor of the mantissa and stored in the exponent register of the data block, which is a memory word attached to the data block. A data block has its fixed length specified by the application. For example, for voice codec, the block length should be the length of a data frame.

After scaling, the solid line in Figure 2.9 is represented by dashed lines called block floating-point data associated with the block scaling factor. The results will be postprocessed to remove the scaling after the signal processing. A scaling factor of a data block must be stored and used all the way during the computing and presenting of the data block. That is the main difference between block floating-point signal

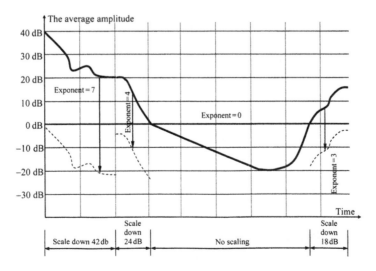

FIGURE 2.9

Block floating-point data.

processing and normal fixed-point signal processing. For normal fixed-point signal processing, a scaling factor is not bounded to a specific data block and may not be stored.

Dynamic range is enhanced by introducing the block floating-point data format to a fixed-point DSP processor. The benefit is to obtain (almost) floating-point data quality using the low-cost fixed-point hardware. The drawback is that the firmware for block floating-point data processing is complicated [14, 15].

In Table 2.10, the HW design costs represent the complexity of the hardware. The HW speed represents the clock speed of the datapath. Acceptable means satisfactory for most embedded applications. Because firmware design costs are very high for signal processing based on block floating-point, it is used for only small and relatively simple applications.

The relation of signal precision versus the signal magnitude is given in Figure 2.10. When using floating-point data formats, the absolute precision always is modified toward the highest hardware resolution because the mantissa is always larger than or equal to 0.5, or less than or equal to -0.5. Data saturation is seldom necessary when using the floating-point data format. Therefore, when the magnitude goes up or down, the resolution of a floating-point data scales to fit the magnitude.

When using fixed-point data representation, for example, using a 16-bit system, the precision of the hardware will not be fully utilized to avoid possible overflow. In a 16-bit system, the hardware dynamic range is 90 dB, and the dynamic range of

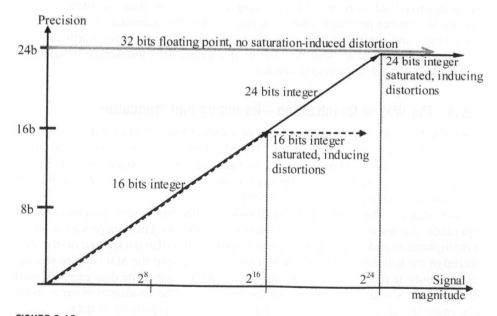

FIGURE 2.10

Precision versus signal magnitude.

data is usually around 70 dB to avoid overflow. For example, if the signal magnitude has values close to the limit of 16 bits, computations can result in numbers larger than (or out of) the representable range of 16-bit hardware. The values will either be truncated or saturated, which will introduce distortions in the output.

When the dynamic range is not enough, distortion will be introduced. When the resolution is not enough, extra noise will be introduced, resulting in lower signal-to-noise ratio. When the resolution is low, extra harmonics of the quantization noise will be introduced in the signal. Therefore we need a DSP processor built with longer data width to reduce the computation error throughout the calculations. However, the longer the data width, the higher the silicon cost. Thus in an embedded system trade-offs of data width and selection of data presentations are always essential.

2.5 DSP BASED ON FINITE PRECISION

In theoretical DSP books, signals and systems could be represented with infinite resolution. However, most real implementations in DSP processors use fixed-point number representations. DSP processors are fixed-point processors with finite data precision and finite computing resolution. When representing data using finite-length data format, quantization errors are introduced. When signals pass through or are processed by a finite precision system (e.g., a fixed-point DSP processor), frequency and amplitude deviations are introduced. To couple with the limited dynamic range of a fixed-point processor, data scaling is necessary to avoid overflow-induced distortion. Further problems arise from data scaling; the quantization error will be even larger, and the worst SNR may not be acceptable. Obviously, hardware and firmware design skills are required. Before discussing the finite-length behavior in detail, some basic definitions are needed.

2.5.1 The Way of Quantization—Rounding and Truncation

Truncation is defined as the conversion of a longer numerical format to a shorter one, simply by discarding bits at the LSB side. The system or signal quantization error is defined as the numerical error introduced when a longer numeric format is converted to a shorter one. The quantization error is induced by the limited precision of a computing or storage device.

For example, the result of multiplication usually has double precision of the operands. The result of an $N \times N$ bit multiplier is $2N$ bits. The data precision inside a multiplication and accumulation unit is usually higher than (twice that of) the data stored in the general register file. When moving data from the MAC unit to storage devices outside it, we need to truncate the extra bits so that the data can be stored into memories with the precision of the bus width. The quantization error to be discussed in this section is introduced due to physical truncation of data.

Equation 2.7 describes truncation on a fixed-point value from N_c bit data to N_b bit data ($N_c > N_b$). E_T is the value of the discarded part and $Q_T[V]$ is the remaining value with quantization error E_T. Therefore, $Q_T[V]$ is always smaller than V. When using signed-magnitude numerical representation, $Q_T[V]$ is less than V when V is larger than zero; and $Q_T[V]$ is larger than V when V is less than zero. Figure 2.11 summarizes the quantization errors of different kinds of binary data formats.

The analysis shows that the quantization error from truncation on a two's complement binary code is $\Delta = 2^{-b}$. The truncated value is thus always smaller than the original value. This introduces a bias when accumulation is performed. In many DSP applications, truncation operations might be conducted many times until the final result is achieved. For example when executing an IIR filter and using the register file as the computing buffer, the accumulating values are truncated after each computing step of a tap. After several steps of accumulation, the negative biases will be increased over time as creeping errors during accumulation. The impact of the bias will keep increasing in a fixed-point DSP system.

The rounding technique reduces the truncation-induced error because the balance between the deviation error is induced by rounding up and cutting down. After iterative computing, positive and negative biases will be compensated during accumulation (see the right part in Figure 2.11). The simplest rounding is the so-called round-to-nearest technique, the conventional type of rounding provided by most fixed-point processors. The rounding arithmetic is to add the smallest value after truncation if the MSB of the discarded part is one. The Verilog code of round operation before truncating M-bits off from N bits data ($N > M$) is:

```
// The N-bit data before truncation is D[N-1:0]
Result[N-M-1:0] = D[N-1:M] + D[M-1];// Rounding and truncation
```

After rounding, the quantization errors become half, and the distribution of positive error and negative error is balanced. The rounding operation is "conditional add" arithmetic. It should be performed before truncating bits from a long word to a short word. From the HW point of view, in order to perform a rounding, one

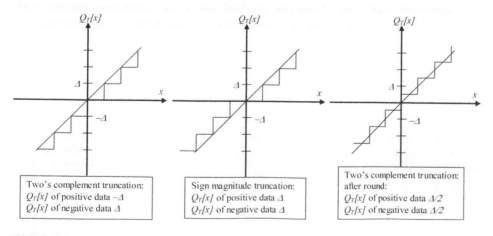

FIGURE 2.11

Quantization errors.

extra addition operation is needed. Therefore, the hardware availability and timing for rounding have to be considered during the instruction set design.

Even the round-to-nearest scheme adds asymmetry due to the rounding up of numbers lying not exactly at the midpoint. Thus on average more values are rounded up than rounded down. This means that the round-to-nearest operation adds a small bias to the signal. This bias is typically many orders of magnitude smaller than the bias introduced by truncation. For IIR and other recursive or feedback algorithms, the small bias might be considered in firmware instead of in hardware. Round-to-nearest is the default rounding technique implemented in hardware, and round-to-nearest arithmetic is given in Figure 2.12.

2.5.2 Overflow Saturation and Guards

Accumulation is used most often in DSP algorithms and is executed in fixed-point DSP processors. When executing accumulations, the value in an accumulator register of a fixed-point DSP processor may exceed the maximum representable value. In this case, computation errors will occur, which is called an overflow. For example, an overflow happens if X is not in the range $-2^{-N-1} \leq X < 2^{N-1}$, where N is the data length of the accumulator register. In a normal computer, overflow can be managed by exception handling and reexecution of the algorithm after scaling input data. In a DSP processor running real-time applications, there is no time to handle exceptions and reexecution of the application. Therefore overflows in a real-time system need to be handled by other means.

There are three ways to deal with an overflow. The first way is to use a floating-point processor to increase the dynamic range. This solution induces higher hardware costs. The second way is to scale down the input data so that the result of the accumulation is less than the maximum representable value. However, the precision of inputs will be lower because of scaling. The third way is to use "guard and saturation arithmetic." With this method, special accumulation circuits with guards will protect the accumulated value from overflow during accumulations, and

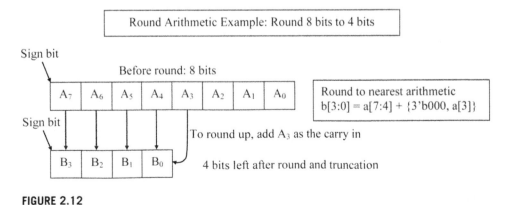

FIGURE 2.12

A round example.

the circuit will detect the overflow on the guarded result when the accumulation is finished. Guarded overflow is the overflow that can be protected by introducing guard bits.

The guard operation is performed implicitly during accumulation by introducing extra sign/guard bits to operands and accumulating circuits. The sign/guard bits are copies of the sign bit before an accumulation. During the accumulation, the sign/guard bits will be involved in computing and will prevent the sign value from overflow. The implicit rule is that the accumulating results will always be within the range of $2^G \times (-2^{N-1}) \le X < 2^G \times (2^{N-1})$; G is the number of guard bits, and N is the significant data length of accumulation. The total data length of accumulation is $G + N$. The dynamic range of the original data is $2^{N-1} - 1$, and the guarded dynamic range is $2^{G+N-1} - 1$.

Saturation will be performed after accumulation, or before moving data from an accumulation register to other places. There is an assumption that all DSP firmware must guarantee to limit results within $2^G \times (-2^{N-1}) \le X < 2^G \times (2^{N-1})$. In most cases, the result will be within $-2^{N-1} \le X < 2^{N-1}$. However, the accumulation result could be outside of $-2^{N-1} \le X < 2^{N-1}$ and within $2^G \times (-2^{N-1}) \le X < 2^G \times (2^{N-1})$. In this case, the result is not very good. If there is no time to recompute the value, the value has to be accepted in a relatively acceptable way. This way is called saturation, replacing the overflow result using the closest acceptable value.

While performing saturation when a positive overflow happens ($X \ge 2^{N-1}$), the erroneous value will be replaced by the largest positive number PMAX $= 2^{N-1} - 1$. In case a negative overflow occurs ($X < -2^{N-1}$), the erroneous value will be replaced by the largest (or most) negative number that can be represented NMAX $= -2^{N-1}$. In some systems, the largest negative number can be NMAX $= 2^{N-1} - 1$.

Although the result is not correct after saturation, it is much better than overflow. The value has to be accepted because there is no time to redo the computation. New data arrives periodically in real-time systems, and processing the new data must be prioritized.

Data quality can be optimized before runtime or during runtime. By detecting the saturation flag, a data scaling subroutine should be called and executed. In some DSP processors, the saturation flag is set when saturation is detected, and the saturation flag will be reset during the running of the data scaling subroutine. During the firmware debugging time, firmware can be improved by checking saturations, inserting data measurement subroutines, and scaling subroutines at required places.

In many applications, the final result is usually within the range of $-2^{N-1} \le X < 2^{N-1}$; values in an accumulation register could temporally exceed $-2^{N-1} \le X < 2^{N-1}$ and within $2^G \times (-2^{N-1}) \le X < 2^G \times (2^{N-1})$. The guards in this case are essential to avoid overflow during the accumulation. Figure 2.13 illustrates how the guard bits are concatenated and overflow can be avoided:

Guards for operand "a" are $ga_4 = ga_3 = ga_2 = ga_1 = a_3$ and for "b" are $gb_4 = gb_3 = gb_2 = gb_1 = b_3$. There are three hardware components in Figure 2.13; the first hardware component is the guard insertion as sign extension of input operands. The second hardware component is the accumulator performing accumulation including guard bits. The third component is the saturation hardware, which

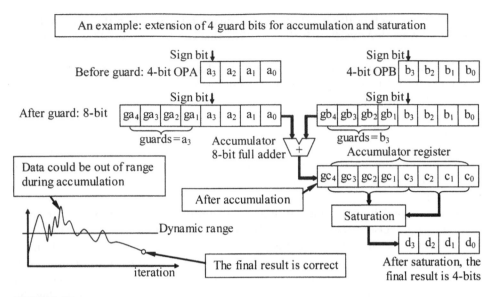

FIGURE 2.13

Saturation and guard bits.

detects for saturation and generates a saturated vector replacing the accumulated result when the result is overflow.

By applying guard arithmetic, the intermediate operands in accumulator are longer. According to this example, a sign bit is copied to four more guard bits. Let us observe the data value in the accumulator register in Figure 2.13 during iteration. During an iteration, the data stored in the accumulator register could exceed the maximum positive ($0111_2 = 2^3 - 1$) or maximum negative ($1000_2 = -2^3$) value that can be represented. The accumulator can present up to the range of $-2^{3+4=7} \leq x < 2^{3+4=7}$. However, the final result is normally within the dynamic range of the native width $-2^3 \leq x < 2^3$ (see the lower left part of Figure 2.13). If we do not use guard arithmetic, the intermediate result in the accumulator register might overflow during iterations. With guard bits, the representable ranges in the accumulator and accumulator register are larger, and the intermediate result will be within the representable range of the guarded hardware.

Finally, if the $G + N$ bits result with G-bit guard bits is $R[G + N - 1, 0]$, the complete saturation arithmetic will be implemented in the following Verilog code:

```
// Check that guard bits are not identical
if ( R[G+N-1:N-1] != {(G+1){R[G+N-1]}} )
// Check sign bit
  if ( R[G+N-1] ) Result = {1'b1,{(N-1){1'b0}}};//MAX Positive
  else Result = {1'b0,{(N-1){1'b1}}};        //MAX Negative
else Result = R[N-1:0];// No saturation
```

2.5.3 Requirements on Guards

To prevent overflow, the dynamic range, scaling, and dynamic range after guarding should be carefully specified. Overflow in a fixed-point processor could happen both in a normal ALU and in the MAC unit. There will be no iterative arithmetic running in ALU so that the sum of the adder will be within $[-2^N, 2^N)$ for fractional data computing based on the operand range of $-1 \leq x < 1$, or for integer data computing based on the operand range of $2^{N-1} \leq x < 2^{N-1}$. In an ALU without iteration support, we need only one guard bit ($G = 1$) to extend the data range to $2^{G+N-1} \leq x < 2^{G+N-1}$ for integer or $2^G \leq x < 2^G$ for fractional data. One guard bit means one more sign bit extension to operands and the arithmetic hardware. At the same time, saturation circuit is required at the output of the arithmetic circuit. The saturation is enabled for single-precision computations and must be disabled, for example, when computing on the lower part data for double-precision. In the following cases, saturation should be disabled:

- For lower subword computing of a double precession algorithm.

- Unsigned arithmetic computing.

- During the execution of an iteration algorithm.

When saturation is disabled, the carry-out bit (flag) should be used to link the upper and the lower part computing of double-precision arithmetic.

During iteration, overflow can be avoided by scaling or guarding. Specifications of scaling factors and the number of guard bits should be given before hardware design. When the requirements on precision and dynamic range are low, the input can be scaled down to prevent the result from overflow during the iteration. The drawback is the increased quantization error. For a high-quality system, we cannot reduce the precision or scaling data before iteration; therefore, we need to guard operands, the arithmetic hardware, and the accumulator register hardware. Equation 2.11 describes the maximum output value of a FIR filter, where x_{max} is the maximum input value. The scaling and guarding factors are calculated according to Equation 2.12. For example, if there are a maximum of 512 iteration steps, all the coefficients are checked and the sum of 512 coefficients is not more than 250, the factor of $2^8 = 256$ or eight guard bits will be enough.

$$|y_k(n)| \leq x_{max} \sum_{m=-\infty}^{\infty} |h_k(m)| \qquad (2.11)$$

$$f_{guarding} = \sum_{m=-\infty}^{\infty} |h_k(m)|$$

$$f_{scaling} = \frac{1}{f_{guarding}} \qquad (2.12)$$

If applications include adaptive filters, values of coefficients are unknown. In this case, we can make a conservative decision based only on the worst-case, that all coefficient values are the positive maximum. The maximum order of the filter will be linked to the number of guards. Continuing with the previous example, the hardware with eight guard bits can support reliable computing for an adaptive filter of 256 taps. If more taps are required by an application, the adaptive filter has to be split up to several adaptive filters [10]. Otherwise, prescaling inputs or accumulated data might be a compensate way.

2.5.4 Execution Order

Finally, execution orders need to be discussed. If the execution order is wrong, there might be a fatal error in the result. The guard operation must be executed before the iterative computing. Before and during the iterative computing, scaling might be required to enhance guards and prevent possible overflow. Rounding, truncation, and saturation shall be executed after the iteration and before storing data to components with different data width. The right order of the computation must be:

1. Guarding operands.

2. Kernel operation and scaling.

3. Round after kernel operation.

4. Saturation and removing guards.

5. Truncation and output.

2.6 EXAMPLES OF CORNER CASES

Identifying corner cases is an essential skill of designing and verifying a processor, especially an ASIP with special data features. A corner case in this book stands for special or irregular inputs, results, and induced computing arithmetic circuits during the design of finite length hardware. A corner case may induce unexpected errors. The following examples expose some corner cases.

Example 2.13

Identify that special saturation arithmetic required for a special case of a fractional multiplication $(-1) \times (-1)$.

According to the definition of fractional data format, the range of fractional data is $-1 \le x < 1$. Recall the example given in Figure 2.4 and the definition of fractional number representation: one bit left shift should be executed after fractional multiplication so that the fractional radix point of the result will be on the right side of MSB. The fractional number system accepts (-1) and does not accept $1 = (-1) \times (-1)$. In this case, saturation must be performed implicitly after the 1-bit left shift on the multiplication result. To obtain correct results from fractional multiplications using the normal two's complement integer multiplier, the operation should be:

```
// A[N-1:0] and B[N-1:0] are signed numbers
Result[2*N-2:0] = Saturation ((A * B) << 1);
```

The order of the execution should also be correct:

```
tmp[2*N-1:0] = A * B;// not real Verilog code.
// Signed numbers aren't that well supported in Verilog yet
if (tmp[2*N-1] != tmp[2*N-2])
  Result [2*N-2:0] = {1'b0, {(2*N-2){1'b1}}}; //saturation
else
Result [2*N-2:0] = tmp[2*N-2:0] << 1;
```

This example exposes one special corner, the only corner of multiplication.

Example 2.14

Absolute operation on −1 in ALU.

Without guard and saturation operations, the absolute operation of A[3:0] = 4'b1000 will become (the pseudocode):

```
ABS (A [3:0]) <= INV (1000) + 0001 <= 0111 + 0001 <= 1000
```

The result is wrong because the result is a negative number −8 after the absolute operation. The right result should be +7 or (4'b0111). The correct operation with guard and saturation should be (the pseudocode):

```
A [3:0] = 4'b1000; // guarded A = GA [4:0] = 5'b11000
If A[3]
{
ABS (GA [4:0]) <= saturation (ABS (GA))
ABS (GA [4:0]) <= saturation (INV(11000)+00001)
             <= saturation (00111 + 00001)
             <= saturation (01000)
             <= 0111// remove the guard bit
}
Else ABS (A [3:0]) <= A [3:0]
```

Remember this typical case of ABS (−1). It requires at least one guard bit in ALU hardware.

2.7 CONCLUSIONS

Basic definitions of data type formats (numerical representations) are the starting point of DSP datapath design. Fixed-point (either fractional or integer) data format will be used dominantly in DSP hardware and firmware development.

DSP applications have different requirements on data precision and dynamic range. To keep low silicon costs, the fixed-point/integer datapath should be used. When high dynamic range is required, a floating-point processor with custom data type might be a solution.

Rounding arithmetic was introduced in this chapter to minimize the quantization error at the lower side of data. Guard and saturation arithmetic was introduced in this chapter to relax the distortion error induced by overflow at the upper side of data.

To keep the relative achievable precision and dynamic range, fixed-point/integer DSP datapaths are supported with special arithmetic such as guard, scaling, rounding, saturation, and truncation. Guard hardware can be found for operands, computing components, and internal storage registers.

EXERCISES

2.1   What is the difference between integer and fractional numerical presentations?

2.2   What are definitions of dynamic range and precision for fixed-point and floating-point representation? What is dynamic range? What is precision?

2.3   When do we need both high resolution and high dynamic range? When do we need higher dynamic range and only reasonable precision?

2.4   Define floating-point corner values following Table 2.9 for a custom floating-point representation system. It represents two's complement data, the floating-point format is 1-4-11 (1 sign bit—4 exponent bits—11 mantissa bits).

2.5   What is quantization error?

2.6   Which operation must be executed first, truncation or rounding?

2.7   What is overflow? What is saturation? What is guard?

2.8   When designing a saturation function at the output of a 16-bit arithmetic unit, in which case does the saturation have to be switched off?

2.9   To design saturation function for a 16-bit full adder without iterative operations, how many guard bits are required?

2.10   While using a MAC unit with eight guard bits for convolution-based iterations, how many iteration steps can be supported?

2.11   What are the dynamic range and precision of a 16-bit signed fractional data? What are the dynamic range and precision of a 16-bit signed integer data?

   a. Dynamic range of a 16-bit signed fractional data.

    b. Precision of a 16-bit signed fractional data.

    c. Dynamic range of a 16-bit signed integer data.

    d. Precision of a 16-bit signed integer data.

2.12 With fractional multiplication and integer multiplication:

    a. Calculate the 4b × 4b fractional multiplication 1000*1000.

    b. Calculate the 4b × 4b integer multiplication 1011*0101.

2.13 Which of the following statement(s) is(are) true?

    a. The MSB of the truncated part is the carry-in bit during round operation.

    b. When the truncated data is positive, the round-up value is 1, when the long data is negative, the round up value is 0.

    c. 32-bit floating-point data has higher dynamic range than 32-bit fixed-point data.

    d. In a two's complement system, 16-bit integer data has higher dynamic range than 16-bit fractional data.

REFERENCES

[1] www.itu.org. ITU R G.711: Pulse code modulation (PCM) of voice frequencies.

[2] Omondi, A.R. (1994). *Computer Arithmetic Systems Algorithms, Architecture and Implementation.* Prentice Hall.

[3] www.ieee.org. IEEE standard 754.1985: IEEE Standard for Binary Floating-Point Arithmetic.

[4] www.ti.com. TI:TMS320C6727.

[5] Eilert, J., Ehliar, A., Liu, D. (2005). Using low precision floating-point numbers to reduce memory cost for MP3 decoding. *Proc of the IEEE Int'l Workshop on Multimedia Signal Processing (MMSP),* Siena, Italy.

[6] Eilert, J., Wu, D., Liu, D. (2007). Efficient Complex Matrix Inversion for MIMO Software Defined Radio. To be presented at IEEE ISCAS, USA.

[7] Wu, D., Eilert, J., Liu, D., Wang, D., Al-Dhahir, N., Minn, H. (2007). Fast Complex Valued Matrix Inversion for Multi-User STBC-MIMO Decoding. To be presented at IEEE ISVLSI, Brazil.

[8] www.coresonic.com.

[9] Maxfield, C., Brown, A. (2005). *The Definitive Guide to How Computers Do Math: Featuring the Virtual DIY Calculator.* John Wiley & Sons Ltd.

[10] Amano, F., Meana, H. P., de Luca, A., Duchen, G. (1995). A multi-rate acoustic echo canceller structure. *IEEE Transactions on Communications* 43(7), 2172–2176.

[11] Embree, P.M. (1995). *C Algorithms for Real-Time DSP.* Prentice Hall Publishing.

[12] Lapsley, P. et al. (1997). *DSP Processor Fundamentals, Architectures and Features.* IEEE Press.

[13] Kuo, S.M., Gan, W-S. (2005). *Digital Signal Processors, Architecture, Implementation, and Applications.* Prentice Hall.

[14] Kobayashi, S., Fettweis, G. P. (1998). A Block-Floating-Point System for Multiple Datapath DSP. *IEEE Workshop on Signal Processing Systems (SIPS'98)*, page(s) 427–436, 08.-10.10.98.

[15] Kobayashi, S., Fettweis, G. (2000). A Hierarchical Block-Floating-Point Arithmetic. *Journal of VLSI Signal Processing–Systems for Signal, Image, and Video Technology*, 24(1):19–30.

DSP Architectures

3

In this chapter, basic architectures will be discussed following a SoC (System on a Chip) design hierarchy. Concepts like system, processor, and processor core will be introduced. The different modules of a DSP core, including memory bus, register bus, ALU, MAC, register file, control path, and address generator, will be described. Advanced architectures, including SIMD (single instruction multiple data) DSP processor, Superscalar DSP, VLIW (very long instruction word) DSP processor, and on-chip multiprocessor DSP, will be introduced.

Architecture design will also be introduced briefly in this chapter. Architecture design is the specification of the top level modules and interconnects between these modules.

The contents of this chapter should be useful for architecture selection. However, techniques for architecture selection cannot be discussed in detail without application analysis techniques and source code profiling techniques. Source code profiling will be discussed in Chapter 6, and the architecture selection in Chapters 19 and 20.

3.1 DSP SUBSYSTEM ARCHITECTURE

A DSP subsystem usually consists of DSP cores, DSP memory subsystems, and peripherals. A processor could play different roles on different hierarchy levels. For example, a processor in a hard disk controller is a peripheral device in a computer system, but the same processor will be the central processor for a hard disk system.

In this book, we discuss only systems for DSP applications. A DSP processor/core is therefore the central processor in our discussion. A DSP core together with its surrounding peripherals and memories forms a DSP subsystem in an embedded system on a chip (SoC). A DSP subsystem in a SoC is shown in Figure 3.1.

Figure 3.1 is a typical SoC for embedded applications. It consists of a DSP subsystem and a MCU (microcontroller unit) subsystem. The MCU subsystem is the task controller of the SoC. It controls all tasks running in the system and executes tasks without real-time requirements. Real-time tasks and tasks requiring high computing performance will be allocated to the DSP subsystem, the executer of the MCU

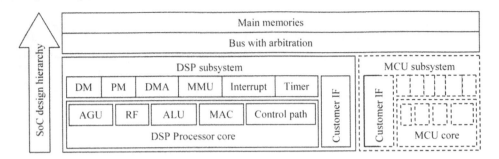

FIGURE 3.1

A System-on-Chip (SoC) and processor hierarchy.

subsystem. The system bus is the communication media between processors and off-chip components.

The definition of a processor core is currently ambiguous. Some cores include memories and peripherals, and some don't. In this book, the DSP core definition is that it consists of a datapath (register file, ALU, MAC unit, and special function units), a control path, and an addressing path (or AGU, Address Generation Unit, including memory bus). The memory bus is the interface between the core and the memories. Peripheral devices, not directly driven by the core instruction set, are not parts of a processor core. However, a DSP core as a SIP (Silicon Intellectual Property) product sometimes also includes IO (input/output) interface circuits, such as interrupt handler, timers, and DMA (direct memory access) controller.

The system bus is the communication medium between processors, main memories, and other external devices. For the case depicted in Figure 3.1, it is the medium between the DSP and the MCU. Note that this bus is an interprocessor bus. Pay attention to the context whenever a bus is mentioned in this book: the bus inside a DSP core is a memory or register bus, whereas the bus between cores is a system or interprocessor bus.

3.2 PROCESSOR ARCHITECTURE

The processor architecture is the hardware organization of the core and its peripherals, including the memory bus architecture. The processor architecture can be divided further into the assembly instruction set architecture (ISA) and the microarchitecture (architecture of each hardware module).

In this book, the ISA specifies all instructions that a processor can execute. Implementation details and performance-related features are not specified by the ISA. For a typical instruction the ISA specifies the operands to be fetched, the operation to be executed, and where the result should be stored. Two operands can be accessed simultaneously by a normal DSP, whereas a four-way SIMD processor (single instruction multiple data) may access up to eight operands simultaneously. The ISA does not specify the number of pipeline steps that are required or how each instruction is mapped to hardware.

The microarchitecture design is the implementation of an ISA specification into hardware modules. Hardware details (functions, input/output, and pipeline) will be specified during the microarchitecture design. The microarchitecture specification therefore is the detailed description of function allocations and implementations in hardware at certain pipeline stages. The microarchitecture design:

1. Exposes and allocates all microoperations from each assembly instruction to hardware modules.

2. Specifies the pipeline and allocates each microoperation to one or several pipeline stages.

3. Specifies the hardware multiplexing.

4. Specifies input/output and intermodule operations.

5. Optimizes the preceding four steps.

Performance, cost (silicon area), and power consumption are key parameters to consider during the microarchitecture design. Enhanced performance by instruction level parallelization (VLIW or superscalar) and deep pipelining has to be traded off against higher area and power consumption. The memory subsystem design (on-chip memory design, memory hierarchy, addressing circuits, memory bus, and DMA) will heavily influence all these parameters.

3.2.1 Inside a DSP Subsystem

A DSP subsystem can be divided into three parts as shown in Figure 3.2: the processor core, the memory subsystem (DM, TM, DMA, and MMU), and the peripheral subsystem (the interrupt handler, the timer, and the custom interface). We further divide the DSP core into the datapath and the control path. The datapath supports basic arithmetic and logic functions as well as dedicated DSP algorithm functions (accelerators).

A DSP usually is based on a RISC architecture with some CISC enhancement. (A discussion of RISC versus CISC can be found later in this chapter.) Similar to a RISC machine, a general-purpose register file can be found as the first-level storage component in the datapath. The register file is used as a computing buffer, and all RISC-type computing gets operands from and sends results to the register file. The register file data is loaded from data memory and results in register files that are stored to data memory.

CISC enhancements are dedicated for DSP computing such as convolution and in special operations such as division. Data array elements are directly loaded from data memories and consumed by the CISC datapath, without using the general register file. FFT works in a similar way: both operands and results are accessed directly from/to the data memories. In general, CISC features will enhance performance in iterative computing.

The DSP core, marked with a fine dotted line in Figure 3.2, is the central part of a DSP. The core can be divided further into three parts: the datapath, the control path,

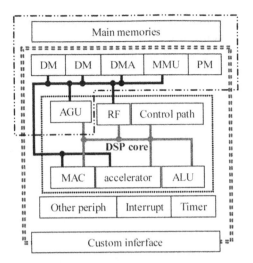

FIGURE 3.2

A DSP subsystem.

and the address generation unit (AGU). The core components are organized around two data buses: the memory bus and the register bus. The memory bus (thick black lines in Figure 3.2) is distributed between the core and the memory subsystem, as well as inside the core. It forms a data connection between on-chip data memories and the register file. The memory bus also is connected to the MAC unit, supplying vectors from data memories. The register bus (thick grey lines in Figure 3.2) connects the register file to all units in the core. This bus is not distributed outside the core. The topology of the memory bus and the register bus can be decided as soon as the assembly instruction set is specified. This is because the assembly instruction set gives an implicit specification of the two buses through the specification of the number of operands, the data source, and the result.

The memory subsystem (marked with a dot-dot-dashed line in Figure 3.2) consists of data memories (DM), program (code) memory (PM), AGU, DMA, and MMU. DM and PM are on-chip memories with limited size that can supply high-speed access. On-chip memories can be scratch-pad memories or cache.

The main memory is not directly accessed by the DSP core; instead data is transferred from main memory to data memories using DMA or cache. If instructions are stored in main memory, the program memory is usually a cache. The main memory could be on-chip, off-chip, or existing both on- and off-chip. Off-chip main memories are large in size with limited speed. Detailed memory issues will be discussed later in Chapters 15 and 16.

There are two levels of peripheral hardware: DSP core peripherals and DSP peripherals. Peripherals can be divided into functional building blocks and nonfunctional building blocks. Functional building blocks can be accessed or configured by programs.

Typical functional peripherals of the DSP core are:

- Timers for counting clock cycles.

- Interrupt controller for handling interrupts.

- DMA (Direct Memory Access) controller for handling data transfers to/from main memory and between other memories/ports.

- MMU (Memory Management Unit) for managing reliable and efficient (address space) memory usage.

Typical functional peripheral devices of the DSP subsystem or a DSP processor are:

- Parallel host control interface for connecting to an off-chip MCU (microcontroller) using a high-speed parallel link.

- Parallel data port for connecting off-chip memory to enhance the on-chip memory bandwidth.

- Serial data port for low-speed communication between chips.

- Dedicated control interfaces for communications of interrupt signals, acknowledgment, and other control and status signals.

Nonfunctional peripherals are not used in normal operation and cannot be accessed by running programs. For example, the JTAG port (Joint Test Acting Group IEEE1149.1) is a port for testing. Peripheral details will be discussed later in Chapter 16.

3.2.2 DSP (Memory Bus) Architecture
History of DSP (Memory Bus) Architecture

Before the detailed description of modern DSP architectures are presented, a review of some classic solutions will be given. One of the first programmable processor architectures was proposed by John von Neumann, and processors using his architecture usually are referred to as von Neumann processors, see Figure 3.3(a). Such a processor loads and executes instructions one by one from a program stored in a memory. Because of memory size limitations in the early years, program and data shared the same memory.

Another solution, called the Harvard architecture, used separate program and data memories, see Figure 3.3(b). The Harvard architecture was introduced by Howard Aiken at Harvard University in the late 1930s. This architecture was used in the Harvard Mark-1 computer, which became operational in 1944. Modern DSPs often use a modified Harvard architecture. The DSP architecture evolution during the last decades is closely related to the rapid progress of silicon technology.

In the following, memory bus architectures will be discussed. A basic Harvard architecture can be found in Figure 3.4(a) (CP is the control path, DP is the datapath,

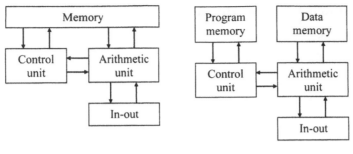

(a) Von Neumann architecture   (b) Harvard architecture

FIGURE 3.3

Classic processor architectures.

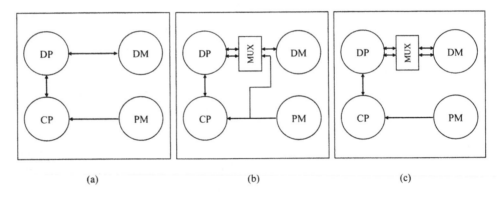

(a)   (b)   (c)

FIGURE 3.4

Basic DSP architectures.

PM is the program memory, and DM is the data memory). This architecture cannot feed multiple data to DP. However, one step of a convolution requires both a data word and a coefficient. Thus two memory accesses are required simultaneously. The architecture in Figure 3.4(a) supplies only one data at a time; it means that a tap of a convolution requires multiple clock cycles.

The circuit in Figure 3.4(a) was improved, resulting in the one in Figure 3.4(b). In Figure 3.4(b), the program memory can also be used as the coefficient memory when a hardware loop convolution is executed. Hardware loop represents hardware acceleration for loop control; for example, loop functionality (for $i = 1$ to $N, i++$) is executed implicitly in hardware. The program for convolution becomes:

```
Instruction of N-Step hardware loop Convolution
Coefficient 1 ; Coefficients are stored directly in the PM
Coefficient 2
...
Coefficient N
The first instruction after the hardware loop convolution
```

In this program, N is the iteration steps (taps) of the convolution. Using the architecture in Figure 3.4(b), we fetch coefficients instead of instructions during a convolution. Simplicity is the advantage, and the drawback is the insufficient support of iterations. The architecture in Figure 3.4(b) can support only simple iterations with one instruction.

The architectures in Figure 3.4(a) and (b) seldom are used in modern DSPs, although many microcontrollers are still using them.

Using the architecture in Figure 3.4(c), more instructions can be fetched while running an iteration loop because the program memory is not used to supply coefficients. Two data words, one operand and one coefficient, are accessed from DM within one clock cycle for one step of convolution. In this case, a dual-port or multiport memory is required. A drawback is that the machine clock rate might be limited by this type of memory. The architecture in Figure 3.4(c) was very much used in the 1980s when logic operations were slower than the speed of memory access.

For highest speed, dual-port memory should be avoided. Architectures in Figures 3.5(d) and (e) were proposed instead. In Figure 3.5(d), a small program memory buffer PMB is used to store a short program when PM is required to access data. The structure in Figure 3.5(d) is better than the one in Figure 3.5(b) because this structure can run only a single instruction in a hardware loop whereas the structure in Figure 3.5(d) supports hardware loops including multiple instructions.

The architecture in Figure 3.5(d) requires the program for iteration to be stored in the PMB in advance. When running a hardware loop using codes from the PMB, the coefficients stored in PM can be accessed. It might be difficult, however, to keep the PMB small if a large number of iterative loops are involved in an application.

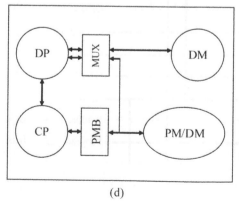

(d)

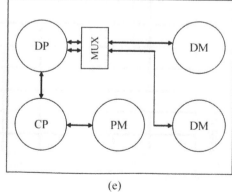

(e)

FIGURE 3.5

More basic architectures.

The architecture shown in Figure 3.5(d) lacks flexibility because the total size of iteration loops is limited by the size of the PMB. Thus it wasn't popular for long. Instead, the most frequently used simple DSP architecture with a single MAC is given in Figure 3.5(e). Two data memories are used. For convolutions and other iterative algorithms, data is stored in one memory and coefficients in the other.

The Most Widely Used DSP (Memory Bus) Architecture

A typical DSP bus architecture is shown in Figure 3.6, which is a detailed view of Figure 3.5(e). For accessing multiple data in parallel, we need a bus that supports multiple connections to multiple memories. The two data memories (DM1 and DM2) are connected to the memory bus and the program memory (PM) to the PM bus. Each data bus can be used to transfer data from the memory to the core or from the core to the memory. Three address buses are associated with the three memories so that three accesses, including two data words and one program word, can be executed independently in parallel.

The bus connections and multiplexing are controlled by the control path. The bus system in a DSP processor is different from that in a normal CPU. In a DSP, the bus system is divided into three parts: the memory bus, the register bus, and the PM bus. Since data allocation and program execution are relatively well known in DSP systems, more parallel and simpler bus architectures can be used. This is an essential concept for DSP design. Following this principle, a DSP chip could be designed with much higher performance compared to a general-purpose CPU.

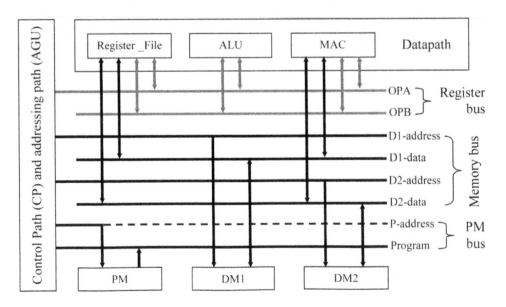

FIGURE 3.6

A typical bus architecture in a DSP datapath.

3.2.3 **Functional Description at Top Architecture Level**

A DSP is a programmable device that fetches, decodes, and executes instructions following a program flow in accordance with the current status (flags and processor configuration) of the processor. In a DSP, there are two state machines, the instruction flow FSM (Finite State Machine) and the data flow FSM.

The instruction flow FSM calculates the address to the instruction to be fetched and decodes the fetched instruction (see Figure 3.7). Machine code in binary format is stored in a program memory (PM). The program counter (PC) points to the next instruction to be fetched from the program memory. The received instruction is decoded by the instruction decoder, which will generate control signals for the datapath (DP) and the control path (CP). Operand addresses are generated and sent to the storage hardware. Operands are then fetched based on the operand addresses. The execution of the instruction generates flags (status), which are used for calculating the address of the next instruction.

The data flow FSM handles the execution of the decoded instruction (see Figure 3.8). The data flow starts at the time when the instruction decoder sends its decoded control signals. These signals specify operand fetching, arithmetic operation, and where to store the result. Operand fetching, in other words, is the same as operand addressing. Operands are fetched according to their calculated addresses and are received by the execution units of the datapath. Arithmetic operations are executed according to the decoded control signals. Results from the operations are then stored according to the decoded result address. Finally, flags are generated and sent to the instruction flow control circuit, which is also called PC FSM (the Program Counter as a Finite State Machine), or to the instruction flow FSM; it will be discussed in detail in Chapter 14.

Figure 3.9 is a further refined description of Figure 3.2. In Figure 3.9, modules are further partitioned, and control signals as black solid lines are exposed (MEM-ctrl means memory control and addressing; Operation ctrl means control signals for execution). By mapping functions in Figures 3.7 and 3.8 to the architecture in Figure 3.9, the basic principles of any processor can be illustrated.

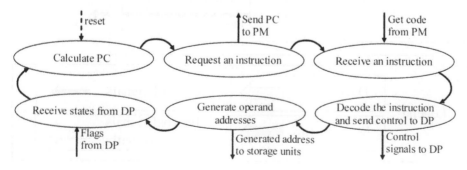

FIGURE 3.7

Instruction flow FSM.

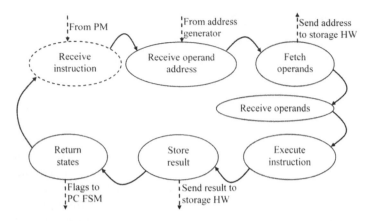

FIGURE 3.8

Data flow FSM.

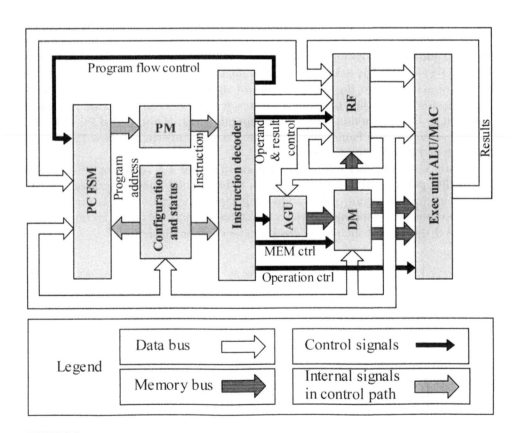

FIGURE 3.9

A basic processor.

The processor starts the execution of an instruction by sending the instruction address to PM. The instruction is fetched from PM and sent to the instruction decoder. The instruction is decoded, and control signals are generated for various components such as the Address Generator (AGU), the Register File (RF), the memories, and the execution units. At the same time, the PC value is updated for fetching the next instruction. The register file and data memory DM send operands to the execution unit, and the result will be available on the register file input.

Memories in a DSP usually are implemented as synchronous Static Random Access Memories (SRAM). If we say "a normal SRAM," it means a single-port synchronous SRAM where there is only one data read out port and one data write in port. Only one access is possible per clock cycle. Memories will be discussed in detail in Chapter 15.

3.2.4 DSP Architecture Design

The design of the architecture is to specify the top-level structure of the processor. The design includes specifications of hardware modules, the top level of the core, and connections among modules. The connections are further divided into two groups, the memory bus (data and address) and the register bus.

The input of the design is the assembly language manual. The output of the design is the top-level block diagram, including components, connections, and component parameters.

Example 3.1

Designing a processor architecture includes the decision of data memories, datapath components, memory buses, register buses, and address generators.

The Assembly Instructions

Assuming there is one assembly instruction set manual that is divided into four groups of instructions as illustrated with Table 3.1 to Table 3.4:

Table 3.1   ALU Instructions.

	Name	Operand	Description	Operation
1	ADD	Rd, Rs	Add	$Rd <= Rd + Rs$
2	ADDI	Rd, K	Add Immediate	$Rd <= Rd + K$
3	ARSH	Rd, K	Arithmetic right shift	$Rd <= Rd >> (K)$

Table 3.2   MAC Instructions.

	Name	Operand	Description	Operation
4	CONV	M1, M2	Convolution: M1 data, M2 Coefficient	ACR <= ACR + M1*M2
5	MAC	Rd, Rs	Multiplication and accumulation	ACR <= ACR + Rd*Rs
6	LADD	Rd, Rs	Long word addition	ACR <= ACR + {Rd, Rs}
7	GDAR	ACR	Move data from AR to a general register	Rd <= SAT(ACRH)

Table 3.3   Load/Store Instructions.

	Name	Operand	Description	Operation
8	Ld M1	Addressing mode	Load a word from DM1	Rd <= M1(address)
9	St M1	Addressing mode	Store a word to DM1	M1(address) <= Rs
10	Ld M2	Addressing mode	Load a word from DM2	Rd <= M2(address)
11	St M2	Addressing mode	Store a word to DM2	M2(address) <= Rs

Table 3.4   Branch and Other Instructions.

	Name	Operand	Description	Operation
12	JUMP	Jump	If condition is true, jump	PC <= Rs
13	CALL	Call	Call, stack the return address	PC <= call address
14	RETURN	Return	Stack pop and return	PC <= stack, pop
15	SET	Constant	Instruction carries a constant	Rd <= constant
16	MOVE	Setting	Configure the processor	Special REG <= Rs

Rs is a source register. Rd is both a source register, and the destination register. K is an immediate value.

ACR is an accumulator register. M1 and M2 stand for data memory 1 and 2.

Addressing modes denote addressing algorithms.

Processor architecture design:

1. Specifies modules and interconnects between modules as well as input/outputs according to the requirements from the instruction set.

2. Specifies module parameters such as memory address space, register file size, and data types.

Analysis of all listed instructions is as follows:

1. ALU instructions are part of the instruction set; therefore an ALU module is required. According to the ALU instruction table, Rd and Rs are general registers. K is a constant carried by the instruction and supplied from the control path. Three ALU inputs are needed: Rd and Rs from the register file and K as immediate data from the control path. The output of the ALU needs a connection to the register file input.

2. MAC instructions are part of the instruction set; therefore a MAC unit is required. According to the MAC instruction table, MAC operands are the internal operand ACR, the external operands {Rd, Rs}, and the external operands from M1 and M2. The CONV instruction requires simultaneous access to M1 and M2. The output of the MAC needs a connection to the register file input. To conclude, MAC inputs are from the register file and the data memories. MAC output goes only to the register file.

3. Load/Store instructions move data between a memory and the register file.

4. In the last table of the example, immediate data can be carried by an instruction. In this case the control path sends the immediate data to the register file and the ALU. The input/output operation and processor configuration operation require data transfers from a general register to special registers (such as loop control registers and address pointers) and vice versa. A link between the register file and the special registers is required. Special registers are allocated in the control path and in AGU.

To summarize, the required components are listed in Table 3.5.

Table 3.5 Required Components.

Modules	Inputs from	Outputs to
AGU	RF, Control path	M1, M2
M1	AGU, RF (Register File)	RF
M2	AGU, RF	RF
CP	ALU (flags), RF	RF, ALU
RF	M1, M2, CP, ALU, MAC, RF	ALU, ALU, MAC, MAC, CP (configure), AGU
ALU	RF, RF, CP,	RF
MAC	RF, RF, M1, M2	RF

Deciding Bus and Component Parameters

Parameters include address spaces, register file size, port size, stack size, etc. Parameters are specified according to the source code profiling, instruction set specification, and HW constraints.

To conclude the discussion, the top-level bus connections can be summarized using Table 3.6.

This table corresponds to a top-level block diagram, which is shown in Figure 3.10. Special registers are allocated in the control path. The thick black lines represent the memory bus, and the gray lines represent the register bus.

Table 3.6 Top-Level Bus Connections.

From memory	M1 → MAC	M2 → MAC	M1 → RF	M2 → RF		
To memory	RF → M1	RF → M2	AGU → M1	AGU → M2		
From register	RF → MAC	RF → MAC	RF → ALU	RF → ALU	RF → CP	RF → AGU
To register	M1 → RF	M2 → RF	MAC → RF	ALU → RF	CP → RF	
Others	CP → AGU	CP → ALU	CP → RF			

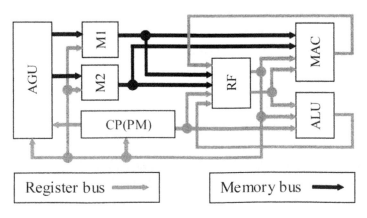

FIGURE 3.10

Top-level block diagram of Example 3.1.

3.3 INSIDE A DSP CORE

As mentioned earlier, a DSP subsystem can be divided into three parts: processor core, memory subsystem, and peripheral subsystem. In this section, we further divide a DSP core into datapath, control path, and address generator. The address generator is discussed sometimes together with the memory subsystem and sometimes as part of the control path. In this book, we will discuss the address generator as a separate unit in a DSP core.

3.3.1 The Datapath and Register Bus

Basic arithmetic and logic functions, part of dedicated DSP algorithms, are calculated in the datapath. The datapath gets its data from the memory subsystem as operands and sends the results back to the memory subsystem. Similar to a RISC machine, many general-purpose registers can be found as first-level storage components in a register file in the datapath. Operations in the datapath are controlled by the control path. Arithmetic and logic computing are done in the datapath. Essential components in a DSP datapath include the ALU, the MAC, the register file, and the register bus.

The register bus connects registers in the register file to all other components in the core. Normally, there are two buses from the register file supplying two operands each clock cycle. These two buses are also called operand buses. Each module in the core, except the address generator, has an output to the register file. Each data memory also has an output to the register file.

3.3.2 MAC

The multiplier and accumulation (MAC) unit is an arithmetic computing block in the datapath. The MAC performs arithmetic functions including multiplications, convolutions (such as FIR, IIR, and correlations), and transformations (such as FFT and DCT) as well as normal arithmetic operations with higher precision than the data width that the memory bus can support. Usually, the precision (data width) inside a MAC is twice the width of the memory bus. A MAC can work in two modes: iterative (CISC) mode and single-step (RISC) mode. In iterative mode, memories supply data to the MAC unit. In single-step mode, data are supplied by the register file and accumulation registers inside the MAC unit.

The most frequently used DSP kernel functions are based on multiplication and accumulation. Therefore the MAC instruction is very essential for a DSP. A complete MAC instruction may take more than one clock cycle to execute. In most DSP processors, the execution of a MAC instruction consumes two clock cycles, one for multiplication and one for accumulation.

Signed and unsigned multiplication as well as integer and fractional multiplication should be supported by the MAC unit. These types of multiplication will be discussed later in Chapters 7 and 13.

A MAC unit includes a multiplication and an accumulation part. The accumulation part consists of an accumulator (a full adder, usually with double precision) and accumulation registers. Accumulation registers are used as computing buffers (especially as iterative computing buffers) in the MAC. Due to the available double precision of the multiplication result, it is convenient to design the accumulator to support double precision. Arithmetic computing with double precision, for example, add/subtract, absolute and compare, will be executed as RISC instructions in one execution cycle. To maximally utilize the high dynamic range and the computing precision of the MAC hardware, other functions such as shift (scaling), rounding, and saturation also should be supported by the MAC unit.

It is sometimes confusing that MAC represents both the name of the hardware module and an instruction. However, this abbreviation has been used in the industry for a long time. In this book, we try to differentiate them in the text by using MAC unit or MAC instruction instead of just MAC.

The Accumulator can have different meanings: long arithmetic circuit with long computing buffers or long arithmetic circuit without computing buffers or addressable registers in MAC unit with long precision. In this book, Accumulator stands for long arithmetic circuit with long computing buffers, and Accumulator register stands for addressable registers in the MAC unit with long precision.

The left part of Figure 3.11 is a structural model illustrating the basic concept of a MAC, the principle of multiplication, and accumulation. Operands A and B are the main inputs of the MAC. The accumulation register is used for iterative computing.

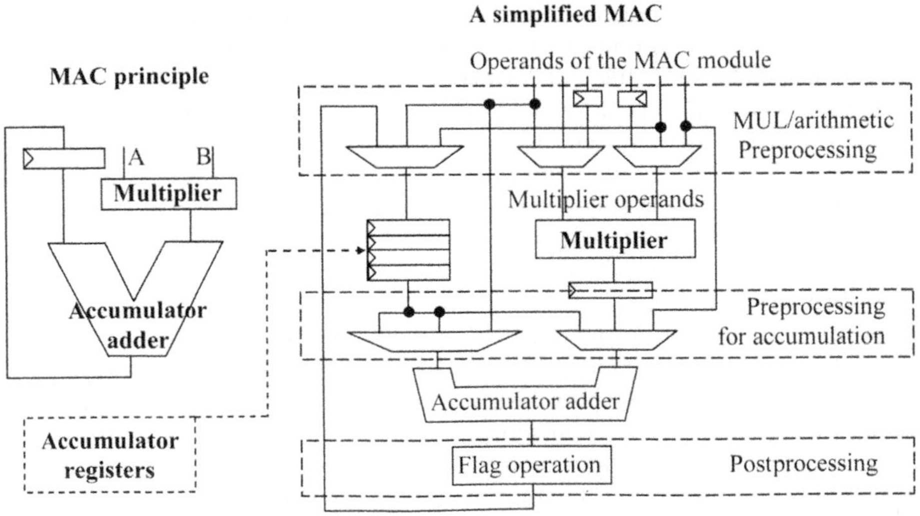

FIGURE 3.11

MAC architecture in a DSP datapath.

The right side of Figure 3.11 depicts a relatively realistic yet simplified MAC. This MAC can be divided into the following functional parts:

- Preprocessing for multiplication and register load: to prepare operands for the multiplier and the accumulator register.

- Multiplication.

- Preprocessing for accumulation, such as guarding.

- Accumulation and result store.

- Postprocessing and flag processing.

The purpose for separating pre- and postprocessing from the kernel computing (multiplication, addition) is to find opportunities for hardware multiplexing of the multiplier and the accumulator adder. By hardware multiplexing, the cost for kernel components can be minimized. Separation of pre- and postprocessing will be discussed in detail in Chapter 10.

Operands for the MAC module are supplied to the preprocessing circuits. Results are collected by the postprocessing circuits and stored in the accumulation registers. Special instructions are used for moving a result from an accumulation register to the general register file. By using preprocessing circuits, functions such as operand selection, operand preprocessing (invert, guard), and accumulator initialization can be specified. MAC module design will be given later in Chapter 13.

3.3.3 ALU

The arithmetic and logic unit (ALU) is a collection of arithmetic and logic computing hardware performing arithmetic operations (plus, minus, absolute, compare, negate, etc.), logic operations (invert, AND, OR, XOR, and bit manipulations), and shift operations. An ALU may also support special hardware operations, such as divide and counting. ALU operations in most DSPs consume one clock cycle.

An ALU is used for single-step computing of RISC instructions, and therefore single precision is enough. In this respect an ALU is different from a MAC, which is used for iterative computing and supports high precision.

The ALU could be a hardware module on its own or part of the hardware inside a MAC module. Usually, the ALU is a separate hardware module. When the DSP ASIP is a slave computing engine running only for iterative computation, the requirements on the ALU will be low. In this case the ALU may be part of the hardware in the MAC. In this book, the ALU is considered separate from the MAC because from a teaching point of view, this is more structured.

The ALU generates flags for program flow control. Flags can also be used as input signals for further ALU operations (for example, the carry-out flag can be the carry-in of the next instruction). An ALU is illustrated in Figure 3.12. Notice that the subblock connections inside an ALU can be different from this figure. The shift unit, for example, can be a component serially connected to an input port

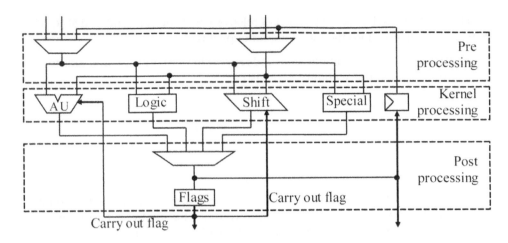

FIGURE 3.12

ALU architecture in a DSP datapath.

of the arithmetic unit (AU). Detailed ALU design techniques will be given later in Chapter 12.

The ALU depicted in Figure 3.12 can be further partitioned into preprocessing, kernel processing, and postprocessing. Separating pre- and postprocessing from kernel processing means that hardware multiplexing or the reusing of kernel components is possible. Operands are supplied through inputs on top of the circuit, and outputs are collected in the postprocessing part. The preprocessing part includes operand selection and operand preprocessing. The postprocessing part includes results selection and flag computing. Kernel processing design will be discussed in detail in Chapter 12.

3.3.4 Register File

The register file is the lowest level computing buffer for RISC instructions, storing temporary operands and results. A register file can receive data from all data memories and all hardware modules inside the DSP core, including data from the register file itself. A register file should be able to supply multiple operands for an instruction; in most cases, it simultaneously supplies two operands. A register file usually accepts only one result at a time in order to simplify the hardware dependency control and the compiler. In a processor with parallel datapaths, multiple results may need to be saved at the same time. Special register files or multiple register files are used for supporting multiple writes.

A register file consists of general registers, special registers, or general registers with implied specific functions. General registers are only used as data buffers. Special registers are designed to support special functions such as addressing, hardware looping, and control. A simplified block diagram of a register file is given in Figure 3.13.

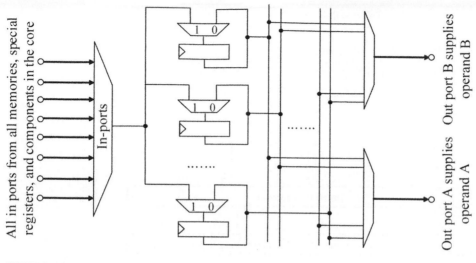

All in ports from all memories, special registers, and components in the core

In-ports

Out port B supplies operand B

Out port A supplies operand A

FIGURE 3.13

General-purpose register file.

The input multiplexer, on the left side in Figure 3.13, is used for selecting the data source;—for instance, the data memory, the ALU, the MAC, the control path, the bus collecting peripheral registers, or the output of the register file itself.

Each register in the register file can be in keep mode or read mode. When the input multiplexer of a register is set to pass data through input pin 0, the register is in keep mode, keeping data unchanged. When the input multiplexer of a register is set to pass data through input pin 1, the register is in read mode, receiving new data. Every register needs a keeper-multiplexer to be able to keep its value. The output multiplexer is used to connect one of the registers in the register file to the output port.

3.3.5 Control Path

Control Path Hardware

The control path is a program control unit that manages the program flow and supplies the control signals. A typical and simplified control path of a DSP core is given in Figure 3.14.

The control path can be divided into the program flow controller, the program memory, and the instruction decoder. The program flow controller or PC FSM decides the next program counter (PC) value according to the current instruction, the current processor configuration, and the flags of the previous execution. Conditions for deciding the next PC will be discussed in detail in Chapter 14. The next PC is the next address for the program memory. The program memory will supply the corresponding instruction to the instruction decoder, which decodes the instruction together with flags and processor status information. The decoded

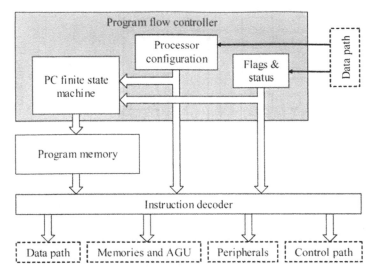

FIGURE 3.14

Simplified control path.

control signals are supplied to the datapath of the core, the memory subsystem, the peripherals, and back to the control path itself.

The Pipeline Concept

Pipeline design will be discussed in detail together with microarchitecture design. This early introduction to the pipeline concept is essential for designing branch (jump) instructions. By partitioning executions of an instruction into several pipeline stages, the system speed can be enhanced because the speed of each stage is faster. The partitioned jobs in each pipeline step can be executed in parallel for different instructions under execution. Pipeline design includes the following steps:

1. Partition the instruction into several steps.

2. Allocate each operation to independent hardware.

3. Put the operations in the right order and assign each operation to a pipeline stage.

4. Run all the operations in parallel.

The pipeline principle is described in Figure 3.15.

As depicted in Figure 3.15, when an instruction is executed without pipelining, it takes time T to finish the execution. Assume the execution can be partitioned into n steps, and the execution time of each stage is T/n. By queuing these steps into n independent hardware pipeline stages, the total execution time of one instruction will still be time T. But the system speed will increase n times. However, imbalanced task partitioning and clock uncertainty induced by physical clock skew and clock jitter will reduce the speed-up. In reality, the speed-up will be between 1 and n. By

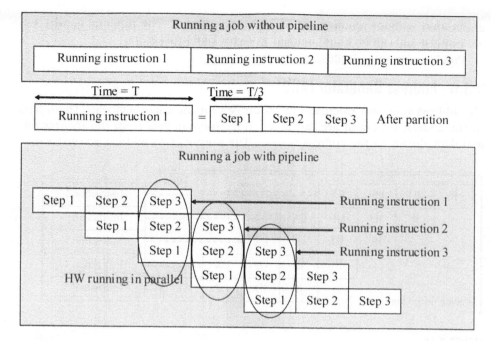

FIGURE 3.15

Pipeline concept.

Table 3.7 Processor Pipelines.

Depth	PST1	PST2	PST3	PST4	PST5	PST6	Notes or cases
2	IF ID	OP EX					Simplest processor
3	IF	ID	EX				Simple RISC MCU
4	IF	ID	EX1	EX2			
4	IF	ID	OP	EX			ALU instructions in DSP
5	IF	ID	OP	EX	ST		Traditional RISC MCU
5	IF	ID	OP	EX1	EX2		MAC instructions in DSP
6	IF	ID	MEM	OP	EX1	EX2	Special instructions

*PST = pipeline stage; IF = instruction fetch; ID = instruction decoding; OP = operand fetch; EX = execution; MEM = memory access, ST = result store.

experience, a five-step pipeline gives 2 to 4 times speed-up. The pipeline depth of a processor could be from 2 to 20 or more. Typical DSPs have a pipeline depth of three to five.

Different pipelining examples are listed in Table 3.7. The architecture with pipeline depth 2 is used only for extreme dedicated architecture in a special application without requirement on high performance. The pipeline depths 4–5 are popular and can be found in many low-cost DSP processors.

Pipeline execution is illustrated in Figure 3.16.

3.3.6 Address Generator (AGU)

The address generator calculates the addresses for the data memories. The inputs to the address generator are the decoded instruction, data from the register file, and data already in the address register. The output is an address toward a data

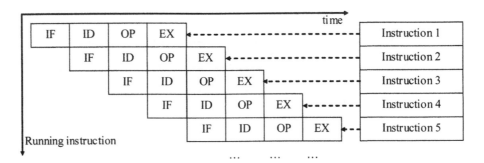

FIGURE 3.16

4-step pipeline in a processor.

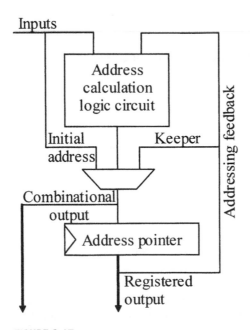

FIGURE 3.17

Simplified addressing circuit.

memory. To supply addresses for DSP computing and program flow control, an address generator should support different and flexible addressing models, such as addressing for convolution.

A simplified addressing circuit is given in Figure 3.17. In this figure, the Input is from the register file or the instruction decoder. The Initial address is used to initialize the addressing algorithm for iterative computing. The Keeper keeps the old address, and the Addressing feedback is used to support iterative computing. The Combinational output is used when the physical distance is long between an address generator and a data memory. Detailed discussion on addressing circuit is given in Chapter 15.

3.4 THE DIFFERENCE BETWEEN GPP AND ASIP DSP

A GPP designer may ask what the difference really is between a GPP and an ASIP DSP. Most students learned fundamentals of computer architectures before reading this book—what is further to learn? When a GPP designer becomes an ASIP DSP designer, he or she needs to know the differences in the way of thinking between designing a GPP and designing an ASIP DSP. In this section, differences will be discussed from several aspects.

3.4.1 The Difference between Designing a GPP and ASIP DSP

The GPP designers think of ultimate performance and ultimate flexibility as well as the compiler-friendly instruction set. The instruction set must be general, close to IP (intermediate representation) or TAC (three address code) style. (IP and TAC will be discussed in Chapter 8.) The application is unknown, and the programmer's behavior is unknown. The branch should be handled efficiently and dynamically using advanced branch prediction hardware associated with speculative execution [9,16,17]. The data dependency should be handled using superscalar (discussed later in this chapter). Addressing must be handled using cache and virtual memory. The GPP designers try to avoid being limited by specific applications (except for the vector instructions for video decoding).

The ASIP DSP designers think of application and cost first, and the challenge is to be efficient. Based on the carefully specified function coverage, the goal of an ASIP design is to reach the highest performance over silicon and the highest performance over power consumption as well the highest performance over the design cost. The flexibility specified according to the class of application is enough, instead of the ultimate flexibility. The performance is application-specific instead of the highest overall.

To reach the greatest efficiency, ASIP DSP designers care more for the understanding of the execution behavior of the application codes from the code analysis (profiling, discussed in Chapter 6). The ASIP DSP designers try to understand the application as much as possible and use the knowledge from the code analysis to fix problems before runtime. Therefore, the static code optimization is important. To minimize the cost, the hardware for dynamic performance enhancement (superscalar, branch predictor, and data cache) should be avoided if possible. In general-purpose DSP processors, program cache can be found. Data cache is not popular, yet it also may be found associated with advanced static data management tools. To minimize the silicon cost, a VLIW (very long instruction word) machine is used more often, and superscalar is not yet popular. Branch predictors never have been used in DSP thanks to the predictable code and data behaviors of DSP application codes.

Moreover, by further utilizing the predictable code and data behaviors of DSP applications, more explicitly parallel features are adapted to ASIP DSP. Algorithms and addressing are further accelerated to enhance the performance over silicon cost and performance over power consumption.

3.4.2 Comparing DSP Processors to Other Processors

Before reading this chapter, you might be curious about the difference between a DSP processor and a general-purpose CPU (central processing unit) or MPU (microprocessor unit). Knowledge of the differences between DSP processor and CPU should be attained after reading this book. In this section, differences between DSP processor and MPU will be discussed.

Actually, a DSP processor is a kind of CPU. CPUs usually are classified into three categories: MPU, DSP, and MCU. A MPU is a kind of GPP. It is the central processor in your desktop or laptop computer. MPUs are very flexible and can also be used for DSP applications. However, the performance over cost ratio will be low and the power consumption will be high.

MCU is a microcontroller for embedded electronics systems. The behavior of a MCU is a comprehensive FSM (finite state machine). MCUs have the largest volume on the processor market. When it comes to total sales of processors, MPU is the dominant, followed by DSP and then MCU. In 2006, DSPs held $22 billion on the global semiconductor market; 65% ($14.3 billion) were sales of embedded ASIP DSP processors [1].

A classification of different CPUs is shown in Figure 3.18.

Table 3.8 compares DSPs and MPUs.

A DSP usually operates as a slave processor, handling digital signal processing. An MPU is usually the master in a system, handling general data manipulations. Tasks running on a DSP processor are relatively known and offline scheduled. Both the functional correctness and timing of running software in a DSP processor should be checked during the development. Most MPUs are designed for general-purpose applications, and it is not known which tasks will be mapped to the MPU during the processor design time. Many software applications running on an MPU have not been designed specifically for this processor.

The assembly instruction set of a DSP is specified for digital signal processing tasks, like filters, transformations, and coding/decoding. Control and peripheral

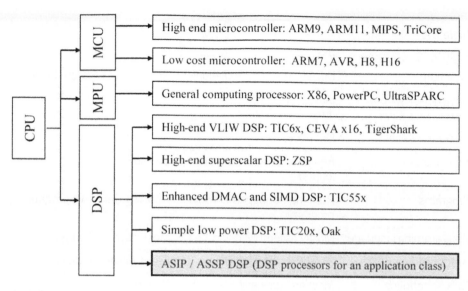

FIGURE 3.18

CPU classification.

functions can be supported by a DSP, but if these functions are dominating, a MPU is a better choice. Through benchmarking, it has been found that the most frequently executed assembly instruction in a DSP is MAC (multiplication and accumulation). The most frequently executed assembly instruction in an MPU is move. Caches are mandatory in an MPU, but they are not used much in DSP processors.

DSP programs are much simpler and compact compared to MPU programs. However, the code efficiency and data memory usage of a DSP must be much better because memory costs are sensitive in most embedded systems. Programming at assembly level is therefore common for DSP applications. However, programming at assembly level on an MPU is not necessary and is unrealistic in most cases.

Most DSP processors support only fixed-point arithmetic whereas an MPU supports both fixed-point and floating-point arithmetic. A DSP is required to support real-time task execution, but this requirement on an MPU is not as strict. In a DSP, fast interrupt without context saving might be required for some real-time applications, but this is not the case for an MPU.

In an MPU, data exception handling is strongly required. If an exception happens, the MPU is interrupted and the abnormal task might be executed once more after the exception handling. In a DSP, tasks must be executed within a fixed period of time and there might be no time left to handle data exception and to reexecute the task. Instead, the real-time exception might be treated as data saturation. A report of saturation will not be given by an interrupt because there might not be interrupt handling time available. Saturation may instead induce data scaling subroutines at a certain point in the streaming data processing flow according to the position of data saturation.

Table 3.8 Comparison between a DSP and an MPU.

	DSP	MPU
Function	Digital signal manipulation	General data manipulation
Behavior in system	Slave device in an embedded system	Master device
Operating system	Not necessary	Mandatory
Typical tasks	Real-time tasks as Infinite loop(s)	Unknown
Programs	Running debugged known programs	Running unknown programs
Programming language	Mostly assembly, could be C	Most high-level languages
Branches	Simple, no prediction	Event-based prediction
Task scheduling	Most are statically scheduled	Dynamic
Interrupts	Fast interrupts and normal interrupts	Comprehensive interrupt system
Most used instructions	MAC instructions executed most of the time	Move/Load/Store
Benchmarking	Based on general DSP kernel algorithms	SPEC and similar
Data type	Custom, integer, and fractional	Standard floating and integer
Datapath	MAC, ALU, data manipulation unit	Floating-point unit, Fixed-point unit
Internal data type	Much higher than the memory bus width	Same as in-out
Overflow management	Guards and saturation	Exception interrupt
Data exception	Skip the current result and use the old one	Interrupt for exception
Memory	Multiple parallel memories	Caches and virtual memory
Data addressing	Dedicated and parallel data addresses	Global and hierarchy
Addressing space	Custom	As large as possible
Memory BUS	Flat and multiple parallel buses	Level hierarchy bus system
Register BUS	Mixed RISC and CISC	RISC
Results	Real-time results	High precision results

3.4.3 **CISC or RISC**

What Is CISC and RISC?

A DSP is just a DSP; it is neither a RISC nor a CISC.

CISC is a Complex Instruction Set Computer described by Figure 3.19(a). RISC is a Reduced Instruction Set Computer, shown in Figure 3.19(b). The primary goal of the CISC architecture is to simplify the assembly code and minimize the code cost by hardware acceleration of general language functions. By executing one single CISC instruction, several low-level (micro) operations, such as a load from memory, an arithmetic operation, and a memory store can be executed. A single CISC instruction might be able to directly support high-level programming constructs, such as procedure calls, loop control, complex addressing modes, and array accesses. The CISC code efficiency is very high, and programming on CISC is easy. However, the hardware implementation of a CISC instruction set is complicated. Because of the integration of multiple low-level (micro) operations into one instruction, hardware design of the instruction decoding becomes very complicated

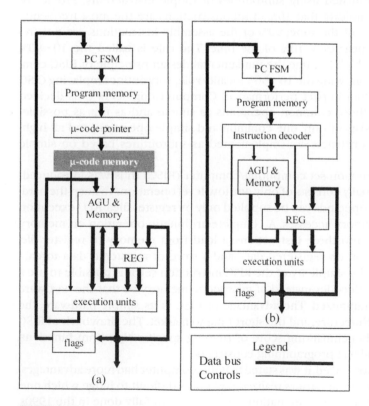

FIGURE 3.19

Comparing CISC and RISC architectures.

and not reliable. To relax the instruction decoding and reduce the cost of hardware verification, the hardware decoding logic circuit was replaced by the microcode. One CISC instruction is equivalent to several lines of microcodes. Each microcode consists of compressed control signals. Because microcodes can be changed after the hardware design, the verification cost is distributed and minimized.

In Figure 3.19(a), the CISC architecture has two execution code memories: the program memory and the microcode memory. The program memory stores the assembly machine code. The microcode stored in the microcode memory consists of both the control signals and addressing signals. Because the execution in a CISC machine is based on microcode, many clock cycles might be used to execute an assembly instruction. The advantages of CISC are the code efficiency and computing performance. The disadvantage of CISC is the expensive hardware design cost.

In the late 1970s and early 1980s, it was discovered that hardware design could be more efficient by reducing the full set of assembly to a simplified set of assembly consisting of only the most frequently used instructions. Seldom used instructions will then be emulated using subroutines of simple instructions. Profiles of various applications indicate that 10% of the instructions are the most frequently used and appear 90% of the time; 90% of the assembly instructions are not frequently used and appear only 10% of the time. The rule is called the 10−90% (code) locality rule, which became the fundamental design principle of RISC computer architectures. Following the 10−90% locality rule, subroutines replacing CISC instructions will seldom impact performance. Computer engineers thus decided to focus on making those common operations as simple and as fast as possible. One instruction consists of only one common and simple operation, and all high-level programming constructs are implemented as subroutines based on simple instructions.

The reduced instruction set computing/computer (RISC) thus was introduced. A RISC instruction usually consists of only one low-level operation. To keep the hardware regularity, RISC operands can be supplied only by register files. ALU execution is separated from memory accesses. A complete arithmetic operation on memory data will be divided into three instructions; a load from memory to register file, an arithmetic operation on registered data, and a store of registered data to data memory. Because of the simplicity of the RISC instruction set, it is possible to use a hardware instruction decoder instead of microcode memory so that the hardware design cost can be minimized. The advantage of RISC is its simple hardware, the compiler friendly architecture, and the short time to market. The drawback of RISC is its relative lower benchmarking score of running specific algorithm functions, especially some high-level programming constructs.

When RISC was introduced, it was stated that this computer had more advantages. Soon computer designers and users realized that it was difficult to judge which one, CISC or RISC, was better. A reexamination of CISC was carefully done in the 1990s, and many modern CISC processors are hybrids, with RISC kernel instruction sets and CISC acceleration features.

Is a DSP a CISC or a RISC?

So far, we know what a DSP should execute. Iterative instructions such as convolution are among 10% of most executed instructions. If a register file is used as the computing buffer for iterative computing, the efficiency of iterative computing will be low. To get efficient iterative instructions, CISC features such as complicated addressing and direct memory (vector) operands should be introduced to DSPs.

Programming of iterative computing on a DSP becomes actually rather simple, and not many iterative instructions will be required. By checking the profiling of DSP applications in Chapter 6, many simple RISC instructions are required by a DSP. By combining CISC and RISC requirements from DSP applications, a DSP can be considered as a RISC processor with CISC enhancement features. It can therefore be simple and with high performance.

The RISC architecture is the fundamental part of a DSP. The register file is the level-one computing buffer for RISC instructions (load/store, ALU, and program flow control instructions) since 1990's. (A register file may not exist in a very early DSP Processor.) CISC enhancements in a DSP are for vector array computing such as convolution, transformation, and special operations such as division. Data arrays are loaded directly from data memories as operands when running iterative array computing. Most or almost all DSPs have instruction decoders based on decoding logic circuits so that control signals toward the datapath are decoded from the instruction decoding hardware instead of microcode.

In general, the instruction set of a DSP is more like a RISC instruction set supporting simplified datapath functions. From another point of view, a DSP accelerates and merges most used DSP instructions, which is called instruction level functional acceleration. A DSP subroutine might be merged into one instruction if it is used frequently. From this point of view (that a DSP has many instruction-level accelerations), a DSP could be considered a CISC. The complicated memory addressing capability in a DSP is also inherited from CISC features. By checking DSP architectures, we find that a DSP, like a CISC, has:

- One execution cycle for ALU and multiple cycles for iteration.

- Complicated data memory addressing modes and circuits.

- Special-purpose registers (accumulator registers).

- Strong instructions for accelerating certain tasks.

However, a DSP, like a RISC, has:

- More general-purpose registers.

- Most instructions as simple instructions.

- Instruction decoding by decoding logic circuit instead of microcode.

- Regular instruction pipelining.

The DSP architecture was formalized in Figure 3.9. Comparing (a) and (b) in Figure 3.19, we find that the control path of the DSP processor is more like a RISC.

Table 3.9 Comparing a DSP with CISC and RISC.

DSP	RISC	CISC
Emphasis on hardware and software	Emphasis on software	Emphasis on hardware
Single and multiclock complex instructions	Single-clock, reduced instruction only	Includes multiclock complex instructions
Operands from registers Operands also from memories	Operands only from registers LOAD and STORE are used to link between register file and data memories	Arithmetic computing based on memory-to-memory variables and register-to-register variables
Small code size	Large code size	Small code size
Most silicon area used for program and data storing	Most silicon area used for program and data storing	Silicon might be used for storing complex instructions (microcode)

The MAC and function acceleration part in the datapath is more like a CISC. The ALU part in the datapath is more like a RISC. Finally, the comparison between DSP and RISC and the comparison between DSP and CISC are summarized in Table 3.9.

3.5 ADVANCED DSP ARCHITECTURE

The DSP architecture can vary in order to support different application requirements, such as processing performance for specific algorithms, ultra low power consumption, or minimized silicon cost.

To classify DSP processors, DSP architectures can be partitioned further into full-custom architecture (DSP ASIC), single-issue DSP, instruction-level parallel architecture DSP (ILP), and on-chip multicore DSP processors (OCMP, sometimes also called CMP). In this chapter, basic advanced architectures will be introduced and discussed. Parallel memory access and architectures, DMA for parallel processing, programming models, task-level compiling, and scheduling will be discussed in Chapters 19 and 20.

3.5.1 DSP with Extreme Specification

A DSP with extreme specifications usually is based on a DSP ASIC architecture. Extreme means that a specific goal must be reached regardless of the flexibility (e.g., extreme performance, extremely low power, or extremely low silicon cost). If extreme flexibility is required, a Pentium might be the right solution. Design methodologies of DSP ASIC have been discussed in detail in Madisetti [5]. The design methods of DSP ASIC can be adapted for designing accelerators of DSP ASIP.

Design of DSP ASIC with Extreme Performance

The first step is to explore possibilities of parallelization of the algorithm. Parallelization and partitioning schemes can be classified into two groups:

> **Group-I:** Pipelined parallelization for streaming signal processing. Several processing tasks/units are connected into a chain. The output of the first task is naturally the input to the second task. All tasks are allocated along a sequential connection of function modules.

The pipelined parallelization mode has been widely used for communication and media signal processing. To conduct streaming signal processing, the execution time of each task must be less than the input data rate.

> **Group-II:** Massive data parallelization for partitionable regular vector signal processing. The parallelism existing in the data is recognized and utilized for parallel computing. Different from the previous case, which is based on task partitioning, Group II is based on data partitioning into multiple parallel modules performing the same function in parallel. Quite often this leads to massively parallel architectures, utilizing homogeneous modules.

Parallelization can be applied at both algorithm level and arithmetic level. On algorithm level, for example, a high-order filter could be partitioned into multiple filters as a filter bank and executed in parallel. Figure 3.20 gives illustrations on pipeline-parallelization and massive-parallelization. By maximum use of all possible parallel opportunities, a high-performance extreme DSP ASIC can be implemented.

In Figure 3.20, the application is partitioned into two streaming chains and two chains running in parallel (top-level parallelization). In streaming chain 1, n modules

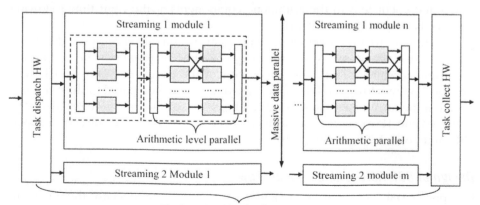

FIGURE 3.20

An example of designing for parallelization of a DSP ASIC.

are partitioned (second-level parallelization). Each module of streaming chain 1 must accept new data and process new data within the time period of data arrival. Inside the modules of streaming chain 1, sublevel pipelined modules are partitioned and can be recognized as dash-line blocks (third-level parallelization). The runtime of each submodule must be faster than the input data rate. Finally, the low-level partitioning is the massive (data) partition-based parallelization. It further divides a pipelined task into parallel computing hardware devices with gray filling-in (fourth-level parallelization).

A special case can be recognized in the second dashed-line module of module 1 in streaming 1, where there is a further partitioning of massive computing components into multiple steps. This lowest level partitioning is necessary if shuffling of device outputs is required by some typical fast algorithm, for example, FFT. Design skill for Extreme Performance DSP ASIC is useful also for accelerators. Accelerator design will be discussed further in Chapter 17.

Design of DSP ASIC with Extremely Low Silicon Cost

Design for low silicon costs is based on maximum hardware multiplexing, lowest control overhead, and memory efficiency. It is just the opposite to design for high performance. By reaching the maximum utilization of parallelism, high performance can be achieved. By reaching the maximum hardware multiplexing, low silicon cost can be reached. Hardware multiplexing technique will be discussed extensively in Chapters 10 through 16. Knowledge from Chapter 10 will help you reach a silicon-efficient solution.

The following example (in Figure 3.21) gives a brief description of hardware multiplexing. The function specified in Figure 3.21(a) can be mapped to the circuit with minimum size in Figure 3.21(b) using hardware multiplexing. In this way five multipliers were merged into one multiplier and four adders were merged into one adder. The address generator is a FSM generating control signals to supply inputs and coefficient to the arithmetic components at the right time following the specification in Figure 3.21(a).

In Figure 3.21(b), the number of components is minimized. Suppose that there is a reference circuit, where the topology is a direct mapping of the computing algorithm in Figure 3.21(a). The silicon cost of the circuit in Figure 3.21(b) will then be 3.6 times lower and the speed five times lower compared to the reference circuit. If the requirement on speed is not high, the solution in Figure 3.21(b) is the preferred one. However, the "performance over silicon cost" of the directly mapped FIR circuit is higher (see Table 3.10).

Design of DSP ASIC with Extremely Low Power

Low-power DSP ASICs can be implemented in many ways because there is more freedom or more dimensions for power minimization. In Equation 3.1, power consumption can be partitioned into dynamic power and static power (V is the supply voltage for full swing logic circuits, α is the activity of a circuit node, C is the total

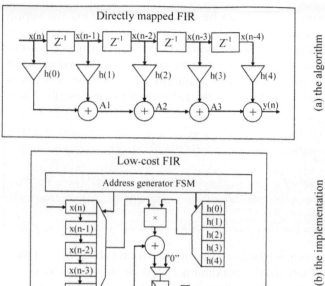

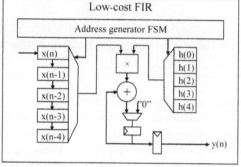

(a) the algorithm

(b) the implementation

FIGURE 3.21

Minimized silicon costs via hardware multiplexing.

Table 3.10 Comparing the Silicon Cost of Two 5-tap FIR Implementations.

	Direct map FIR		Low-cost FIR (Figure 3.21(b))	
	Number	Gate counts (Kilo-gates)	Number	Gate counts (Kilo-gates)
16b Multiplier	5	25	1	5
32b Adder	4	1.4	1	0.35
Flip-flops	64	0.64	208	2.08
16b Multiplexer	0	0	2	0.1
Total silicon cost		27.04		7.53
Performance over silicon		5 / 27.04 = 0.185		1 / 7.53 = 0.133

capacity of the circuit node, and f is the clock frequency of the digital circuit). The static power consumption is the accumulation of leakage power of each circuit path from the power supply to the ground.

$$P = V^2 \sum_i \alpha_i f C_i + P_{\text{static}} \tag{3.1}$$

The most significant power savings can be achieved by reducing the supply voltage. Reducing the circuit capacitance and speed enables linear power reduction. However, mandatory constraints such as the required circuit performance must be met during the power minimization. In this section, we assume that the power supply voltage can be an optimization parameter and requirements on silicon cost can be flexible. Low-power design methods at high levels can be simplified as described next.

> The first step of power minimization is to identify which part of the circuit consumes significant power. The power optimization will start from the circuit modules that consume most of the power.
>
>   The second step is to reduce the supply voltage. As the circuit speed is reduced, the performance of the circuit should be maintained by mapping algorithms and arithmetic functions to parallel hardware.

When the supply voltage is much higher than the threshold voltage of the CMOS transistor in the integrated circuit, the maximum speed of the circuit does not decrease significantly when the supply voltage is reduced. When doubling the circuits for parallelization, the value of C is doubled and the power consumption is doubled as well. If the supply voltage can be reduced to half, the reduction factor of the power consumption will be four, $(1/2)^2 = 1/4$. The total reduction factor of power consumption will be $4/2 = 2$. However, double leakage current is introduced because the circuit size is double. Using ultrasubmicron silicon technology, the leakage current could be significant. Carefully trading off dynamic and static power consumption is required [3].

Many other low-power design techniques can be found in Nurmi [3]. For example, the hardware multiplexing technique could be used to reduce power consumption by minimizing leakage paths. Both memory power and interconnect power can be reduced by minimizing the number of memory accesses.

3.5.2 ILP DSP Processors

The Instruction Level Parallel (ILP) processor has capabilities to issue and execute multiple instructions or multiple arithmetic operations in one processor core during one clock cycle. Typical ILP DSPs are Dual MAC processor, SIMD processor, VLIW processor, Superscalar processor, and custom structure DSP.

In this book ILP DSP stands for a single processor. Complexity issues thus are managed within one program flow. All data and control-induced hazards can be managed by modifying the program flow. Because the DSP applications are known before runtime, instruction-level parallelization is relatively easy to do and the efficiency of parallelization can be high.

Performance Measurements

The absolute performance of a general processor can be measured using the term IPC, or instructions (issued/executed) per clock cycle. Another term that is used for performance measurement is CPI—cycles per instruction. CPI is the inverse of IPC.

Efficiency of an ILP machine is measured by IPC/W, where W is the width of the ILP, the maximum number of simultaneously issued instructions. If a four-way ILP processor ($W = 4$) executes 2.1 instructions on average per clock cycle, then IPC is 2.1. This means that the efficiency of the ILP architecture is $2.1/4 = 0.525$.

IPC is not as useful for performance measurements of ILP DSPs. IPC cannot give the real performance of a DSP because all redundant operations are included. For example, the cost of memory and IO accesses could be high, which does not directly contribute to the arithmetic computing. A more accurate way to measure the performance of an ILP DSP is to measure APC, the number of arithmetic instructions per clock cycle. APC is a real figure and can be achieved via profiling as the real performance. If one-third of the instructions are used for data access, $APC = (1 - 1/3) \times ILP$ and the efficiency of APC is APC/W. To continue on the former example, the APC is $2.1 \times (1 - 1/3) = 1.4$ and the efficiency of $APC = (1 - 1/3) \times 2.1/4 = 0.35$.

To simplify measuring the APC of an ILP DSP, Mega MAC operations per second can be a measure of the performance. For example, if 1200 Mega MAC operations are measured in a second for a four-way ILP DSP, and the processor is running at 500 MHz, then there are $1200/500 = 2.4$ MAC operations executed in the four-way ILP machine per clock cycle.

Problems

Except for programming on a superscalar processor, programming an ILP processor is harder compared to programming a single-issue DSP. However, programming an ILP DSP is manageable by a skilled programmer without requiring special hardware architecture knowledge. This is in contrast to programming a multicore DSP.

IPC could be much less than W when using a W-issue ILP processor. A problem that has been recognized is operation dependency, which is the case when a processor has to wait (stalls). Operation dependency refers to the case when an operation cannot be issued. In this book, dependencies are categorized into data (operand) dependency, control (branch) dependency, and hardware structural dependency (hardware resources are not available).

Data dependency happens when a data/operand is not available for an instruction to be executed. A typical data dependency problem that can happen in a single issue processor is the following case:

```
...
ADD    R0,R1,R2 ;   (I1) R0 = R1+R2
ADD    R3,R0,R1 ;   (I2) R3 = R0+R1
...
```

The execution of I1 takes two clock cycles, and therefore I2 has to wait until the operand R0 from I1 is available. Another example is:

```
... R0←R1+R2 || R3←R0+R1 ...
```

Here "||" stands for parallel execution of two instructions in one clock cycle. Two instructions are distributed to two execution units and will be executed simultaneously. The second instruction cannot be executed in parallel if the second instruction R3 ← R0 + R1 is logically the following instruction using the result of R0 ← R1 + R2 in the program flow.

Control dependency happens when the fetched instructions cannot be executed. In this case, extra stall cycles are needed. For example, a jump that is taken will not be confirmed until the jump instruction is executed. When a jump is taken, already fetched instructions following the jump instruction should be killed. The address of the next instruction to be executed after the jump is contained in the jump instruction. Fetching and execution of the right instruction after the jump takes several clock cycles, and these cycles are wasted.

3.5.3 Dual MAC and SIMD

Dual MAC DSP

The datapath of a Dual MAC machine contains two MAC units. Dual MAC means two different architectures: one is the enhanced dual MAC, two MAC units driven by one instruction, which will be discussed here; the other is the two-way VLIW DSP, two MAC units driven by two simultaneously executed instructions.

The simplest ILP DSP is the enhanced Dual MAC processor. In this machine, two arithmetic computations are executed in two MAC units in parallel by one instruction. Later we will see that an enhanced Dual MAC processor is actually a simplified SIMD (single instruction multidata machine). The enhanced dual MAC processor often is used as an ASIP for applications where the usage of a dual MAC unit can be predefined as explicitly parallel. Because we know those cases when two MAC units run in parallel, the instruction set can be classified into the six groups in Table 3.11.

A Dual MAC DSP can execute two MAC operations in parallel at the same time, indicating the possibility of doubling the performance for convolution-dominated

Table 3.11 Dual MAC Instructions.

	Group	Function specification
1	Dual MAC	Driving two MAC units using one instruction for dedicated algorithms
2	Single MAC	As conventional single-issue instructions, one instruction drives one MAC
3	ALU	As conventional single-issue instructions, one instruction drives one ALU
4	Parallel load/store	Load or store two or more words simultaneously
5	Load/store	Load or store one word as a conventional instruction
6	Controls	As conventional branch, call, and other control instructions

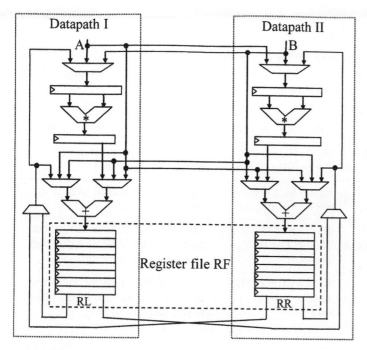

FIGURE 3.22

Dual MAC datapath.

DSP applications. The datapath of a typical Dual MAC processor is presented in Figure 3.22. By just looking into the datapath, we cannot tell if the architecture is a Dual MAC or a two-way VLIW.

In Figure 3.22, RL stands for register for the left datapath cluster, and RR stands for the register file of the right datapath cluster. Figure 3.22 will be analyzed further in Chapter 11. In this figure, both operands A and B are long operands consisting of two data words. A and B will be split into four operands for two multipliers or used as two operands for one or two accumulators (full adders in Figure 3.22). The result of the left path can be saved only in the register file at the left side, RL. The result of the right path can be saved only in the register file at the right side, RR. Therefore, write dependency can be avoided in hardware. Both RL and RR can supply operands to either their own computing path or to the one on the opposite side. Therefore, all data can be shared by the two MAC units. By using the register file in Figure 3.22, a parallel datapath can be divided into two clusters. A clustering technique can relax data dependency by isolating write operations and keeping the datapath busy. Table 3.12 lists six typical Dual MAC instructions.

For example, the following instruction drives both the left and right paths:

DICONV m RL1, RR1 //instruction 1 in Table 1−12

Table 3.12 Typical parallel Instructions for Dual MAC.

	Instructions	Function specification	Applications
1	Dual parallel MAC or convolution	RL<=RL±AH*BH; RR<=RR±AL*BL.	Normal two FIR in parallel
2	Dual Pipeline MAC or convolution	RL<=RL±AH*BH; RR<=RR±AH*RL.	Echo canceller or other algorithms for automatic control
3	Dual MAC using one coefficient	RL<=RL±AH*BH; RR<=RR±AL*BH.	Two taps of symmetrical FIR using one coefficient
4	Dual iterative accumulation	RL<=RL±AH; RR<=RR±AL.	Sum of one vector or two vectors
5	Single MAC or convolution without Pipeline stall	RL<=RL±AH*BH; MAC-R can do others	The left MAC runs a convolution and the right MAC can do other single-step computing
6	Single MAC or convolution without Pipeline stall	RR<=RR± AL*BL; MAC-L can do others	The right MAC runs a convolution and the left MAC can do other single-step computing

This instruction, Dual Integer Convolutions, initializes two integer convolutions, both running m iterative steps in two MAC units simultaneously. This instruction is useful in many cases; for example, a N-tap FIR filter can be partitioned to two $N/2$ tap filters and running them in parallel. The instruction is specified in the following pseudocode:

```
for(N = 0; N < M; N++){
{
RL1 [39:0] = RL1 [39:0] + 8guards {A [31:16]* B [31:16]};
RR1 [39:0] = RR1 [39:0] + 8guards {A [15:0] * B [15:0]};
}
```

When two MAC units are driven by a single instruction, the instruction will be long if all parameters are carried explicitly by the instruction. For example, 40 or more bits might be needed to carry all codes for two convolutions in two MAC modules. Implicit functions or parameters might be used in the instruction. For example the operand accesses of instruction DICONV might be implicitly specified. Unless the programmer reads the manual of the assembly instruction set, it is impossible to guess which part of operand A or B will come to which multiplier. That is one drawback of enhanced Dual MAC DSP.

As soon as the implementation of a dual MAC instruction is decided and carried out, functions in the two MAC modules will be fixed and must be executed as a predefined parallel instruction. When using a two-way VLIW, there are always two instructions controlling the two MAC modules, and flexibility is guaranteed by combining the two instructions.

A processor based on the enhanced Dual MAC architecture brings more complexity, which requires extra programming skills. However, there are also advantages such as low silicon cost because of the small code (program) size.

Many DSP algorithms are rather regular, especially 2D algorithms, such as image/video signal processing and radio baseband signal processing. If a processor is designed only for a class of applications requiring very high performance, a Dual MAC machine might be a good solution.

SIMD DSP

In a single instruction multiple data (SIMD) machine, only one instruction is fetched and executed per clock cycle. A SIMD instruction controls multiple datapaths and executes multiple arithmetic functions in parallel. When all datapaths are the same and execute the same function, we call this a homogeneous SIMD machine, or systolic SIMD machine. Otherwise, a normal SIMD or a heterogeneous SIMD contains different hardware in each datapath and it can run different functions in parallel. Normally, a SIMD machine is a slave machine or a coprocessor. It is a relatively simple design and often is used for vector signal processing.

A SIMD machine usually is designed for running regular and parallel tasks based on massive (data) parallelism. Image filtering and image transformations are typical applications. For such tasks, the vector data array can be divided into several small data arrays, and the algorithms running on each small data array are exactly the same. All datapaths can therefore be driven by the same program flow. The SIMD machine is a very efficient machine for these kinds of applications. In Figure 3.23, a typical SIMD task is illustrated.

In this example, four loops can be executed in parallel if DM(i), DM(i+20), DM(i+40), and DM(i+60) and if TM(i), TM(i+20), TM(i+40), and TM(i+60) can be accessed in parallel. DM is the on-chip data memory, and TM is the on-chip coefficient memory. To prepare for the parallel execution in a four-way SIMD machine, four data words and four coefficients are stored in parallel in DM and TM. After

```
For (i=0; i<20; i++)          For (i=20; i<40; i++)
{                             {
// Convolution:               // Convolution:
A=A+DM[i]*TM[i];              B=B+ DM[i]*TM[i];
} // loop1                    } // loop2

For (i=40; i<60; i++)         For (i=60; i<80; i++)
{                             {
// Convolution:               // Convolution:
C=C+ DM[i]*TM[i];             D=D+ DM[i]*TM[i];
} // loop3                    } // loop4

A=A+B+C+D //merge partial accumulated values
```

FIGURE 3.23

A typical SIMD example.

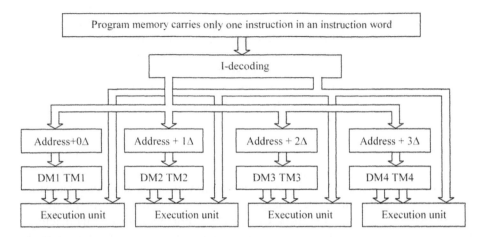

FIGURE 3.24

Systolic array SIMD DSP processor.

the parallel execution, results from the four datapaths in the SIMD machine will be collected and accumulated.

The homogeneous datapath SIMD (or systolic array SIMD) machine is depicted in the block diagram of Figure 3.24. A four-way SIMD machine is shown here, and with $\Delta = 20$ it corresponds to the example in Figure 3.23.

In Figure 3.24, the same set of control signals come to all addressing generators and all execution units. The addresses for the data memories are generated in parallel. The same set of control signals generated from the instruction decoder will be sent to all datapaths to execute the same function.

The homogeneous SIMD machine can handle only systolic massive parallel functions. To accelerate complex yet regular computing, heterogeneous arithmetic functions can always be merged into an instruction, if the merged functions can be executed in parallel at the same time. This is typical for many embedded applications. Features of heterogeneous SIMD are:

- Different functions can be scheduled in parallel because there is no explicit data dependency.

- All these parallel functions consume about the same amount of execution time.

- These parallel functions use different computing resources.

- These functions do not use multiple data words from one data memory at the same time (resource dependence).

In a heterogeneous SIMD machine, the datapath could be very complicated, but the controller is simple or regular. Much control information is implicitly coded, inducing a complicated instruction decoder. A block diagram of a heterogeneous SIMD machine is depicted in Figure 3.25.

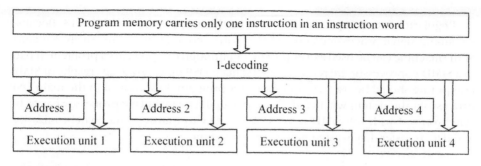

FIGURE 3.25

A SIMD DSP processor with heterogeneous datapath clusters.

	Table 3.13 Typical Systolic Array SIMD Instructions.		
	SIMD Instructions	**Function specification**	**Applications**
1	Parallel convolution	$\sum\sum a(n)b(k-n)$	Filters
2	Matrix multiplication	$C(n) <= a(n)*b(n)$; $n = 0$ to $k - 1$	2D, matrix, vector computing
3	Matrix accumulation	$\sum\sum a(n)$	2D, matrix, vector computing
4	Scale a row	$[a1, a2, \ldots an]xk$	2D, matrix, vector computing
4	A row divided by a variable	$[a1, a2, \ldots an]/k$	2D, matrix, vector computing
5	Add value to a row	$[a1 + k, a2 + k, \ldots an + k]$	2D, matrix, vector computing

In this block diagram, the execution units 1, 2, 3, and 4 could be different or the same, yet running different functions. The address generators provide the addresses to the memory blocks containing the data vectors. However, all heterogeneous functions are carried by one instruction. Coding of the functions represented by the instruction is done mostly implicitly. An explicit code for a heterogeneous instruction may require hundreds of bits, which would be an inefficient way of coding.

Function mapping and execution must be predefined during the instruction set design if the datapaths in a SIMD machine execute heterogeneous algorithms in parallel. The coding flexibility of a heterogeneous SIMD will be low, because SIMD machines are intended for vector signal processing of simple algorithms with massive parallel features. When new algorithms are introduced, the drawback of SIMD will be exposed. That is why companies introduced a VLIW (very long instruction word)/SIMD mixed architecture around year 2000. In a VLIW/SIMD mixed machine, SIMD instructions are used for known applications, and VLIW instructions are used for new applications. High code efficiency can be achieved by maximizing the use of SIMD instructions, and the high flexibility is guaranteed by the VLIW instructions. Instructions for a typical SIMD machine are listed in Table 3.13.

Regularities are recognized by observing the instructions in Table 3.13. Because streaming signal processing and massive matrix computing are relatively regular, a SIMD machine can be used as a coprocessor, although this presents a problem. If data in a SIMD coprocessor are required by the main DSP processor (e.g., results of SIMD computing should be used as conditions for program flow control in the main DSP processor), the system will become complicated. The main DSP processor has to be synchronized with the SIMD processor in order to transfer data. Synchronization and data transfer may waste much computing time.

3.5.4 **VLIW and Superscalar**

Superscalar DSP

By analyzing the problems induced by SIMD machines, we recognize that a SIMD machine is weak in processing control flows and running irregular programs. A SIMD machine should be used as a coprocessor for vector processing, and a master DSP processor should handle DSP program flow control, data quality control, and other miscellaneous functions. The complexity of programming multiple processors will be discussed in the next subsection. A rule of thumb is to avoid multiple processors if it is not necessary. If the performance of a single ILP machine is acceptable, this solution is preferred because of the lower complexity of programming.

Fortunately there are ILP architectures that support both parallel vector signal processing and can run irregular programs. An architecture that can execute multiple instructions in parallel while handling the operation dependences in hardware is called superscalar.

The first step is to handle the data dependences. The hardware must recognize whether an operand is available or not. For example, it may not be available because there is another earlier fetched instruction that is going to write to the operand. There are three types of data dependencies, as seen in Table 3.14.

As soon as all operands are fetched, an instruction should be executed. The next step is to check whether the hardware resource is available or not. Let's assume that there are four MAC instructions in the queue and the datapath contains two ALU

Table 3.14   Instruction Dispatch and Data Dependence Management.

Situation	HW implementation
The operand is available and in the register file	Fetch the operand in register file
The operand is available in the result register and not yet written to the register file	Fetch the operand from the result register
The operand is not available, locked by a previous fetched instruction in the execution queue	Wait for the execution of the dependent instruction, fetch it right after the execution

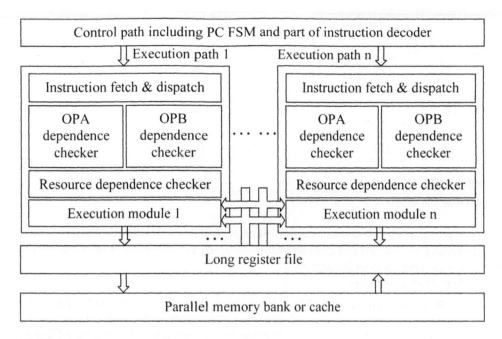

FIGURE 3.26

A simplified superscalar architecture.

and two MAC modules. In this case, the first two MAC instructions can be issued and executed. The second two MAC instructions will not be issued until the two MAC modules are available.

So far, two types of dependences (data and resource dependences) have been discussed. The control dependency will be covered in detail in the chapter about the control path.

A simple superscalar architecture based on Table 3.14 is shown in Figure 3.26. Three dependence checkers, for operand A, operand B, and the resource for execution, can be found in each execution path. The execution of an instruction starts at the instruction fetch and instruction dispatch in the superscalar.

Example 3.2

Specify a four-way instruction dispatcher accepting four instructions per cycle.

Before we can do anything else, we need to describe the architecture further. Four instructions can be fetched to the fetch-buffer in each clock cycle. If instructions cannot be allocated to execution paths, there should be a temp-buffer to store up to six instructions in execution order for the four-way superscalar. The hardware is shown in Figure 3.27. Four instructions can be fetched in a clock cycle. An execution path, EP, could be an ALU or a MAC unit. A MAC instruction consumes two clock cycles, and other instructions consume one clock cycle.

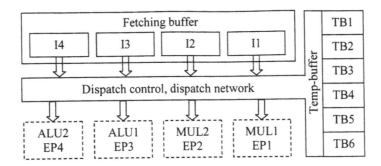

FIGURE 3.27

A 4-way dispatch unit.

The dispatch circuit should include a dispatch controller, a dispatch connection network connecting any fetched and buffered instruction to any EP. There are six temp-buffers that temporally store up to six instructions.

Specifications include:

1. Program memory and PC. Always fetch four instructions per cycle. That means an instruction word contains four instructions in the program memory. For every PC <= PC + 1 operation, four new instructions will be fetched. Here PC is the program address counter.

2. Instruction fetch. If there are more than two instructions stored in the temp-buffer in the dispatch circuit, the dispatch circuit will assign the next PC as PC <= PC and there will be no further fetch of instructions in the next clock cycle. Otherwise, PC <= PC + 1 and four new instructions are fetched.

3. The order definition. The instructions dispatched and queued in the temp-buffer should be reorganized in dispatch circuit in each clock cycle. The oldest instruction is in TB1 in the temp-buffer, and I4 in the fetching-buffer is the youngest instruction. The execution order should be kept both in the temp-buffer and in the fetching buffer.

4. Dispatch. A fetched instruction can be dispatched either to an execution path (issued if there is no dependence) or to a temp-buffer (data dependence or HW dependence).

5. The order of dispatch. The oldest one in the queue will be dispatched first. The dispatch will follow the order of the queue. If an instruction cannot be dispatched, the following instructions shall not be dispatched.

6. Rule of hardware allocation. The first dispatched MAC instruction will be routed to MAC1, and the first dispatched ALU instruction will be routed to ALU1.

7. Temp-buffer shifting. If an instruction in temp-buffer is issued, the following instructions in the temp-buffer will immediately shift and fill in all empty buffer spaces following the fetched one.

8. The rule of temporally storing not yet issued instructions. Temporal store shall follow the fetching order. If the oldest instruction in the fetching-buffer (I1) cannot be issued,

it will be temporally dispatched to the first of the available temp-buffers with the lowest number (e.g., TB1 is the first buffer if it is empty).

9. Rule for the next fetch. Always fetch four instructions per cycle. If there are more than two instructions in the temp-buffer, the dispatch circuit will not give the fetch permission, PC <= PC instead of PC <= PC + 1.

10. If there are less than four spaces in I1-I4 and TB1 to TB6 and there is no instruction that can be issued in the dispatcher, there will be an exception of "locked by data dependence."

Example 3.3

Specify a dependence checker for one incoming instruction line in the in-buffer (I1 to I4) or a waiting instruction in the temp-buffer (TB1 to TB6) that can check the dependencies of two operands.

Based on circuits in Figure 3.26 and Figure 3.27, an instruction can be executed if operand A, operand B, and the execution unit of the EP are all available. Operations during the dispatching cycle include:

1. Check that operand A address is not equal to any destination addresses of other instructions dispatched and not yet issued.

2. Check that operand B address is not equal to any destination addresses of other instructions dispatched and not yet issued.

3. Check that the instruction can be executed in the next clock cycle.

As soon as the instruction can be executed, we need to check and decide where to fetch the operand for the execution. Operations include:

1. If operand A address is equal to a destination address of an instruction running in an EP (if it is a MAC EP, it must be at the second execution cycle of the MAC instruction), the operand A will be fetched from the out port of the EP; otherwise, the operand will be fetched from the register file.

2. If operand B address is equal to a destination address of an instruction running in an EP (if it is a MAC EP, it must be at the second execution cycle of the MAC instruction), the operand B will be fetched from the out port of the EP; otherwise, the operand will be fetched from the register file.

Readers might immediately find a problem. If there is data dependence, the fetched and dispatched instruction will not be executed until the dependence is gone. A waiting instruction occupies the execution module, which will decrease the performance of the superscalar. To manage the problem, so-called reservation stations should be added to each execution path. A superscalar with reservation stations is illustrated in Figure 3.28.

A reservation station consists of several reservation buffers. A reservation buffer consists of a long register and control circuits. Data stored in the long register include the fetched instruction, the execution order code, operands, dependence

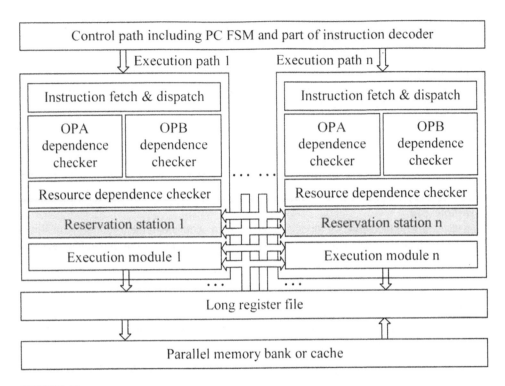

FIGURE 3.28

Superscalar with reservation station.

status, and some decoded control signals. A fetched instruction will be sent directly to the execution module if operands are available, the execution unit is available, and the execution order is correct. Otherwise, the instruction will be temporarily stored in a long register in a reservation station until the dependence is gone and the execution order is correct. Issuing an execution follows the fetching order.

If data dependence cannot be resolved during several clock cycles, the reservation buffer of an execution path will be full. Fetched instructions then have to be dispatched to other execution paths. A worst case happens if there is a cache mismatch when fetching an operand. Many clock cycles are required to access data from the main memory. In this case, all reservation stations might be full and the superscalar stalls.

Example 3.4

Specify functions around a reservation buffer of a reservation station of an EP, execution path.

Based on the circuits in Figure 3.28, an instruction can be reserved in a buffer of the reservation station in an execution path (EP). The reserved instruction can be executed if operand A, operand B, and the execution unit of the EP are or will be available in the next clock cycle. As soon as this instruction is issued for execution, the buffer in the reservation station will be available to accept a new instruction in the next clock cycle. Operations include:

1. Check that operand A address is not equal to any destination addresses of other instructions dispatched and not yet issued.

2. Check that operand B address is not equal to any destination addresses of other instructions dispatched and not yet issued.

3. Check that the execution unit is available or will be available the next clock cycle.

4. Release the instruction from the reservation station for execution if conditions from three previous checks are true.

5. Release the buffer containing the issued instruction; it will be available to accept a new instruction the next clock cycle.

6. Otherwise, reserve the instruction in the reservation buffer. When the instruction waits for an operand, the result address and the name of the instruction lock the operand fetching for later fetched instructions.

As soon as the instruction is released from the reservation station, we need to fetch the operands for the execution:

1. If operand A address is equal to a destination address of an instruction running (its last execution cycle) in an EP, the operand A will be fetched from the out port of the EP; otherwise, the operand will be fetched from the register file.

2. If operand B address is equal to a destination address of an instruction running (its last execution cycle) in an EP, the operand A will be fetched from the out port of the EP; otherwise, the operand will be fetched from the register file.

Comparing a hardware dependence checker to a C-compiler dependence checker or scheduler, the C-compiler is superior to the superscalar hardware. In the C-compiler, there is a scheduler managing out-of-order (OOO) executions. The OOO scheduler checks data dependences and collects independent instructions in a subroutine. While waiting for dependent data, these independent instructions can be scheduled and executed to save runtime. To emulate the function of the OOO scheduler in the C-compiler, OOO execution hardware can be implemented in a superscalar. It then becomes an advanced superscalar with OOO execution as shown in Figure 3.29.

The superscalar with OOO execution holds the IPC record among ILP architectures. The principle of the architecture is its out-of-order execution and in-order results released to the register file. As soon as operands and resources are available, an instruction is executed regardless of its execution order.

The result will first be stored in the OOO buffer instead of in the long general register file. The result can be released to the general register file when the instruction is the oldest in the reorder buffer. In summary, OOO execution is a technology for separating the writing of a result from instruction execution so that the execution and dependence handling can be optimized separately.

Instructions temporarily inside the superscalar should be named with numbers following the fetching order so that the order of execution and the order of releasing

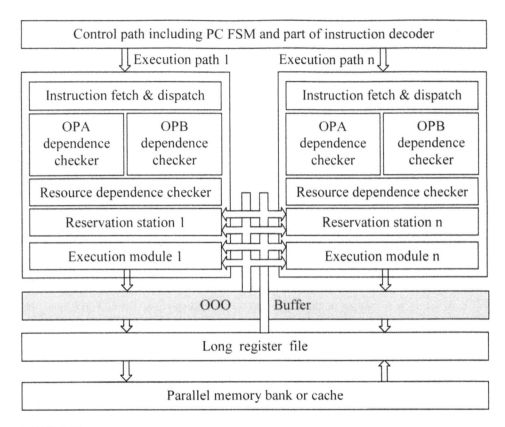

FIGURE 3.29

Superscalar with out-of-order execution.

results (to the general register file) can be exposed and controlled. The name (order) of each instruction can also be used for tracking and matching dependency between operands and result so that the data dependence can be managed right after the execution of the instruction writing to the dependent register. This technique is called renaming, an essential technique for designing a superscalar. It renames the long name (PC value) to a short name in the superscalar.

Example 3.5

Specify a four-way reorder buffer that can accept four instructions in parallel and release four results in parallel.

Specification

1. A reorder buffer is a conditional FIFO. When a register space is available, an instruction is shifted in as soon as the instruction is fetched. An instruction is shifted out (released) following the execution order when the flag (executed) is true. A result cannot be

released if a result of any earlier fetched instruction is not yet available. If there is no space available, the fetching of instructions will be stalled.

2. Suppose that there is a renaming circuit, RN, allocated in the reorder buffer. The RN assigns a name of an instruction using binary code following the fetching order as soon as the instruction is fetched.

3. The reorder buffer accepts a fetched instruction in the fetched order so that all results of these instructions can be released in order. For each clock cycle, the reorder buffer can accept up to four instructions.

4. The reorder buffer contains the name, the running flag, the destination address of the instruction, and the result of the instruction.

5. The reorder buffer keeps track of the execution of each instruction. As soon as an instruction is executed, it has the right to be released.

6. The release of a result must follow the fetching order instead of the execution order.

7. In one clock cycle, up to four results can be released (written to) the general register file.

8. In one clock cycle, if multiple released instructions write to one register in the general register file, the latest fetched instruction (with higher number in RN) will write to the register.

There are 16 reorder buffer registers, RBR, in the reorder buffer (see Figure 3.30). In a RBR, the 5-bit Rename is the name of the instruction fetched and not yet released (5 bits instead of 4 bits are used for naming instructions inside a superscalar; the reason can be found in the following example or from any superscalar book). If an instruction is executed, the flag F[0]

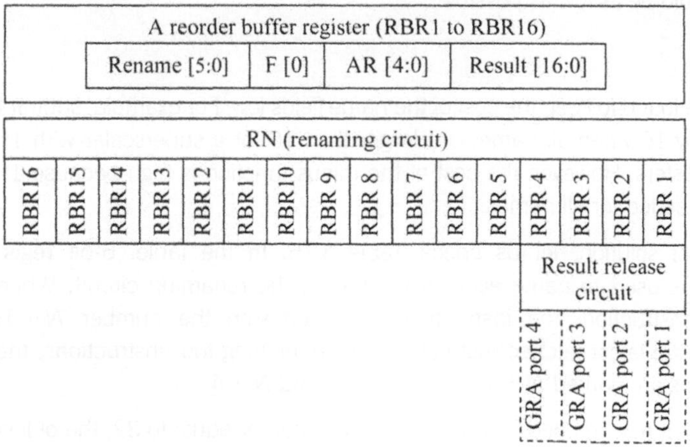

FIGURE 3.30

Reorder buffer.

will be 1. AR is the address of the result in the general register, the address in the general register file for the result.

A result will be written to the register in the register file with the same address. There are four ports (port 1–4) linking four reorder buffer registers to four registers in the register file. The function of the result release circuit is:

If F of RBR1=1, the AR1 and the result 1 will be sent to port 1;
If F of RBR2=1, the AR2 and the result 2 will be sent to port 2;
If F of RBR3=1, the AR3 and the result 3 will be sent to port 3;
If F of RBR4=1, the AR4 and the result 4 will be sent to port 4;
In the same clock cycle,
    if AR4=AR3, Result 4 will over-write result3;
    if AR3=AR2, Result 3 will over-write result2;
    if AR2=AR1, Result 2 will over-write result1.

After writing results, the positions RBR1 to RBR4 are available and RBR5 to RBR8 will be shifted to replace, and become, RBR1 to RBR4. (The implementation is actually a circulating buffer instead of shifting. Design of circulating buffers will be discussed in Chapter 15.)

Example 3.6

Design a renaming circuit RN for the reorder buffer of a four-way superscalar, containing 16 instruction buffers.

Requirements

The renaming circuit shall give a name to each fetched instruction following the fetching order. The cost of the code should be as low as possible.

Low-cost solution

1. Many ways can be found to keep the cost of the name fields low. For example, subtract all name values by 16 when all names are larger than 16 for a superscalar with 16 reorder buffer registers. However, the cost of the minus function is high because 16 5-bit adders are needed for all names.

2. To find a low-cost solution, let us check Table 3.15. In the table, 6-bit register Rename[5:0] is used to name each instruction in the renaming circuit. When fetching a new instruction, the instruction is named with the number $N + 1$; N is the name of the latest fetched instruction. When fetching four instructions, the fetched instructions are named $N + 1$, $N + 2$, $N + 3$, and $N + 4$.

3. When the name of a new fetched instruction is larger than or equal to 32, the oldest instruction in the reorder buffer must be larger than or equal to 16 (because there are only 16 buffer positions).

4. After logic right-shifting the two MSBs (Rename[5:4]), the order of the instruction names is not changed.

This is an exciting result. We actually do not care about the absolute value of each Rename[5:4] in each RBR. We need only the correct order of Rename[5:4]. The renaming circuit therefore is designed as follows:

1. Each instruction in the reorder buffer holds 6-bit rename register Rename[5:0].

2. Rename the fetched instruction in the fetched order as Rename[5:0] = $N + 1$ for the first fetched instruction, Rename[5:0] = $N + 2$ for the second fetched instruction, $N + 3$, and $N + 4$ for the third and fourth fetched instruction.

Table 3.15 Checking Renaming Order when Logic-Shifting Rename[5:4].

Before logic shift [5:4]			After logic shift [5:4]		
[5:4]	[3:0]	Order	[5:4]	[3:0]	Order
01	0010	18	00	0010	2
01	0011	19	00	0011	3
01	0100	20	00	0100	4
01	0101	21	00	0101	5
01	0110	22	00	0110	6
01	0111	23	00	0111	7
01	1000	24	00	1000	8
01	1001	25	00	1001	9
01	1010	26	00	1010	10
01	1011	27	00	1011	11
01	1100	28	00	1100	12
01	1101	29	00	1101	13
01	1110	30	00	1110	14
01	1111	31	00	1111	15
10	0000	32	01	0000	16
10	0001	33	01	0001	17

3. When $N >= 32$, assign all Rename[4] $<=$ Rename[5] and all Rename[5] $<=0$ for all rename registers of all RBR in the reorder buffer.

Superscalar architectures are attractive for DSP applications when silicon cost is not sensitive or when backward binary code compatibility is essential between different versions of the product. The TIM (time in market) of a superscalar processor can be long. That matches typical product requirements of an in-house semiconductor vendor.

The datapath silicon cost of a four-way superscalar is estimated in Table 3.16.

The gate counts in Table 3.16 are under the following assumptions: An execution path of a four-way superscalar includes one MAC, one ALU, and one register file (RF). The MAC has four accumulator registers and one pipeline register between the multiplier and the accumulator. The size of the register file for four execution paths is 64×16 bits. To estimate the silicon costs including connection wires, the silicon cost from wires is translated to gates.

In this example, a 4×4 reservation station means four reservation station buffers in each execution path. Each reservation buffer uses 80 flip-flops (for the instruction, the operands, the instruction name, the dependence and running status, etc.). The total number of flip-flops used by all reservation stations in four execution paths will be $80 \times 4 \times 4 = 1280$. In the last column of Table 3.16, 16 000 execution buffers mean that there are a maximum of 16 temporal result registers in the reorder buffer.

Table 3.16 Estimated Silicon Cost of 4-Way Datapath.

Cost from	Simple datapath		Simple superscalar		Plus 4X4 reservation stations		Plus 16 000 execution buffers	
	Number	Gates	Number	Gates	Number	Gates	Number	Gates
4XMAC	4	30 k	4	30 k	4	30 k	4	30 k
4XALU	4	10 k	4	10 k	4	10 k	4	10 k
Flip-flops in 64 × 16 RF	1024	24 k	1024	24 k	1024	24 k	1024	24 k
Other Flip-Flops	400	4 k	500	5 k	1800	18 k	2400	24 k
Other logic		10 k		28 k		35 k		50 k
Equivalent wire cost		78 k		97 k		117 k		138 k
Total equivalent area		156 k		194 k		234 k		276 k

Each result register consists of 32 flip-flops including 16 result data bits and 16 status bits (for instruction name, RF address of the result, running status, release status, etc.). The total number of registers used by the reorder buffer will be $32 \times 16 = 512$. The silicon cost of a superscalar datapath is 1.3 to 1.5 times higher than a simple datapath. The absolute and relative performance over silicon cost of different solutions is estimated and given in Table 3.17.

It is interesting to see that the IPC value increases when hardware and compiler become more advanced. However, the performance over silicon does not improve for advanced superscalar architectures. The conclusion is, if advanced compiler technology is available, the advantage of using superscalar is not high.

Silicon parameters and performance figures used in Tables 3.16 and 3.17 are from reviewing different designs. A direct comparison of these designs is meaningless, because design conditions are so different. However, after extrapolations and adaptations of the available information according to my experiences, the information in the tables should be relevant.

Figure 3.31 contains a complete schematic of a four-way superscalar machine with OOO execution. Superscalar circuit modules include instruction dispatcher, the reservation station, and the reorder buffer. The schematic exposes heavy interconnection cost. There are more than 50 data buses consuming significant silicon and power. Around the register file, the reservation station, and the reorder buffer, unexpected high fan-in and fan-out will impact the VLSI implementation.

Table 3.17 Estimated Performance over Silicon Cost.

Silicon and performance	Simple 4-way datapath + simple compiler	Simple 4-way datapath + good compiler	Superscalar with reservation station	Superscalar with OOO execution
Datapath	156 k	194 k	234 k	276 k
AGU + DMA	110 k	110 k	110 k	110 k
Control path + interrupt HW	40 k	40 k	60 k	80 k
Total silicon of the DSP core	306 k	344 k	404 k	466
IPC of DSP applications	1.4	2.2	2.4	2.7
IPC over silicon cost	0.4575	0.6395	0.5941	0.5794
Relative IPC over silicon	1	1.398	1.299	1.267

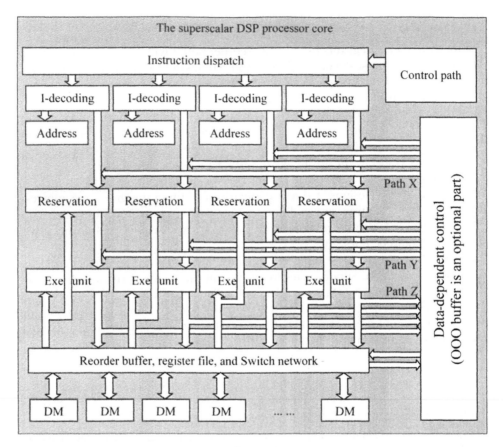

FIGURE 3.31

4-way superscalar DSP.

A review of the superscalar principle is given in Figure 3.32. The PC FSM can fetch one, two, three, or four instructions according to the availability of the dispatch, the reservation station status of each execution path, and the reorder buffer status. As soon as new instructions are fetched, they are dispatched to the corresponding execution unit. Next, the instruction decoder checks for availability of operands. If operands are available and execution hardware will be available in the next clock cycle, the instruction will be issued and executed. Otherwise, the instruction will be stored in the reservation station. Availabilities of all dependent operands of all waiting instructions in the reservation station will be checked in parallel in each clock cycle. An instruction in a reservation station will be executed as soon as operands and the execution unit are available.

After execution, if there are no older unexecuted instructions in the reorder buffer, the result will be sent to the register file immediately. Otherwise, the result

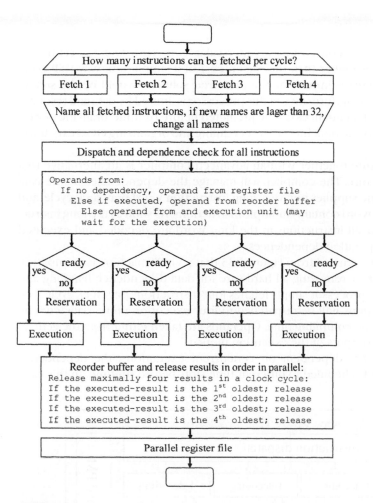

FIGURE 3.32

Data flow of Figure 3.31.

will be stored in the reorder buffer. A result can be released to the register file when the corresponding instruction becomes the oldest in reorder buffer. An instruction can be retired as soon as its result is stored in the register file.

There are a lot of "check and decide" functions in a superscalar in order to manage data dependencies and execution orders. Implementing this in hardware means a lot of extra circuitry using extra power. Therefore, to design low-power yet high-performance DSPs, designers should use a compiler instead of a superscalar processor to reach the required performance. In other words, if compiler design competence is available, a VLIW architecture might be superior to a superscalar architecture.

VLIW DSP

From Table 3.17, the silicon cost of a superscalar core with OOO execution is $466/306 = 1.52$ times higher than the silicon cost of a simple four-way DSP core. When small silicon area is important, the preferred solution would be the simple four-way DSP core. The performance of this processor is limited by data and resource dependences. If the dependences can be predicted before runtime, dramatic performance improvements can be achieved by scheduling and organizing OOO execution before runtime.

A VLIW architecture together with an advanced compiler is an alternative to a superscalar architecture. The compiler will manage the dependencies before runtime. A VLIW machine supplies one very long instruction word per clock cycle, and the long instruction word contains multiple instructions. As soon as the long instruction word is fetched, all instructions in the long word are decoded and executed without checking operation dependences.

It can easily be recognized from Figure 3.33 that a VLIW machine is a simplification of a superscalar by removing all hardware for data dependence handling and OOO execution. The processor simply fetches four instructions per clock cycle. The four fetched instructions will be dispatched to four execution units. Data fetching was defined during the compiling time. Operands are taken either from the general register file or from outputs of the execution units.

When operands are dependent on concurrent executions, the dependence will be identified and scheduled to different time slots before runtime. In this case,

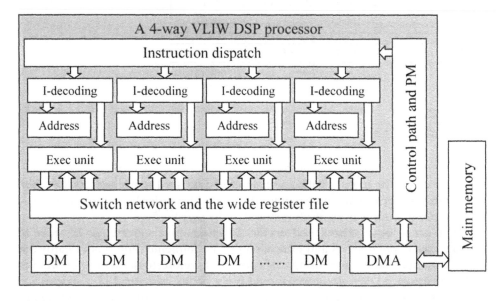

FIGURE 3.33

VLIW DSP processor architecture.

it is very important to identify all possible independent instructions, because independent instructions can be scheduled for OOO execution. The available time slots for avoiding data dependencies can be utilized by these independent instructions. This is a very good way to improve the quality of parallelization. However, when there are not enough independent instructions available, NOP (no operation) instructions have to be executed. Resource dependences are handled in the same way. Many NOP instructions can be found in VLIW programs if the compiler is not good enough or if the application is not "enough explicitly parallel."

To conclude, as soon as we understand the design of a superscalar, the VLIW hardware design is relative easy to do. The essential part of the VLIW concept is to understand ILP compiler design. In Chapter 8, a brief introduction to ILP compiler design will be given. Control dependences will be further discussed in Chapter 7. Handling of interrupt, exception, and multitask threading for ILP machines will be discussed in detail in Chapter 14. More hardware design details for VLIW machines can be found in Fisher et al. [6].

Superscalar or VLIW Machine?

It was soon found out that writing programs on a VLIW DSP or on an enhanced Dual MAC DSP is not so easy. Reasons are that data dependency problems have to be handled during assembly coding. Many NOP instructions also were discovered in VLIW programs when applications are relatively complicated. Thus, many VLIW solutions have a low code (program) efficiency, resulting in a large program memory. Table 3.18 lists and discusses the pros and cons for VLIW and superscalar.

In general, for a volume- and cost-sensitive product requiring both ILP and low-power consumption, VLIW is the choice. When the value of backward compatibility of binary executable codes becomes essential, superscalar is the only choice. When programming on assembly level is required, superscalar is also the choice. Superscalar is always the preferred solution for general computing when applications are complicated.

Using standard compiler, programming a superscalar is easier and more efficient than a VLIW machine. We have to use explicit information to denote parallel execution in assembly code for VLIW. For example, the || symbol must be used to indicate that two instructions will be executed in parallel, which means that one bit of the binary code must be used for indicating the parallelization. Instead, it is not necessary to use explicit code for parallelization in a superscalar processor because data dependency and order of execution can be managed by the superscalar hardware during runtime. In VLIW programs, NOP instructions are frequently used to avoid data dependencies. The code efficiency of a VLIW program therefore could be lower than a superscalar program [7, 8]. For COTS general DSPs, the size of the program memory is fixed so the code efficiency does not give much impact to users. However, running NOPs will consume extra power and that cannot be hidden from users.

Combining VLIW compiler technique and the superscalar circuit technique, most execution ordering and data dependencies can be scheduled in the compiler.

Table 3.18   Compare Superscalar with VLIW.

Behavior	Superscalar	VLIW
Performance	Relative high	Relative low
Silicon cost	High	Relative low
Performance over silicon	Not high	High
Power consumption	High	Low (for simple parallel algorithms)
Code efficiency	High, all instructions will be useful	Low because of many inserted NOP
Parallel execution	Implicitly coded, cannot see from the code	Explicitly noted by compiler, can see the parallelization from ASM and BIN
Order of parallel code	Achieve high performance by scheduling, it is not mandatory	The order must be exactly correct and managed by compiler
Data dependence	Managed by hardware, flexible	Managed by compiler, must be explicit during off-line compiling
Resource dependence	By hardware	Managed before fetching instruction, not efficient
OOO execution	By hardware	Scheduled by compiler
Program ASM	Easy to program assembly code	Very difficult to program the ILP code
Binary code compatibility	Binary code backward compatible	Not between hardware versions

Using a superscalar in this way, all NOP instructions can be eliminated from the program. If the compiler is good enough, the size of the reservation stations in the superscalar can be small and the OOO execution technique will not be necessary. That is why the hardware of a superscalar DSP is relatively simple and the silicon cost overhead of a DSP superscalar can actually be acceptable [8].

Low code efficiency is a drawback of VLIW machine. To partly use advanced superscalar features on VILW machine based on predictability features of DSP applications, different solutions to reach higher code efficiency were proposed to enhance the performance over silicon of VLIW machines. Researchers proposed modified VLIW architectures such as TVLIW (Tagged VLIW) [18], and CLIW (Configurable Long Instruction Word) of Camel [19].

3.5.5 On-Chip Multicore DSP

In this subsection, programs will be partitioned and executed in several DSP cores. Both multicore solutions and previously discussed ILP solutions are based on parallel processing, executing multiple instructions in one clock cycle. Both solutions support high performance, and their IPCs are usually larger than one.

The Reliability Challenge

Programming a VLIW machine is not easy, yet it can be managed using VLIW compilers. However, programming tools for multicore processors are not well developed, and programming is more difficult. When running a program on an ILP machine, the control dependence can be managed explicitly in the control path, and the data and resource dependencies can be managed either explicitly using compilers or implicitly in superscalar hardware. Programs can be debugged and can be bug-free in principle. However, when running a program on a multicore processor, both reliability and efficiency problems are new challenges that were not seen in ILP machines.

To utilize the parallel resources efficiently, tasks should be partitioned into subtasks and allocated to different DSP cores on a chip. Each core has its own local memory, and all cores have access to a common memory (or main memory). Different subtasks might share the same data in the main memory. When local data is changed, we sometimes do not know if this will affect other subtasks. If we do not know the dependence, we have to inform all others sharing the same data as soon as the data is changed. A potential problem is if two or more tasks are changing data in a common data block at exactly the same time. Another problem is the latency of communication: while informing about change of the data, the data may have been changed again from other sources.

The reliability problem can be managed by synchronization and protocols for memory/data sharing. Using and writing data is regulated and performed in a predefined way during a fixed time period. In this way, all subtasks sharing data will use the latest changed data. Waiting for specific time sharing data and running dependence control protocols, however, will induce overhead.

The Efficiency Challenge

To manage parallel task execution, different types of overhead cannot be avoided:

Parallelization overhead: This overhead is due to the execution of programs for parallel task management. It includes task identification, task initialization, data initialization and allocation, setting up parallelization environment, and terminating parallel tasks.

Synchronization overhead: Synchronization is required to handle data or task dependencies. Synchronization can be implemented by setting barriers, locks, and events. When a running program reaches a synchronization point, or it is interrupted by a synchronization point, a synchronization program should be executed.

Communication overhead: Data have to be shared and transferred between tasks allocated in different cores. The cost of intertask communication includes the cost of direct communication between tasks and the cost of indirect communication via the operating system. The cost can be divided further into the cost for setting up a communication and the cost for payload transfer. The communication cost can be minimized by task repartitioning and the elimination of unnecessary communications.

Load imbalance overhead: Tasks can never be partitioned with exactly the same runtime, because some tasks run faster and finish in advance and some tasks run slow. To synchronize tasks, a core running a fast task has to wait for cores running slow tasks until the slowest one reaches the synchronization point.

The challenge of parallel system design is to minimize all overhead while keeping the reliability high and the hardware cost of the system low. This will be discussed in the following paragraphs.

Classification

Before dealing with the challenges just listed, a classification of parallel systems should be discussed briefly. In general, parallel computing systems can be divided into multicores, multiprocessors, and multicomputers.

The multicore solution is an integration of multiple processor cores on a chip. The integration is based on an on-chip connection network. Chip peripherals and the main memory are shared by multiple cores, and each core may have its own core peripherals (interrupt handler, timer, etc.).

The multiprocessor solution is an integration of multiple processor chips in a system. The integration is based on an interchip connection network (mother board bus or backbone bus) and shared data stored in an external memory (DRAM, disk, etc.). A system is not a multiprocessor system if tasks are not partitioned and shared by multiprocessors, though there are multiple processors on a system board or in a system cabinet. For example, there are multiple processors on a board for public phone local access equipment. Each processor handles data conversion for several phones. However, these processors neither share tasks nor share data. The system is therefore not a multiprocessor system. Table 3.19 compares multicore and multiprocessor systems.

The multicomputer solution is an integration of multiple computers as a distributed system based on an open network (Ethernet or Internet, for example). Each computer is a piece of equipment in a box. The system does not supply any media to share data and data coherency control is on the user's side.

A multicore system can be further classified as a memory-sharing system, a distributed-memory system, or a parallel stream processing system.

Shared Memory Architecture

In general, shared memory architecture allows cores to access all memory contents under a global address space. Multiple cores can operate independently but share the same memory resources and address space. Data changes in a memory location

Table 3.19 Comparing Multicore with Multiprocessor and Multicomputer.

Specification	Multicore	Multiprocessor	Multicomputer
Interconnections	On-chip network	Motherboard/backbone bus	Open network
Data sharing	Based on main memory	Disk and main memory	Via file server or no sharing
Communication	Memory sharing	Message passing	Message passing
Granularity	Fine and coarse	Coarse	System level
Overhead statistics	Can be very low	Depends on type of task	High
Real-time behavior	Suitable for real-time applications	Not suitable to run one stream flow in parallel	Not for real-time
Programming environment	Hardware dependent C coding	MPI, or OpenMP	MPI

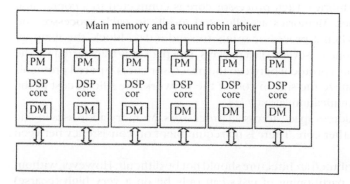

FIGURE 3.34

Shared memory architecture.

by one core must be visible to all other cores sharing that data. The memory or cache coherency can be implemented using a memory-coherent protocol. This protocol can be simplified as follows: If one of the cores modifies any shared data (i.e., performs a write operation), then a copy of this data is sent to all other cores sharing the same data. Based on this protocol, shared local data remain consistent.

Shared memory architecture usually is designed as a master–slave system. A master core controls a number of slave cores. The master synchronizes and distributes tasks, configures the on-chip connections, and manages communication to external components. Slave cores are usually application-specific datapath machines, like vector machines and other computing engines. A simplified shared memory architecture is depicted in Figure 3.34.

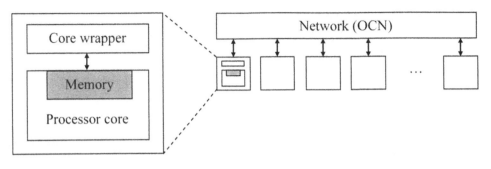

FIGURE 3.35

Distributed memory architecture.

Distributed Memory Architecture

Instead of using a main memory as the sharing and communication media, the distributed memory system in Figure 3.35 uses a communication network to connect interprocessor memories. Each processor core is connected to a connection network via its wrapper. Memories are allocated locally in each processor and behind the wrapper. When a processor wants to send a data block, the wrapper packs it and sends it as one or several data packets. The packet is transferred via the shared network based on a protocol between the processor cores. Special protocols should be used to minimize the silicon cost of the network, the silicon cost of the wrapper, and the communication latency [11].

There is no global address space across all processors. Memory addresses in one core do not map to another core. There is no requirement on consistency between memories.

Programming a distributed architecture should not be difficult. However, without dependence checking, partitioning of tasks can only be on a very high (coarse) level. The load imbalance overhead will be high. Lack of memory coherency means that either tasks are independent or tasks must be independent of each other.

A distributed memory architecture is suitable for multistreaming signal processing. Different channels are processed almost independently with only occasional communication. A typical application is a voice gateway, where multivoice channels are processed independently. Another example is the baseband subsystem in a radio base station.

Parallel Streaming Architecture

The shared memory architecture and the distributed memory architecture are not very useful for parallel DSPs. Instead, the most common solution for parallel DSPs is the parallel streaming DSP architecture, simply called streaming DSP. In a stream-

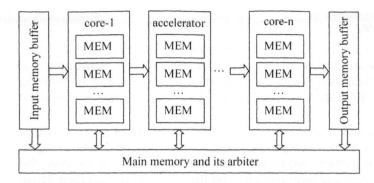

FIGURE 3.36

Streaming DSP architecture.

ing DSP system, each processor has its own local memories. Parallel cores operate independently without memory coherency control. From a hardware point of view, a main memory usually is shared by streaming processing cores. Without memory coherency control during programming, the streaming DSP system is actually not a real shared memory system. The architecture of a streaming DSP is given in Figure 3.36.

A streaming DSP is intended for real-time applications. Streaming signal processing must be faster than the rate of the arriving streaming data. The input data, usually a data packet, arrives periodically to the input memory buffer. As soon as a data packet is received, it is sent to the processing chain of the streaming DSP. The first core in the streaming chain processes the first step of the task. The output of the first core will be the input of the second core. The complete application therefore is partitioned into tasks and allocated to processor cores connected sequentially. Finally, the output of the streaming DSP is available to the output memory buffer as the last stage in the streaming DSP chain.

The main memory is required mostly as an extension to the in-core memories and as the media for data and control message passing among DSP cores. To minimize the total silicon cost, the size of in-core memories should be small and the main memory should be shared as much as possible. In this case, the main memory can be used as a data swapping buffer for different cores at different times. For example, in a video signal processing system, a core cannot contain several picture frames. The main memory may then store all recently used picture frames while each core is processing only part of a picture frame. Streaming connections can be static as shown in the figure, or dynamic using an on-chip connection network (to be discussed in Chapters 19 and 20).

Streaming DSPs can be homogeneous solutions consisting of only one type of DSP core. But most streaming DSP solutions are heterogeneous, consisting of different DSP cores. Some cores in the streaming flow could even be a SIMD machine handling data-level parallelization.

Parallel Programming

Parallel programming is actually a high-level specification for parallelization based on a specific hardware architecture. It includes specification, evaluation, and optimization of parallelization. A simplified parallel programming flow is shown in Figure 3.37. After careful analysis of an application, the application task is partitioned into DSP cores.

After partitioning, functions are allocated to different processor cores. Relations between tasks and cores are thus exposed. By checking the memory resource capacity of a core and the memory requirement from tasks allocated to the core, the communication between the main memory and the core will be specified. These communication subroutines are scheduled following the order of task execution. By checking the intertask relations, two types of communication tasks are further added to the parallel program: the task of passing data from one core to another core, and the synchronization task to start, monitor, and stop certain tasks in the cores due to dependencies.

The parallelization tasks are summarized and listed in Table 3.20. These management tasks provide reliable parallel computing. By scheduling and synchronizing all tasks running in the DSP cores, the execution and waiting times in each core can be exposed. Imbalance of the parallelization will be recognized, and the quality of the partitioning can be evaluated. Based on the evaluation of the parallelization quality, optimization may be needed. Optimization of the partitioning includes repartitioning, regrouping of tasks, reorganization of communication, and add–cancel management of tasks for handling parallelization. The repartitioning induces

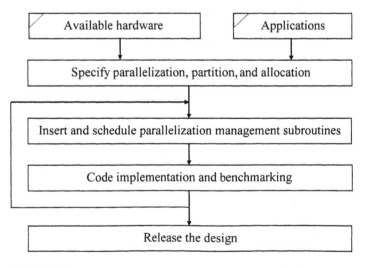

FIGURE 3.37

Parallel programming.

Table 3.20 Frequently Used Tasks for Parallelization.

Tasks and function description	Cost estimation—The overhead cost is:
Packet or depacket a data block	Low when there is no error control coding
Load data packets from main memory to one memory in the core	Packet size, or negligible because of using DMA
Store data packets from a memory in the core to the main memory	Packet size, or negligible because of using DMA
Pass one data packet from one core to another core	1 or 2x packet size, or negligible behind DMA
Assess request and set up an access to/from the main memory	Response time if not busy
Set up communication links	Response time if not busy
Terminate communication links	Negligible
Synchronization	
Other arbitrations and interruptions	

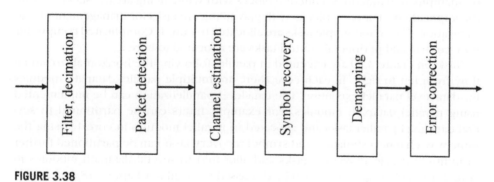

FIGURE 3.38

Functional partitioning of a radio baseband stream.

further scheduling and further optimization. The iterative optimization is finished when tasks are balanced and parallelization overhead is minimized.

A parallel programming example for radio baseband streaming signal processing is illustrated in Figure 3.38. Functions are partitioned into six task-level pipeline steps.

The six steps are allocated into three heterogeneous cores (see Figure 3.39). Core 1 is a simple SIMD processor for flexible filters, core 2 is a processor with advanced parallel computing and control features, and core 3 is a processor for programmable

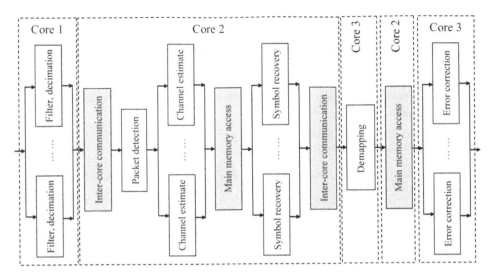

FIGURE 3.39

Parallelization of streaming signal processing in Figure 3.38 [20, 21].

bit manipulation functions. Function blocks with white filling are the six tasks from the applications. Function blocks with gray filling are tasks for management of parallelization. Heavy and simple tasks are allocated to core 1. Complicated computing, flow control and peripheral control tasks are allocated to core 2.

Tasks in Figure 3.39 are executed in parallel following the horizontal streaming flow from left to right. In each core, there are multiple parallel datapath modules for data-level parallel execution. The data-level parallelization can be implemented using parallel datapath modules. For example, filters can be partitioned to several data-level parallel tasks and allocated to parallel modules in core 1. Using the same way, channel estimation and symbol recovery tasks can be partitioned further to multiple data-level parallel tasks and allocated to several datapath modules in core 2. Parallel programming will be discussed in detail in Chapters 18 and 20.

Amdahl's Law

Parallelization cannot be perfectly balanced. Datapaths can be made parallel in an easy way. But memories, especially the off chip main memory, usually is not accessed in a parallel way. Sometimes, special iterative algorithms in an application can give constraints for parallelization. Input or output functions may also affect the performance of parallel computing. Gene Amdahl [10] described the maximum expected improvement to an overall system when only part of it is improved through parallel computing.

By defining the overall speed up = (old runtime/new runtime), the performance improvement can be measured. Amdahl's law describes the overall speed up as

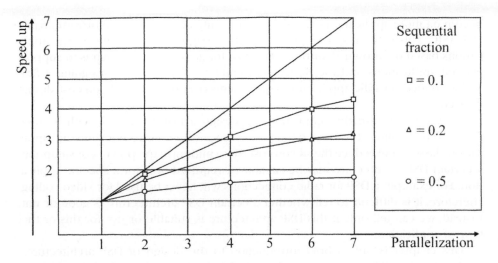

FIGURE 3.40

Amdahl's law: Speed up versus sequential fraction.

a function of the fraction of parallelization (if only part of the system can be parallelized). Amdahl's law gives speed up A:

$$A = \cfrac{1}{\text{Fraction}_{\text{sequential}} + \cfrac{\text{Fraction}_{\text{enhanced}}}{\text{Speedup}_{\text{enhanced}}}} \qquad (3.2)$$

Here $\text{Fraction}_{\text{sequential}} = 1 - \text{Fraction}_{\text{enhanced}}$, and $\text{Fraction}_{\text{sequential}}$ is the part that cannot be sped up.

For example, 50% of the runtime is used for accessing the main memory in a router core, and the access of the main memory cannot speed up because of the irregularity of the data. The $\text{Fraction}_{\text{sequential}}$ is therefore $1 - \text{Fraction}_{\text{enhanced}} = 0.5$. Assume there are six processor cores used and the speed up from the cores are four times without considering the memory availability. The overall speed up $A = [0.5 + (0.5/4)]^{-1} = 1.6$. By adding more cores, the speed up will never be up to two according to Equation 3.2. When designing a system and applying parallelization, the decision of realistic parallel factors is essential. Figure 3.40 shows the feasibility of parallelization.

3.6 CONCLUSIONS

Architectures of general-purpose CPUs (MPUs) are still converging in order to manage all possible applications. However, DSP architectures are diverging and becoming more application-specific by continually taking over non-programmable DSP ASIC. This is because DSP processors are designed for domain-specific applications based on known algorithms. The goal for general CPU design is to support general applications and the design of compilers. The goal for designing a DSP ASIP is to meet specific application requirements (performance, silicon cost, design cost, etc.).

We therefore have the opportunity to design a DSP processor with its own features, for example, to enhance the performance for certain algorithms and application classes, or to reduce the silicon costs or to minimize the power consumption. A certain DSP architecture may be suitable or required only for a specific application. For example, a DSP for radio connections will never be used for video coding. Therefore, it is difficult to tell whether a certain DSP architecture is good or not. Instead, we can tell only if this DSP architecture is suitable or not for this or that type of application.

This chapter is only a brief introduction to the design of DSP architectures. Design details will be discussed in Chapters 7, 9, 10, 14–16, 19, and 20. Architecture design is a specification of the top level of a processor, which includes defining hardware modules of the core and connections among them. The connections are divided further into two groups: the memory bus and the register bus. The input to architecture design is the assembly language, manual, and the output is the top-level block diagram including components and their connections. Principles for handling operation dependencies should also be specified during the architecture design.

All top-level components in a DSP processor have been addressed in this chapter including the datapath (e.g., MAC, ALU, and the register file), the address generator, the control path, and the memory subsystem. The microarchitecture design details of these modules will be discussed in Chapters 10 through 15.

Parallel architectures also were introduced in this chapter. Different product objectives induce different parallel architectures. ILP architectures were discussed, and the superscalar DSP was introduced in order to describe the concept of operation dependencies and how to deal with them. To move these methods from hardware to compiler, the VLIW DSP was introduced. The VLIW machine in this chapter is only used as basic concept, and further discussions on its instruction set, compiler, and control dependencies will be covered in Chapters 7, 8, and 14. On-chip multicore architectures were also introduced. A detailed discussion on advanced issues will be given in Chapter 20.

EXERCISES

3.1   What are differences between the memory bus and the register bus?

3.2   Describe functions of all five modules in a DSP core.

3.3 Why is the architecture in Figure 3.5(e) the most popular architecture for a single-issue DSP processor?

3.4 Describe the instruction flow FSM in Figure 3.7 and describe the data flow FSM in Figure 3.8.

3.5 Try to merge the instruction flow FSM in Figure 3.7 and data flow FSM in Figure 3.8 into one FSM.

3.6 Describe the advantages and drawbacks of architectures in Figure 3.4.

3.7 Design a processor architecture including the decision of number of data memories, the decision of datapath components, the decision of the memory bus, the decision of the register bus, and the decision of the address generator. The processor is a SIMD machine issuing one instruction per clock cycle. The SIMD machine has one ALU and two MAC units. The ALU executes instructions using the register file as the computing buffer. The two MAC units execute only iterative instructions from data memories M1, M2, M3, and M4. M1 and M3 supply data. M2 and M4 supply coefficients. Results from both MAC units can be loaded to the register file. Four address generators generate four addresses in parallel. Address generators require inputs from both the register file and immediate data from the instruction decoder. Register files get inputs from M1, M2, M3, M4, ALU, MAC1, MAC2, and the control path. Register files can supply data to ALU, address generators, and the control path.

3.8 List preprocessing and postprocessing of ALU and MAC.

3.9 What is an accumulator, and what is an accumulation register in a MAC?

3.10 Why can a MAC usually support double precision arithmetic computing?

3.11 What are keep mode and read mode of a register in a register file?

3.12 Describe the circuit in Figure 3.17 and explain why both combinational output and registered output are required.

3.13 List five modules inside a DSP core requiring output data (operand) from the general register file. (Inside a DSP core means memories are not taken into account.)

3.14 What are the main differences between a general CPU and a DSP processor?

3.15 Describe the differences between a CISC machine and a RISC machine.

3.16 Which features of a DSP processor are similar to a CISC machine which features of a DSP processor are similar to a RISC machine?

3.17 Describe the basic method to design a performance extreme DSP ASIC.

3.18 Design a Biquad IIR filter using the minimum silicon cost. There is no requirement on the runtime cost. The precision of results are single precision.

3.19 What are three kinds of operation dependences? Give three examples.

3.20 What is an enhanced Dual MAC DSP processor? What is the difference between an enhanced Dual MAC DSP processor and a two-way VLIW DSP processor?

3.21 What are the main differences between a VLIW machine and a SIMD machine?

3.22 What is the main difference between a VLIW machine and a superscalar?

3.23 Respecify the four-way dispatcher in Example 3.2 based on the architecture in Figure 3.29. The difference is the OOO execution.

3.24 Respecify the four-way dependency checker in Example 3.3 based on the architecture in Figure 3.29. The difference is the OOO execution.

3.25 Respecify the four-way reservation station in Example 3.4 based on the architecture in Figure 3.29. The difference is the OOO execution.

3.26 Use a small assembly program with a branch (taken) as an example to demonstrate that the PC (program counter) cannot be used to keep the order in a superscalar, and the superscalar must use an internal renamed instruction name instead of the PC to keep track of the order of fetching-dependence-execution-release. (*Hint*: PC can jump backward when the branch is taken.)

3.27 Design an RN circuit following Example 3.6.

3.28 What is memory sharing parallel architecture, and what is distributed memory parallel architecture?

3.29 What is streaming parallel architecture?

3.30 Use a flowchart to summarize the discussions of parallel programming in Section 3.5.5.

3.31 What are the typical parallelization management subroutines mentioned in Figure 3.37 parallel programming?

3.32 Which of the following statement(s) is(are) true?

a. A DSP processor does not need a MAC unit.

b. A DSP processor always has the same data width in the datapath and on the bus.

c. A four-way superscalar has higher performance than a four-way VLIW.

d. Superscalar can schedule instructions in hardware and VLIW cannot.

3.33 Are following statements true? The main difference(s) between a DSP processor and a microcontroller unit is/are:

a. The DSP is used mainly for connection protocol processing.

    b. The DSP is a master device and the MCU is a slave device.

    c. The DSP has instruction level acceleration for data processing.

3.34   Redraw Figure 3.40 to further quantitatively understand Amadahl's law. The sequential fractions are 0.05, 0.1, 0.2, 0.3, 0.4, and 0.5.

REFERENCES

[1] www.fwdconcepts.com/DSP06.htm.

[2] Furber, S. (2000). *ARM System-on-Chip Architecture*, 2nd ed. Pearson Education Limited.

[3] Nurmi, J. (ed.). (2007). *Processor Design, System-On-Chip Computing for ASICs and FPGAs*. Springer.

[4] Lapsley, P., Bier, J., Shoham, A., Lee, E. A. (1997). *DSP Processor Fundamentals, Architectures and Features*. IEEE Press.

[5] Madisetti, V. K. (1995). *VLSI Digital Signal Processors, An Introduction to Rapid Prototyping and Design Synthesis*. IEEE Press.

[6] Fisher, J. A., Faraboschi, P., Young, C. (2005). *Embedded Computing: A VLIW Approach to Architecture, Compilers, and Tools*. Morgan Kaufmann, Elsevier.

[7] Kozyrakis, C., Patterson, D. (2002). Vector vs. superscalar and VLIW architectures for embedded multimedia benchmarks. *Proceedings of MICRO-35*.

[8] www.zsp.com.

[9] Shen, J. P., Lipasti, M. H. (2004). *Modern Processor Design: Fundamentals of Superscalar Processors*. McGraw-Hill.

[10] Amdahl, G. (1967). Validity of the single processor approach to achieving large-scale computing capabilities. *AFIPS Conference Proceedings* 30, 483–485.

[11] Wiklund, D., Liu, D. (2005). Design, mapping, and simulations of a 3G WCDMA/FDD base-station using network on chip. *Proc of the International Workshop on SoC for Real-Time Applications*, Banff, Canada.

[12] Ohsawa, T. Takagi, M. Kawahara, S. Matsushita, S. PINOT: (2005). Speculative multi-threading processor architecture exploiting parallelism over a wide range of granularities. *38th Annual IEEE/ACM International Symposium on Micro architecture (MICRO'05)*.

[13] Pirsch, P. (1998). *Architectures for Digital Signal Processing*. John Wiley & Sons.

[14] Culler, D. E. (1998). *Parallel Computer Architecture, A Hardware/Software Approach*. MK.

[15] Liu, D., Tell, E. (2005). *Low Power Baseband Processors for Communications, Low Power Electronics Design*, C. Piguet (ed.), Ch. 23. CRC Press.

[16] Hennessy, J., Patterson, D. A. (1990). *Computer Architecture: A Quantitative Approach*. Morgan Kaufmann.

[17] Hennessy, J., Patterson, D. A. (2003). *Computer Architecture: A Quantitative Approach, 3rd ed.* Morgan Kaufmann.

[18] Weiss, M., Fettweis, G. P. (1996). Dynamic Codewidth Reduction for VLIW Instruction Set Architectures in Digital Signal Processors. *3rd International Workshop on Signal and Image Processing (IWSIP'96)*, page(s) 517–520, 04.-07.01.96.

[19] Sucher, R., Niggebaum, R., Fettweis, G., Rom, A. (1998). Carmel - A New High Performance DSP Core Using CLIW. *International Conference on Signal Processing and Techniques (ICSPAT'98)*, page(s) 499-504, 13.-16.09.98.

[20] Nilsson, A., Tell, E., and Liu, D. (2008). "An 11mm^2, 70mW fully programmable baseband processor for mobile WiMAX and DVB-T/H in 0.12μm CMOS," ISSCC, San Francisco, February 2008.

[21] Kneip, J., Weiss, M., Drescher, W., Aue, V., Strobel, J., Oberthuer, T., Bolle, M., and Fettweis, G. (2002). Single Chip Programmable Baseband ASSP for 5 GHz Wireless LAN Applications. *IEICE Transactions on Electronics*, E85-C(2):359-367, February 2002.

[22] Limberg, T., Winter, M., Bimberg, M., Klemm, R., Matus, E., Ahlendorf, H., Robelly, P., Tavares, M., and Fettweis, G. (2008). A Fully Programmable Single-Chip 45GOPS SDR Baseband For LTE/WiMAX Terminals, ESSCIRC 2008.

DSP ASIP Design Flow

4

In this chapter, ASIP design flow is introduced briefly. This chapter focuses on the discussion of what and why rather than how. Chapters 5 through 15 will give detailed discussions on how to carry out each step of the design flow introduced in Chapter 4. The focus of this chapter is the design flow of embedded DSP processors (ASIP). Nevertheless, the same design flow is also suitable for designing other hardware.

4.1 DESIGN AND USE OF ASIP

4.1.1 What Is ASIP?

The definition of ASIP (application-specific instruction set processor) was used since the late 1980s [1]. Alternatively, ASIP can stand for application-specific integrated processor in other books. ASIP also is considered to be embedded processors, but this is not really true because Pentium can also be used as embedded processors. The definition of ASIP in this book is to differentiate architectures and applications from general-purpose microprocessors.

The main difference between a general-purpose processor and an ASIP is the application domain. A general-purpose processor is not for a specific application class so that it should be optimized based on the performance of the application mix. The application mix has been formalized and specified using general benchmarks such as SPEC (Standard Performance Evaluation Corporation) [24]. The application domain of an ASIP usually is limited to a class of specific applications, for example, video/audio decoding, or digital radio baseband. The design focus of an ASIP is aimed at specific performance and specific flexibility with low costs for solving problems in a specific domain. A general-purpose microprocessor aims for the maximum average performance instead of specific performance.

There were two major families of integrated digital semiconductor components: microprocessors and ASIC. Microprocessors take the role of flexible computing of all programs running on a desktop or laptop. An ASIC supplies application-specific function, its functional coverage, and performance is fixed during the hardware design time. When requiring higher performance, lower silicon area, and lower power consumption, ASIC was preferred because it consumes about two orders lower power as well as the silicon area compared to general-purpose processors. In the 1980s to 1990s, ASIC played dominant roles in communications and consumer electronics products; for example, the radio baseband ASIC in the GSM mobile phone, IC in the telephone switching system, IC in cable terminals, and IC in the DECT (Digital European Cordless Telephone) phone.

Since 2000, more standards in wireless communications and audio/video processing have emerged, which need to be supported in a single product; therefore, more and more ASIC circuit modules have been integrated in one product to support these standards with multiple working modes. For example, nowadays it is quite common that a high-end mobile phone supports wireless connections including GSM in dual frequency bands, WCDMA, WLAN including IEEE802.11a/g/b, and Bluetooth. At the same time, the mobile phone should also support different voice CODEC standards, audio MP3 decoder, image CODEC, and different video CODEC. Therefore, the design cost and silicon cost of ASIC will surge if a mobile phone is based on ASIC. Actually, as the requirements on product flexibility become mandatory, ASIC is gradually giving way to programmable devices such as ASIP, because using ASIC becomes less realistic. ASIP is so far the best flexible solution for volume products, for the following reasons [2]:

- ASIP supplies with sufficient flexibility.

- ASIP can be used for a class of multimode applications.

- ASIP is silicon efficient.

- ASIP can reach relatively low power.

- Design cost of ASIP is much lower than the cost of ASIC design.

- New generations and new applications can be adapted by changing the ASIP code.

Multimode means the functional coverage of multiple applications or the support of an application of several standards. For example, a multimode video decoder handles video coding standards such as MPEG2, MPEG4, H.261, H.263, H.264, WMV, and JPEG.

An ASIP is an essential electronic component for a class of specific functions with programmability. An ASIP could be a DSP processor, a network processor, a microcontroller, or any programmable device in a specific system. In this book, we discuss ASIP only as DSP processors.

4.1.2 **DSP ASIP Design Flow**

DSP ASIP design flow can be simplified as shown in Figure 4.1. In later sections and from Chapters 5 through 15, Figure 4.1 will be discussed in detail. This flowchart gives a guidance of ASIP design from the marketing strategic decision down to the microarchitecture specification, as a complete ASIP design flow. After specifying

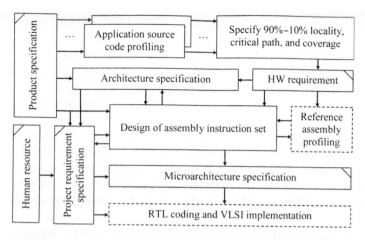

FIGURE 4.1

DSP ASIP design flow.

the microarchitecture of the ASIP, the output of the microarchitecture design will be the input of VLSI implementation, which is the same as all other backend designs and which can be found in any back-end VLSI book.

Before designing an ASIP (either DSP or other kind of processors), the product specification should be supplied from the marketing strategy responsible. A product specification includes product descriptions and requirements. The descriptions include the type of the processor, applications to be supported, customers, and the role of the ASIP in OEM (original equipment manufacturer) products. The requirements include the range of the ASIP price, the TTM (time-to-market), design cost limit, fabrication cost limit, performance, and physical constraints. The product specification also includes management issues such as available competence, the long-term strategy of the company, customer relations, cooperator relations, and so on.

Design of an ASIP is a risky and expensive business, but rewarding when the design is successfully implemented. An ASIP is a system component. The decision of ASIP could be from the system. The ASIP provider will decide the design of the product after the market analysis (profit, product life time, risk, competitors, etc.). An ASIP product also could be initialized by an ASIP provider. In this case, an ASIP is a kind of ASSP (application-specific standard product) or COTS (commercial off-the-shelf) component. Writing a product specification of an ASIP is a complicated decision process based on multiple objectives and is beyond the scope of this book.

After collecting enough application documents and source codes, the first step of the design is source code profiling [3, 4]. Source code profiling is to understand and predict the running statistics of applications by running or parsing the source code or assembly code of the reference assembly language, the assembly language of another machine used as the reference of the ASIP design.

Based on the result of source code profiling, there should be sufficient understanding of the performance critical path in the source code, and the function coverage. Based on the understanding, the performance requirements, and the cost limitation, the early architecture should be decided through the design space exploration of the ASIP architecture (architecture selection). At the same time, hardware requirements can be specified for assembly instruction set design [5-9].

If the expected performance or power consumption is close to the physical limitation, the code freedom (the freedom to write assembly code) is low. In this situation, a reference assembly language should be used to assist the profiling of miscellaneous functions, for example, the extra cost of handling fixed-point data, multithreads, and interrupts. Otherwise, the step of profiling the reference assembly code can be skipped.

Assembly language design (instruction set design) is an iterative process. By using results from the source code profiling, an ASIP assembly language design can be divided into the design of basic assembly instructions and the design of accelerated instructions. The quality of designed assembly language should be evaluated through benchmarking.

The architecture selection includes the selections of processor structure, available modules, and interconnection between modules. Microarchitecture design will further specify modules. In particular, ASIP microarchitecture design maps the function of every assembly instruction into each hardware module.

RTL coding and VLSI implementation is the last step of ASIP design. However, the last design step for ASIP will not be discussed in this book, because there are already enough books available on that topic.

ASIP design requires multiple disciplines and comprehensive knowledge structure. The knowledge of ASIP design cannot simply be partitioned and isolated in different design steps. Therefore it is impossible to formulate a perfect pedagogical knowledge structure of ASIP design. We therefore try to introduce the practical design skills using iterative and progressive ways of learning through the textbook. In the following sections of this chapter, the ASIP design flow will be briefly introduced to provide an overview of the complete flow. The overview will help readers to understand every step of the design flow presented in the following chapters.

4.2 UNDERSTANDING APPLICATIONS THROUGH PROFILING

In the academic community, understanding applications usually means DSE (design space exploration) at the system level. In industry, understanding applications is called source code profiling or source code analysis. Source code profiling for ASIP design is to achieve enough knowledge for designing the assembly instruction set. The profiling is to analyze the source code itself or to analyze its running behavior [2-4].

A complete knowledge system of an application might be developed and accumulated by hundreds of people through many years. Application engineers take years to learn and accumulate fundamentals of system design experiences. Instead of application engineers, ASIP designers are hardware engineers living on knowledge of circuit specification and implementation. It is impossible that ASIP engineers can really understand all design details of an application. However, a hardware engineer may have no future without function design experiences.

Source code profiling technique therefore is developed to fill in the gap between system design and hardware design. Profiling is a good technique, isolating the ASIP design from the design of applications.

> **BOX 4.1** The purpose of profiling is not to understand the system design; rather it is to understand the cost of the design, including execution (dynamic, meaning data-dependent) behavior, the code (static, meaning data independent) structure, hardware cost, and runtime cost.

Profiling exposes the code structure and execution behaviors, such as the coverage of arithmetic operations, costs of algorithm execution, and costs of memory accesses. The source code profiling can also expose opportunities of parallelization for further performance enhancement. *Coverage* is the capability of running different operations by the ASIP, and *performance* is the computing capacity of the ASIP for certain algorithms.

The profiling input is a list of applications and a collection of the source code. The profiling output is the code analysis and execution statistics. Profiling results will be used as the input of the architecture and assembly instruction set design. Profiling techniques will be discussed in detail in Chapter 6.

4.3 ARCHITECTURE SELECTION

The purpose of architecture design is to select a suitable ASIP architecture for the class of applications. The decision includes how many function modules are required, how to interconnect these modules (relations between modules), and how to connect the ASIP to the embedded system.

In the academic community, exploring the architecture usually is called DSE at the architecture level, and in industry, it usually is called ISA (instruction set architecture) selection. The architecture can be decided by selecting and reforming reference architecture or by directly mapping functional task flow to circuits.

4.3.1 General Methodology

The traditional embedded system design flow is given in Figure 4.2. To simplify a design, system design inputs are partitioned and assigned to the HW design team

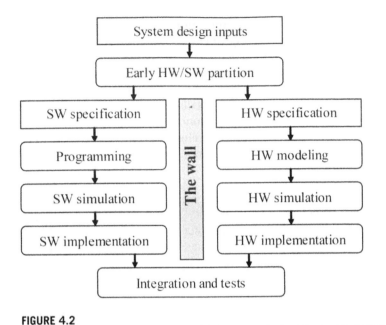

FIGURE 4.2

Traditional embedded system design flow.

and the SW design team at the beginning of the system function design. The SW design team and HW design team design their SW and HW independently without much interwork. In the SW design team, functions mapped to SW will be implemented in programs using high-level behavior language. The implemented functions will be verified during simulations, and finally the simulated programs are translated to machine language during the SW implementation.

From the HW design perspective, functions assigned to the HW team are allocated to HW modules. Modules could be either processors or functional circuits. Processors could be standard components or ASIP. The behaviors of programmable HW modules are described by an assembly language simulator. The behaviors of nonprogrammable HW modules are described by hardware description languages. These models must be carefully and completely simulated and validated. HW can be implemented only after the validation of functions and performance of the HW modules. Finally, the implemented SW and HW will be integrated. The implemented binary code of the assembly programs will be executed on the implemented HW.

The design flow in Figure 4.2 is correct because it follows the golden rule of design: divide-and-conquer. However, when designing a high-quality embedded system, we actually do not know how to make a correct or optimized early partition. In other words, the early partition usually is not good enough without iterative optimizations through the design. To optimize a design, we have to modify the basic principle of divide-and-conquer and interactively exchange functions and specifications between SW and HW design teams during the embedded system design. Under such a challenge, HW/SW codesign appeared in the early 1990s associated with many research job opportunities and conferences [10]. This is because HW/SW codesign is not an easy method. From this chapter until Chapter 9, we will discuss ways of HW/SW codesign for ASIP. The discussion will be based on engineering and case studies because the formal way of HW/SW codesign is not yet practical (as of 2007).

The HW/SW codesign flow is depicted in Figure 4.3. Following the figure, the idea of the new design flow is to optimize the partition of HW and SW functions cooperatively at each design step during the embedded system design. HW/SW codesign trades off function partition and implementation using either SW or HW during all design steps through the embedded system design. Eventually, the results will be optimized following certain goals.

Repartitioning is not difficult. The difficulty lies in the fast modeling of the performance and the cost of the new design after each repartition. Eventually, the decision of repartition will be based on the quantitative analysis and trade-off between the costs and performance before and after the repartition.

The HW design cost is much higher than the design cost of the same function in SW, whereas the silicon costs of a dedicated HW module can be much lower than the silicon costs of a programmable processor for the same function. The

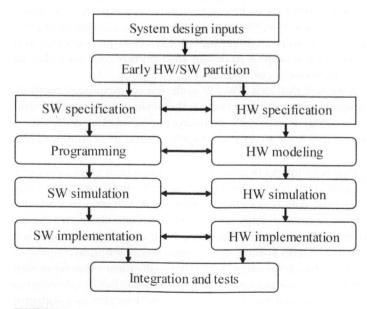

FIGURE 4.3

Hardware/software codesign flow.

performance of a HW module of a dedicated function can be much higher than that of SW, whereas the flexibility of a SW solution is always higher than that of a HW module. Therefore, the design decision will be based on:

- Definition of the product performance and flexibility.

- Function modeling.

- Modeling of performance, design costs, silicon costs, and power consumption.

- Weighing factors of performance, design costs, silicon costs, and power consumption for the design.

The decision can be achieved by running optimizations step-by-step following the requirements of performance, design costs, silicon costs, and power consumption, under constraints of flexibility.

To reach different objectives—such as low design costs, higher performance, lower silicon costs, or higher flexibility—an embedded system can be implemented in different ways based on different kinds of HW and SW partitions. Through HW/SW codesign, the optimization for certain objectives can be reached by the repartitioning of HW and SW through all steps of embedded system design.

As we discussed before, HW/SW codesign is not an easy task at all. The way to observe and extract design parameters for optimization cannot be modeled easily. Many basic design criteria have not been touched at all, such as the modeling of flexibility and the modeling of functional coverage. There are so many design freedoms and so many possible solutions to investigate and reach a perfect solution. From the engineer's point of view, HW/SW codesign has remained an infant since it was born in the early 1990s. Usually, engineering HW/SW codesign is divided into several design steps. One of these steps is to design and optimize ASIP for a class of specific applications, which is one focus of this book.

Later in this section we will find that HW/SW codesign is a necessary methodology for processor design because the processor is the device combining HW and SW. The codesign methodology is especially mandatory for designing ASIP. This is simply because the HW and SW of an ASIP are tightly related to accommodate a class of applications.

The identity of an ASIP is its instruction set architecture, which gives processor features such as functional coverage, performance, parallelization, data addressing space, program memory cost, data types, data memory cost, silicon area cost, and power consumption. Applications running on a DSP ASIP are relatively known during the processor design time. The designers therefore have opportunities to reach specific performance at low silicon costs with low power consumption.

Implementing part of the functions using ASIP HW usually means that we design a single or several new instructions performing a specific function. Implementing part of the SW functions means that we design the specific function as a software subroutine using available instructions. The SW of an ASIP is an executable program stored in the program memory in binary format. The hardware of ASIP is a specific

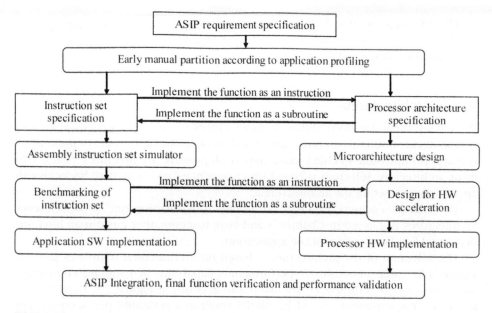

FIGURE 4.4

Hardware/software codesign for an ASIP.

piece of hardware that executes instructions. HW/SW codesign moves functions or tasks between SW and HW to meet the specific design constraints.

In the early design phase (source code profiling) of an ASIP, applications are carefully analyzed. An instruction set architecture can be proposed based on the profiling and early partition, together with the allocation of functions to SW or HW. HW/SW codesign of ASIP in Figure 4.4 is to further modify the proposal by trading off HW/SW partition through all design steps of an ASIP. Readers may notice that the HW/SW codesign discussed in Figure 4.4 is just about the partition. Instruction design will be discussed later in detail.

An assembly instruction set shall efficiently supply enough performance for the most frequently appearing functions and be general enough to cover all functions in the specified application domains. When the performance is not enough, the most used (kernel) subroutines or algorithms will be hardware-accelerated. Those subroutines or algorithms that are seldom used will be implemented as SW subroutines. Performance and coverage will be fulfilled by carefully trading off the HW/SW partition.

However, to evaluate functional coverage is not easy. So far, there is no better way than running either the benchmarks or the real applications. The cost is high, and it takes a long time to evaluate an assembly instruction set by running benchmarks and applications. Experiences in programming DSP assembly code will allow the user to speed up the evaluation.

HW/SW codesign flow in Figure 4.4 is part of the design flow in Figure 4.1. The HW/SW codesign starts from source code profiling down to the release of assembly instruction set design.

4.3.2 Architectures

Principally, there are many different architectures to select, and it is not an easy task to decide a suitable architecture for a class of applications. One of the reasons is that it is impossible to check and compare all possible architectures. That is why most architecture selection is based on experiences and why drawbacks usually appear after selecting the architecture.

There are two general ways to make architecture selection: one, to use reference architectures discussed in Chapter 3 and two, to generate a custom architecture dedicated for the task flow of the application.

The selection of the architecture is based on architectural templates given in Figure 4.5. Three dimensions—performance, capability of handling complexity, and power/silicon efficiency—are taken into account in Figure 4.5. Here the performance measurement should be MOPS (million operations per second). The mathematic operations are required by DSP algorithms and can be counted explicitly by source code profiling. Other operations such as memory access, program flow control, data quality control, and hardware resource control are redundant

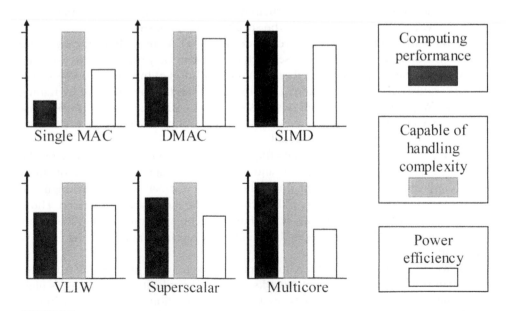

FIGURE 4.5

Intuitive architecture selection.

operations. These redundant operations may not be exposed during the source code profiling.

Complexities in Figure 4.5 can be divided further into data complexity and control complexity. Data complexity is the irregularity of custom data types, data storage, and data access. For general computing, the impact of the differences can be hidden by using high-precision data types. Data features in embedded applications are relatively known during the system design phase. The knowledge of data features should be used to minimize the silicon costs, thus increasing the design complexity. For example, extra complexity can be from precisions and dynamic range control of monitoring and scaling data. Data complexities cannot be found in a functional behavior model. In Chapter 2, problems were discussed; it is beyond the scope of this book to discuss their impact in detail. To design a high-quality system with low silicon costs, the way to handle the data complexity issues can be found in references [11, 12]. To summarize the data complexity problems in precision-limited hardware, the following hints can be used to select architecture.

- Efficient data quality control instructions can decrease the cycle costs of running data quality monitoring and scaling.

- Data quality control usually is executed in the RISC datapath, and iterative computing will be executed in the iterative or vector datapath. An efficient architecture can share data between vector datapath and RISC datapath, so that the vector parallel computing and data quality control can be executed without extra overhead.

- Architecture to support conditional execution is useful when data-dependent executions are involved in the kernel subroutines. To eliminate branches in kernel subroutines, conditional execution instruction should be available.

The data complexity includes also the complexity of memory addressing and access. Design of data storage and data access is based on the knowledge of memory subsystem (HW) and data structure (SW). Complexities of data storage and data access of DSP applications are different from the complexity of memory subsystem and the data structure of general computing. For general computing, the complexity comes from the requirements of flexible addressing, large addressing space, and short latency induced by cache mismatch. For embedded DSP applications, the complexity of memory addressing and access comes from:

- Running memory accesses in parallel with DSP algorithms by adapting the memory accesses to custom algorithms for most used or MOPS extensive algorithms.

- Accessing data in parallel for flexible parallel computing.

- Other custom addressing algorithms.

Control complexity is required by program flow control and hardware resource management. On a subroutine level, the control complexity is mixed up with data complexity, such as managing data dependencies. At the top level of an application program, the complexity can be algorithm selection, data precision control, working mode (data rate, levels, profiles, etc.) selection, and asynchronous event handling. At the super level of interapplication handling, the complexity can be the hardware resource management through threading control and control for interprocessor communications. The management of control can be efficient if instructions for I/O control, conditional branches, and bit manipulations are carefully designed. The SIMD machine cannot supply comprehensive control functions because it was only designed for accelerating vector computing. The multicore solution introduced extra control complexities from intertask dependencies between processors.

Computing performance is obviously the main goal of the architecture selection. It is the measure of million operations per second (MOPS). DSP instructions must be much more efficient (can be less flexible) compared to the instruction set of GPP. The performance measure therefore is based only on arithmetic computing or MOPS instead of MIPS (million instructions per second), and the data access instructions must not be included.

Power efficiency is measured by performance over power consumption. Low-power design is equivalent to eliminating redundant operations (to minimize the dynamic power) and minimizing the circuit size (to minimize the static power). Power efficiency and silicon efficiency are more related because the static power is no longer negligible and even becomes dominant. In a modern high-end ASIP design, since the memory consumes most of the power, the design for power efficiency is almost a problem of designing memory efficient architecture in order to optimize the memory partition, to minimize the memory size, and to minimize the memory transaction.

The complexity handling and performance enhancement based on limited silicon costs has been the focus of embedded electronics. In Chapter 20 questions posed here will be discussed. Others not yet discussed are silicon costs, customer preference (customers might have unexpected requirements), HW design costs, and SW design costs. Visit this book's web page for more design references from our research.

How to Use a Reference Architecture

A typical method of ASIP design is to modify a selected hardware template—for example, one template in Figure 4.5. The decision of a selection is based on requirements on performance, complexity, power consumption, and silicon costs. Architecture selection will be discussed in later chapters of the book.

The HW design cost of a superscalar is high, and the compiler design costs for VLIW could be much higher. The SW design costs for a VLIW processor or a multicore solution are usually higher. It might be troublesome to select VLIW architecture without very good compiler knowledge. A superscalar is the preferred solution when higher performance (VLIW cannot handle) and code compatibility between product generations are required.

When the control complexity cannot be separated from the vector computing, VLIW or superscalar should be used. If the control complexity can be separated from the vector computing and if there is an available processor that can execute the top-level DSP application program, SIMD architecture is preferred to enhance the performance of vector/iterative processing.

How to Generate a Task-Flow Architecture

If the programming cost, the hardware design cost, and the complexity of hardware and system verification can be manageable, generating a custom (task flow) architecture may induce very good results. Task flow architecture is also called function mapping hardware or algorithm mapping architecture. The architecture is from the direct implementation of the DSP stream or its control flow graph. In Figure 4.6, the method of hardware implementation of control flow graph is depicted.

The left part of Figure 4.6 is the behavior task flow of the DSP application. This architecture is useful when the input data rate is high and the function cannot be executed by any conventional architecture in Figure 4.5. If the control flow graph is relatively simple and will not be changed through the product's lifetime, the

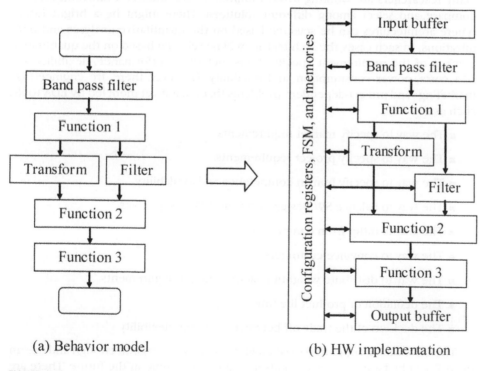

(a) Behavior model   (b) HW implementation

FIGURE 4.6

Implementation of task flow architecture.

control flow graph can be implemented using HW on the right part of Figure 4.6. In Figure 4.6 (b), the datapath is the mapping of the control flow graph in Figure 4.6 (a). The FSM controls the execution of the DSP task in each hardware block. Memories are used as the computing data buffer passing data between the hardware function blocks [12].

The control flow graph architecture is preferred when the requirement on hardware multiplexing is not high, configurability is required, and programmability is not mandatory. The task flow architecture is used mostly for designing accelerators.

Here the configurability stands for the ability of the system functionality to be changed by external configuration. The programmability stands for the ability of the hardware to run programs. The difference between configurability and programmability is the way the hardware is controlled; configuration control is relatively stable, it is not changed in each clock cycle, and the program changes the hardware function in each clock cycle.

4.3.3 Quantitative Approach

Many researchers are working toward multiobjective-oriented evaluation tools to compare and select among different solutions. There might be a bright future where architectures can be selected based on the quantitative analyses and optimizations. In such cases, the architecture will be selected based on the quantitative reaches of the requirement specifications including performance, flexibility, silicon costs, power consumption, and reliability. However, there are engineering-, marketing-, and project-dependent problems that are not yet modeled by EDA tools, such as:

- The way to specify market requirements.

- The way to specify project requirements.

- The way to specify human competence and availability.

- The way to balance SW design costs and HW design costs.

- The way to differ from competitors.

- The way to convince customers.

- The way to distribute weighting factors among requirements.

- The definition of product life time.

- The decision of the trade-off between costs and flexibility.

The quantitative approach based on EDA tools is not yet good enough and design of ASIP will be based on design skills in a rather long time in the future. There are other nontechnical decision factors. For example, the life time of ASIP may not be "the longer the better." A longer life time may block sales of new products.

From the technology point of view, the trade-off between the engineering costs of assembly coding and the hardware efficiency or hardware cost will be different in different projects running in different countries. The trade-off issues will be discussed more in the book and will be available from future publications on this book's web page. Nevertheless, design skills introduced in this book will be essential. The EDA tools for design space explorations of ASIP architecture will be used only as references instead of being used to release a design, which means that the release of an ASIP design will be based on your design skill!

4.4 DESIGNING INSTRUCTION SETS

As soon as the architecture is decided, the instruction set of the architecture can be designed to meet the balance of requirements. The input of assembly instruction set design is the result of source code profiling (coverage and performance) and architecture selection. The design task is to map the functions to the instruction set and achieve balance between the performance and cost. The design output will be the assembly instruction set manual. The design of an instruction set includes the design of:

- Arithmetic instructions and accelerated arithmetic instructions for DSP algorithms and data quality controls.

- Memory access and addressing algorithms to supply sufficient data with acceptable access latency for all arithmetic computing.

- Program flow control.

- Instructions for I/O access and supporting accelerators.

The quality of the designed instruction set cannot be evaluated before a complete benchmarking and essential assembly code profiling. The toolchain for assembly programming should be prepared for assembly-level benchmarking and profiling as soon as the assembly instructions are specified. The iterative design flow of an assembly instruction set is given in Figure 4.7.

The gray part in Figure 4.7 denotes the basic activities of assembly instruction set design. The white part with solid lines denotes the activities involved in the processor firmware design. The processor firmware design supports assembly-level code library and code profiling, which is part of the assembly instruction set design. The design activities within the dashed lined blocks are the remaining parts of ASIP design, related to those not yet included in the assembly instruction set design.

The processor hardware design starts from the designing of the assembly instruction set. The design of an assembly instruction set can be the modification on a selected instruction set template or on the generation of an instruction set by directly mapping functions to instructions based on results of source code profiling. Assembly instruction set design will be discussed in Chapter 7.

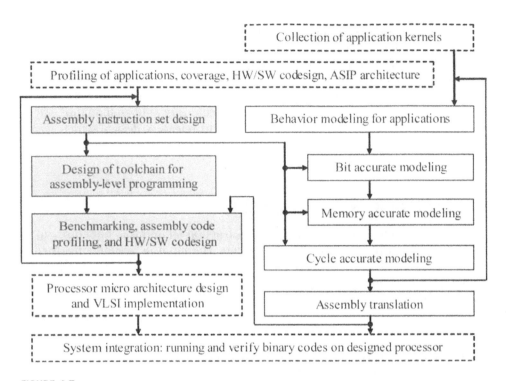

FIGURE 4.7

Simplified DSP ASIP design flow.

4.5 DESIGNING THE TOOLCHAIN

After proposing an assembly instruction set, the assembly programming toolchain should be designed and provided for benchmarking. An assembly programming toolchain includes the C-compiler, the assembler, the linker, the ISS (Instruction Set Simulator), and the debugger. The complete set of the programmer toolchain is called the integrated design environment (IDE). Design of a toolchain will be discussed in Chapter 8. Benchmarking will check the performance and costs of kernel DSP programs running on the processor. Assembly code profiling will further expose the statistics of the instruction usage and the SW costs of application kernels. Benchmarking and assembly code profiling will be discussed in Chapter 9.

Toolchain given on the right part in Figure 4.8 will support firmware design on the left part in Figure 4.8 for an application from specification down to the binary machine code.

Seven main design steps are used to translate the behavior code to the qualified executable binary code, and eight tools in the programmer's toolchain on the right side of Figure 4.8 are used for the code translation. Each tool in the tool chain in Figure 4.8 (b) is identified by a number. The way that tools are used in each design

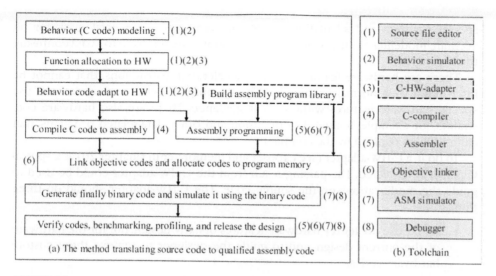

FIGURE 4.8

Firmware design flow using programmer's toolchain.

step are annotated by the numbers marked within each design step in Figure 4.8 (a), indicating the usage of tools in the toolchain as depicted in Figure 4.8 (b).

Toolchains for assembly programming will be discussed in detail in Chapter 8. In this chapter, the toolchain will be described briefly to provide the knowledge necessary for designing assembly programming tools, including assemblers, linkers, and assembly instruction set simulators.

Source file editor as the first tool in the toolchain can be any available text editor. It helps the programmers to create, analyze, and modify the source files. Some editors can help the programmers to organize the source code hierarchically, such as printing statements and comments in a decent and tidy way and statically checking and discovering potential bugs.

The behavioral simulator is the source code (C or MATLAB) simulation environment. In some cases, special tools and a development environment such as Simulink are used as the behavioral simulator.

Before translating the C code to assembly code, the firmware designers should adapt or rewrite part of the C code to:

- Remove the legacy hardware adaptation of early designs (if there is any) from the behavior C code.

- Adapt to the current ASIP hardware (finite-length data types, parallel and accelerated hardware) features and follow our ASIP hardware constraints (memory size, for example).

Rather often, the source C code contains hardware features used for other products or for the previous version of the ASIP. The hardware features, such as special partition and adaptation for vector computing, or special mask for bit accurate modeling of the previous hardware, should be removed to avoid confusing the C compiler.

There are hardware features, such as accelerated instructions, that cannot be interpreted by a normal C compiler. The C compiler cannot recognize and use special hardware features without clear guidance. There are also hardware constraints, for example, limited data precision and other data-related problems, that cannot be recognized by the compiler.

To do it, a C-HW-adapter (see exposer and memorizer in Chapter 20) as a special C parser is expected. The parsing result from the C-HW-adapter could be used to guide the modifications on the source code. The C-HW-adapter should do three things:

1. To expose the legacy hardware features of early design, not for the hardware of the current design, and to guide the designer to remove these custom hardware codes.

2. To expose opportunities to use compiler features, and to guide the code designers to use the acceleration features available from the selected hardware.

3. To expose the opportunities for hardware parallelization and parallel memory accesses. With hardware configurations, the C-HW-adapter may also guide programmers to code the parallel programs.

So far, there has not been any good and universally accepted method to develop a C-HW-adapter for all listed requirements. Part of these requirements could be identified by tools from university researchers.

There might be a special C library adapting some ASIP hardware features—for example, the special data types, special operations, special functional acceleration, and special addressing algorithms. The distance between C and assembly language can be reduced further by including the special library and using library functions in the C code. In this case, the C-HW-adapter might be used to identify the places requiring special C library features. The compiler may need special annotations to guide the compilation for certain specifically accelerated code. In this case, the C-HW-adapter might be required to insert and to interpret annotations.

After the C code qualification and hardware adaptation, the C code will be compiled to assembly code by the C compiler. Usually, the compiler will translate C code to intermediate representation (IR) first. IR is the language inside the compiler. IR will be optimized further and finally translated to target assembly language.

High quality and simple software can also be translated by firmware designers without using a C compiler (see the right part of Figure 4.8 (a)). However, usually a complex application has to be translated to assembly language by the C compiler due to the size of code. In most firmware design projects, the design methods on both the right and left parts of Figure 4.8 are used. The C compiler is used to generate the top-level assembly code, and the assembly programmers are responsible for making high-quality assembly subroutines much used.

The assembler translates assembly code into binary code. The translation process has three major steps. The first step is to parse the code and set up symbol tables. The second step is to find symbol locations in the program and data memories so that the relationship between symbolic names and addresses can be known when translating instructions. The third step is to translate each assembly statement by combining the numeric equivalents of op-code, registers, specifiers, and labels into binary code. The input of an assembler is the source assembly code. The output of an assembler is the relocatable binary code, called the object file. The output file from an assembler is the input file to a linker, which contains machine (binary) instructions, data, and bookkeeping information (symbol table for the linker).

It is usually a tedious task to design an embedded firmware system based on the assembly language. A firmware system usually is developed by a team of many designers, including library designers, subsystems designers, and firmware integration engineers. Therefore, a firmware system consists of many separate subsystems or subroutines. These subroutines can be modified independently so that the cost of the design iterations can be minimized locally. A firmware system is generated finally by a linker that links subroutines and libraries when all sublevel designs are ready. The linker collects all related application and library subroutines, links them together, and assigns global relative distances to the addresses of the code and data memories.

All assembly code of the application including subroutines in libraries must be linked together, and all program code and data should be allocated with fixed distance to the starting address of the code and the starting address of the data. Finally, the code must be executed and debugged by the instruction set simulator before it can be loaded to the hardware program memory.

The assembly instruction set simulator executes assembly instructions (in binary machine code format) and behaves the same way as a processor. It therefore is called the behavioral model of the DSP processor. A simulator can be a DSP core simulator or a simulator of the whole DSP chip. The simulator of the core covers the functions of the basic assembly instructions, which should be "bit and cycle accurate," meaning the result from the simulator matches the result from the hardware of the processor core. The simulator of the chip covers the functions of both the core and the peripherals, which should be "bit, cycle, and pin accurate," meaning the result from the simulator should match the result from the whole processor hardware.

Once the assembly firmware is developed, the programmer must debug it and make sure it works as expected. Firmware debugging can be based on a software simulator, or a hardware development board. In any case, the programmer will need a way to load the program and data, run the program, and observe the outputs. Signal processing applications tend to require large amounts of streaming data. It is helpful if the tools can easily deal with I/O operations. By integrating all tools into the toolchain in Figure 4.8, the integrated design environment (IDE) is obtained.

4.6 MICROARCHITECTURE DESIGN

The microarchitecture design of an ASIP is to specify the hardware implementation of the assembly instruction set into function modules of the core and its peripheral modules. The input of the microarchitecture design is the ASIP architecture specification and the assembly instruction set manual. The output of the microarchitecture design is the microarchitecture specification for RTL coding, including module specification of the core and interconnections between modules. The microarchitecture design of an ASIP can be divided into three steps. The first step is:

> **BOX 4.2** Partition each assembly instruction into microoperations, and allocate each microoperation into corresponding hardware modules.

Microoperation is the lowest level hardware operation; for example, guard an operand, invert an operand, addition, and others. The second step is:

> **BOX 4.3** For each hardware module, collect all microoperations allocated in the module and specify hardware multiplexing for RTL coding of the module.

The third step is:

> **BOX 4.4** Fine-tune intermodule specifications of the ASIP architecture specification and finalize the top-level connections and pipeline of the core.

Similar to architecture design, there are also two ways to design microarchitecture. One way is to use reference microarchitecture, which is the available design of a hardware module from publications, other design teams, and available IP. The reference microarchitecture is close to the requirements of the module to be designed and can execute most microoperations allocated to the module. Several reference microarchitecture modules are presented in this book.

Another method is called custom microarchitecture, which is to generate a custom architecture dedicated for a task flow of an application. This method is used when there is no reference microarchitecture, or when the reference microarchitecture is not good enough. The method is to map one or several CFGs (Control Flow Graphs, the behavior flowchart of an application) to hardware and to generate a microarchitecture. A typical case is to design accelerator modules for special algorithms, such as lossless compression algorithm [20], forward error correction algorithms [21], or protocol packet processing [22]. This method will be described in detail in Chapter 10.

Methodologies, basic skills, and examples of ASIP microarchitecture design will be discussed in detail in Chapter 10. As case studies, microarchitecture design details of register file, ALU, MAC, the control path, the memory subsystem, and peripherals will be discussed in Chapters 11 through 16.

4.7 FIRMWARE DESIGN

Firmware by definition is the fixed software in products. The software will not be changed when using the system. When firmware is discussed, it is implicitly meant as the firmware in embedded systems. Typical firmware can be a voice codec in a mobile phone, which can be recognized by its user. On the other hand, firmware can also implicitly exist in a system not exposed to the user, like the firmware for radio baseband signal processing. Firmware can be application firmware or system firmware. The latter is used to manage the use of system hardware and the running of application software. A typical system firmware is the real-time operating system, RTOS.

Qualified DSP firmware design is based on the deep understanding of DSP processor architectures. The design of DSP firmware will be discussed in detail in Chapter 18, after the discussion on hardware. The purpose of this chapter is to provide enough knowledge for the assembly instruction set specification and microarchitecture design. The iterative way of learning follows the pedagogy for undergraduate teaching.

Different from application engineers, processor designers are hardware designers and might not understand much of DSP algorithms or applications. However, during the design of an ASIP instruction set, hardware designers need to have a certain level of understanding of applications or software in order to coordinate with application engineers. The question is how to define the level of understanding of algorithms and applications. Since the time to run a project always is limited, deep understanding of algorithms and applications is not feasible. In this chapter, we try to guide you as hardware designers to reach the right level of understanding of involved applications for instruction set and hardware design.

Some basic applications have been introduced in Chapter 1. To understand applications means to know what, why, and how: *What* is the algorithm and the application, including inputs, outputs, and how inputs are processed; *why* is the deep understanding of the reason to choose such applications or algorithms, including the market decision, the technology decision, and the product strategy, performance and cost trade-off; and *how* is the implementation, which includes the algorithm design, the hardware design, the system design and optimizations, as well as fabrication.

An intuitive view of DSP firmware design is given in Figure 4.9. There are two main constraints existing throughout all design phases: the real-time constraints of the application and the hardware constraints—for example, the timing period of the coming data packet, the performance, and the hardware precision. Correct

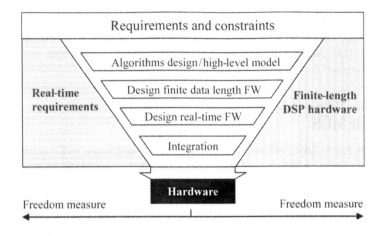

FIGURE 4.9

An intuitive view of firmware design.

implementation of DSP firmware includes the correctness of functionality, the correctness of the runtime behavior, the correct handling of precision and dynamic range for finite-length computing and storage hardware.

4.7.1 Real-time Firmware

Requirements and programming of DSP firmware for real-time applications can be much different from offline signal processing. Real-time signal also is called streaming data and is the system being able to finish all processing tasks within a certain period of time (before the new signal arrives), which means the speed of signal processing is higher than the arriving speed of the signal.

Figure 4.10 (a) gives the definition of parameters of real-time tasks. After the arrival of the data packet at time (1), the signal processing task can start at time (2). The time interval between (1) and (2) can be negligible if the start-up time of the task is small. The computing time is (3) and it finishes at time (4). Time (5) is the deadline by which the execution of all tasks must be finished. The time interval between (4) and (5) is reserved for possible data-dependent uncertainty and asynchronous events such as interrupts. Figure 4.10 (b) gives an example of a relative practical scheduling. It achieves both the minimum execution time and the minimum computing latency. Here the computing latency is the time interval between (1) and (5) in Figure 4.10. The minimum computing latency can be achieved by scheduling the data packet reception and task execution in parallel. Detailed real-time knowledge can be found in references [13, 14].

A hard real-time system cannot tolerate any missing of time deadline. For example, a decision to switch a locomotive to a track must be based on hard real-time computing. If we miss the time deadline, the locomotive may come to the wrong track. In a normal real-time system, the time requirement may be less stringent.

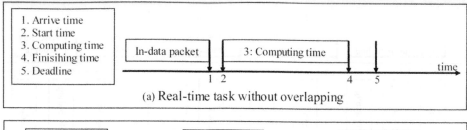

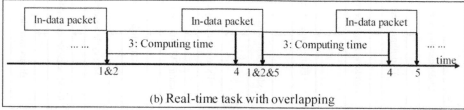

FIGURE 4.10

Real-time parameters.

Deadline miss with low probability is acceptable sometimes. For example, during the voice packet reception of VoIP (Voice over IP), losing one data packet out of a thousand may not be a significant problem. To design a real-time system, we need to specify and model the system having time as a significant parameter.

4.7.2 Firmware with Finite Precision

Finite precision firmware is the firmware running data with finite precision on fixed-point hardware. The data quality of finite precision firmware is optimized by running special subroutines for the data quality measurement at different processing steps and scaling data during computing to reach the highest possible data quality. The purpose is to best utilize the low-cost hardware with finite precision to reach the best reachable data quality. The way is to dynamically adapt to the data feature during signal processing by monitoring and scaling data to be processed. Fundamentals regarding the finite precision problem was discussed in Chapter 2. Extra firmware for promoting the data precision using fixed-point hardware will be discussed in Chapter 18.

4.7.3 Firmware Design Flow for One Application

The firmware design flow was introduced briefly as part of Figure 4.7. The firmware design starts from the behavioral modeling of all involved applications. The behavioral modeling includes algorithm selection and coding using high-level behavior language. Based on the behavioral model, the firmware shall be implemented following hardware constraints. The hardware constraints are divided into three groups: the finite-length hardware constraints, the memory constraints, and the speed or

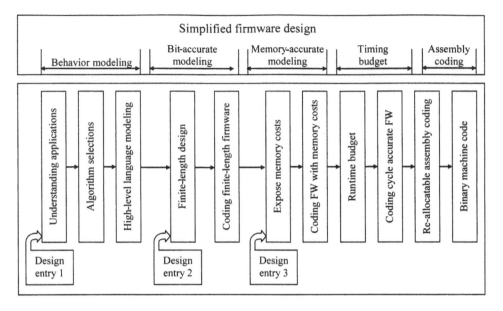

FIGURE 4.11

Detailed firmware design flow.

execution time constraints. To follow these constraints, the behavioral model is translated down to a bit-accurate model to adapt to the finite-length hardware. A memory-accurate model may also be developed to adapt to the memory resources and memory-bus architecture of the processor if the memory addressing for parallel computing is complicated.

The firmware design flow is further refined to Figure 4.11 based on Figure 4.7. The firmware design can have different starting points. Examples starting from design entry 1 to 3 in Figure 4.11 are discussed in the following text.

Following design entry 1, a design starts from paper specification. In this case, we have the freedom to select algorithms. For example, during the implementation of radio baseband signal processing, we have the freedom to select either frequency-domain algorithms or time-domain algorithms under constraints of a standard.

However, if there is no freedom to select algorithms, a design starts from design entry 2. A typical example is video decoding for H.264 or audio decoding for MP3. Algorithms and C code are available from the standard. The C code is based on floating-point with excessive high data precision. For embedded applications, low precision hardware will be used. The design following entry 2 starts from bit-accurate modeling. In this case, the freedom to decide the data precision is available; data can be scaled and masked during processing. For example, freedom to decide the data quality is available for noise suppression and echo cancellation algorithms.

Freedom of data precision control may not be available from some designs. When the freedom is not available, the firmware design entry is (3) in Figure 4.11. In this case, the bit-accurate model is available from the standard committee; for example the implementation of some voice CODEC (coder and decoder) starts from available bit-accurate C code.

A memory-accurate model is essential for parallel computing using parallel datapath. The memory costs are exposed after inserting hardware constraints (limited sizes of physical memories assigned for the application) into the source code during the profiling. The memory-accurate model specifies the parallel addressing of parallel computing. The size of the memory is also specified during memory-accurate modeling.

The firmware has runtime constraints when processing for streaming data. From profiling of the source code, runtime cost can be estimated. If the runtime cost is less than the arriving time period of input signal, the system implementation is feasible or finally the C code is qualified. Otherwise, enhancing hardware performance is required.

Figure 4.8 presents two ways of translating C code to assembly code. Assembly code implementation and verification will be discussed in Chapter 18.

4.7.4 Firmware Design Flow for Multiapplications

Running two or more real-time signal processing applications in one DSP processor, the usage of the hardware, and time resource are decided by running a program above the application programs. In a GPP, this program is called the operating system (OS). In a DSP as a slave processor, this program is relatively simple and usually is not called an OS, since it contains only a task scheduler. When a DSP is running the task scheduler instead of an application, we say the processor is running under super mode.

When running multiple applications, the application with the shorter streaming period holds higher priority than the priorities of the applications with longer streaming periods. The application with higher priority can interrupt applications with lower priority. The signal processing task without a streaming timing feature has the lowest priority. The total computing load of applications plus the computing load of the super mode must be less than the computing capacity of the processor.

As soon as an interrupt comes to the task scheduler, the scheduler (or the interrupt handler) compares the priority of the interrupt with the priority of the current running application. If the interrupt has higher priority, the scheduler accepts the interrupt and runs the application as an interrupt service. Meanwhile, if there is a new interrupt with even higher priority, the new interrupt will be served immediately. After the interrupt service, the highly prioritized application releases the hardware resource to the interrupted application program with lower priority.

Most schedulers for streaming signal processing are based on static scheduling. In such cases, a static scheduler (offline scheduler) schedules and optimizes the execution of multiple applications before runtime. The scheduling result is a timing budget that will be used as the input of the runtime scheduler.

4.8 CONCLUSIONS

The ASIP design flow and DSP ASIP HW/SW codesign are introduced briefly in this chapter. The flow can be divided further into the flow of architecture selection, the flow of assembly instruction set design, (including design of programming tools), the flow of microarchitecture design, and the flow of firmware design. These four flows briefly introduced in this chapter will be discussed further in later chapters throughout the book.

EXERCISES

4.1 What are the inputs and outputs of the following design steps?

    a. Source code profiling for ASIP design.

    b. Architecture design.

    c. Assembly instruction set design.

    d. Microarchitecture design.

4.2 What is HW/SW codesign, and why is HW/SW codesign difficult?

4.3 When codesigning HW/SW to optimize an ASIP based on the flow in Figure 4.4, what happens when moving a piece of HW to SW or moving a piece of SW to HW?

4.4 What are the two general ways to make an architecture-level decision?

4.5 When selecting the reference architecture depicted in Figure 4.5, what is the definition of data complexity, and what is the definition of control complexity?

4.6 What should be done on the behavior source code before the C compilation?

4.7 Discuss the relation between software design flow and software design tools in Figure 4.8. What design tools are required to design an assembly library?

4.8 There are three steps to specify the microarchitecture of an ASIP; what are they?

4.9 There are several design entries in Figure 4.11; discuss the reason. Select a correct design entry for designing a voice CODEC based on a standard with bit-accurate specifications.

4.10 Which of the following statements are correct? Architecture design of a DSP core:

    a. Should be conducted in parallel with the instruction set design.

    b. Decides the memory bus architecture according to move instructions.

   c. Decides modules and interconnection of modules inside the core.

   d. Specifies the way of implementation of each module.

4.11 Which of the following statements are correct? Instruction set design of a DSP core:

   a. Is before application code profiling and after ASM benchmarking.

   b. Is before ASM benchmarking and after application code profiling.

   c. Includes also the specification of hardware module functions.

   d. Gives decisions on how many operands can be carried by an instruction.

4.12 Further reading for Ph.D. students: reference papers [9] and [19].

REFERENCES

[1] Wolfe, A., Bretemitz, M., Jr., Stephens, C., Ting, A. L., Kirk, D. B., Bianchini, R. P., Jr., Shen, J. P. (1988). The White Dwarf: A High-Performance Application-Specific Processor. 15 Computer Architecture Conference.

[2] Malik, K. K. S., Newton, A. R. (2002). From ASIC to ASIP: The next design discontinuity. 2002 IEEE International Conference on Computer Design: VLSI in Computers and Processors ICCD '02.

[3] Balsamo, S., Di Marco, A., Inverardi, P., Simeoni, M. (2004). Model-based performance prediction in software development: A survey. *IEEE Trans. on Software Engineering* **30**(5).

[4] Karuri, K., Al Faruque, M. A., Kraemer, S., Leupers, R., Ascheid, G., Meyr, H. (2005). Fine-grained application source code profiling for ASIP design. DAC 2005, June 13–17, 2005, Anaheim, California, USA.

[5] Qin, W., Rajagopalan, S., Vachharajani, M., Wang, H., Zhu, X., August, D., Keutzer, K., Malik, S., Peh, L -S. (2002). Design tools for application specific embedded processors. EMSOFT '02.

[6] Sun, F., Ravi, S., Raghunathan, A., Jha, N. K. (2003). A scalable application-specific processor synthesis methodology. ICCAD 2003, San Jose, CA. USA.

[7] Kucukcakar, K. (1999). An ASIP design methodology for embedded systems. CODES 99, Rome, Italy.

[8] Cousin, J. -G., Sentieys, O., Chillet, D. (2000). Multi-algorithm ASIP synthesis and power estimation for DSP applications. IEEE ISCAS 2000, Geneva, Switzerland.

[9] Fisher, J. A. (1999). Customized instruction-sets for embedded processors. DAC 99, New Orleans, Louisiana, USA.

[10] Wolf, W. (1994). Hardware-software co-design of embedded systems. *Proceedings of the IEEE, July 1994.*

[11] Eilert, J., Wu, D., Liu, D. (2007). Efficient complex matrix inversion for MIMO software defined radio. IEEE ISCAS, USA.

[12] Liu, D., Tell, E. (2005). *Low power Baseband Processors for Communications, Low Power Electronics Design,* C. Piguet, ed., Chapter 23. CRC Press.

[13] Krishna, C. M., Shin, K. G. (1997). *Real-time systems.* McGraw-Hill.

[14] Laplante, P. A. (1997). *Real-time systems design and analysis*, 3rd ed. IEEE Press and Wiley.

[15] Embree, P. M. (1995). *C Algorithms for real-time DSP*. Prentice Hall.

[16] Kuo, S. M., Gao, W. -S. (2005). *Digital signal processors, architectures, implementations, and applications*. Prentice Hall.

[17] Potkonjak, M., Rabaey, J. (1999). Algorithm selection: A quantitative optimization-intensive approach. *IEEE Transactions on Computer-Aided Design of Integrated Circuits and Systems*, 524–532.

[18] Paulin, P. G., Liem, C., Cornero, M., Nacabal, F., Gussens, G. (1997). Embedded software in real-time signal processing systems: Application and architecture trends. *Proceedings of the IEEE, March 1997*.

[19] Vachharajani, M. (2004). Microarchitecture modeling for design space exploration. Ph.D. thesis from Department of Electrical Engineering, Princeton University.

[20] Flodal, O., Wu, D., Liu, D. (2006). Configurable CABAC encoder for multi-standard media compression. *Proc. RAW06*.

[21] Kamuf, M., Anderson, J. B., Öwall, V. (2004). A simplified computational kernel for trellis-based decoding. *IEEE Communications Letters* 8(3), 156–158.

[22] Nordqvist, U. (2004). Protocol processing in network terminals. Ph.D. thesis, Linköping Studies in Science and Technology, Dissertation No. 865. Linköping, Sweden.

[23] http://www.adi.com.

[24] www.spec.org.

A Simple DSP Core—The **Junior** Processor

5

Many students and young engineers may bring forward questions such as: How is an instruction set designed, and why is it designed in that way? In which circumstances should a function be implemented using an instruction instead of a subroutine? Designing an instruction set is based on experience, and first we should identify a suitable starting point.

The assembly instruction set of a simple DSP processor, **Junior**, will be designed in this chapter. The **Junior** instruction set is based on fundamental C operators, the language most frequently used for behavior modeling of embedded systems. After the benchmarking of this instruction set, its drawbacks will be identified, which will provide us with hints to design better DSP processors.

5.1 JUNIOR—**A SIMPLE DSP PROCESSOR**

Basic knowledge of logic design and computer architecture is necessary before reading this book. You also should have basic knowledge of RISC architecture and instruction sets, as well as a basic understanding of functional units in a RISC processor core.

Let us start from the architecture discussed in Figure 3.8 in Chapter 3. Following this architecture and the fundamental knowledge that can be obtained from classic computer architecture books, a DSP processor based on a simple instruction set will be proposed in this chapter as the starting point of our grand tour in the area of DSP design. This simple DSP processor is named **Junior**, and is only a preliminary one with obvious weaknesses. Therefore, the identified weakness of **Junior** can be used to guide our further learning through the book. Of course, this chapter is written following the pedagogy point of view; therefore it is not prepared, nor is it necessary, for experienced DSP processor designers.

Since most DSP behavior models (the source codes of applications) are based on the C language, the **Junior** instruction set intuitively resembles the primitive operators of C. We will find what is missing of the instruction set just from the mapping of C-primitive operators. The missing parts or weaknesses of **Junior** will be identified after benchmarking the **Junior** instruction set. The weaknesses identified will be used to promote the quality of DSP processor design.

5.2 INSTRUCTION SET AND OPERATIONS

Following a conventional RISC processor architecture, assembly instructions are classified into three groups: load/store and move instructions, arithmetic instructions, and program flow control instructions. The definition depicted in Table 5.1 is used to regulate the instruction format. Each row in the table tells how to describe a group of instruction. Each column in the table describes part of the specification of an instruction.

In Table 5.1, instructions are classified into three instruction groups. CC is the execution cycle count. The mnemonics of an instruction (not available in the table) is part of the verb of the operation. In the operands part, either register or data memory addressing (or simply addressing) mode shall be described. The operands part also describes how to fetch and calculate the target address for jump instructions. The function of an instruction will be described in plaintext and mathematical notations. Flag computing should be described for each instruction together with cycle cost and the usage of pipeline.

We simply assume that the pipeline depth of **Junior** is three including instruction fetch, instruction decoding, and execution. The jump taken cost is therefore three clock cycles (two clock cycles without computing the result after jump).

5.2.1 Load/Store Instructions

Let us assume that we are going to use the most popular architecture in Figure 3.2 of Chapter 3 to build a simple DSP core issuing one instruction per clock cycle. Figure 5.1 is the simplification of Figure 3.2. Two data memories are needed for digital signal processing: one for supplying data and the other for supplying coefficients.

Table 5.1 Instruction Groups.

Instruction group/type	Operand	Operation	Mathematical description	Flags	CC
Load, store, and move	Register name and memory addressing	Data transfer and addressing modes	DST (ADR) <= SRC (ADR)	No flag	1
ALU instructions	Register names or immediate data	Arithmetic and logic operations	DST (ADR) <= OA op OB	ALU Flag	1
Flow control	Way to get target address	Jump taken decision	If condition true PC <= target address	No flags	1 or 3

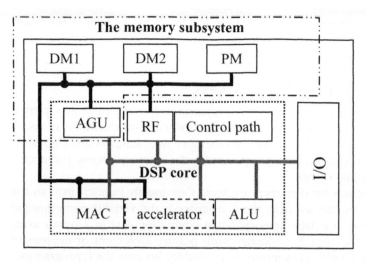

FIGURE 5.1

A simple single-issue DSP architecture.

	Mnem	**Operand**	**Description**	**Operation**	**Flags**	**CC**
Table 5.2			Load/Store and Move Instructions.			
1	Load	Rd, DA	Load data from memory 0/1	Rd ← DM(DA)	no	1
2	Store	DA, Rs	Store data to memory 0/1	DM(DA)← Rs	no	1
3	move	Rd, Rs	Move between two registers	Rd ← Rs	no	1
4	move	Rd, K	Move immediate data to a register	Rd ← immediate	no	1

Obviously, in order to access two data memories simultaneously, we need two address generator units (AGU).

Being a RISC processor keeps the architecture simple. Data and parameters of a subroutine are loaded to the register file first. Operands of RISC arithmetic instructions are supplied by the register file or immediate data carried by an instruction. After data processing, results in the register file need to be moved back to the data memory in order to save the storage space of the register file for further data computing. Since load/store instructions are responsible for setting up the link between the register file and data memories, they are essential for a RISC processor to prepare data for computing and to store the results.

A register in Table 5.2 could be either a destination register (Rd) or a source register (Rs). Both Rd and Rs are registers in the general register file. We simply specify that the processor core contains 32 general registers requiring 5 bits to address each of them. The suitable size of a register file will be discussed later in the book. DA in Table 5.2 stands for data memory address; we assume that DA is available when running a load/store instruction.

5.2.2 Addressing for Data Memory Access

Data memory addressing (or simply addressing) is an addressing algorithm carried by an assembly instruction. The algorithm specifies the way to calculate the memory address (the location) of the data in the data memory. According to this algorithm, a unique location in the data memory is pointed to for memory access. The access could be either a read or a write.

Addressing algorithms for data memory access are mostly implicit in C to the programmers. What we can see from the C language is the variable name instead of the variable address in a data memory. When computing data in a vector array, the data position in the array is associated with pointer increment or decrement operations. This means address computing in C is hidden because the C programmer does not need to manually manage the physical address of memories.

However, in a DSP processor, the physical address must be managed by the assembly language programmer. An addressing algorithm will be specified according to the requirements of DSP algorithms such as the size of the data memory and of the program memory. DSP algorithms to be executed on **Junior** processor were described in Chapter 1 following BDTI benchmarks (Berkeley Design Technology Incorporation, a company supplying benchmarks of DSP processors to the DSP market [3]). To run all these algorithms, we may need advanced addressing algorithms to enhance the performance. So far, however, we do not yet know what kind of advanced addressing algorithm should be used. At least, we need direct addressing if the address is already known; we need register indirect addressing because usually the result of address computing is stored in a general register; we need postincremental addressing for array data access during vector computing and stack push operations; and we finally need predecrement for stack pop. The minimum set of addressing algorithms is listed in Table 5.3.

In Table 5.3, DA denotes the name of the addressing algorithm in the assembly code; D is an immediate address value; R is a register in the general register file. The third column in the table exposes the code cost for memory addressing; the fourth column denotes which of the data memory is addressed; the fifth column describes the addressing algorithm to execute.

5.2.3 Instructions for Basic Arithmetic Operations

By checking the basic arithmetic operations in C, we find that those essential arithmetic operations are $+$, $-$, $\times$, $/$, and %. The modulo operation % is not used very often for DSP arithmetic computing; therefore we decide to implement the modulo function using a subroutine. From basic logic design books, we know that the division operation "/" is not easy to implement in hardware. To simplify the instruction set, the division function will be implemented using software.

Table 5.3 Minimum Set of Addressing Modes.

Name	DA	DA code cost (b)	Memory	Algorithm
Direct	D	16	DM0/1	16-bit constant as the direct memory address
Register indirect	R	5	DM0/1	A register containing the memory address
Register incremental	R++	5	DM0/1	R gives address, and R = R + 1 after addressing
Register decrement	−−R	5	DM0/1	R = R−1 before addressing, R gives address

In C, there is no dedicated MAC (multiplication and accumulation) operation. However, MAC is intuitively the most used DSP arithmetic operation, which will be carried out by a single instruction of **Junior**. A 32-bit accumulation register A is therefore necessary to store the result of the 16-bit multiplication. Recall what was discussed in Chapter 2—guard bits are required for the iterative MAC computing. In **Junior**, eight guard bits are used to support long iteration. The length of the accumulation register A is therefore 40 bits. Finally, round and saturation instruction is necessary when moving the result from the accumulator register A to a 16-bit general register. Behavior functions of guarding, saturation, and round can be found in Chapter 2.

Nine fundamental arithmetic operations are given in Table 5.4.

In both Table 5.4 and Table 5.5, flags and other notations are specified in the following list:

- V: Overflow Flag.

- Z: Zero Flag.

- N: Negative Flag (the guard bit of the result).

- Rd: Destination register in the register file.

- Ra: Operand A from the register file.

- Rb: Operand B from the register file.

- Rs: Source register.

5.2.4 Logic and Shift Operations

Shift operations are part of the arithmetic and logic operations. Again by checking the basic logic and shift operations in C, we can find that the essential logic operations are &(and), |(or), ~(not), ^ (xor), << (left shift), and >> (right shift).

Table 5.4 Basic Arithmetic Instructions.

	Mnem	Operand	Description	Operation	Flags	CC
1	ADD	Rd, Rr	Add	Rd ← Ra + Rb	Z, N, V	1
2	SUB	Rd, Rr	Subtract	Rd ← Ra − Rb	Z, N, V	1
3	ABS	Rd, Rr	Absolute operation	Rd ← ABS(Ra)	Z, N, V	1
4	INC	Rd	Increment	Rd ← Ra + 1	Z, N, V	1
5	DEC	Rd	Decrement	Rd ← Ra − 1	Z, N, V	1
6	MPL	A, Rd, Rr	Multiplication	A ← Ra × Rb	Z, N, V	1
7	MAC	A, Rd, Rr	Multiplication and accumulation	A ← A + Ra × Rb	Z, N, V	2
8	RND	Rd, A	Round, saturate, and truncate	Rd ← Saturate(Round(A))	Z, N, V	1
9	CAC	A	Clear an accumulator	A ← 0	Z, N, V	1

Table 5.5 Logic and Shift Operations.

	Mnem	Operand	Description	Operation	Flags	CC
1	AND	Ra, Rb	A logic-and B	Rd ← Ra and Rb	C, Z	1
2	OR	Ra, Rb	A logic-or B	Rd ← Ra or Rb	C, Z	1
3	NOT	Ra, Rb	Invert A	Rd ← INV (Ra)	C, Z	1
4	XOR	Ra, Rb	A logic-xor B	Rd ← Ra xor Rb	C, Z	1
5	LS	Ra, Rb	Logic left shift	Rd ← Ra left shifted by Rb [3:0]	C, Z	1
6	RS	Ra, Rb	Logic right shift	Rd ← Ra right shifted by Rb [3:0]	C, Z	1

Here "and" operates on each bit of operand A and B; that is, $C[0]=A[0]$ & $B[0]$, $C[1]=A[1]$ & $B[1]$, ... $C[15]=A[15]$ & $B[15]$. In the same way, or, not, and xor are bit-level operations on the operand word. Zeros will be filled in for logic shift operations; for example, logic right-shift 4 bits is: $A[15:0] = \{4'b0, A[15:4]\}$. 4'b0 stands for four binary zeros. {A, B} stands for a concatenation of A and B.

5.2.5 Program Flow Control Instructions

There are conditional and unconditional program flow controls in C. Unconditional program flow control is simply goto operations. Conditional program flow control depends on the test of conditions. Condition test and jump operations in

C are usually integrated, for example, if A then B else C. In an assembly language, condition test and condition jump usually are separated into two instructions: the first instruction offers arithmetic and flag computation, and the second instruction is the conditional jump with jump decision based on flags and the jump operation, see Table 5.7. The arithmetic computation of flags are separated from jump condition tests because of hardware limitations. Condition tests in C include the logic operators in Table 5.6.

Table 5.6 Logic Operators.

	Condition symbol	Conditions
1	<	Less than
2	<=	Less than or equal to
3	==	Equal to
4	>=	Greater than or equal to
5	>	Greater than
6	!=	Not equal to
7	&&	Boolean AND
8	||	Boolean OR
9	!	Boolean NOT

Table 5.7 Logic and Shift Operations.

	Mnem	Description	Conditions	Flags meet	CC
1	JLT	Jump when Less than	<	$N = 1$	1
2	JLE	Jump when Less than or Equal to	<=	$N = 1$ or $Z = 1$	1
3	JEQ	Jump when Equal to	==	$Z = 1$	1
4	JGE	Jump when Greater than or Equal to	>=	$N = 0$	1
5	JGT	Jump when Greater than	>	$N = 0$ and $Z = 0$	1
6	JNE	Jump when Not Equal to	!=	$Z = 0$	1
7	JUMP	Unconditional jump			1
8	CALL	Jump, push return address into stack			1
9	Return	Return to the stacked address			1

Table 5.8 Target Address (TA) of a Jump.	
TA	**Algorithm**
Absolute	16 bits constant
Relative	In a general register

Program flow control instructions can be divided further into unconditional branches, conditional branches, and others. Unconditional branch instructions include JUMP, CALL, and RETURN. Conditional branch instructions are jumps when the condition flags fulfill certain specified values.

For example, JLT is a conditional jump instruction; the jump is taken when flag $N = 1$. JLT stands for a jump when the flags of previous result meet the condition "less than." In the same way, JGE stands for a jump when the flags of the previous result meet the condition "greater than or equal to" or the flag "$N = 0$." In the same way, JNE stands for a jump when the flags of the previous result meet the condition "not equal to" or the flag "$Z = 0$."

A call instruction jumps to the target address and pushes the return address as the current PC + 1 to the stack. The last instruction in the call subroutine is "return." After executing the call, the subroutine pops out the stacked address to PC, and the program continues at the break point PC + 1.

TA is the target address of a jump. When a jump is taken, the next PC will be TA instead of PC + 1. Two essential operations are included in a jump instruction: the jump decision (decide whether the jump is taken by checking flags) and the computing of the target address. Checking the C language, we find that the jump TA could be a predefined address before runtime (when the label as the target address is a constant) or a variable (the label is a variable). The programmer uses predefined TA (called absolute TA) if the TA can be decided before runtime. A TA can be a variable, for example, if the algorithm selection is based on the input data or the result see Table 5.8.

Because the word size in a general register is 16 bits, it is easier to see that the addressing space of the program memory is $64k = 2^{16}$.

5.3 ASSEMBLY CODING

Assembly coding includes coding of the assembly language and coding of the binary machine codes. The coding of the **Junior** instruction set shall be given intuitively, and the assembly language of **Junior** will be used only for benchmarking. The coding format therefore is simplified to:

```
Mnemonic Destination, Operand A, Operand B, Others
```

The assembly coding of all **Junior** assembly instructions can therefore be specified using Table 5.9.

Table 5.9 Junior Assembly Instruction Set Summary.

No.	Mnemonic	Operation	No.	Mnemonic	Operation
1	Load	Rd ← DM(DA)	15	OR	A logic-or B
2	Store	DM(DA) ← Rd	16	NOT	Invert bits in A
3	Move	Rd ← Rr	17	XOR	A logic-xor B
4	Move	Rd ← constant	18	LS	Logic left shift
5	ADD	Add	19	RS	Logic right shift
6	SUB	Subtract	20	JLT	Jump when Less than
7	ABS	Absolute operation	21	JLE	Jump when Less than or equal to
8	INC	Increment	22	JEQ	Jump when Equal to
9	DEC	Decrement	23	JGE	Jump when Greater than or equal to
10	MPL	Multiplication	24	JGT	Jump when Greater than
11	MAC	Multiply & accumulate	25	JNE	Jump when Not equal to
12	RND	Round and saturate	26	JUMP	Unconditional jump
13	CAC	Clear an accumulator	27	CALL	Call
14	AND	A logic-and B	28	Return	Return using stacked PC address

Instruction binary coding is to give each assembly instruction a unique binary code so that the machine hardware can store it, recognize it, and execute it. Intuitively, the longest binary code will be identified and will be the size of the binary code. Here the longest binary code is the machine code of a specific instruction. It requires the longest binary string, compared to other instructions, to carry all information of the instruction necessary for instruction decoding and execution.

To decide the memory size in **Junior**, we try to make things as simple as possible. Since a data word in **Junior** is 16 bits in both the memory and the register file, the addressing cost will be minimized if the memory space of each data memory is not more than $64k = 2^{16}$. Based on all the assumptions just given, the direct memory addressing algorithms can be specified in Table 5.10.

Based on the address space limitation of 64k, the longest binary machine code might be from the instruction of load/store using direct memory addressing mode. The code cost for "type" is 2 bits covering three kinds of instructions (load/store, arithmetic, and flow control). The OP in Table 5.10 is "operation", and the code cost is 2 bits because the number of load/store instructions in Table 5.2 is $4 = 2^2$. Because there are four addressing modes in Table 5.3, the code cost of the DA is 2 bits. To point to one of the two data memories, one binary bit is required. Finally, a general register should be pointed to as the source register for loading the input data and the destination register for storing the result, consuming 5 bits. The binary code length of a load/store instruction is $= 2 + 2 + 2 + 1 + 5 + 16 = 28$ bits in Table 5.10.

Another possible longest binary code is the instruction to move a 16-bit immediate data to a general register in the register file. In this instruction, the immediate data consumes 16 bits. The possible code is shown in Table 5.11.

Obviously, the code cost is $2 + 2 + 5 + 16 = 25$, which is not longer than the binary code in Table 5.10. Yet another possible longest binary code is the conditional instruction using absolute target address. The possible code is shown in Table 5.12.

In Table 5.12, the OP code uses 4 bits because the number of flow control instructions is 9 and $2^3 < 9 < 2^4$. TA represents the target addressing mode. It could be absolute TA $= 0$ or relative TA $= 1$. The code cost of the instruction in Table 5.12 is $2 + 4 + 1 + 3 + 16 = 26$ bits.

Since other instructions use shorter binary code than that of the load/store instruction with direct memory addressing mode, the length of the binary code of **Junior** is 28 bits.

However, the length of the machine binary code could be predefined. For example, if the longest binary code cannot be more than 24b, the length of the immediate data for direct memory addressing will not be longer than $24 - 5 - 1 - 2 - 2 - 2 = 12$

Table 5.10 Coding a Load/Store Instruction Using Direct Memory Addressing.

	Type	OP	DA	Data memory 0/1	Destination/source	Immediate
Number of bits required	2	2	2	1	5	16

Table 5.11 Coding of Move Immediate.

Type	OP	Destination	Immediate
2	2	5	16

Table 5.12 Coding of a Conditional Jump to an Absolute TA.

Type	OP	TA	Jump conditions	Immediate
2	4	1	3	16

Table 5.13 Binary Machine Code Coding Table.

Instructions	Mnemonics (type+OP)	Memory (target)	DA (condition)	Destination (address R)	Source a	Source b	Immediate	Total
Load	4b	1b	2b	5b			16b	28b
Store	4b	1b	2b		5b		16b	28b
Move	4b	1b	2b	5b	5b			17b
Move	4b			5b			16b	25b
ALU	6b			5b	5b	5b		21b
JUMP	6b	1b	3b				16b	26b
JUMP	6b	1b	3b	5b				15b
CALL	6b	1b					16b	23b
CALL	6b			5b				11b
Return	6b							6b

bits. In this case, the direct memory addressing will not be able to cover the entire addressing space.

There are 28 instructions listed in Table 5.9, and the binary coding cost is listed in Table 5.13.

5.4 ASSEMBLY BENCHMARKING

Benchmarking or benchmark test is a process to test the performance of the HW, SW, or the quality of the compiler targeting the processor based on a specified set of program, data, and given configurations. DSP benchmarking generates information such as the cycle cost and code size of the assembly code of frequently used DSP algorithms with single-precision data. Following the convention of DSP benchmarking, round operation is required before moving long data from an accumulation register to a general register. Benchmarking details will be discussed in Chapter 9. Professional DSP benchmarking can be found in reference [3], and the essential DSP algorithms are listed in Table 5.14.

The benchmarking of the **Junior** assembly instruction set follows the benchmarking convention proposed by BDTI. It measures the execution time (cycle cost), the code size (program memory cost), and the cost of data memories. The cycle cost of the benchmarking will be divided further into the cycle cost of prologue-epilogue and that of the kernel subroutines. Prologue is the part of the program preparing for running a program, and epilogue is the part of the program terminating the

Table 5.14 Essential DSP Benchmarking Algorithms.

Algorithm Kernel	Description
Block transfer	Transfer a data block from one memory to another memory
Single FIR	N-tab FIR filter running one data sample
Frame FIR	N-tab FIR filter running K data samples
IIR	Biquad IIR (2nd order IIR) running one data sample
16-bit division	A positive 16-bit value divided by another positive 16-bit value
Vector add	$C[i] = A[i] + B[i]$, $i = 0 \ldots N - 1$ (not included in this chapter)
Windowing	$C[i] = A[i] * B[i]$, $i = 0 \ldots N - 1$ (not included in this chapter)
Vector Max	$R <= MAX \{A[i]\}$, $i = 0 \ldots N - 1$
DCT	8X8 2D Discrete Cosine Transform
256p complex FFT	256 point FFT including all computing and addressing

program after its execution. In the same way, the code cost (program memory cost measured by number of lines of instructions) will also be measured including the kernel code cost and code cost of the prologue-epilogue. To separate the prologue-epilogue cost from the total cost, overheads of running algorithms will be exposed. The result of benchmarking will be shown in later tables in the chapter. Complete assembly code for benchmarking can be found on the web page of this book.

The benchmarking in the following subsections will give the exact cycle cost of several DSP kernel algorithms. The cycle cost of prologue and epilogue will only be roughly estimated because it is relatively low. The benchmarking follows these definitions:

- The size of a data frame is 40 data words (40 samples).

- The number of taps of a FIR will be 16.

- The cycle cost is one clock cycle per instruction except for jump taken, which consumes three cycles.

- MAC takes one clock cycle if the following instruction does not use the data in an accumulator register.

- The **Junior** benchmark will be compared with the benchmark of a typical single MAC DSP (TSMD) processor available as a COTS (commercial off-the-shelf) component.

- Finally, the difference exposed from comparing benchmarks will be discussed.

Table 5.15   Junior Block Transfer (BT) Benchmark Results.

Processor	Algorithm	Total cycle cost	Pro-epilogue cycle cost	Kernel cycle cost	Total code cost	Code for pro-epilogue	DM Cost
Junior	BT	242	4	238	9	4	84
TSMD		47	4	43	7	4	84

5.4.1 Benchmarking of Block Transfer

The benchmark moves a data array from one place in a memory to another place in another memory. The purpose of running this benchmark is to test the load/store performance of a data array. This benchmark frequently is used when preparing for an autocorrelation algorithm.

In this benchmark, 40 data words will be moved from DM0 to DM1 by running the following subroutine. The moving process is shown in Listing 5.1.

LISTING 5.1   Junior block transfer kernel.

```
Loop:   Load R0,DM0(R1++)   ; R1: Source address pointer
        Store DM1(R2++),R0   ; R2: Destination address pointer
        DEC R3               ; R3: loop counter, initial value is 40
JGT Loop                     ; Jump cost 3 cycles, no jump cost 1
                             ; cycle
```

There are 39 jumps taken while transferring the first 39 data words. After transferring the last data word, there will be no jump. Therefore the cycle cost of the kernel is $39 \times 6 + 4 = 238$ clock cycles. The prologue of the subroutine includes loading the program parameters (segment address of the parameters, address pointer 1, address pointer 2, and loop size), and the cost of the prologue and epilogue is four clock cycles using four instructions.

The cycle cost of data block transfer between data memories in a typical single MAC DSP (TSMD) processor will be $7 + N$ clock cycles. When $N = 40$ in this example, the cost will be 47 clock cycles, see Table 5.15. Opportunities for improvement are:

- The loop: The extra cost of each jump taken and DEC of the loop counter consumes four clock cycles. HW loop may eliminate the cost.

- Load and store can be merged to a memory move to memory instruction.

5.4.2 Benchmarking of Single-Sample FIR

The benchmark of single-sample FIR is used mostly for measuring basic DSP performance. The benchmarking runs a single-precision data sample using a 16-tap FIR filter to get one result sample with single precision. The kernel program is shown in Listing 5.2.

LISTING 5.2 Junior Single-Sample FIR kernel.

```
Loop:    SUB R6,R0,R5      ; R5 is the top limiter of the data FIFO
         JNE FIFO          ; Check if data pointer R0 reached FIFO
                             top
         MOVE R0,R7        ; Move bottom pointer R7 to R0
FIFO:    Load R1,DM0(R0++); R0:data address pointer, R1:data
         Load R3,DM0(R2++); R2;data address pointer, R3:tap
         MAC A,R1,R3       ; A=A+R1*R3
         DEC R4            ; R4: loop counter
         JGT Loop          ; Jump cost 3 cycles, no jump cost 1
                             cycle
```

In the subroutine, there is a FIFO as the FIR computing buffer defined by its top limiter register R5 (R5 = top address + 1) and bottom limiter register R7 (the bottom address). The data pointer R0 must be R7 ≤ R0 < R5. Before loading data, we have to check if the postincrement operation of the last loading has reached the boundary or not. In case R0 = R5, it means the data address pointer reached the top limiter of the FIFO; the bottom pointer should be assigned to the data pointer.

Let us define that a convolution is to run all the instructions once from the top to the bottom of the kernel code. Sixteen convolutions consist of a 16-tap FIR filter. The longest execution time of a step of convolution is 11 cycles when both jumps JNE and JGT are taken. There will be one convolution in which the data pointer reaches the top limiter while JNE is not taken. The cycle cost of a step of convolution is 10 clock cycles. Finally, the last convolution consumes nine clock cycles. To run a 16-tap convolution kernel without counting of the prologue and the epilogue, the cycle cost will be $14 \times 11 + 10 + 9 = 173$ clock cycles.

Another 12 (prologue) +7 (epilogue) cycles are required to prepare the execution of the filter including clearing the accumulation register and initializing filter parameters. Finally, to run a 16-tap single-sample FIR filter, the total cycle cost is $12 + 173 + 7 = 192$ cycles.

A TSMD processor requires $A + T + C$ clock cycles to run a T-tap FIR filter. A is the cycle cost of the filter initialization, which is about 10 cycles; T is 16 cycles; and C is the epilogue with five cycles, including storing the data pointer, round-saturating the result, sending the result, and storing the result pointer. The total cycle cost of running one sample on a 16-tap FIR filter using TSMD is $10 + 16 + 5 = 31$ cycles, see Table 5.16.

To conclude, the extra cycle cost is $192 - 31 = 161$ cycles. The cycle cost of **Junior** is 6.2 times higher than the benchmark of a TSMD. Opportunities for improvement are:

- The cost of SW emulated circular buffer and modulo addressing is high. HW circular buffer and modulo addressing is essential.

- Data and coefficient loading, MAC, and the loop control can be merged into one instruction, convolution, which is one of the most frequently used instructions in DSP.

5.4.3 Benchmarking of Frame FIR

A frame FIR processes a frame of k words data (producing k words results) using an N-tap FIR filter. Simply saying, running a frame FIR is like running a single-sample FIR k times. Because the principle of emulating modulo FIFO buffer using a random access memory is not yet introduced, the data memory organization has to be introduced briefly in Figure 5.2. A data FIFO consists 16 data samples from

Table 5.16 Junior Single-Sample FIR Benchmark Results.

Processor	Algorithm	Total cycle cost	Pro-epilogue cycle cost	Kernel cycle cost	Total code cost	Code for pro-epilogue	DM cost
Junior	16-tapFIR	192	19	173	26	18	25
TSMD	16-tapFIR	31	15	16	15	14	25

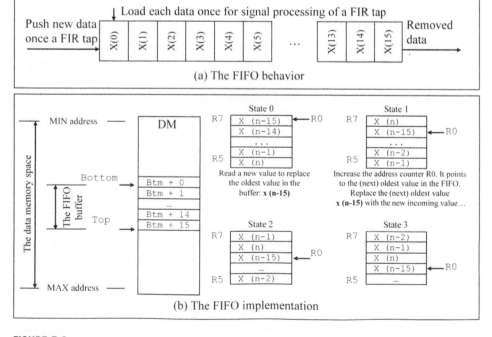

(a) The FIFO behavior

(b) The FIFO implementation

FIGURE 5.2

The procedure to insert a new data sample into the FIFO (16 data).

Btm + 0 to Btm + 15 limited by the bottom register R7 and the top register R5. R0 is the data pointer, which can point only data within the limits of R5 and R7, meaning the address range of R0 is (R7) ≤ R(0) ≤ (R5). The new incoming data replaces the oldest data in the FIFO buffer when loading it from R31 to the data memory pointed by R0. After one tap signal processing, the R0 as the data pointer points the oldest data to be replaced by the latest data. If the pointer R0 reaches the buffer limit stored in R7, the pointer R0 will jump to the top of the FIFO buffer. For each tap of FIR computing, all data in the FIFO are pointed to by R0, loaded, and processed.

The kernel code can be as shown in Listing 5.3.

LISTING 5.3 Junior frame FIR kernel.

```
Top_loop:
        SUB    R6,R0,R5       ; R5 is the top limiter of the input data
                                 FIFO
        JNE    LD             ; Check if data pointer R0 reached
                                 FIFO top
        MOVE   R0,R7          ; Move bottom pointer R7 to R0
LD:     Store  DM0(R0),R31    ; Load data from port R31 to DM0
        CAC    A              ; Clear accumulator
        MOVE   R4,16          ; N = 16 (R4: Inner loop counter)
        MOVE   R2,TAPSTART    ; TAPSTART points to the memory location
                             ; where the filter coefficients are
                             ; stored
Inner_loop:                  ; This loop is the same as the single
                             ; sample FIR example
        SUB    R6,R0,R5       ; R5 is the top limiter of the data FIFO
        JNE    FIFO           ; Check if data pointer R0 reached FIFO
                                 top
        MOVE   R0,R7          ; Move bottom pointer R7 to R0
FIFO:   Load   R1,DM0(R0++);  R0:data address pointer , R1:data
        Load   R3,DM0(R2++);  R2;data address pointer , R3:tap
        MAC    A,R1,R3        ; A=A+R1xR3
        DEC    R4             ; Decrease the loop counter
        JGT    Inner_loop

                             ; We are now done with one sample,
                             ; store it
        RND    R8,A           ; R8 = saturate(round(A))
        Store  DM0(R9++),R8;  Store Y(n) to DM0 (R9:Output buffer
        pointer) DEC R10      ; R10 is the k, the length of the data
                                 frame
        JGT    Top_loop       ; Run the next sample until k runs
```

Suppose that a data sample in the frame is available in the port register R31 (the input buffer) when a sample processing starts. While running k=40 data samples in this example, the jump JNE LD will not be taken in every 16 data samples (the FIFO buffer size). In fact, JNE LD will be taken about three times while processing 40 data samples. When the jump JNE LD is taken, the in-loop prologue cycle cost is eight cycles for each data sample. When the jump JNE LD is not taken, the in-loop prologue cycle cost is seven cycles for each data sample. The total cycle cost of the in-loop prologue will be $(40-3) \times 8 + 3 \times 7 = 317$. In a similar way, the cycle cost of the epilogue will be $6 \times 39 + 4 = 238$. The top costs level prologue and epilogue are 11 and 6 clock cycles.

The inner loop cost of a single-sample FIR is 173 cycles. The total cycle cost of the benchmark will be $173 \times 40 + 317 + 238 + 11 + 6 = 7492$ cycles, and the kernel consumes $173 \times 40 = 6920$ cycles. Here 11 and 6 are the prologue and epilogue of the top-level FIR (which are not included in Listing 5.3.). The typical BDTI cycle cost of running 40 data samples using a 16-tap FIR is:

```
CC = A + 40 ×  (B + T + C) + D
   = 9 + 40 ×  (2 + 16 + 4) + 4 = 893 cycles
```

A represents the cycles of top-level prologue; B is the in-loop prologue cycle cost of loading a data sample; T is the number of taps; C is the epilogue cycle cost of a data sample processing; and D is the top-level epilogue cycle cost of the FIR filter. TSMD processors can use only 893 cycles to process 40 data samples using a 16-tap FIR.

Comparing the cycle cost of a TSMD processor to the cost of **Junior**, the difference in cycle cost ratio is 7373/893—roughly eight times more clock cycles are consumed, see Table 5.17. The reason is obviously that the assembly instruction set of **Junior** has neither loop instruction nor modulo addressing acceleration (addressing to eliminate the overhead of FIFO dada access).

Opportunities for improvement are:

- The same as the single-sample FIR; a convolution instruction and the HW circulation buffer and modulo addressing are essential.

- The top-level loop also should be accelerated because the program inside the loop is simple.

Table 5.17 Junior Frame FIR Benchmark Results.

Processor	Algorithm	Total cycle cost	Pro-epilogue cycle cost	Kernel cycle cost	Total code cost	Code for pro-epilogue	DM cost
Junior	40 sample 16-tap FIR	7492	572	6920	19 + 17 = 36	17	65
TSMD	40 sample 16-tap FIR	893	253	640	8 + 17 = 25	17	65

5.4.4 Benchmarking of Single-Sample Biquad IIR

The function of a Biquad IIR filter is:

$$y(n) = a_0 x[0] + a_1 x[1] + a_2 x[2] + b_1 y[1] + b_2 y[2]$$

The kernel part of Biquad IIR based on the **Junior** assembly instruction set is shown in Listing 5.4.

LISTING 5.4 Junior Biquad IIR filter.

```
; For a biquad filter we can afford to move memory contents
; around by hand which will be much better on Junior than
; to try to emulate modulo addressing by hand.
;
; R0=a0; R1=a1; R2=a2; R3=b1; R4=b2
; R5=DM0 address pointer for x vector
; R6=DM1 address pointer for y vector
; 5 cycles to move coefficients to R0 ... R4.
CAC    A                     ; Must clear accumulator first
Load   R7,DM0(R5++)          ; Load x0
MAC    A,R7,R0               ; A = x0*a0
Load   R8,DM0(R5++)          ; Load x1
MAC    A,R8,R1               ; A += x1*a1
Load   R9,DM0(R5++)          ; Load x2
MAC    A,R9,R2               ; A += x2*a2

; At this point we save the x values shifted one position
Save   DM0(--R5),R8          ; Save previous x1 into x2
Save   DM0(--R5),R7          ; Save previous x0 into x1
; Old x2 is scrapped

Load   R7,DM0(R6++)          ; Load y0 (not used in this iteration)
Load   R8,DM0(R6++)          ; Load y1
MAC    A,R8,R3               ; A += y1*b1
Load   R9,DM0(R6++)          ; Load y2
MAC    A,R9,R4               ; A += y2*b2

; Now we must shift y one step'
Store  DM0(--R6),R8          ; Save previous y1 into y2
Store  DM0(--R6),R7          ; Save previous y0 into y1
RND    R7,A
Store  DM0(--R6),R7          ; Save new y0
```

By merging prologue and epilogue into the kernel code, the cycle cost of the Biquard IIR filter is **23 clock cycles**.

5.4.5 Benchmarking of 16-bit Division

The behavioral model of binary division is based on restoring division and operates on positive fixed-point fractional numbers and $N < D$, where N is the numerator (dividend) and D is the denominator (divisor). The quotient is Q and its digits are binary q. The basic algorithm for binary (radix 2) restoring division is:

```
P = N; i = n; q = 0;          // n is the number of bits
while(i > 0){
    P = 2*P - D;
    q = q « 1;
    if(P >= 0) { q += 1; }    // Shift in a one from the left
    else { P = P + D; }       // Restore P
    i-;
}
```

$2\times$ P[i] is a logic left-shift. To prevent overflow of the shift, a guard bit is required in the shift hardware. Otherwise, N must be less than 1/2. The implementation of the behavior model into **Junior** assembly code is shown in Listing 5.5.

LISTING 5.5 **Junior** assembly code for division.

```
; R0 is P;  R1 is i; R2 is D;
; R4 is Q
        MOVE   R1,16           ; Loop counter
Loop:
        ADD    R4,R4,R4        ; Left shift of q
        ADD    R0,R0,R0        ; 2*P
        SUB    R0,R0,R2        ; R0 contains TP = 2*P - D
        JLT    Q0              ; jump if negative (TP < 0)

        INC    R4              ; q += 1
        JMP    Next
Q0:
        ADD    R0,R0,R2        ; Restore P (TP = TP + D
Next:
        DEC    R1
        JNZ    Loop
```

The kernel cycle cost of the 16-bit positive fractional division is between $16 \times 12 = 192$ and $16 \times 11 = 176$ cycles. The prologue is only one instruction, and there is no epilogue.

The benchmarking result from a TSMD processor can be found from BDTI or other sources. The kernel cycle cost of running 16-bit division on a TSMD processor is about 49 clock cycles see Table 5.18. **Junior** is about a factor of 3.49 times behind a normal commercially available DSP processor. Reasons and opportunities are:

Table 5.18 Junior Positive 16-bit/8-bit Division Benchmark Results.

Processor	Algorithm	Total cycle cost	Pro-epilogue cycle cost	Kernel cycle cost	Total code cost	Code for pro-epilogue	DM cost
Junior	16b division	182–199	5	176–192	10	5	5
TSMD	16b division	57	6	49	∼ 10	6	5

- As usual, hardware acceleration of running a small simple loop can dramatically enhance the performance.

- Conditional filling in of the shift-in bit for the logic left-shift is essential for division computing. This instruction involves the following operations: if greater than, shift in "1" without changing flags, else shift in "0" without changing flags.

- If the left-shift instruction does not change the flags, the flag from the computing of $P = 2 \times P - D$ can be used and the instruction $P = P + D$ can be executed using one clock cycle; that is: $P = P + D$ if LT. (LT: less than)

- If the operand preoperation, left-shift one bit on operand A is available, two instructions will be merged and $P = 2 \times P - D$ will be executed using one instruction.

5.4.6 Benchmarking of Vector Maximum Tracking

The behavior of maximum tracking is to find the maximum value in a data array. Let us check the kernel cycle of running maximum data tracking of 40 data words on **Junior**.

LISTING 5.6 **Junior** vector maximum tracking.

```
      ; Inputs: R1 points to data array
      ; Outputs: R2 contains address of maximum value
      ;          R4 contains the maximum value
      Move  R6,40         ; initialize the loop counter R6 with 40
      Move  R4,0x0000     ; initialize R4 with Hexadecimal ''zero''
LOOP:
      Move  R3,DMO(R1++); load new value
      SUB   R5,R4,R3      ; Check if the new value is larger
      JGT   Jp_1          ; Jump if not larger
      Move  R4,R3         ; Replace max with new value
      Move  R2,R1         ; Save the address of the new value
```

```
Jp_1:
        DEC     R6          ; R6 is loop counter
        JGT     LOOP        ; loop of vector addition
```

The kernel cycle cost is $9 \times 39 + 7 = 358$ clock cycles for running maximum data tracking from an array of 40 data words. Running the same kernel algorithm on a TSMD processor, the cycle cost is about 41 clock cycles. **Junior** is about $358/41 = 8.7$ times slower, see Table 5.19. The reasons and opportunities are:

- Hardware loop is essential for any vector computing.

- Conditional execution is essential for If True A else B.

- A <= max{A, B} and A <= max{ABS(A), ABS(B)} are essential.

5.4.7 Benchmarking of 8 × 8 DCT

The 8×8 DCT translates an 8×8 point two-dimensional data array from time-domain to frequency-domain or vice versa. 8×8 DCT is based on an 8-point DCT kernel. 8×8 DCT is executed by running the one-dimensional DCT kernel eight times horizontally and eight times vertically. The definition for one-dimensional DCT is:

$$F(u) = \frac{2c(u)}{N} \sum_{m=0}^{N-1} f(m) \cos\left(\frac{(2m+1)u\pi}{2N}\right) \text{ for } u = 0, 1, 2 \ldots N-1 \qquad (5.1)$$

where

$$c(u) = \begin{cases} \sqrt{2} & \text{if } u = 0 \\ 1 & \text{otherwise} \end{cases} \qquad (5.2)$$

The two-dimensional DCT transforms a two-dimensional data array into another two-dimensional data array as the output by performing multiple one-dimensional transforms.

Table 5.19   Junior Vector Maximum Tracking Benchmark Results.

Processor	Algorithm	Total cycle cost	Pro-epilogue cycle cost	Kernel cycle cost	Total code cost	Code for pro-epilogue	DM cost
Junior	MAX track	361	3	358	10	2	43
TSMD	MAX track	44	3	41	4	3	43

It is mathematically defined as

$$F(u, v) = \frac{4c(u, v)}{N^2} \sum_{m=0}^{N-1} \sum_{n=0}^{N-1} f(m, n) \cos\left(\frac{(2m+1)u\pi}{2N}\right) \cos\left(\frac{(2n+1)u\pi}{2N}\right)$$

for

$u, v = 1, 2, 3, \ldots N - 1$

where

$$c(u, v) = \begin{cases} \frac{1}{2} & \text{if } u = v = 0 \\ 1 & \text{otherwise} \end{cases} \tag{5.3}$$

Many algorithms for accelerating the computation of DCT exist. In this chapter, we use a fast DCT algorithm [5] [6], which is available in most video DSP books. Figure 5.3 is the 8-point DCT computation algorithm. In Figure 5.4:

```
xi is the out-of-order input
yi is the Bit-reversal output
ci is cos(i*pi/16)
Solid line means addition
Dash line means subtraction
```

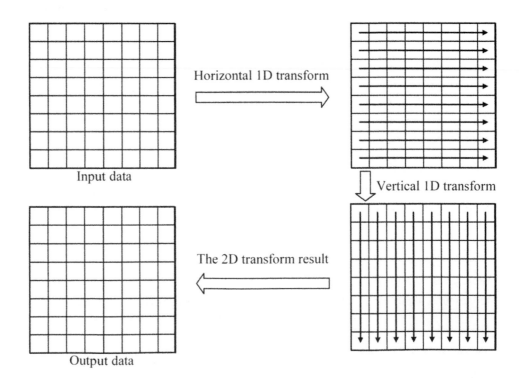

Input data

Horizontal 1D transform

Vertical 1D transform

The 2D transform result

Output data

FIGURE 5.3

2D DCT transform for image and video signal processing.

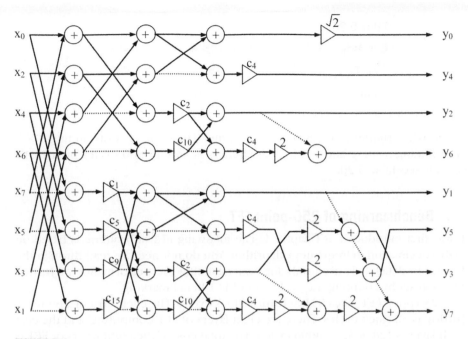

FIGURE 5.4

Signal flow diagram of 8-point 1D DSC algorithm.

The inputs are assumed to be 16-bit fractional numbers, taken from the data memory, row by row. The coefficients are stored in the tap memory in the order $c1$, $c5$, $c9$, $c13$, $c2$, $c10$, $c4$.

The program starts with the setup of the status buffer and address pointers. The DCT algorithm is executed 16 times, following Figure 5.4. The first eight DCT runs row by row, and the second eight DCT runs column by column. The kernel part of the program in Figure 5.4 starts from fetching data from the memory to the registers. The computation is performed, and the result of each row will be written back to the memory. All eight rows of the DCT kernels are calculated according to row-based addressing. Then the address pointers will be changed to perform column-based computation. The complete assembly code of the DCT is long and takes time to understand. Readers especially interested in the assembly implementation of DCT can check this book's web page.

By observing the unrolled 8-point DCT assembly code, the cycle count of the kernel operation is 123 cycles, including reading in and writing out 8-point data buffers. When running an 8×8 DCT for a DU (data unit of image and video signal processing), including eight horizontal DCT and eight vertical DCT, the total cycle cost of running kernels is $123 \times 16 = 1968$ cycles.

The overhead of 8×8 DCT can be observed. If the butterfly instruction BUTTFLY (subtraction and addition in parallel) is supported, about 50 cycles can

Table 5.20   Junior 8 × 8 DCT Benchmark Results.

Processor	Algorithm	Kernel cycle cost	DM cost
Junior	8 × 8 DCT	1986	∼ 150
TSMD	8 × 8 DCT	860	∼ 150

be saved while running each 8-point DCT kernel. The cycle cost of running 8 × 8 DCT including reading and writing data buffers is around 860 cycles on a TSMD processor, see Table 5.20.

5.4.8 Benchmarking of 256-point FFT

FFT was first introduced in Chapter 1. The following FFT subroutine is based on the DIF (decimation-in-frequency) algorithm. You do not need to read through the FFT code in detail; it is used only for understanding the performance of **Junior** instruction set by counting the cycle cost of FFT benchmarking.

The 256-point FFT takes 52329 **Junior** clock cycles. The cycle cost of a butterfly in **Junior** is 32 clock cycles. There are eight layers of FFT computing, and the cost of each layer is $128 \times 32 = 4096$ cycles. The total cost of the eight layers of FFT is $4096 \times 8 = 32768$ cycles. The overhead is $52329 - 32768 = 19601$ cycles. Extra cycle cost will be mainly consumed by the extra address computing cost of 16384 cycles including:

- A bit-reversal addressing, which uses 34 cycles; the total is $256 \times 34 = 8704$ cycles.

- Extra cost to run loops, which is around $3 \times 128 \times 8 = 3072$ clock cycles.

The minimum cost of 256-point FFT kernel running on a TSMD processor is estimated, and it can be as low as $(128 \times 14 + 10) \times 8 + 256 \times 2 + 10 = 14938$ clock cycles, see Table 5.21. Ten clock cycles are consumed by the prologue of each layer and the top of FFT. Fourteen cycles are required for processing a butterfly:

- Two data loads instructions (loading 4 data).

- One data load instruction (loading 2 coefficient).

- Eight arithmetic operations; four MAC and four long add operations.

- Two data writes instructions (write 4 data).

- One address computing instruction (distance computing).

Obviously, the cycle cost of **Junior** is $52329/14938 \approx 3.5$ times higher. To conclude, we first need to enhance memory addressing and accessing instructions; for

Table 5.21 Junior 256 Points FFT Benchmark Results.

Processor	Algorithm	Total cycle cost	Pro-epilogue cycle cost	Kernel cycle cost	DM cost
Junior	256 point FFT	52329	19601	32768	~600
TSMD	256 point FFT	14938	522	14416	~600

example, the variable size postincremental addressing and bit-reversal addressing. Nonoverlap loop is also important.

Opportunities to improve the performance of FFT and DCT are discussed and finally listed:

- Hardware loop for vector computing.

- Postincremental memory addressing with variable step size for butterfly computing.

- Bit-reversal addressing for input or output of FFT and DCT.

- Read from two memories simultaneously.

- Write to two memories simultaneously.

- Two ALU operations (+ and + or + and −) in parallel.

5.4.9 Benchmarking of Windowing

A finite-length time-domain signal gives an infinite response in frequency-domain, and a finite-length frequency-domain signal gives an infinite-length time-domain response. To avoid problems induced by cutting a signal to finite length, windowing algorithms are used. For a data packet with N samples, the windowing algorithm multiplies each sample of the input with the corresponding part of the window signal:

```
Signal_w(i)<= Window(i) * Signal_in(i) 0 < i <(N-1).
```

The cycle cost of the windowing algorithm kernel is $9 \times N - 2$. The total cycle cost is $5 + 9 \times N - 2$. In this example, N is 40 and the total cycle cost is $5 + 9 \times 40 - 2 = 363$ clock cycles. The major extra cycle cost is from the data memory addressing, data memory access, and cost of conditional jump. By using nonoverlapped loop and advanced memory addressing algorithms, the score of a TSMD processor of the windowing algorithm is about $5 + 4 \times 40 = 165$ clock cycles, see Table 5.22.

Table 5.22   Junior Windowing Benchmark Results.

Processor	Algorithm	Total cycle cost	Pro-epilogue cycle cost	Kernel cycle cost	DM cost
Junior	Windowing	363	5	358	82
TSMD	Windowing	165	5	160	82

5.5   DISCUSSION OF JUNIOR DSP

By comparing **Junior** with the state-of-the-art benchmarking from BDTI, and by comparing the performance difference between **Junior** and TSMD, we found that the benchmarking of **Junior** is very poor. Obviously, we cannot define an assembly instruction set by simply transferring all operators from C. C is a hardware-independent language, and hardware execution, such as physical addressing, shall be hidden. An assembly language is a hardware language, and it should include all hardware features.

The misusing of C as the entry for an assembly instruction set design will be summarized in the following text. The lacking of hardware features discussed in the following list is not the drawback of C. Actually, the hiding of hardware features is an advantage of C. The following discussion only exposes the distance between C and AIS (Assembly Instruction Set), a hardware assembly language.

- We cannot find the nonoverlapped loop instruction in C. Nonoverlapped loop is very important for DSP processors because most DSP tasks are based on iterative computing. For example, with nonoverlapped loop instruction, the performance of convolution can be one tap per clock cycle. Without nonoverlapped loop instruction, to compute a tap of convolution will need up to 11 clock cycles in **Junior**.

- C does not expose RISC features and does not have load/store between the registers and memories because both the register variables and the array data can be directly used as operands in C. In DSP, most assembly instructions are RISC-like instructions requiring operands only from the register file instead of from the memories. Therefore, memory load and store must be executed explicitly on an assembly level to support RISC-like instructions.

- Computing of memory addresses is a physical operation not appearing in the behavior C code. C hides the computing cost of memory addressing from users. At assembly language level, memory addressing must be explicitly executed, and the cost of memory addressing and data access must be minimized. The first step of memory design is to expose the addressing and memory access. The goal or the second step of memory design is to minimize all the exposed memory addressing and data access. The way to minimize the memory access cost including computing of memory addresses is to hide the cost by parallelizing and pipelining the correspondent operations. From the benchmarking of **Junior**, we can conclude that most memory access and addressing can be pipelined and executed in parallel with the arithmetic operations. One essential ASIP design technique will therefore be grouping the arithmetic and memory operations into one specific instruction if they are used together all the time.

- To hide the cost of memory addressing and data access is to design smart addressing models by finding and using regularities of addressing and memory access. During the benchmarking of the **Junior** instruction set, we have recognized and discussed regularities such as the postincremental addressing, modulo addressing, postincremental with variable step size, and bit-reversal addressing. We need to formalize the addressing activities, allocate the addressing to other hardware in parallel, and schedule the addressing before using operands. Finally, these addressing operations will be coded to corresponding instructions or will be specified as specially accelerated addressing instructions.

- Each operator in C is simple, and simple stands for the meaning of being general or flexible. A simple operator set cannot support comprehensive computing so that math libraries must be used when compiling the C code. While using C, dedicated DSP operations, such as bit manipulation, arithmetic shift, and rotate are realized by the invocation of library calls. However, if we implement these dedicated DSP operations using assembly subroutines, the performance of a DSP processor will not be sufficient. The conclusion is that several C-level subroutines should be accelerated. Each of the subroutines should be a dedicated instruction if the related hardware cost and complexity is acceptable.

- Low silicon cost is the main requirement of ASIP, which was discussed in early chapters. The first way to minimize silicon cost is to use a fixed-point DSP processor. Operations dedicated for fixed-point computing and hardware-dependent corner case management cannot be found in the original operator list in C and they are implemented using C subroutines. Some C subroutines are hidden to C users, such as subroutines that check overflow and underflow. However, at assembly level, dedicated instructions or microoperations (as parts of an instruction) for corner case handling must be exposed and designed. Special algorithms for corner cases of fixed-point computing shall be supported, for example, round, saturation, and truncation by dedicated assembly instructions.

- Conditional test and conditional branch were not separated in C. The mixup of conditional test and conditional branch is suitable for software design. However, at assembly level, computing for checking conditions requires extra code and execution time. Actually, most of the computing of DSP algorithms implicitly generates flags so that in most cases, the dedicated flag computing is not necessary if the assembly code is scheduled well enough. The explicit separation of flag computing and conditional branch in assembly language is a main difference between C and assembly code.

- Target address computing for a jump instruction is implicitly executed in C because the physical address of program memory is not exposed to C users. However, assembly language is the lowest level language of a specific processor, and the target addressing eventually must be executed on assembly level. The explicit execution of target addressing algorithms must be specified in the assembly instruction set.

- Many DSP arithmetic operations are tightly connected because of the high regularity of algorithms and data types. Especially, streaming signal processing is the dominant part of DSP. Arithmetic operations in a streaming signal processing chain is tightly related. In most cases, multiple arithmetic computing steps can be merged into one instruction, which can dramatically enhance the performance as well make the program more compact. For example, several operations around an addition can be collected into one instruction. The addition instruction should be configured and executed with or without carry-in, saturation, and the computation of other flags. This is another main difference between C and assembly.

- C does not give any parallel features; for example, a convolution must be executed using many lines of C code. Since the convolution is one of the most used DSP operations, we want very high efficiency by having the memory addressing, arithmetic computing, result store, and program flow control carried out in parallel in one instruction. Indeed this is possible because the parallel hardware can be organized in a pipeline. Other most frequently used and iterative DSP operations can also be specified into one instruction. We should find opportunities to pack these operations into one instruction or a limited number of instructions.

- Other differences between the C and assembly languages can also be found in later chapters. To conclude: An assembly language instruction set must be more efficient. Acceleration could be implemented at arithmetic and algorithmic levels. Addressing and memory accesses should be executed in parallel to support arithmetic computing with the minimum extra cost. Program flow control such as loop or conditional execution shall also be accelerated.

5.6 CONCLUSIONS

A simple instruction set, the **Junior** assembly instruction set, was proposed. If students do not have enough knowledge and experience, they can simply specify the **Junior** instruction set by mapping C to the assembly instruction set. After carefully benchmarking DSP kernel algorithms using the **Junior** instruction set, students will be able to realize why the **Junior** instruction set cannot give enough performance. The lesson learned during the design of **Junior** can be used as input knowledge to design a more qualified instruction set, **Senior**, in Chapter 7.

EXERCISES

5.1 Try to list a number of instructions to enhance the **Junior** instruction set according to the discussion and proposals in Section 5.5.

5.2 Design the prologue and epilogue for the benchmarking of a single sample FIR and a frame FIR filter based on the **Junior** instruction set.

5.3 Design the prologue and epilogue for the benchmarking of a Biquad IIR filter.

5.4 Design a piece of the **Junior** benchmarking program for the "vector add" specified in Table 5.14.

REFERENCES

[1] Barton, G. (1984). Towards an assembly language standard. *IEEE Micro*.

[2] IEEE standard for microprocessor assembly language. IEEE Std 694–1985.

[3] Buyer's guide to DSP processors, Edition 2001. Berkeley Design Technologies Inc. CA, USA. (www.bdti.com)

[4] Embree, P. M. (1995). *C Algorithms for Real-Time DSP*. Prentice Hall.

[5] Lennartsson, P., Nordlander, L. (2002). Benchmarking a DSP processor. Master Thesis, LiTH-ISY-Ex-3261, Linkoping University.

[6] Hou, H. (1987). A fast recursive algorithm for computing the discrete cosine transform. *IEEE Trans. Acoustics, Speech, and Signal Processing* 35(10), 1455–1461.

Code Profiling for ASIP Design

6

This chapter will introduce source code profiling for ASIP design and will supply profiling examples of advanced applications. The intention is to introduce more about profiling skills than discussing applications. You will gain an understanding of where the *cost* is from and how to estimate it, rather than the algorithm *principle* of the applications.

6.1 SOURCE CODE PROFILING

From Chapter 5, we understood that the instruction set of a processor can never be created directly from the operators of a high-level language (e.g., C). An immediate question pops up—what should be the input of an instruction set design for embedded DSP processors? Before answering the question, let us check what the goal of the design of embedded DSP processors is. It can neither be the ultimate functional coverage nor the ultimate performance. This is easy to understand; the processor is a part of a system, and the requirement of the system is usually specific instead of ultimate. The ultimate performance and functional coverage are required only when the purpose or users of the processor are not specified; in other words, ultimate is equivalent to general.

> **BOX 6.1** The goal of (ASIP) DSP processor design is to implement programmable application-specific components with defined performance (instead of the highest possible performance), while minimizing the silicon cost, power consumption, and design cost.

The ASIP design can be divided into two steps. The first step is to understand the functions of the application and its requirements on performance. The second step is to find ways to minimize all cost while meeting the requirements on functions (flexibility) and performance. In other words, the first step is to understand the application and propose and initialize a design, and the second step is to apply design skills to optimize the design.

There are ways to understand applications behind the processor. Reading related textbooks and standards is the necessary and the first step. However, reading is not

217

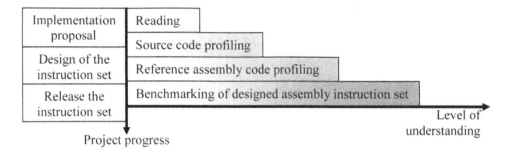

FIGURE 6.1

Understanding applications via profiling.

sufficient because the knowledge achieved from reading cannot be quantitatively accurate.

A widely accepted way to understand applications is through code profiling. Code profiling analyzes the application code structure and measures its execution time and memory cost. Source code profiling has been accepted as an accurate way to understand a system. In industry, the acceptable inputs for designing an instruction set should be from the profiling of source code and reference assembly code. Normally, an assembly instruction set has not yet been designed during the early profiling phase. A similar and available assembly instruction set could be used as a reference for early assembly-level profiling. The reference could come from the early design or from a third party, see Figure 6.1.

In this chapter, we will expose how industry gets inputs for designing an ASIP assembly instruction set, including organizing and running profiling of source and assembly code. There are many publications on source code profiling for software optimization. Code profiling was mentioned for instruction set design of general processors [19]. Not many publications can be found for the profiling for ASIP design. Source code profiling to be discussed in this book is only for instruction set design, including early HW/SW partitioning, computing costs, memory access costs, and code costs. In this chapter, we will discuss what profiling is and how to profile source code for ASIP design. The design and implementation of source code profiler will be discussed in Chapter 8.

6.1.1 What Is Source Code Profiling?

Source code profiling is a technique to estimate the execution cost and memory cost of source code. Source code in this book stands for the behavior description of a DSP application using high-level language. While executing or parsing the source code, the profiling gives records and summaries of time cost of subroutines, algorithms, number of calls, and hardware usage. Profiling also exposes the behavior of program entities including functions, loops, basic subroutines, and kernels. Most source code profiling is for software design and optimization [1]. Recently, discussion on profiling for ASIP design can be found in Karuri et al. (2005, 2006) [2, 3]. To optimize software, profiling is conducted on a coarse-grained level because software designers are more interested in software organization and algorithm selection. To design an ASIP, profiling is conducted both on a coarse-grained level for architecture selection and on a fine-grained level for the instruction set design. Fine-grained level profiling exposes the cost of arithmetic and logic operations, low-level algorithm operations, and memory accesses.

Code profiling can be hardware-dependent or hardware-independent. The hardware-independent code profiling gives basic arithmetic computing costs of the source code [1]. Hardware-dependent profiling is used more for performance and cost estimation of parallel systems or systems with parallel features. Parallel features can be found in any DSP processor whether or not it is simple, for example, the parallel features for convolution. Hardware-dependent profiling exposes the execution time and/or memory cost under conditions of using specific hardware. For example, if the convolution hardware is available, several steps of convolution functions will be merged.

Example 6.1

Counting computing costs of running N-step convolution when convolution hardware is available.

Acceleration of convolution function

```
for(i=0;i<N;i++){
{
        Accumulator <= Accumulator + x[SEG1 + i] * Y[SEG2 + i]
}
```

Solution

To accelerate the convolution, the parallel hardware shall include at least data memory and its address generator, coefficient memory and its address generator, multiplier, accumulator, and the loop control logic. The purpose is to finish a complete step of convolution in one clock cycle. The order of execution is:

Addressing and memory loading
Multiplication
Accumulation
Loop control

In this case, the computing cost is N steps of accelerated convolution (not including the extra pipeline cost).

Discussion

The profiling for ASIP design is much different from the profiling for software optimization.

However, hardware-independent profiling is very different from hardware-dependent profiling. In the following example, we will discuss hardware-independent profiling and compare it with the profiling of hardware-dependent convolution, given in Example 6.1.

Table 6.1   Computing Costs of a Convolution.

Operation	Costs
Data memory addressing (including test of FIFO boundary)	2N
Data memory load	N
Coefficient memory addressing	N
Coefficient memory load	N
Multiplication	N
Accumulation	N
Conditional Jump	3N
Total	10N

Example 6.2

Hardware-independent profiling of running N steps of convolution.

```
for(i=0;i<N;i++){
{
       Accumulator <= Accumulator + x[SEG1 + i] * Y[SEG2 + i]
}
```

Solution

The cost can be summarized in Table 6.1.

In this case, the computing cost of the convolution is (at least) 10N operations including the extra pipeline stall from conditional jumps.

Conclusion

The profiling is much different from the hardware-dependent profiling.

Usually, source code profiling counts the number of operations instead of the number of clock cycles. Hardware designers translate the number of operations to the number of clock cycles according to the architecture of the hardware to be used.

6.1.2   Why Profiling?

Let us recall the HW/SW codesign introduced in Chapter 4. HW/SW codesign for the ASIP instruction set is to decide what function will be implemented using a single instruction and what function will be implemented using a subroutine. However, we haven't discussed issues such as how to conduct the HW/SW codesign

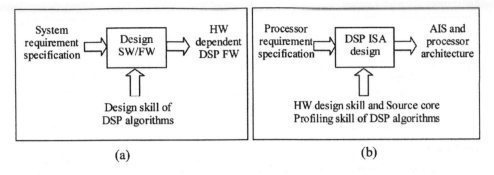

FIGURE 6.2

Differences between SW design and ASIP design.

and how to quantitatively guide the HW/SW codesign. These questions will be answered while discussing the source code profiling.

Understanding the application is to know what, why, and how. "What" stands for what the application is, including the inputs and outputs, and how the inputs are processed. "Why" stands for the purpose of the application or the driving force behind it. "Why" also explains the market behind the design, as well as its significance and use. "How" corresponds to the implementation details. It includes the engineering implementation of the algorithm, the hardware, and the system.

Here, the understanding of applications does not mean the deep understanding of the complete principle of DSP algorithms. For an ASIP hardware designer, understanding does not mean to fully hold the knowledge and skills of SW design. Instead, understanding means enough knowledge for hardware or firmware design; in other words, how the software *works* instead of how it was *designed*. The difference can be explained using Figure 6.2.

Figure 6.2 (a) describes the design of a DSP system or its SW, which requires deep understanding of the applications. Figure 6.2 (b) describes the design of an assembly instruction set, which requires just sufficient understanding of the applications. The differences in design inputs and design outputs as well as requirements on design skills are illustrated by comparing (a) to (b) in Figure 6.2. In Figure 6.2 (b), AIS stands for Assembly Instruction Set and ISA stands for Instruction Set Architecture. A simplified assembly instruction set design flow is given in Figure 6.3 and was discussed in Chapter 4.

6.1.3 What to Profile

Functional Coverage

Before source code profiling, collection and selection of source code is an essential step. Too many kinds of source code may induce excessive functional coverage. On the other hand, the flexibility of the ASIP might not be enough without enough

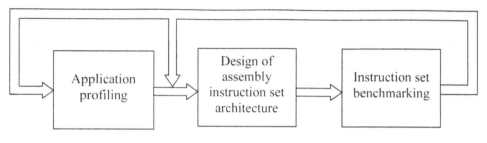

FIGURE 6.3

Simplified assembly instruction set design flow.

source code being profiled. The decision of the source code collection is beyond the scope of this book. Readers interested in product strategic plan can read books or papers of technical marketing of ASSP.

Computing Cost

Computing cost includes the cost of algorithm computing and the cost of overhead. Overhead costs are those that cannot be directly recognized during the high-level source code profiling, and include the following:

- Memory access cost.

- The cost of running program flow control.

- The execution cost for hardware and timing resource management.

- Cost to serve asynchronous tasks.

- Synchronization and communication as well other cost of parallelization.

The overhead costs cannot be recognized because directly the cost might or might not be visible while reading or profiling the source code. By using a reference architecture, experienced engineers can estimate the listed overheads. To get high performance, parallel opportunities and ways to minimize overheads will be investigated. The following questions will be investigated by the instruction set designers during the source code profiling.

1. Which are the most MIPS consuming functions that should be accelerated?

2. Which are the most frequently appearing functions that should be accelerated?

3. What will be the datapath architecture after identifying questions 1 and 2?

4. What will be the memory and bus architecture to support questions 1 and 2 and reach the minimum on-chip memory cost?

Memory Cost

The memory cost is measured as the number of bits used by a program. It is architecture dependent and run-time dependent. The memory cost analysis of a program depends on variable lifetime and the accumulation of memory usage. The accumulation includes variable sizes (in, out, and computing buffers) and constant parameter sizes (coefficient, program parameters). The variable lifetime is the period from the first time the variable is used to the last time using the variable by a program. It should be considered during the memory cost analysis. Memory space might be specified based on the accumulation of data with long and short lifetimes to minimize the runtime cost. If not all variables can be kept in the on-chip memory while running a program, the runtime might be relatively longer, including context swapping. Trade-off of runtime and memory cost will be discussed in Chapter 18.

The memory access cost analysis includes (1) the accumulation of memory access (read and write) operations while running an application (a task) and (2) analysis on memory access models (addressing algorithms or parallel addressing algorithms). The profiling result from the memory cost analysis and memory access cost analysis will be used when designing for addressing, and for the memory subsystem when designing the assembly instruction set architecture.

The memory cost analysis and memory access cost analysis from the source code are usually inaccurate because the lifetime of a variable in a DSP processor depends on the size of its register file and the size of its on-chip memory, algorithms, and design skills. If the size of the register file is small, useful variables can be overwritten by others. The overwritten variables need to be loaded from memory again, which induces extra cycle cost and memory cost.

If the size of the on-chip memory is small, the on-chip memory may not be enough to support all data of all functions. Very often, useful data of one function can be overwritten by other functions. Before overwriting useful data, a DMA transaction is required to temporarily back up the data to off-chip memory. When it is needed again, another DMA transaction is required to load the data. Extra DMA requires extra cost for DMA management operations. Memory cost analysis will be discussed in detail in Chapters 18 and 20.

Hidden Cost

Hidden cost includes hidden computing cost, hidden memory access cost, and hidden memory cost. We will find that some hardware-dependent computing cost will not be recognized during the source code profiling. We have to use reference architecture and reference assembly code to expose and estimate these costs. Here, the reference architecture is the architecture of an available ASIP or the architecture proposed. The reference assembly code is based on the instruction set of the reference architecture. The hidden costs could be from:

- The cost of preparing for a subroutine call and the termination of a subroutine.

- The cost of preparing and running HW (DMA and other IO transactions).

- The cost of the top-level management of a program.
- The cost of thread and interrupt handling.
- The extra cost of parallelization (synchronization, communication, etc.).

6.1.4 How to Profile

Selecting or designing a profiler is essential for profiling. A profiler is based on a source code analyzer. As a software development tool, it analyzes the source code, and collects and reports execution statistics of the source code. A profiling tool usually includes three parts: the code analyzer (static profiling), the instrumentation part, and the part for execution time statistics (dynamic profiling). Discussion on the three steps of profiling will be the focus of this chapter.

The first step of source code profiling is to analyze the source code (also called static profiling). The analysis exposes the code structure using a task flow graph (TFG). A TFG is a tree including arithmetic operations in each branch of the tree and program flow control on each node of the tree. A fine granularity flowchart can be used to expose the computing arithmetic in detail. A coarse granularity flow graph exposes control behaviors of the source code by hiding the arithmetic computing into basic blocks. After accumulating arithmetic operations in each basic block, the worst-case execution time can be further identified by accumulating the total computing cost of each path in the flowchart.

Dynamic behavior of the source code execution can be further exposed by running the source code with typical inputs (dynamic profiling). Instrumentation should be carried out before dynamic profiling. The instrumentation (or probe) is to mark which part of the execution should be counted by inserting additional counters into the original source code. The inserted counters shall not change the program function. Instead, the inserted code is used only to monitor the program behavior by counting the interesting execution and accumulating the gathered information to the profiling score, which is usually a log file.

Runtime profiling is data dependent because some subroutines may or may not be executed depending on features of the input data. Statistics are required to get both the maximum and average execution time. The statistics also expose the most frequently executed algorithm kernels, the most demanded arithmetic operations, and the total time cost.

After the first step of profiling, the source code is analyzed, and a profiling strategy should be decided, including what to observe and where the observation points are. A dynamic profiling is thus prepared and source code will be executed. After the dynamic profiling, results of both static and dynamic profiling will be analyzed. A profiling process can be divided further into five parts:

1. Decide the scope of a profiler.

2. Analyze source code (static profiling).

3. Prepare and configure the profiling tool (instrumentation).

4. Run dynamic profiling.

5. Analyze the results of static/dynamic profiling.

Deciding what to profile is actually not that easy. We may not know what to profile until the results of the first iterative profiling is available. If the source code is large and complicated, identification of critical path might not be easy. We may not know the computing cost until several profiling iterations have been completed.

6.1.5 The Language to Profile

Profiling can be performed on source code or on assembly code. Profiling on source code is fast but in general not accurate enough. A single line of C code may consist of several different operations mapped to several assembly instructions. In C/C++, some assembly level operations are not visible to the source-level profiler;—for example, load/store between registers and memories, computation of physical data memory addresses, and target addresses in the program memory. Profiling on assembly code is accurate but slow and sometimes even impossible during the early ASIP design phase.

The IP level profiling was investigated in recent years. Intermediate representation (IR) is used inside a compiler as an intermediate language [2, 3]. During the compilation, a compiler will translate C first to IR. After the code optimization, the compiler will finally translate the optimized IR to target assembly language. While profiling at IR level, the source code is parsed and analyzed by a compiler. Since the assembly language is not yet specified, the compiler will stop after the IR generation and optimization. The optimized IR is then profiled. Static IR profiling is not difficult because the IR level task flow graph is actually available. Dynamic profiling of IR is possible if there is an executable simulator of IR. However, IR simulator is usually not available.

Since all micro operations must be exposed when translating C to IR, and at the same time in IR there is only one operation within each line, accurate profiling is feasible and the memory cost will be exposed during the profiling at IR level. I believe that IR level profiling will be popular in the future for embedded processor design.

However, the memory cost might be excessively counted during IR profiling. For example, using a good assembly language, certain regular vector memory access can be eliminated because of the pipelined parallel execution of datapath computing and memory access. As with source code profiling, hardware information cannot be sufficiently exposed during the profiling at IR level.

Profiling efficiency and accuracy based on different languages is intuitively depicted in Figure 6.4.

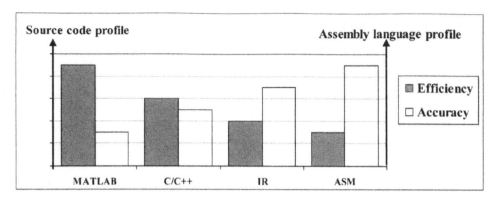

FIGURE 6.4

Different kinds of profiling.

6.2 STATIC PROFILING

6.2.1 Dynamic and Static Profiling

By definition, dynamic profiling is performed by executing the source program and accumulating the execution time. Many profiling tools are for dynamic profiling. Dynamic profiling is used to identify the specific cost of execution time and memory under a specific set of data stimuli. If proper data stimuli are used, the profiling results can be considered to be valid for most types of input data. However, how to find the right stimuli is an open and experience-based question, and the evaluation of the trust on stimuli selection might be required.

Static profiling is used to identify the limit of costs, the upper limit of execution time, and the upper limit of memory cost. Static profiling is the same as the analysis on WCET (worst-case execution time). Static profiling gives the worst execution and is used for reliable design of real-time systems. Results from static profiling are data independent.

6.2.2 Static Profiling

Static profiling analyzes the source code instead of running it; in other words, it is conducted by parsing the source code, presenting the software structure using Control Flow Graph (CFG), and counting the profiling score by parsing instead of running the source code.

The innermost loop is the implementation of an iterative algorithm kernel in an application. It is implemented as the lowest nest loop. The typical code size is less than 10%, and all innermost loops take 90% of the execution time overall. The innermost loops will be identified while parsing the program hierarchy. The accumulation of computing cost on basic block will be conducted from the innermost loop to the top level of the tree. Each basic block will be annotated with its execution cost, and total execution cost of the program will be accumulated. After the accumulation, the critical execution path with the highest cost will be identified. If the critical path does not exist in reality, it can be annotated as a false path that should be ignored. Finally, the true critical path will be identified as the worst case (the longest execution time).

To summarize, static profiling can be divided into four steps:

1. Generate CFG, identify the basic blocks that are the lowest component.

2. Perform cost annotation on each branch of the CFG.

3. Identify cost accumulation based on annotations on all branches of the CFG.

4. Analyze the result and find the cost of the true critical path.

The principle of code parsing and CFG extraction will be discussed in Chapter 8. Total cost of each branch should be accumulated, and finally the total cost of each path (accumulation costs of all branches from top to the bottom) of the CFG will be accumulated.

6.2.3 Fine-grained Static Profiling

Fine-grained static profiling is to parse the source code deeply inside each basic block at the arithmetic level. Fine-grained static profiling should be conducted in each branch (a basic block in a CFG). The purpose is to understand the computing behavior in detail, including the costs of arithmetic and logic computing for selection and decision of assembly instructions. Except for arithmetic and logic operations, test operators, memory access, and communications shall also be profiled.

Counting operations in each basic block of the source code can help the designers decide the datapath hardware and instructions. Basic arithmetic operators in C-language for DSP were listed in reference [4] including $+, -, *, /,$ and %. Necessary logic operations are &&, ||, and !. Bitwise operations are $\sim$ (bitwise not), $\hat{}$ (bitwise xor), & (bitwise and), | (bitwise or), << (shift left), and >> (shift right). Relation operators are $>, >=, <, <=, ==,$ and $!=$. The fine-grained profiling discussed in this section counts only the individual operations [4].

An example of a fine-grained static profiling is given in Figure 6.5.

Figure 6.5 includes both fine-grained profiling of each B and coarse-grained profiling on the top level (to be discussed in the following subsection). In Figure 6.5, the result of a21 = B2 (a15, a12, a16) or a21 = B2 (a12, B1(a11), B3(a13)). A "B" is a basic block that is a straight-line sequence of code that can be entered only at the beginning and exited only at the end [6]. There might be more than one entry at both the beginning and the end of the block. The computing cost of blocks in Figure 6.5 is accumulated and listed in Table 6.2.

Some costs such as load, store, and move are not included in the table. This group of costs depends very much on the target architecture (the assembly instruction set, the size of the register file, the size of the data memories). If the profiling on load,

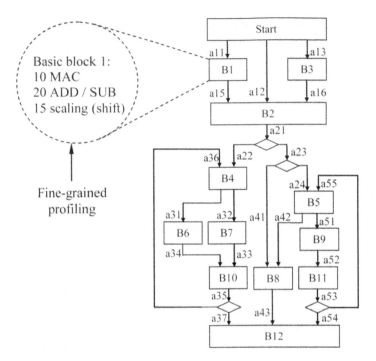

FIGURE 6.5

Block arc-based profiling.

store, and move is further exposed, the architecture independent load/store usually happens when

- A variable is used by a subroutine.

- An output of an algorithm is generated.

- A global variable is changed.

In this case, we assume that the size of the register file is large enough and the computing is buffered using the register file while running a subroutine. The hidden load/store will be actually more.

The memory access cost of RISC computing is not very explicit. However, at least, load operations can be identified when loading new variables and initializing subroutines. Store operations can be identified when the outputs of algorithms are generated, when the global variables are changed, and when the subroutines are finished. While profiling source code for ASIP design, the assumption on the size of register file is usually large enough, and extra data swapping between the register and data memories is not counted in.

Table 6.2 An Example of Computing Cost Profiling.

| Coarse-grained profiling | | Fine-grained profiling | | |
Basic block	Total	Arithmetic operations	Logic / Shift	MAC
B1	45	20	15	10
B2	12	10	2	0
B3	3	2	1	0
B4	140	10	30	100
B5	8	4	2	2
B6	145	25	0	120
B7	122	2	10	110
B8	53	50	3	0
B9	14	10	2	2
B10	214	12	2	200
B11	11	8	1	2
B12	6	5	1	0

6.2.4 Coarse-grained Static Profiling

Coarse-grained profiling analyzes and accumulates cost of basic blocks (a branch) on the system level instead of the cost inside a branch. Coarse-grained profiling is used for architecture decision instead of instruction set design. During coarse-grained profiling, the basic blocks consuming the most computing time will be identified. The cost of these basic blocks and the relation between these basic blocks will be analyzed. The basic blocks that do not consume most of the computing time, but instead appear very often, will also be recognized. There may also be other basic blocks that instead of consuming most of the execution time, consume a significant amount of program memory or data memory. These shall also be identified.

Example 6.3

Estimate the computational workload of 802.11a.

IEEE 802.11a [7, 8, 13] is a standard of wireless local area network access. Baseband signal processing for IEEE 802.11a is a typical real-time DSP application. In this example, the coarse-grained profiling and the counting of the execution time of those most computationally intensive basic blocks will be given and discussed. The profile is based on MOPS (million operations per second). The memory cost will not be included. The example will include only the profiling of the channel estimation and reception algorithms of IEEE 802.11a.

Solution

The example will give an estimation of the computing cost of the digital signal processing for radio baseband. Instrumentation probes are inserted at the top of every basic block or subroutine to monitor the execution time of the signal processing for a symbol. By converting the computing cost of a symbol to MOPS, the final score is depicted in Table 6.3.

Discussion

Readers might not understand the algorithms. Again, to design hardware, we need to understand only the cost instead of the principle of algorithms. The deep understanding of algorithms can be found in references such as [7] and [8]. In this example, one step of basic algorithm execution is counted as one operation; for example, a step of convolution, a butterfly

Table 6.3   Profiling of the Main Tasks in 802.11a Wireless LAN.

Tasks	Function	Algorithms	MOPS	Data types
Viterbi decoding	Viterbi decoding	Viterbi decoding algorithm	1700	Short integer
Receiving Filter	Anti-alias and band pass	Complex FIR convolution	960	Complex
De-interleaving	Permutation	Permutation algorithm	260	Bit
Track fast fading	Track Pilot distortion	FFT and vector diff	160	Complex
Fading compensation	Interpolation	F-domain LMS	200	Complex
Sampling phase offset	Get rotor coefficient	FFT / phase decision	100	Complex
Demapping	Symbols to bits	Decision FSM	80	Integer
Descrambling	Bit permutation	Polynomial algorithm	80	Bit
Packet detection	Synchronization	Autocorrelation	80	Complex
Energy detection	Antenna diversity	Autocorrelation	80	Complex
IFFT/FFT	IFFT/FFT	64 points FFT and IFFT	60	Complex
Payload Rotor	Rotor	Vector product	40	Complex
Normalization	Normalization	1/x and vector product	40	Complex
TOTAL			3840	

computing, a step of ACS (add-compare-selection for Viterbi decoding), and a subcarrier demapping of an imagery or quadrature value. The equivalent MIPS cost in a general RISC could be much higher.

6.3 DYNAMIC PROFILING

Dynamic profiling has been discussed by many publications, and readers can get further details of dynamic profiling from the references listed in [1]. The instrumentation for computing and memory access shall be implemented as probes in the interesting path in the source code before running it.

After specifying and inserting probes, dynamic profiling can be realized. An essential work should be done before the dynamic profiling, which is the selection and decision of stimuli. The stimuli should be selected to represent either typical cases or the worst case. Right stimuli can be found from the standard specification or from the knowledge of the application. For example, for video encoding, typical references can be found at www.mpeg.org.

6.3.1 Instrumentation for Coarse-grained Profiling

Instrumentation (also called probe) is used to define and insert a measurement point in a program to measure and accumulate the appearance of the measured parameter during runtime. A probe is a subroutine identifying the appearance of interests and counting on it. Instrumentation for coarse-grained profiling checks the appearance of a basic block on the top-level during the execution time of the source code. Probes or instrumentations are inserted to:

1. Check the frequency to reach each program mode (behavior of the high-level CFG).

2. Check the cycle cost of each basic block.

3. Check the frequency of invocation of a basic block (low-level paths).

The purpose of instrumentation of a coarse-grained profiling is to understand the software system and the structure of the code at a higher level. It identifies the critical paths, the scope of the algorithms, the general behaviors of kernel algorithms, and the use of hardware in general. The objectives of instrumentation include application specific algorithms such as data transforms, filters, correlations, and all other basic blocks.

6.3.2 Instrumentation for Fine-grained Profiling

Instrumentation for fine-grained dynamic profiling is used to count the running of the source code deeply inside each basic block at the arithmetic level.

Example 6.4

Accumulate datapath operations of the voice coder and decoder specified in the ITU G.723 and ITU G.729 standards.

Solution

The bit-accurate C code as the behavior mode of voice CODEC G.723, G.729 can be found from ITU (International Telecommunications Union) [9–12]. By parsing the source code, the arithmetic operations, shift, logic, and normalization are carefully specified, listed, and instrumented.

Memory access and addressing algorithms cannot be included in the instrumentation because the processor architecture and instruction set are not yet specified. The example is only part of profiling, and the test of conditions is not included. The profiling results are depicted in Table 6.4. The dynamic profiling result is from the counters of probes (instrumentation), and the static profiling result is from the accumulation of the execution costs of all basic blocks in the longest true path.

Discussion

The profiling includes both the coder and decoder of two standards. The static profiling is based on the analysis of the source code provided by ITU. The dynamic profile uses data stimuli recommended by the standards committee. A data frame of G.723 is 30ms including 240 voice samples. A data frame of G.729 is 10ms including 80 voice samples. The profile in detail can be found in references [11] and [12].

Table 6.4 gives the counts of operations. Finally, the counts of operations should be translated to the counts of cycle costs. Some operations, such as division and double-precision computing might use multiple clock cycles. The final MIPS cost based on an assembly instruction set will be much higher than the MOPS in Table 6.4. The invocation frequency of conditional branches shall also be profiled to support the design of program flow control instructions, loops, and the control path hardware; for example, what kind of nonoverhead loops are essential, and what kind of jumps are most used. Details can be found in references [11] and [12].

6.3.3 Implement Instrumentation

Instrumentation is used to annotate probes into the interesting places in the source code. The probe code usually is inserted into a basic block in the source code as a probe subroutine. The following example is the pseudocode of an instrumentation subroutine. The probe subroutine is inserted at the entry of the basic block. Every time the basic block is executed, the call of the instrumentation subroutine will increase the counter by a certain constant value (1 or a larger integer). After the profiling, the accumulated parameters as the profiling scores will be stored in data log.

Table 6.4 Profiling of Arithmetic Computing of Voice CODEC G723 and G729.

Standards Operations in a voice frame	G.723 Dynamic	6.3 kb/s Static	G.723 Dynamic	5.3 kb/s Static	G.729 Dynamic	8.0 kb/s Static
16 bits Fractional MAC	222373	264810	130872	141129	38734	42384
16 bits Integer MAC	58565	74768	71251	73104	5081	5383
32 bits arithmetic left shift	8550	10242	5210	5693	3068	3097
16 bits absolute value	6975	8945	1189	1201	21	21
32 bits absolute value	6937	9068	646	652	100	100
16 bits multiplication with 16 bits rounded result	3308	3478	3436	3478	240	240
32 bits arithmetic right shift	2625	4075	4540	6794	3150	4081
16 bits multiplication 32 bits result	2126	7076	5729	10653	5923	6852
32 bits subtraction	638	841	366	841	832	845
32 bits addition	557	682	489	597	596	597
32 bits normalization	378	782	400	790	70	71
32 bits by 16 bits multiplication	339	1406	312	1406	972	1002
16 bits multiplication 16 bits result	329	7544	5347	10856	6564	7544
32 bits by 16 bits division	47	50	49	50	10	10
16 bits by 16 bits division	11	24	15	24	23	24
32 bits by 32 bits multiplication	8	166	1	166	166	166
16 bits normalization	6	11	6	11	11	11
TOTAL main arithmetic operations in a frame	313772	393968	229858	257445	65561	73428
The frame length (milliseconds)	30	30	30	30	10	10
TOTAL main arithmetic operations per voice sample	1307	1642	958	1073	820	918
TOTAL arithmetic MOPS	10.46	13.14	7.75	8.59	6.56	7.34

Example 6.5

Write an instrumentation subroutine and use it in source code.

```
//Instrumentation counter code
Function instrument (basic_block, counter)
  {
  If basic_block then Counter <= counter + 1
  Return.
  }
//Source code after annotating parsing information
{
  ...
  //insert the instrumentation code here.
  Instrument (basic_block_n, counter_basic_block_n)
}
```

6.4 USE OF REFERENCE ASSEMBLY CODES

6.4.1 Expose Hidden Costs

Hidden computing costs cannot be exposed before analyzing assembly-level codes. It was discussed that the extra execution time cost such as the prologue and epilogue of a subroutine, the task threading, the asynchronous task handling, and the IO handling will not be exposed during the profiling of source code. Normally, these extra costs are relatively low and negligible. However, there are exceptions. For example, the thread handling cost could be significant when threads appear too many times.

Example 6.6

Schedule and run two radio baseband signal processing tasks simultaneously using one baseband signal processor. Task one is IEEE802.11 a/g baseband DSP, and task 2 is Bluetooth baseband DSP [13, 14].

The processing period of a Bluetooth data packet is one millisecond. The processing period of an IEEE802.11a baseband OFDM symbol is four microseconds. To run two tasks using the same DSP core, the threading technique based on the EDF (Earliest Deadline First) algorithm is used [15]. The WLAN has the earliest deadline, and it holds higher priority. When a data packet of WLAN arrives, it will interrupt the signal processing of Bluetooth.

The scheduling of the two flows of streaming signal processing on one processor is given in Figure 6.6.

In every four microseconds, a data packet comes from the WLAN radio channel. Suppose the system clock frequency is 250 MHz and the thread handling cost is about 50 clock cycles

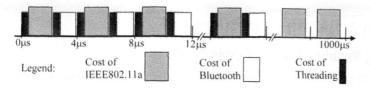

FIGURE 6.6

Scheduling Bluetooth and IEEE802.11a on one processor.

for all context saving and task switching, and that there are two thread control slots consuming in total 100 clock cycles during each period of four microseconds. Therefore there are more than $100/(250 \times 4) = 10\%$ computing time used for thread handling.

Discussion

Obviously, the runtime cost of thread handling is too high. Unfortunately, the thread handling cost cannot be exposed while profiling the source code.

The conclusion from the example shows that some execution time cost cannot be exposed during the source code profiling, for example, the cost of thread handling. To get better understanding of extra execution time cost, early assembly code profiling is helpful. However, as mentioned, the assembly language instruction set is not yet specified during the time of profiling. We therefore have to use some reference assembly languages.

6.4.2 Understanding Assembly Codes

Reference assembly language instruction sets could be used for the early assembly-level profiling. References could be:

- Assembly instruction sets of early product versions.
- Similar assembly instruction sets in the market.
- Assembly instruction sets from research case studies.
- Pseudo assembly instructions specified by experienced engineers.

After selecting a reference assembly instruction set, the gap between the expected instruction set and the reference instruction set can be identified. For example, since we hope that the expected instruction set is as simple as possible, the reference assembly instruction set might be too advanced, which will bring too high a cost. Therefore, the minimum number of instructions in the reference instruction set can be selected and the rest of them are not used. If the reference instruction set is less advanced than the expected instruction set to be designed, pseudo instructions should be added. Adding instructions to a reference instruction set simulator might be impossible (the reason will be exposed in Chapter 17).

6.5 QUALITY EVALUATION OF RESULTS

6.5.1 Evaluating Results of Source Code Profiling

The quality of profiling stands for accuracy and coverage. Accuracy can be measured using the difference between the profiling results (discussed in this chapter) and benchmarking score of the produced assembly instruction set (discussed in Chapter 9), including the execution time cost and memory cost. If the difference between the profiling results and the benchmarking score is not significant, we may consider the accuracy to be acceptable. Coverage here is the measurement of the completeness of the profiling.

The accuracy of a profiling cannot be measured until an assembly instruction set is designed and the benchmarking is performed using the assembly code. However, this kind of evaluation of accuracy is usually too late. We need an early evaluation of the accuracy of profiling. Intuitively, the more algorithms and operations are instrumented, the higher the accuracy can be reached. The lower the program hierarchy and the more operators exposed, the better the accuracy will be. Fine-grained source code profiling was discussed in this chapter, which can give higher accuracy by including both DSP algorithms and memory access algorithms. If the execution time is very much dependent on the data stimuli patterns, the trust of dynamic profiling is low. For example, the execution time using the data pattern one is twice that of using data pattern two; this extreme case could happen when running Huffman coding.

6.5.2 Using Profiling Results

The creditable profiling results can be used to design an assembly instruction set for an ASIP. There are two sets of profiling results, coarse-grained and fine-grained. The coarse-grained profiling result includes statistics on the execution of basic blocks. It is used for planning an instruction set architecture in general. The fine-grained profiling result includes the counts of arithmetic and logic operations of DSP algorithms and memory addressing. The fine-grained profiling will be used further to design the instruction set details.

In the previous chapter, the rule of 90%–10% code locality was introduced. By analyzing the results of profiling, we could ask:

- Which 10% of the operations take 90% of the total execution time (invocation frequency)?

- Which path requires the most intensive computing power (critical path)?

After finding the critical path, the opportunities of optimization or acceleration can be further exposed. The first opportunity is to accelerate regular and volume tasks, such as simple array computing, convolution, and division arithmetic operation. Complete computing on a data sample in a data array may require many steps of operations: address computing, operand loading from the memories, kernels processing, results storing. If all steps of the computation can be allocated to hardware modules in parallel and be executed in a pipelined mode, all steps of the computation can be merged into one instruction and the cost of the array computing will be lower. The second opportunity is to identify the correlation between instructions. If one instruction is always following another instruction and the instruction sequence can be allocated in parallel in available hardware, the sequence of the instructions can be merged as a new instruction.

In a program, a prologue is a part of the program initializing a subroutine, for example, to load the subroutine parameters. An epilogue is a part of the program terminating a subroutine, for example, to store the results, parameters, and context. Execution time of prologue and epilogue is not visible during the source code profiling. If the execution times of most subroutines are short, the prologue and epilogue costs might be significant.

6.6 CONCLUSIONS

The quantitative design of an assembly instruction set requires deep understanding of applications. Profiling is a way to understand applications. In this chapter, source code profiling for designing ASIP assembly instruction sets was discussed. Static and dynamic profiling was introduced through examples. The way to estimate extra execution time using the reference assembly instruction set was also introduced.

Static profiling can be the starting point of ASIP profiling, which gives the worst execution time. Dynamic profiling gives typical results if right data stimuli can be specified. The instrumentation for both coarse-grained and fine-grained profiling was also discussed.

EXERCISES

6.1 To design an ASIP instruction set, profiling technique is essential. What is coarse-grained profiling and what is fine-grained profiling?

6.2 What is source code profiling, and what is profiling on the IR level?

6.3 What are differences between source code profiling for software optimization and source code profiling for ASIP design?

6.4 What is dynamic profiling, and what is static profiling?

REFERENCES

[1] Balsamo, S., Di Marco, A., Inverardi, P., Simeoni, M. (2004). Model-based performance prediction in software development: A survey. *IEEE Trans. Software Engineering* **30**(5).

[2] Karuri, K., Al Faruque, M. A., Kraemer, S., Leupers, R., Ascheid, G., Meyr, H. (2005). Fine-grained application source code profiling for ASIP design. DAC 2005, June 13–17. Anaheim, California, USA.

[3] Karuri, K., Huben, C., Leupers, R., Ascheid, G., Meyr, H. (2005). Memory access micro-profiling for ASIP design. DAC 2005, Proceedings of the Third IEEE International Workshop on Electronic Design, Test and Applications (DELTA'06).

[4] Embree, P. (1995). *C Algorithms for Real-Time DSP*. Prentice Hall.

[5] Eilert, J., Andreas Ehliar, A., Liu, D. (2004). Using low precision floating-point numbers to reduce memory cost for MP3 decoding. *Proc. of the IEEE Int'l Workshop on Multimedia Signal Processing (MMSP)*. Siena, Italy.

[6] Muchnick, S. (1997). *Advanced Compiler Design and Implementation*. Morgan Kaufmann.

[7] Liu, D., Tell. E., Nilsson, A., Söderquist, I. (2005). Fully flexible baseband DSP processors for future SDR/JTRS. CEPA2 Workshop, Digital platform for defense, Brussels Belgium.

[8] Heiskala, H., Terry, J. T. (2002). *OFDM Wireless LANs: A Theoretical and Practical Guide*. Sams Publishing.

[9] ITU-T Recommendation g.723.1. (1988). Dual rate speech coder for multimedia communications transmitting at 5.3 and 6.3 kbit/s.

[10] ITU-T Recommendation g.729. (1996). Coding of speech at 8 kbit/s using conjugate structure algebraic-code-excited-linear-prediction (cs-acelp).

[11] Olausson, M., Liu, D. (2001). Instruction and hardware accelerations in G.723.1(6.3/5.3) and G.729. Proceedings of the 1st IEEE International Symposium on Signal Processing and Information Technology (ISSPIT), Cairo, Egypt, 34–39.

[12] Olausson, M., Liu, D. (2002). Instruction and hardware acceleration for MP-MLQ in G.723.1. Proceedings of IEEE Workshop on Signal Processing Systems (SIPS'02), San Diego, California, USA, 235–239.

[13] IEEE Std. 802.11a-1999. High speed physical layer in the 5 GHz band.

[14] http://www.bluetooth.com. Specification of the Bluetooth System V2.0+EDR-Core.

[15] Krishna, C. M., Shin, K. G. (1997). *Real-time Systems*. McGraw-Hill.

[16] Deb, A. K. (2004). System design for DSP applications with the MASIC methodology. Doctoral thesis of Royal Institute of Technology, Stockholm, Sweden.

[17] Ravasi, M., Mattavelli, M. (2003). High-level algorithmic complexity evaluation for system design. *Journal of Systems Architecture* **48**.

[18] www.bdti.com. (2001). Buyer's guide to DSP processors. Berkeley Design Technologies, Inc. California, USA.

[19] Hennessy, J., Patterson, D. A. (1990). Computer Architecture: A Quantitative Approach. Morgan Kaufmann.

Assembly Instruction Set Design

7

In this chapter, instruction set design will be discussed in detail based on case studies and experience accumulated from Chapters 2 through 6. The focus of this chapter will be principles for assembly instruction design. Inputs to the assembly instruction design process are source code profiling of the application class; the proposed architecture; and requirements including flexibility, silicon cost, power consumption, and design cost. Output of the assembly instruction design process is the assembly instruction set manual. Knowledge provided in this chapter is essential for both processor and system designers using processors as components.

7.1 METHODOLOGY

In Chapter 4, general methodologies for HW/SW codesign were introduced. From designing **Junior** in Chapter 5, it became apparent that assembly instruction set design experience is essential in order to provide a good assembly instruction set. According to Chapter 4, a prerequisite is also a deep understanding of applications. Profiling, discussed in Chapter 6, is the essential method to understand applications. In this chapter, accumulated experiences from Chapters 4, 5, and 6 will be used to design an ASIP assembly instruction set.

7.1.1 Opportunities and Constraints

Opportunities and the Design Spaces

It is always expected that an assembly instruction set should be optimized; however, no instruction set can be the best in all aspects. An optimization promotes some features and restrains other features. Figure 7.1 exposes some imaginable basic design dimensions. Obviously, there are many dimensions to handle or many requirements to follow while designing an instruction set.

At a certain time, an instruction set could be "the perfect one" for one application. However, this instruction set may not be suitable at all for another application. An assembly instruction set can be regarded as state-of-the-art, but not as "the best."

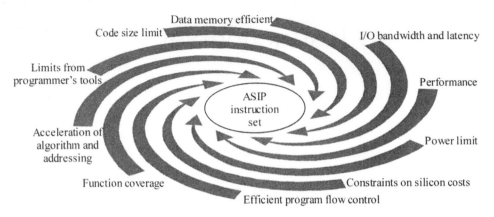

FIGURE 7.1

Design space of an ASIP assembly instruction set.

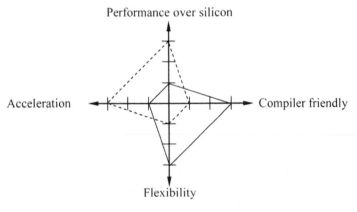

FIGURE 7.2

Trade-off among design dimensions assuming similar design efforts.

When designing a processor, we can focus on and optimize some features or parameters. At the same time, other features will be restrained. The trade-off among the four most important dimensions is illustrated in Figure 7.2.

Figure 7.2 shows that different designs have different focuses. Designs with high performance will lose flexibility; designs for general digital signal processing goals of high flexibility will lose performance.

For each design, profiling on these four mandatory dimensions in Figure 7.2 shall be conducted. Performance over silicon is always a requirement from semiconductor companies. It is obvious that necessary function acceleration should be used and unnecessary functions should be eliminated.

The concept of a compiler-friendly instruction set should be discussed briefly before reading the chapter about toolchain for assembly language design. Compiler-friendly means that the compiler (usually C compiler) can translate high-level source code to assembly code and use the assembly instruction set in an efficient way. A compiler-friendly instruction set usually is characterized by generality and orthogonality (for example, any instructions can use any register). Unfortunately, in ASIP instruction sets there are by definition application-specific instructions where generality and orthogonality may have been sacrificed for performance reasons. The compiler may need extensive help in order to use these instructions.

After code processing, a compiler translates the source C code to an intermediate representation (IR) code. The IR code is the bridge between the C and the target assembly code. To adapt a compiler generator to an assembly language, first an assembly language template table should be created, containing all assembly instruction patterns (templates). The IR code then is analyzed, and a task flow graph tree is generated from the IR code. The task graph is checked step by step, and the assembly templates are matched to each part of the IR task graph. Until the complete task graph is replaced by assembly templates, the IR code (the translation of the source code) is mapped to the assembly code [2]. During the template replacing process, if the compiler cannot match the corresponding code to CISC instructions, the design of the CISC instructions will not be compiler-friendly. Reasons could be that the compiler is not advanced enough or that the CISC instruction is so complicated that it cannot be supported by the compiler. Later in this chapter we'll see that an ASIP design usually includes special and not-compiler-friendly instructions.

By linking the marked points for a design in Figure 7.2, a quadrilateral can be formed. The quadrilateral can be interpreted as an effort space with fixed design costs. Different designs will have different quadrilaterals. However, the areas of all quadrilaterals will be about the same if the design costs are in the same range. The area of a quadrilateral can be larger by putting in more design effort; for example, introducing full custom silicon design.

If the quadrilateral is moved toward upper-left, performance will be enhanced, but the flexibility will be lower. Figure 7.2 illustrates a rule of thumb for processor design: improving one feature will make another worse. A focus of performance over silicon induces lower flexibility. High flexibility and high performance cannot be achieved at the same time.

10%–90% Code Locality Rule

Recall the discussions in Chapters 1 and 3: The primary result of source code profiling is the identification of the 10%–90% code locality of the application. The 10%–90% code locality rule is [7]:

> **BOX 7.1** 10% of instructions take 90% of the runtime, and 90% of instructions take 10% of the runtime.

This means that 90% of the instructions are for function coverage and are seldom used. The 10% most used instructions should be optimized to enhance the performance of the processor. Finally, what to accelerate by hardware and what to leave in software will be decided according to the rule of HW/SW codesign, described in Chapter 4.

ASIP Design Methodology

The architecture and instruction set of an ASIP should be optimized for a class of specific applications. The only required flexibility is within the application domain. Performance over silicon cost should be maximized under the required flexibility. For designing an ASIP DSP, the seven inputs shown in Figure 7.3 should be taken into account.

The coverage of source code profiling comes from the definition of application coverage. Requirements on flexibility and performance can be predicted by profiling the source code, checking necessary control features, and estimating memory as well as I/O costs. Before the instruction set design has started, the architecture should be proposed based on requirements on flexibility, performance, and cost from product requirements. A basic template of a RISC instruction set can be specified based on available references of the proposed architecture.

Design of an assembly instruction set is an experience-based activity because of the enormous decision factors. Many factors are not yet modeled and analyzed. This chapter intends to provide the necessary information in order to design an instruction set based on limited experience. Architecture should be proposed as the initial point of the design activity as soon as the product is specified. The templates discussed in Chapter 3 should be useful for the architecture decision.

As shown in Chapter 3, a DSP processor usually is based on a RISC core plus CISC (accelerated) features. For a DSP processor without strong requirements on program flow control, the basic RISC instructions can be simplified further. For a DSP processor with many control features, more instructions can be added to the basic RISC instruction set for more flexibility.

Experience from **Junior** was collected in Chapter 5. Eleven conclusion points indicate the distance between C code and an assembly instruction set of a DSP

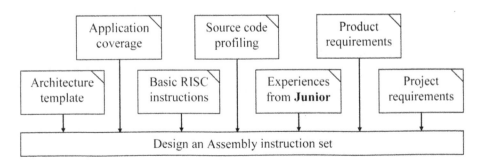

FIGURE 7.3

Inputs of an instruction set design.

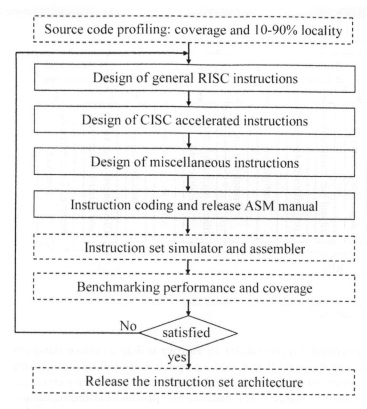

FIGURE 7.4

Instruction set design flow.

processor. The collected ten points from Chapter 5 can be a good starting point for designing a qualified instruction set. The basic instruction set design flow is shown in Figure 7.4.

In Figure 7.4, ASM stands for Assembly language. Design of an instruction set will be divided into the design of general (RISC) instructions and the design of accelerated (CISC) instructions. The output from the instruction set design activity is an assembly instruction set manual and requirements for microarchitecture design. The instruction set manual contains the functional specification of each instruction and assembly code, and the binary coding of each assembly instruction.

The first step of an ASIP instruction set design is to select the architecture and its instruction set template (default RISC ISA). By profiling the applications, simplification and enhancement of the RISC instruction set can be specified. According to the 10%–90% code locality rule, an instruction from the selected template can be cancelled if it is infrequently used. Accelerated functions should also be identified during profiling. The most frequently used subroutines may need to be accelerated. After specifying RISC and CISC instructions, the available components of the

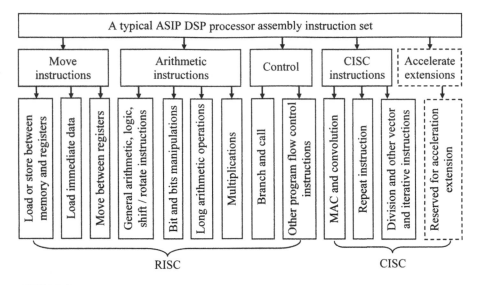

FIGURE 7.5

Classification of an instruction set[1].

datapath can be enumerated. Opportunities to further utilize available datapath functions should be investigated. If the coding space is not fully used, extra instructions could be implemented on identified opportunities without extra cost.

Product requirements (see Figure 7.3) constitute the product definition, including the contents of the product packet (the chip or the RTL core, the assembly programming toolchain, the assembly subroutine library, and the software debugging board), the function coverage, the performance, the expected volume, and the range of the price.

Project requirements (see Figure 7.3) constitute the project management, including the requirements of the team competence and experience, the design flow, the profiling tool, the design entry, and the simulation and verification tools.

7.1.2 Classification of General Instructions

A DSP instruction set usually is divided into two groups: the RISC subset and the CISC subset, see Figure 7.5. All DSP processors, both ASIP DSP processors and general-purpose DSP processors, need RISC subset instructions for handling general arithmetic and control functions. CISC subset instructions are used for special arithmetic functions and low-level algorithms.

---

[1]Actually, to improve code efficiency, more functions are added to the 10% frequently used instructions (running 90% time). CISC instructions can be found in arithmetic and control instructions of most DSP processors.

If a processor is required to supply a rather wide set of complex control functions, the RISC instruction subset could be similar to an instruction set of a general RISC processor.

The RISC instruction subset usually is divided into three groups: instructions for basic load and store, instructions for basic arithmetic and logic operation, and instructions for program flow control.

CISC subset instructions can be divided into two categories: normal CISC instructions and instructions for accelerated extensions. The normal CISC instructions are specified in the assembly instruction set manual during the design of the processor core. The normal CISC instructions are completely decoded by the control path of the core. The arithmetic circuits for the CISC functions are part of the datapath in the core.

Instructions for accelerated extensions (accelerator instructions) are reserved for hardware accelerators. They usually are specified after the design of the core. The format of these instructions is regulated during the design of the processor core, and the accelerator instructions are recognized and partly decoded by the core. The final decoding is done within each hardware accelerator. Design of instruction for accelerated extensions will be discussed in detail in Chapter 17.

7.1.3 Design of General RISC Subset Instructions

Assembly instruction set designs are based on design experience. Basic assembly instructions can be found from textbooks and available instruction sets. Starting from these basic instructions, a processor design can be speeded up. Design for general RISC instructions can be simply illustrated as in Figure 7.6.

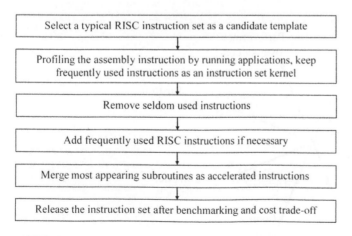

FIGURE 7.6

Designing an instruction set based on templates.

Select a Template for RISC Subset Instructions

To design a fixed-point ASIP DSP, it is preferable to select an available DSP instruction set close to the requirements instead of a general RISC instruction set as the template. There are two main reasons for this:

1. The special data-type features of a fixed-point DSP processor cannot be found in a general RISC instruction set.

2. A DSP requires special DSP arithmetic operations, such as multiplication and accumulation.

Reference instruction sets can be found from publications and open sources. For example, the instruction set, the simulator, and the RTL code of the **Senior** processor core is available on the web page of this book. There are also instruction sets from large DSP semiconductor companies, which are released and free to use. Using an available instruction set kernel could also mean access to programming tools and reference designs. It is beneficial to use an available instruction set if the requirements of an ASIP design are not special. Actually, the kernel part of a DSP RISC instruction set has strong common features, and these common features will be discussed in this chapter. Several DSP instruction sets are free to use (for example **Senior** and some old DSP processors from ADI). Several RISC instruction sets are also open and free to use, such as the early MIPS instruction set and OPENRISC [4].

Following Figure 7.7, the principle of instruction set selection depends on the application requirements. If more control features are required, for example, an ASIP DSP for automatic control system, a RISC template is preferred. The final ASIP product will be rather like a RISC with DSP features. If an ASIP DSP is a slave processor for computing acceleration (e.g., an audio CODEC controlled by a master MCU), a simple DSP template is the choice [5]. The final ASIP product will be more application specific.

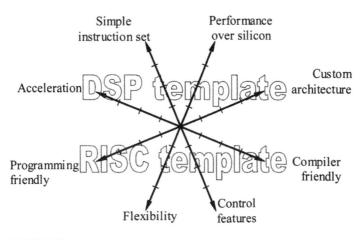

FIGURE 7.7

How to select an instruction set template.

The design process will be:

1. Get and analyze several available instruction sets.

2. Select one instruction set that is the most suitable solution.

3. Benchmark the instruction set (following BDTI to evaluate general DSP instructions and following applications to evaluate special instructions).

4. Decide to use it or not. If you decide to use it, a further decision on the degree of modification will be specified.

In this book, an ASIP DSP is assumed to be a slave processor without strong requirements on control features. **Senior**, designed and introduced by my research team, is suggested as a template of a DSP instruction set, see Table 7.1. In order to adapt **Senior** to different applications, acceleration features will be added.

To select a template, it is important first to evaluate the toolchain for assembly programming. For example, the simulator and assembler source code are essential if new instructions should be added. The way to evaluate an assembly programming toolchain will be discussed in the next chapter.

Remove Infrequently Used Instructions

Some RISC instructions are intended for microcontrollers and seldom are used by DSP algorithms. For example, DSP processors do not use unsigned arithmetic except for special cases like data packing and depacking. Also, instructions for string processing can be cancelled from a DSP instruction set. If an infrequently used instruction consumes a significant amount of logic gates and instruction encoding space, this instruction should be removed.

When an instruction is removed, the cancelled function shall be replaced by an assembly subroutine in order to keep the function coverage of the instruction set. For example, NEG is an instruction performing negation of an operand. The function is $R <= INV(R) + 1$. The NEG instruction can be replaced by the following instruction: $R <= 0 - R$ (0 is immediate data).

There is another example: The instruction "rotate right or left N bits" seldom is used. This instruction can be replaced by rotating one bit at a time. A subroutine of multibits rotation based on the instruction of rotating one bit should be coded.

Add New RISC Instructions

Each ASIP has its own features. A frequently used small subroutine may not be available as an instruction of a normal RISC instruction set. To make an instruction of this subroutine may significantly improve the performance of the processor.

A function "count leading identical bits from MSB and return the result to a register" is very useful for software-based floating-point implementations. It is also required by lossless coding and decoding algorithms. For example, if the input is %0000 0101 1001 0011, the result will be 5. This function is used often in variable-length coding and Huffman coding [6]. However, this function is usually not available in a normal instruction set.

Table 7.1 Typical Template Choices.

Template	Availability	Applications	Design goal	Source
Openrisc	Free	Control plus simple DSP	Flexible, control features	www.opencores.org
Senior	Free	Voice, Audio, DSP control	Low power, low silicon	www.da.isy.liu.se

Bundling Operations into One Instruction

When operations or instructions mostly appear together at the same time in the same order, combining instructions or bundling operations into one instruction can significantly improve the performance and shrink the code size.

Some pre- and postoperations can be bundled together to become a frequently used instruction. To fully utilize the available datapath, possible pre- and postoperations may be bundled into one custom instruction in order to minimize the code cost.

Some arithmetic operations also frequently appear together. Bundling these instructions into one instruction is a good choice if the bundled instruction is used frequently. For example (this example will be further discussed in Chapter 12), a frequently used DSP instruction is R $\le$ ABS(A − B). It can be implemented using an ALU with two full adders. Another frequently used instruction is accumulation of absolute values, R $\le$ R + ABS(A). Designers should aware the increased delay of logic circuit after bundling functions into one instruction.

7.1.4 Specify CISC Instructions

The decision of a CISC instruction is based on 10%–90% code locality identified during profiling. A CISC instruction usually is specified by bundling a chain of multiple arithmetic operations from the most frequently appearing subroutine into one or several instructions. A subroutine identified by a profiling tool is a branch of a tree in a task flow graph, consisting of a piece of code without branch. A subroutine could be one or several lines of C code. If the subroutine code is long, several instructions are needed in order to accelerate it.

The preferred bundling of function acceleration is to assign many arithmetic operations, either a long operation chain or massive parallel operations, into one instruction and then map this instruction to an existing datapath. The possibility for this, however, is low. Normally, only a short arithmetic operation chain or part of a chain can be mapped to an existing datapath. If a long operation chain has to be implemented, extra hardware components must be added. Trading off instruction granularities is an ASIP design challenge during the design for function acceleration. If an accelerated instruction contains many arithmetic operations, the performance will be much improved. At the same time, the compiler may not be able to recognize the opportunity to map such a long chain of functions to one instruction.

Application-specific instructions can be semiautomatically generated by analyzing and extracting operations from a control flow graph. Such solutions are available from Tensilica [9] and LISA [9] design flows. This is a new design method exposing a new design dimension for future embedded systems.

It is not difficult to identify hot subroutines from source code profiling and to bundle arithmetic operations in these subroutines into CISC instructions. The problem is compiling for acceleration. This is how to match C or IR code to accelerated CISC instructions. There are two methods to help the compiler.

One way is to keep the level of instruction granularity low so that the compiler can match the intermediate representation code with the template representing function acceleration. This is an extension of the current method for instruction selection in a compiler. The trend of the academy is to accept longer and more complicated templates. So far, an accelerated template cannot be long. There is still a long way to go in order to meet the requirements from the industry.

The second way is to annotate the C code so that the compiler can be guided to use accelerated instructions or assembly libraries following the annotation. In this case, the compiler will replace the annotated C code with a precoded assembly subroutine. This method is called compiler-known functions. To use this method, the C code must be analyzed and modified for the methodology. Because the cost of source code adaptation is acceptable, this method is widely used in the industry today.

7.1.5 **For Undergraduates: From** Junior **to** Senior

Most DSP source code is written in C; therefore the **Junior** assembly instruction set was implemented based on the primitives of the C language. The performance of this instruction set was benchmarked using several typical DSP kernel algorithms. The weaknesses of the **Junior** assembly instruction set were identified. Important conclusions in the previous chapter were that an assembly instruction set should be much more comprehensive, more linked to hardware, more capable of utilizing parallelism, and more supportive for array and loop computations.

By knowing and avoiding the weaknesses of **Junior**, **Senior** is implemented with a compact yet comprehensive DSP instruction set. Through the book, **Senior** will be used for various signal processing applications such as voice and audio signal processing. With extendible instruction set features, **Senior** will support hardware acceleration. It can also act as a task manager for vector and parallel processors.

The starting point of the instruction design in this chapter will be the analysis of the weaknesses of the **Junior** instruction set. Basic instructions of **Junior** and the discussion on **Junior** in Chapter 5 will be used as design references. The 10 drawbacks listed in the previous chapter will be quoted again. Advanced instructions or advanced microoperation will be added one by one to some instructions to eliminate the listed drawbacks. The design methodology for the **Senior** assembly instruction set is given in Figure 7.8.

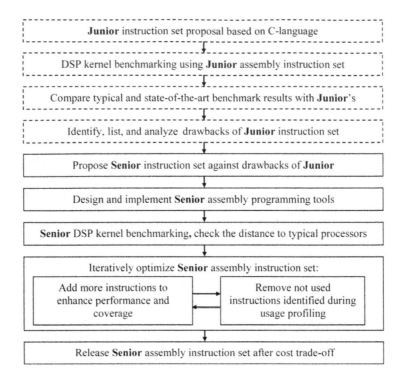

FIGURE 7.8

Senior assembly instruction set design process.

7.2 DESIGNING RISC SUBSET INSTRUCTIONS

7.2.1 Data Access Instructions

A data access instruction moves data from one place to another. To simplify and speed up instruction execution, RISC instructions use operands only from the register file or the instruction code (immediate data). Data stored in data memories must be loaded to the register file before RISC computing, and results should be stored to data memories after computing. In most DSP processors, data in special registers cannot be directly used as arithmetic operands. In order to manipulate data in a special register, the data must be moved to the general register file first. After the data manipulation, the result must be moved back from the general register file to the special register.

Requirements on data access instructions include the bandwidth, the latency, the address space, and addressing modes. The bandwidth is the number of bits that can be transferred in one clock cycle on average. Latency is the time interval from the start of addressing until a complete data word is transferred. Bandwidth and latency are determined from the processor requirements. It is an advantage if the address space enables data to be moved from any addressable storage component to any other addressable storage component, including all data memories, all I/O ports, and all addressable registers. Some pipeline registers do not need to be addressable.

DSP Data Types

Data types are used to classify operands; a data word (operand) is classified according to its numerical representation and word width. Data involved in mathematical computing in DSP hardware is a two's complement data. In most fixed-point DSP processors, the width of a data word is 16 bits, and the most used data type is the 16-bit signed two's complement data. 8-bit data in a fixed-point DSP processor is called a byte or half-word. A 32-bit word in a fixed-point DSP processor is called double-precision data. A data precision of 64 bits seldom is used in embedded computing. Most DSP processors are fixed-point processors. If floating-point operations are necessary in a fixed-point processor, floating-point computing can be emulated in software.

Special attention should be paid to internal data types of a fixed-point DSP processor. Internal data types do not have to be fully exposed to the programmer, but hardware designers should understand all details of the internal data types. Different data types are used in the MAC unit. For example, in a 16-bit DSP processor, the internal MAC operands could be 40 bits wide including 8 guard bits, which may not be observable except in the case of context switching. After saturation and truncation operations, only 32-bit data will be observed by the programmer. In a 24-bit processor, the internal MAC operands could be 56 bits, including 8 guard bits. After saturation and truncation operations, only 48-bit data can be observed. In a 16-bit ALU, the internal data type is usually 17 bits.

A DSP processor seldom handles text (strings) [7] since such processing usually is done by a microcontroller. A low-end DSP processor could support 4-bit data called binary-coded decimal (BCD) or packet decimal.

Data Access in *Junior* and Other RISC Machines

By benchmarking DSP kernels on the **Junior** instruction set, inefficiencies were identified from addressing modes and load/store instructions. The missing instructions from the **Junior** instruction set include moving data between the general registers and the special registers. For example, it is essential to be able to move data from the general registers to an accumulation register for concatenation of two or three data words to double-precision data in this register. When loading data into an accumulator register in the MAC, guard bits should be added implicitly. For context switching, data swapping is required between a 40-bit long word in the MAC unit and three general registers. After double-precision computing or iterative computing, 40-bit data shall be transformed to 16-bit data and saved in a 16-bit general register. None of these functions were covered by the **Junior** instruction set. The reason is that the C language does not adapt the data format to the specific underlying hardware.

General Load/Store Instructions

In order to design a complete instruction subset for data access, all data access paths shall be explicitly marked between addressable storage components. Suppose there are two data memories in an ASIP DSP core; then data transfers can be specified using Figure 7.9.

According to the specified data access channels in Figure 7.9, load, store, and move instructions of **Senior** shall include:

1. Load data from DM0 to GRF (general register file).

2. Load data from DM1 to GRF.

3. Store data from GRF to DM0.

4. Store data from GRF to DM1.

5. Load data from I/O port to DM0.

6. Store data from DM0 to I/O port.

7. Load data from I/O port to GRF.

8. Store data from GRF to I/O ports.

9. Move data from GRF to ACR.

10. Move data from ACR to GRF.

11. Move data from GRF to SRF in AGU (address generator unit).

12. Move data from SRF in AGU to GRF.

13. Move data from GRF to SRF in control path.

14. Move data from SRF in control path to GRF.

15. Load immediate data from control path to GRF.

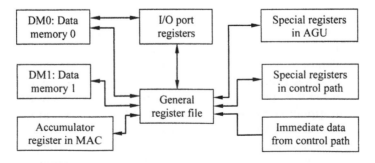

FIGURE 7.9

Data access channels in a typical DSP processor.

GRF is the general register file, SRF is the special register file, and ACR is an accumulator register. It is essential to be able to load immediate data into a general register. Immediate data is a constant such as a control vector (for configuring an interface port, configuring a control register, etc.), a program parameter (data allocation, data buffer margin, start and stop address, and program flow control parameters such as the number of iterations), or a data value to be processed.

If data accesses between some hardware modules appear infrequently, the corresponding hardware links can be eliminated to save silicon area. In order to allow data accesses between all addressable storage components, these accesses must be possible to emulate by running the assembly instruction emulation. Otherwise, the processor will not be flexible enough. All emulations use the register file as the media (see Table 7.2; note that emulation requires multiple instructions).

By listing all possible data accesses with emulated solutions, it has been verified that all possible data accesses are feasible and that data can be moved between any two addressable storage units. For example, in most DSP processors, data in the special registers cannot be processed directly by the ALU, because the ALU accepts only operands from the general registers. An ALU operation on data in a special register will be executed using the following code:

```
MOVE R0,SR5; Read from special register 5
OR R0,8    ; Set a bit
MOVE SR5,R0
```

Table 7.2 Data Access Instruction Emulation.

Operation	From	To	Emulation
Move	DM0	DM1	Load DM0 to GRF; Store GRF to DM1
Move	DM1	DM0	Load DM1 to GRF; Store GRF to DM0
Load	I/O port	DM1	Load I/O port to GRF; Store GRF to DM1
Store	DM1	I/O port	Load DM1 to GRF; Store GRF to I/O port
Load	DM0	ACR	Load DM0 to GRF; Move GRF to ACR
Store	ACR	DM0	Move ACR to GRF; Store GRF to DM0
Load	DM1	ACR	Load DM1 to GRF; Move GRF to ACR
Store	ACR	DM1	Move ACR to GRF; Store GRF to DM1
Load	DM0	SRF	Load DM0 to GRF; Move GRF to SRF
Store	SRF	DM0	Move SRF to GRF; Store GRF to DM0
Load	DM1	SRF	Load DM1 to GRF; Move GRF to SRF
Store	SRF	DM1	Move SRF to GRF; Store GRF to DM1

If data widths are the same, moving data between two registers is easy. As described in Chapters 2 and 3, the accumulator register is wider than a general register. In a typical 16-bit fixed-point processor, the data width of a register in GRF is 16 bits. The data width of an ACR is typically 40 bits including the guard bits. Data-type conversion will be implicitly executed while accessing data between GRF and ACR, according to Table 7.3.

In Table 7.3, $\{X[n-1, 0], Y[m-1, 0]\}$ is a concatenation operation to form a longer word. It uses X as the higher part of the longer word and Y as the lower part. The new word length is $n + m$. The operation $\{8\{X[15]\}\}$ creates an 8-bit new word by copying $X[15]$ eight times. Rd is the destination register in GRF. Ra and Rb are the operand registers in GRF.

Moving a result from an accumulation register to a general register usually includes the implicit operations rounding, saturation, and truncation. The behavior of this move instruction is:

```
TMP_Signal [23:0] = Truncate (Round (ACR [39:15]));
R [15:0] = Saturate (TMP_Signal [23:0]);
```

For supporting asynchronous tasks and executions of multiple tasks, a stack can be used for saving the context. In this way a program can be interrupted temporarily and hardware resources can be used by other tasks. Data accesses for stack

Table 7.3 Implicit Data Format Conversion While Accessing ACR.

Instruction	Operation	Application
Load ACR lower part	ACR <= {ACR[39:16], Ra[15:0]}	Restore ACR low
Load ACR high part	ACR <= {ACR[39:32], Ra[15:0]; ACR[15:0]}	Restore ACR high
Load ACR guard	ACR <= {Ra[7:0], ACR[31:0]}	Restore ACR guard
Load ACR high and guard	ACR <= { {8{Ra[15]}}, Ra[15:0]; 16'b0 }	Prepare an operand
Load ACR	ACR <= { {8{Ra[15]}}, Ra[15:0]; Rb[15:0]}	Prepare a long operand
Store ACR higher part	Rd <= ACR[31:16]	Push high
Store ACR lower part	Rd <= ACR[15:0]	Push low
Store guard bits in ACR	Rd [7:0] <= ACR[39:32]	Push guard
Store result	Rd <= Saturation (round(ACR[31:16]))	Store result

Table 7.4   Stack Instructions.

Instruction	Load/Store operation	Address operation
Push R	Store R to stack memory	Postoperation: Address <= Address −1
Pop R	Load stack memory to R	Preoperation: Address <= Address +1

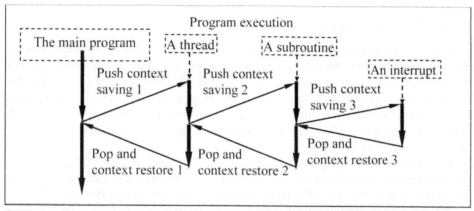

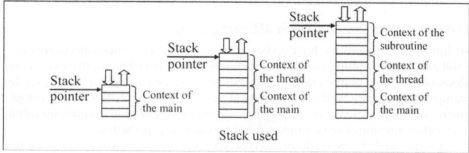

FIGURE 7.10

How the stack is used.

operations include stack push and stack pop, see Table 7.4. The stack is a FILO (first in/last out) buffer, meaning that the latest stored data shall be used first.

The principle of context saving using a stack is shown in Figure 7.10. Following the "last in/first out" principle, the context will be temporally stored and exactly restored. Stack instructions can be emulated using normal load/store instructions with special memory addressing modes.

While saving the context, we usually move an accumulation register value to the general registers first and then push them to the stack memory. Three general registers are used for storing an accumulator register, including the guard bits [39:32], the high word [31:16], and the low word [15:0] of an accumulator register. Likewise, while restoring the context, these three words in the stack will be popped to the general registers first and finally moved to an accumulation register. Three move operations between general registers and an accumulator register can be in serial using normal instructions or in parallel using accelerated instructions if the time cost of context saving is critical.

Addressing algorithms for load, store, and move instructions will be discussed later in this chapter.

7.2.2 Basic Arithmetic Instructions

Basic arithmetic instructions are also called ALU instructions. They include all arithmetic, logic, and shift operations executed in the ALU. The ALU instructions get their operands from the general register file or from immediate data carried in the machine code. Since a MAC unit can be found in all DSP processors for iterative computing, an ALU is not required to support iterative computing. In a 16-bit fixed-point DSP processor, 16-bit two's complement integer data is the default data type for all ALU instructions.

Two types of basic arithmetic operations can be found in most DSP processors. The first type is executed in the ALU and performs arithmetic operations based on native data length. The second type is executed in the MAC and performs arithmetic operations with double precision (twice the native width).

Microoperations Can Be Used in ALU Instructions

Arithmetic instructions are built up based on arithmetic computing primitives, like a full adder. The function of this primitive is addition, ADD. Due to the two's complement data type, subtraction, $A - B$, can be executed using the full adder by manipulating one operand and the carry in, $A - B = A + INV(B) + 1$. INV is logic invert. This indicates that by selecting microoperations around computing primitives, other functions can be implemented based on this primitive.

By analyzing ALU instructions in a typical RISC instruction set, several microoperations can be found, such as kernel operations, preoperations, and postoperations (see Figure 7.11). Opportunities exposed in Figure 7.11 include the number of operands that can be selected, the number of preoperations that are available, what the kernel operations are, and the number of postoperations available.

By fetching data from different operand sources; selecting different preoperations on operands; and selecting different carry-in bits, flag operations, different postoperations, and ways of writing the result, $7 \times 3 \times 3 \times 2 \times 4 \times 4 = 2016$ choices of possible ALU operations can be found in Figure 7.11. By using 12 control bits, it is possible to make 2016 arithmetic instructions. Of course most of these instructions are not practically useful. However, the choices in Figure 7.11 indicate that an instruction set exposes only a few percent of all possible datapath functions.

For executing the "absolute operation ABS(A)," the condition for conditional invert in Figure 7.11 is used. The MSB is the sign bit of operand A (OPA). If the sign bit is 1, then input OPA is negative data, and inversion will be performed on OPA.

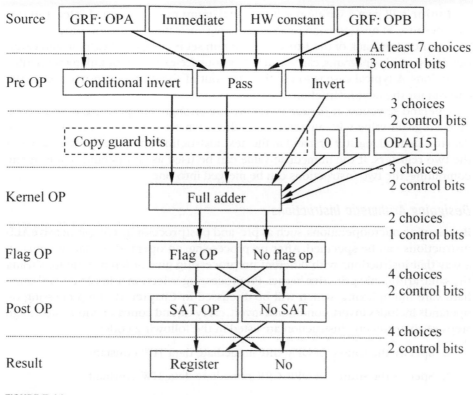

FIGURE 7.11

Microoperations in an arithmetic unit.

The MSB, OPA[15], will be used as carry in. If OPA is negative, INV(OPA) + 1 will be conducted. If OPA is positive, the sign bit is 0 and OPA + 0 will be the result. The absolute instruction R <= ABS(OPA) can thus be implemented using conditional invert on OPA and OPA[15] as carry in.

A saturation operation is essential after an absolute operation. If OPA is the maximum negative value, then the output should be the maximum positive value. According to the range definition of two's complement data, the maximum positive value is:

```
R = ABS (16'h8000) = 16'h7FFF;
```

If a saturation operation on the result is not available, then the result will be wrong because:

```
R = (INV (16'h8000) + 1
  = 16'h7FFF + 1
  = 16'h8000;
```

Junior will give a wrong result on a maximum negative signed integer in this case because of missing configuration on carry-out/saturation computing.

Another drawback of the **Junior** instruction set is the lack of carry-in and carry-out. In assembly programs, carry-in and carry-out are essential for linking sequential operations. A typical case is to use the carry-out of the previous instruction as the carry-in for the next instruction.

Operand selection in **Junior** is not flexible enough. Suppose we want to compare a register value with an immediate value; then we have to load the immediate data to a general register first and in the next instruction compare the two values in the two registers. If the immediate value can be used as an operand of arithmetic computing, the two instructions can be merged into one.

Designing Arithmetic Instructions

By selecting microoperations, such as pre- and postprocessing, comprehensive ALU instructions can be specified. After preprocessing, an operand is changed in such a way that instructions with different preprocessing give different functions using the same arithmetic primitive kernel. Preprocessing selection is implemented using hardware multiplexing, which will be discussed in Chapter 10. Preprocessing on operands includes invert, conditional invert, carry-in, and concatenation. The design steps of an arithmetic instruction are listed in the following code:

1. Specify the source of OPA: GRF, immediate data, HW constant.

2. Specify the source of OPB: GRF, immediate data, HW constant.

3. Specify preprocessing on OPA: INV, conditional-INV, or NOP.

4. Specify preprocessing on OPB: INV, conditional-INV, or NOP.

5. Copy sign bit as the guard bit for OPA and OPB.

6. Prepare for Carry-in: 0, 1, Carry flag, or OPA[15].

7. Kernel operation: 17bits full adder with carry in.

8. Flag operation: enable or disable.

9. Saturation operation: enable or disable.

10. Result store: to register file or NOP.

Following this code, arithmetic instructions can be selected from the 2016 choices in Figure 7.11. After profiling, the most used arithmetic instructions are listed in Table 7.5.

There are four instructions consisting of $2 \times 4 + 3 \times 4 + 2 \times 1 + 1 = 23$ of 2016 combinations in Table 7.5. Arithmetic instructions in Table 7.5 consist of four instructions, eight operations, and 23 combinations.

Other useful arithmetic operations were identified from the source code profiling. Four instructions are specified according to these operations in Table 7.6.

Table 7.5 General Arithmetic Instructions.

No.	Mnemonic	Carry/ borrow in	Saturation	Description	Operation
1	ADDN	0/1	yes/no	Addition without flag change	Rd <= ra + rb
	ADD	0/1	yes/no	Addition	Rd <= ra + 16'b imm
2	SUBN	0/1	yes/no	Subtraction without flag change	Rd <= rb − ra Rd <= 16'b imm − ra
	SUB	0/1	yes/no	Subtraction	Rd <= rb − 16'b imm
3	CMP	0	yes	Compare	Compare (ra, rb) Compare (ra, 16'b imm)
4	ABS	0	yes	Absolute	Rd <= abs(ra)

Table 7.6 Other Arithmetic Instructions.

No.	Mnemonic	Description	Operation
5	DTA	Difference of two absolute values	Rd <= |Rd| − |Rs|
6	ATD	Absolute value of the difference	Rd <= |Rd − Rs|
7	MAX	Maximum	Rd <= max (ra, rb) Rd <= max (ra, 16'b imm)
8	MIN	Minimum	Rd <= min (ra, rb) Rd <= min(ra,16'b imm)

DTA is an instruction examining the difference between two absolute values, which is used for checking the difference in energy between two variables. ATD is used for calculating the absolute difference of two values. One use for this is in motion detection.

There are four instructions consisting of $2 + 2 \times 2 = 6$ operation combinations in Table 7.6. MAX/MIN is used to find the maximum/minimum value in a group of data. Rather often, a DSP program needs to decide the gain or the scaling factor at certain processing points. MAX will be used for checking and avoiding possible overflow, and MIN will be used for checking the limit of the computing precision.

Arithmetic Instruction Emulation

Sometimes an instruction can be emulated by another arithmetic instruction. There are two advantages in emulating arithmetic instructions; the hardware design can

Table 7.7 Emulations of Arithmetic Instructions.

No.	Instruction to emulate	Emulation
1	NEG	SUB Rd IMM, Rs; IMM = 16'b0; the operation is: Rd <= 16'b0 − Rs
2	INC	ADD Rd Rd IMM; IMM = 16'b1; the operation is: Rd <= Rd + 16'b1
3	DEC	SUB Rd Rd, IMM; IMM = 16'b1; the operation is: Rd <= Rd − 16'b1

be simplified, and the binary coding space can be better utilized. Some frequently used arithmetic instructions are listed and emulated in Table 7.7.

To minimize the silicon cost, the emulated instructions are not implemented in hardware. However, emulated instructions can be implemented in the assembler. For example, if INC Rd is used in the assembly code, the assembler will generate ADD Rd, Rd, 1.

Single-Precision Data Manipulation

Data manipulation functions include word-level and bit-level logic operations. A word-level logic operation applies the same logic operation to each aligned bit of operand A and operand B. For example, the instruction AND OPA OPB means a bitwise AND operation on each bit of operand A and operand B:

```
Result [15] <= OPA [15] & OPB [15];
Result [14] <= OPA [14] & OPB [14];
......
Result [1]  <= OPA [1]  & OPB [1];
Result [0]  <= OPA [0]  & OPB [0];
```

Logic OR and XOR operations work in the same way. The most frequently used logic operations are specified in Table 7.8.

Flag manipulations are used for supporting the precomputing of branches and conditional executions. Examples of flag manipulation instructions are given in Table 7.9.

For program flow control and data frame packaging, functions such as setting, clearing, and reading bits are useful. Instructions including load/store one bit, test one bit, and set/reset one bit, are shown in Table 7.10.

Here the bit position pointer can be carried by the immediate $K[3:0]$ or the register $Rs[3:0]$. If the position of the bit is known during programming, $K[3:0]$ can be used for specifying it. If the position is a variable, general register $Rs[3:0]$ is needed for carrying the bit position.

Table 7.8   Logic Operations.

No.	Mnemonic	Description	Operation
1	ANDN	And without flag change	rd <= and (ra, rb)
	AND	And with flag change	rd <= and (ra, 16'b imm)
2	ORN	Or without flag change	rd <= or (ra, rb)
	OR	Or with flag change	rd <= and (ra, 16'b imm)
3	XORN	Xor without flag change	rd <= xor (ra, rb)
	XOR	Xor with flag change	rd <= and (ra, 16'b imm)

Table 7.9   Flag Manipulations.

No.	Mnemonic	Description	Operation
1	ANDF	Reset flags	flags <= (flags) logic and (K);
2	ORF	Set flags	flags <= (flags) logic or (K);
3	XORF	Toggle flags	flags <= (flags) logic xor (K);

Table 7.10   Basic Bit Manipulation Instructions.

No.	Mnemonic	Description	Operation
1	VWB	View one bit in Rd at position Rs or K	Carry_out <= Rd [Rs]; Carry_out <= Rd [K]
2	STB	Set one bit in Rd at position Rs or K	Rd [Rs] <= "1"; Rd [K] <= "1"
3	CLB	Clear one bit in Rd at position Rs or K	Rd [Rs] <= "0"; Rd [K] <= "0"
4	IVB	Invert one bit in Rd at position Rs or K	Rd [Rs] <= INV(Rd [Rs]); Rd [K] <= INV(Rd [K]);

In order to run lossless coding and decoding algorithms, especially variable-length coding, the instructions in Table 7.11 are useful.

For example, LED with Rs = 16'b0000101010100101, then Rd = 4. These instructions can be also very useful for software-based floating-point arithmetic.

Emulations of Data Manipulation Instructions

The data manipulation instructions in Table 7.12 can be emulated using other data manipulation instructions.

Table 7.11 Logic Operations Counting Leading Identical Bits.

No.	Mnemonic	Description	Operation
1	LED0	Count left leading 0's in Rs and save in Rd	
2	LED1	Count left leading 1's in Rs and save in Rd	Rd <= led (Rs)
3	LED	Count left leading identical x in Rs and save in Rd	

Table 7.12 Emulations of Arithmetic Instructions.

No.	Instruction to emulate	Function	Emulation
1	INV	Invert	XOR Rd Rs $FFFF; Rd <= Rs $\oplus$ $FFFF
2	CLN	Clean	XOR Rd Rs, Rs; Rd <= Rs $\oplus$ Rs
3	SET	Set	OR Rd Rs $FFFF; Rd <= Rs logic or $FFFF

Table 7.13 Behavior of Shift and Rotate.

No.	Function	Result specification
1	Arithmetic right shift	$\{N'A[15], A[15:N]\}$; $C <= A[N-1]$
3	Logic right shift	$\{N'0, A[15:N]\}$; $C <= A[N-1]$
4	Logic left shift	$\{A[15-N:0], N'0\}$; $C <= A[15-N+1]$
5	Rotate right with carry	$\{C, A[15:N]\}$ if $N=1$; $\{A[N-2:0], C, A[15:N]\}$ if $N>1$; $C <= A[N-1]$
6	Rotate right without carry	$\{A[N-1:0], A[15:N]\}$
7	Rotate left with carry	$\{A[15-N:0], C\}$ if $N=1$; $\{A[15-N:0], C, A[15:15-N+2]\}$ if $N>1$; $C <= A[15-N+1]$
8	Rotate left without carry	$\{A[15-N:0], A[15:15-N+1]\}$

N = the number of shift or rotate steps;
{ } = concatenation.

Single-Precision Shift and Rotate Instructions

The **Junior** instruction set does not have enough shift and rotate instructions. This is because shift operations in C are simple and hardware independent. Shift and rotate are used for scaling, bit test, program flow control, data packaging, depackaging (protocol processing), efficient coding, and redundant coding. It can also be used for software division.

While shifting, fill-in bits to vacant bit-positions must be specified. Sign bits are filled in for arithmetic right-shifts and zeros are filled in for logic-shifts. A rotate operation shifts one bit from one end of a register to the other end of the same register. As a special case, the carry-in C can be used as a fill-in bit while rotating.

Shift and rotate are controlled by either operand B or immediate K. Table 7.13 and Figure 7.12 describes the shift and rotate instructions.

The hardware cost of a barrel shifter is almost negligible in a DSP processor; therefore a barrel shifter is always used. This means that the number of execution cycles for a shift instruction is always one.

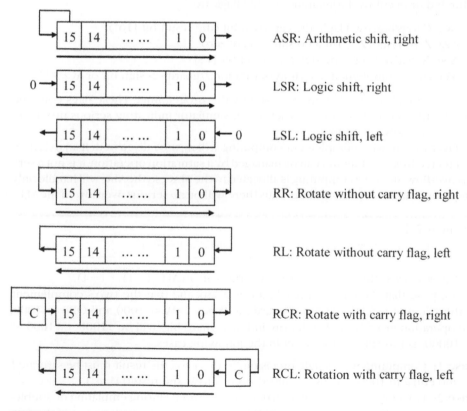

FIGURE 7.12

Shift and rotate operations.

7.2.3 Unsigned ALU Instructions

There are no dedicated instructions for unsigned arithmetic. The ALU performs unsigned arithmetic when input operands are unsigned. Unsigned data could be recognized on behavior level, but there is no way to find out if data is signed or unsigned by observing it in binary format. Unsigned arithmetic operations can be executed in an arithmetic unit based on two's complement computing hardware when MSB of operands are 0. It is up to the programmer to define if the arithmetic computing is based on 2's complement data or unsigned data. If unsigned arithmetic is executed, flags will be interpreted and used differently. Logic shift, rotate, and logic operations are based on unsigned data.

Designing ALU Flags

When arithmetic instructions are executed, flags are used in order to specify certain features of the result. They typically indicate if the result is zero or not, if the result is positive or negative, what the carry-out value of the result is, and if the result is saturated or overflow. Programmer visible flags are:

AC: Carry/Borrow Flag;//the carry-out bit (the result bit [16]).
AZ: Zero Flag;//Z = 1 if all result bits are zero.
AN: Negative Flag;//the sign bit (result bit [15]).
AV: Two's complement overflow/saturation;//guard <> sign bit or AC<>AN.

There are also flags that are not used by the programmer. These flags could be used by the hardware or the instruction set simulator indicating serious problems, such as overflow exception.

Overflow in two's complement computing will now be discussed. If an overflow is detected by guard bits and can be managed by a saturation operation, it is not a serious overflow. Iterative computing is allocated to the MAC, so the ALU will handle only simple and single-step computing. Thus there will be no serious overflow in the ALU.

Example 7.1

Discussion on an extreme case MAX-NEG + MAX-NEG in the ALU.

Let us discuss the extreme case of overflow in an ALU, $(-1) + (-1)$.

Suppose that the datapath resolution is 4 bits and that the two's complement arithmetic unit is 5 bits with one guard bit. When OPA = 1000 and OPB = 1000, the operation in a 5-bit full adder (including a guard bit) will be 11000 + 11000 = (1)10000. Let us analyze the result in the following cases.

Case 1: Computing with single precision. In this case, the result must be saturated and equal to 1000 because result[4] <> result[3] and result[4] = 1.

Case 2: Lower part computing with double precision. During emulating of double-precision computing, the carry-out bit will be used for concatenation of the two steps of computing. 1000 stands for unsigned 8 and 1000 + 1000 = 10000 stands for 16. The carry flag is 1 and the result is 0000. It is correct!

Case 3: Single precision without saturation. In this case, we can only say that the programmer is not experienced. This is a bug due to poor understanding of fixed-point arithmetic and wrong use of fixed-point computing hardware!

Suggestion: If single precision is used and there is no further computing requiring the carry flag, the result should be saturated.

7.2.4 Program Flow Control Instructions

Program flow control instructions include conditional and unconditional jumps, function calls, and other control instructions. For example, RETURN is a program flow control instruction associated with call and interrupt. Loop control instructions, stack instructions, and debug instructions are all program flow control instructions.

In Chapter 5, the features of conditional test and branch operations in C were reviewed. Mixing conditional test and conditional branch in C makes programming easy. However, the conditional branch in C requires extra computing for testing conditions. When DSP ALU instructions are executed, flags are generated implicitly so the extra cost of condition test can be eliminated. If flags are not available, the instruction "compare" could be used to set the flags. In an assembly language, we implicitly accept the fact that flags generated by previous instruction are conditions for the instruction to be executed. The behavioral model of a conditional branch is:

```
If condition is true, then jump to T else continue
The following instructions are executed if the condition is
not true
...
T: The first instruction pointed by the target address "T"
```

Basic requirements on branch instructions include the capability to test all conditions and to jump to any position in the program memory. In order to design a high performance processor, the time for conditional test and the time for making a jump must be minimized. No time is consumed for conditional testing if it is implicitly executed by previous arithmetic instructions.

In a pipelined processor, instructions following a jump instruction, already loaded in the processor, must be flushed when the jump is taken. The reason is simple: instructions following the jump instruction should not be executed in this case. The next instruction to be executed after the jump is pointed to by the target address of the jump instruction. The target address is calculated when the jump instruction is executed, and this address is immediately used for fetching the next instruction. If a processor pipeline consists of IF (Instruction Fetch), ID (Instruction Decoding), OP (Operand Fetch), and EX (Execution), the extra cycle cost of a jump taken will be three clock cycles (one clock cycle each for IF, ID, and OP). The following rule can be formulated: The extra clock cycle cost of a jump taken is the number of pipeline steps before the execution pipeline stage.

By analyzing a typical DSP program, conditional jumps can be classified into four cases:

Case 1: A simple conditional jump back to the top of an iteration loop (the most frequently used jump instruction in a DSP).

Case 2: A branch to select between two instructions: "if zero then A = B else A = C." Data-dependent execution is common in DSP subroutines.

Case 3: A branch to select between "continue" and "go to a subroutine."

Case 4: Call and return.

Repeat and Jump Hint

Case 1 is used in iterative algorithms, for example, running an FIR kernel program with N data samples. Suppose that K cycles are required to process one data sample; the total runtime will be $K \times N + P \times (N - 1)$ cycles. The number of conditional jumps will be $N - 1$. P is the cycles of pipeline stall while a jump is taken. P of the **Senior** DSP processor in the appendix is 3. The extra cycle cost, $P \times (N - 1)$ from the jumps, was recognized as the major drawback of **Junior**. To minimize this extra cost, a repeat instruction or a JUMP hint can be used instead.

Example 7.2

Calculate the cycle cost of an FIR execution of 40 sample data. The kernel cycle cost of FIR on each sample is $K = 16$ clock cycles, and the pipeline of the processor is IF, ID, OP, EX, and EX2.

In this processor, the cycles of pipeline stall is $P = 3$ (there are three pipeline stages before execution). The total cycle cost will be $K \times N + P \times (N - 1) = 16 \times 40 + 3 \times 39 = 657$ clock cycles.

Iterative computing consumes most of the runtime in DSP programs. Therefore repeat instructions are very useful in DSP processors. A repeat instruction takes M subsequent instructions and executes them N times. During the execution of a repeat instruction, the processor jumps back to the top of the iteration loop after executing the last instruction of the loop, without extra cycle cost from the jump taken. When the last instruction in the last iteration loop has been executed, the processor will continue with the next instruction after the loop. The repeat instruction is given in Table 7.14. The cycle cost of this repeat instruction is $CC = N * M + O$. O is the top-level overhead of the loop, usually only one clock cycle. The repeat instruction is not a RISC instruction.

Table 7.14   No Overhead Loop Control Instruction.

Operation	N	M
REPEAT	Number of iterations	Number of instructions carried by REPEAT

Example 7.3

Use Example 7.2 with the REPEAT instruction.

The total cycle cost will be $M \times N + 1 = 14 \times 40 + 1 = 561$ clock cycles. The extra cycle is consumed by the REPEAT instruction.

A Jump hint is usually a configuration bit of a conditional jump instruction. The jump hint indicates that a jump will be taken or not taken. When a jump instruction is fetched, the hint directs the fetching of the next instruction without pipeline stall. If the hint is taken, the next PC will be the target address instead of PC + 1. In this case, the target address must be available and precalculated. Hints can be added during programming or during debugging. For all iterative loops, taken is always the hint to lead the execution back to the top of the loop.

Example 7.4

Use Example 7.2 with JUMP HINT.

The total cycle cost will be $M \times N + P = 14 \times 40 + 3 = 562$ clock cycles. The three extra cycles are consumed by the pipeline stall of the last jump in the iteration.

Conditional Execution

While selecting a subroutine during execution, a conditional jump is needed, and several clock cycles will be lost when the jump is taken. For a long subroutine, the relative overhead of a conditional jump is durable. However, in case the conditional jump is part of a short inner loop and it is used for selecting one of two instructions (e.g., if C is true then A else B), then the overhead of the conditional jump will be significant. Conditional execution is a way to reduce this overhead.

Case 2 (earlier) appears quite often inside a subroutine of a DSP algorithm kernel. The corresponding pseudocode is:

```
JUMP to T if Zero;
      A = C;
      JUMP to Next;
 T: A = B;
 Next: ... ...
```

The cycle cost of this program in **Junior** is $3 + 3$ (not true) or $2 + 3$ (true), including pipeline stall, $P = 3$, when the jump is taken. If conditional execution is available, the following pseudocode is much simpler, and the cycle cost is only two clock cycles no matter what the condition is:

```
A=C;
A=B if Zero;
```

However, including the condition to test, the binary machine code will be longer.

Branch Instructions

Unconditional branch instructions are JUMP and CALL. For a conditional branch instruction, the jump is taken only when the condition is true. A conditional jump executes NOP (no operation) if the condition is not true.

When a jump is taken, the processor cannot execute the following instructions that are already fetched and not yet executed in the hardware pipeline. Instead, the processor has to load a new instruction from the new PC target address and flush all instructions already fetched. A stall of P clock cycles will occur while waiting for the fetching and decoding operands of this new instruction. The target address is the position in the program memory from where the new instruction is fetched after the jump is taken. The target address is calculated by the branch instruction.

The usage of flags must be specified for conditional jumps. Table 7.15 depicts how the flag combinations are used as jump conditions:

AC: Carry Flag of ALU
AZ: Zero Flag of ALU

Table 7.15 Jump Conditions.

CDT	Code	Description specification	Flag tested
–	0000	Unconditionally true	None
EQ	0010	ALU equal/zero	$AZ = 1$
NE	0011	ALU not equal/not zero	$AZ = 0$
UGT	0100	ALU unsigned greater than	$AC = 0$ and $AZ = 0$
UGE/CC	0101	ALU unsigned greater than or equal	$AC = 0$
ULE	0110	ALU unsigned less than or equal	$AC = 1$ or $AZ = 1$
ULT/CS	0111	ALU unsigned less than	$AC = 1$
SGT	1000	ALU signed greater than	$AN = AV$ and $AZ = 0$
SGE	1001	ALU signed greater than or equal	$AN = AV$
SLE	1010	ALU signed less than or equal	$AZ = 1$ or $AN! = AV$
SLT	1011	ALU signed less than	$AN! = AV$
MI	1100	ALU minus/less than	$AN = 1$
PL	1101	ALU positive/greater or equal	$AN = 0$
VS	1110	ALU has overflowed	$AV = 1$
VC	1111	ALU has not overflowed	$AV = 0$

AN: Negative Flag of ALU

AV: Two's complement overflow/saturation

It is not enough to check greater/less by using only the AN flag, because this flag will not be correct if the result is overflow. Instead the AV flag must be used together with the AN flag for checking greater/less. The overflow flag is set if the guard bit is not equal to the sign bit, AV = result[16] XOR result[15]. Checking greater/less using both AV and AN will be correct regardless if saturation is used on the result or not.

Jump instructions and call instructions are described in Table 7.16.

A flow control instruction carries binary coded information including the type of instruction, the operation code (the main operation of the instruction), the condition code (what conditions to check), the target address specifier (ways to decode the target address), and the target address (or part of the target address). Suppose the target address is 20-bit immediate data covering an addressing space of little more than one million instructions. Then the length of the machine code could be up to 34 bits (see Table 7.17). A 34-bit instruction is usually too long for an embedded processor, though.

R/D is the target addressing mode; R represents PC relative addressing (PC <= PC + instruction carried TA) and D represents direct addressing (PC <= instruction carried TA). The Target Address (TA) can be immediate data (R/I = I) or a register address (R/I = R). If TA is a register address, then TA = the register value, see Figure 7.13.

Other Flow Control Instructions

Other frequently used flow control instructions are listed in Table 7.18.

NOP is used for waiting and synchronization of operation dependencies.

Table 7.16 Flow Control Instructions.

No.	Mnemonics	Description	Operation
1	JUMP CDT I TA	Jump to indirect target address (TA)	If true PC <= PC + Target address
	JUMP CDT D TA	Jump to direct target address (TA)	If true PC <= Target address
2	CALL I TA	Call a subroutine indirect, stack PC	PC <= PC + Target address
	CALL D TA	Call a subroutine direct, stack PC	PC <= Target address

Table 7.17 Machine Binary Code Cost for Jump Instructions.

Type	Operation	CDT	R/D	R/I	TA	Total
3 bits	4 bits	5 bits	1 bit	1 bit	20 bits	34 bits

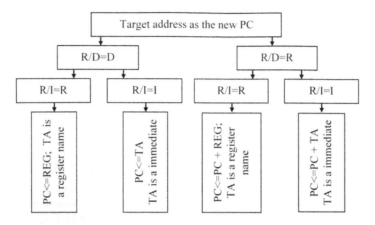

FIGURE 7.13

Target addressing modes for a jump.

Table 7.18 Other Instructions for Program Flow Control.

No.	Operation	Configuration	Description
3	NOP	No	No operation
4	RET RETI		Return from subroutine Return from interrupt
5	SLEEP	Pin or Timer	Let the core sleep, waiting for timer or I/O activity Different sleep modes based on immediate data

There are two types of return instructions: return with flag and return without flag. When calling a subroutine, flags will be used continually and not saved. After executing a subroutine of a CALL, flags from the subroutine execution will also be used continually. However, interrupts may happen at any time, and an interrupt may be accepted just before a conditional jump instruction is loaded. In this case, if the flag is not saved as contexts and restored after the interrupt service, the conditional jump right after the interrupt service will be executed based on flags computed by the interrupt service routine. To avoid this, both the next PC and the flags must be pushed to the stack as contexts, when an interrupt is accepted. After the interrupt service, the RETI instruction will pop both the PC value and the flag pushed to the stack.

A sleep instruction is required for reducing the power consumption during longer periods of inactivity. The difference between "sleep by stopping the clock" and "sleep by turning off the power supply" must be specified during the instruction set design. One clock cycle latency is enough for waking up the processor from clock sleep. To wake up from "sleep of the power supply," tens of thousands of clock cycles might be required.

7.3 CISC SUBSET INSTRUCTIONS

CISC instructions, in general, are irregular instructions for function acceleration, run-time minimization, code cost reduction, silicon cost reduction, and minimization of power consumption. Advantages are promotion of performance and reduction of relevant costs. Disadvantages are compiler unfriendly instructions and increased hardware complexity.

There are two types of irregularities in CISC instructions: data irregularity and control irregularity. Data irregularities are instructions based on irregular data format or data format conversions that are implicitly executed. (For example, there are data irregular instructions with implicitly execution of saturation or rounding or truncations, also for example, special instructions directly handling complex data.) Control irregularities are special control features and pipelines beyond the RISC control functions. (For example, interactive instructions and conditional arithmetic executions).

By adding more irregular CISC instructions, we achieve the higher performance over silicon (higher silicon efficiency). At the same time, the more irregular CISC instructions, the processor core will be less general and the portability will be lower. The trade off decision must be taken according to the long and short term product strategy. Because the method of modeling and quantitative analysis on ASIP product strategy is not yet well investigated, it is again an experience based activity.

CISC instructions can be classified as:

- Multiplication- and accumulation-based instructions.

- Long arithmetic instructions.

- Iterative instructions.

- Special hardware instructions.

7.3.1 MAC and Multiplication Instructions

From DSP source code profiling, the MAC operation was identified as the most frequently used arithmetic operation. The MAC instruction is therefore the most important CISC instruction in any DSP processor. The MAC operation is:

```
ACR = ACR + OPA * OPB;
```

A MAC instruction contains both data and control irregularities. The internal data format is irregular. The instruction might require more than one execution cycle (two execution cycles are used by the MAC in the Senior core). MAC instructions are shown in Table 7.19.

Accumulation can be configured as add or subtract (diminish) in order to support certain algorithms. More instructions can be supported by the MAC hardware module with no or very few hardware modifications. For example, by adding

Table 7.19 MAC Instructions.

No.	Mnemonics	Description	Operation
1	MAC	Multiply and accumulate	ACRD <= ACRD + Scaling ((((S/U) RA * (S/U) RB))
2	MDM	Multiply and diminish	ACRD <= ACRD - Scaling(((S/U) RA * (S/U) RB))

U = unsigned; S = signed; RA = operand A; RB = operand B; ACRD = the destination accumulator register.

preoperations before the pipeline registers of the multiplier inputs, and by adding microoperations at the multiplier output, multiplication of different arithmetic operations can be implemented.

From the **Junior** discussion, the following two concluding points were identified at the end of Chapter 5: Eliminate the loop overhead, hide the memory access cost. The loop overhead can be eliminated by using a zero overhead hardware loop unit. Most DSP processors have this type of hardware unit in order to speed up iterative computing. The Repeat instruction, which was introduced in Table 7.14 and in Example 7.3, relies on this hardware. The zero overhead hardware loop unit can also be used for convolution:

$$y(n) = \sum_{k=0}^{N-1} x(n-k)h(k) \tag{7.1}$$

The iterative behavior of a convolution instruction for each n value in Equation 7.1 can be:

```
for(i=0; i< N; i++){
/* N-tap convolution for a sample with modulo addressing*/
ACR = ACR + DM0(AP0)* DM1(AP1);/* A step of convolution */
AP0 = MODULOINC(AP0);/* post increment modulo addressing*/
AP1 = MODULOINC(AP1);/* AP1++ modulo or not modulo*/
}
```

Here $x(n-k)$ is the input stored in DM0 and addressed by the modulo address pointer AP0; $h(k)$ is the coefficient of the filter stored in DM1 and addressed by the normal address pointer AP1; $y(n)$ is the output and $y(n) = $ACR after convolution. Here modulo addressing is also called *circular addressing*. It is an addressing mode using modulo addressing arithmetic for postincrement or decrement addressing. By using modulo addressing, an efficient and low-power FIFO data buffer can be emulated using a single-port data memory. Modulo addressing will be discussed later in this chapter. We should assume that a FIFO in DM0 supplies N data samples from the latest sample $x(n-0)$ to the oldest sample $x(n-N+1)$. The convolution instruction is specified in Table 7.20.

Table 7.20	Iterative Convolution without Loop Control Overhead.	
Mnemonics	**Description**	**Operation**
CONV	Convolution	ACR <= ACR +/− Scaling (DM0 (modulo(AP0 + +)) * DM1 (modulo or not moduloAP1 + +))

The convolution CONV instruction will be executed in N steps. N can be carried by the instruction or preloaded using a previous instruction. The accumulation buffer is the ACR. Accumulation can be plus or minus (minus for transformation algorithms). The precision of the result from the multiplier was discussed in Chapter 3. A typical solution for a 16-bit fixed-point processor is 32-bits output precision. Modulo addressing is mandatory for DM0 and optional for DM1. Modulo addressing is needed for DM1 when running autocorrelation.

Occasionally multiplication with ultra high precision is needed, for example, for multiplication with double-precision operands. If a coefficient of IIR has double precision, then a 32b * 16b multiplication is required. Full precision for a 32b * 16b result is 48 bits. The accumulator in a MAC module of a 16-bit fixed-point processor supports only 32-bit precision, and in most applications 32-bit results are good enough. 32b * 16b multiplication with a 32-bit result is called pseudo-double-precision multiplication. The lower 16 bits are truncated during computing.

```
A32[31:0] <= Truncate lower 16bits(OPA[31:0] * OPB[15:0]);
```

To emulate this long multiplication using an assembly program, both signed and unsigned multiplications are needed. The long multiplication is partitioned into multiple steps. The RTL code of the operation is:

```
wire signed [31:0] mul1 = $signed(X[31:16]) *
  $signed(Y[15:0]);
    wire signed [31:0] mul2 = $signed({1'b0,X[15:0]}) * '
  $signed(Y[15:0]);;
    assign ACR[31:0] = $signed(mul1[31:0])+
  $signed(mul2[31:16]);
```

In extreme cases, for example, when implementing the advanced audio codec (AAC), a 32b * 32b multiplication is needed with a 32-bit precision for the result (a typical overspecified standard). For supporting this operation using a 16-bit multiplier and a 64-bit accumulator, both signed and unsigned multiplications are needed in order to emulate the long multiplication. The behavior of the operation is:

```
ACR[63:0] =
($signed({X[31],X[31:16]}) * $signed({Y[31],Y[31:16]}) << 32) +
($signed({ 1'b0,X[15:0]}) * $signed({Y[31],Y[31:16]}) << 16) +
```

```
($signed({X[31],X[31:16]}) * $signed({ 1'b0,Y[15:0]}) << 16) +
($signed({ 1'b0,X[15:0]}) * $signed({1'b0,Y[15:0]}) << 0);
```

In the pseudocode, $>> 16$ is equivalent to $\times 2^{-16}$ and $>> 32$ is equivalent to $\times 2^{-32}$. The implementation of unsigned multiplication using unsigned or signed multipliers will be discussed in Chapters 10 and 13.

Multiplication is shown in Table 7.21. The result of the multiplication must be scaled. Integer multiplication will be executed if the scaling factor is 1 (without scaling). If the scaling factor is 2, fractional multiplication will be performed. As discussed in Chapter 2, scaling factors of 0.5, 0.25, and 0.125 ($2^{-1}, 2^{-2}$, and 2^{-3}) will enlarge the range of guarding and allow more iteration steps. Different configurations for operand A, operand B, and the result may provide different functions for a multiplication instruction (see Table 7.22).

When the scaling factor is greater than 1, the implementation of scaling is a left-shift operation. A left-shift might induce an overflow. For left-shifting, two hardware operations must be specified. The first is what to fill in on the LSB side. Following the history of integer computing, the LSB side will be filled in with 0. The second operation is the overflow control. This is to conduct saturation if shifting out significant bits. If multiguard bits are used, the most significant guard bit will be kept as a reference bit and not be involved in scaling (not be shifted). After scaling, saturation will be conducted using the reference bit.

7.3.2 Double-Precision Arithmetic Instructions

Following the code profiling on audio and voice signal processing, the most used double-precision arithmetic instructions are listed in Table 7.23.

In Table 7.23, {ra:rb} = {8'ra[15], ra[15:0], rb[15:0]} and H/L means that RA is used either as the high part or the low part in ACR. Pre- and postoperations around the accumulator adder provide multiple choices for double-precision arithmetic operations. Some of these choices are listed in Table 7.24.

The register file supplies data with single precision. Long-precision data can be supplied by the accumulation register ACR in the MAC. Because concatenating a long data requires several clock cycles, implicitly concatenating operands for double-precision computing will simplify the code.

Postprocessing of the result includes "flag operation" and "saturation on result." Carry-out is not necessary for double-precision computing because 64-bit precision often is not used. However, for some extreme cases, for example, the multiplication of two 32-bit data, a carry-out flag is required.

Requested by iteration and double-precision computing, 32-bit data types have to be converted to internal data types with guards (copies of the sign bit), for

Table 7.21	Multiplication Instruction.	
Mnemonics	**Description**	**Operation**
MUL	Multiplication	ACR <= Scaling((((S/U) RA * (S/U) RB))

Table 7.22 Configuring a Multiplication.

No.	OPA	OPB	Scaling	Operation	Arithmetic
1	U	U	1	UU integer	$ACR <= \{0, OPA[15:0]\} * \{0, OPB[15:0]\}$
2	U	S	1	US integer	$ACR <= \{0, OPA[15:0]\}$ $* \{OPB[15], OPB[15:0]\}$
3	S	U	1	SU integer	$ACR <= \{OPA[15], OPA[15:0]\}$ $* \{0, OPB[15:0]\}$
4	S	S	1	SS integer	$ACR <= \{OPA[15], OPA[15:0]\}$ $* \{OPB[15], OPB[15:0]\}$
5	S	S	2	SS fractional	$ACR <= 2 * \{OPA[15], OPA[15:0]\}$ $* \{OPB[15], OPB[15:0]\}$
6	S	S	0.5	SS/2	$ACR <= 2^{-1} * \{OPA[15], OPA[15:0]\}$ $* \{OPB[15], OPB[15:0]\}$
7	S	S	0.25	SS/4	$ACR <= 2^{-2} * \{OPA[15], OPA[15:0]\}$ $* \{OPB[15], OPB[15:0]\}$
...	...	...	...	...	...

Table 7.23 Long Arithmetic Computing Instructions.

No.	Mnemonics	Description	Operation
1	ADDL	Long Addition	ACRD <= ACRA + ACRB ACRD <= ACRA + RA:0 ACRD <= ACRA + sign: RA ACRD <= ACRA + RA:RB
2	SUBL	Long Subtraction	ACRD <= ACRA − ACRB ACRD <= ACRA − RA:0 ACRD <= ACRA − sign: RA ACRD <= ACRA − RA:RB
3	CMPL	Long Compare	Compare ACRA, ACRB Compare ACRA, RA:0 Compare ACRA, sign: RA Compare ACRA, RA:RB
4	ABSL	Long Absolute	ACRD <= abs(ACRA) ACRD <= abs(sat(H/L RA))
5	NEGL	Long Negation	ACRD <= neg(ACRA) ACRD <= neg(sat(H/L RA))
6	MOVEL	Long move	ACRD <= RA:RB ACRD <= sat(H/L RA)

Table 7.24 Configurations for Long Arithmetic Computing.

No.	Instruction	Long operand B	Carry in	Saturation	Arithmetic operations
1	ADDL	ACRy	0	yes/no	ACR <= ACRx + ACRy
2	SUBL	INV(ACRy)	1	yes/no	ACR <= ACRx + INV(ACRy) + 1
3	ADDL	{OPA, OPB}	0	yes/no	ACR <= ACRx + 8'OPA[15], OPA[15:0], OPB[15:0]
4	SUBL	INV{OPA, OPB}	1	yes/no	ACR <= ACRx + INV {8'OPA[15], OPA[15:0], OPB[15:0]} + 1
5	ADDL	OPA-H	0	yes/no	ACR <= ACRx + {8'OPA[15], OPA[15:0], 16'b}
6	ADDL	OPA-L	0	yes/no	ACR <= ACRx + {24'OPA[15], OPA[15:0]}
7	SUBL	INV(OPA-H)	1	yes/no	ACR <= ACRx+ INV{8'OPA[15], OPA[15:0], 16'b0}+1
8	ADDL	INV(OPA-L)	1	yes/no	ACR <= ACRx + INV {24'OPA[15], OPA[15:0]} +1
9	NEGL	INV(ACRy)	1	yes/no	ACR <= 0 + INV(ACRy) + 1
10	ABSL	Sign of ACRx	0	yes	ACR<=conditional invert (ACRx)+{39'b0, ACRx[39]}
...	...	...	...	...	...

preventing overflow. The most popular data format inside the MAC is the 40-bit format including 8 guard bits (discussed in Chapter 2). After computation, the result should be converted to 32-bit precision by a saturation operation (saturation was discussed in Chapter 2).

```
if((ACR[39:31]!=9'b000000000) && (ACR[39:31]!=9'b111111111))
    if(ACR[39]) Result[31:0] <= 32'h80000000;
    else Result[31:0] <= 32'h7FFFFFFF;
else Result[31:0] <= ACR [31:0];
```

Saturation is needed for the NEGL operation when the operand holds the maximum negative value. With saturation, the result of NEGL (-1) is $1-2^{-2N+1}$. Without saturation the result will overflow.

Table 7.25 Processing on Accumulation Registers.

No.	Mnemonics	Operand	Operation
1	CLR	ACR	ACR <=0
2	PostOP	ACR	ACR <= sat (round (scale(ACR)))

While moving data from an accumulation register to a general register, the data type is changed from 40 bits to 16 bits. Changing data type is a kind of postprocessing that includes three operations: scaling, rounding, and saturation. Scaling can be based on predefined scaling factors—for example, 2, 0.5, 0.25, and 0.125. Rounding and saturation were discussed in detail in Chapter 2. The order of executing these three operations is very important. The right order is:

1. Scaling.

2. Rounding.

3. Saturation.

If any other order is used, the result might be wrong. We encourage you to investigate postprocessing using the wrong order for corner cases.

As a preoperation for iteration, it is mandatory to clean an accumulation register. Pre-and post-ACR operations are given in Table 7.25.

CLR ACR can also be done with SUBL—for example, SUBL ACR0, ACR0.

7.3.3 Other CISC Instructions

The purpose of introducing more CISC instructions is to decrease the runtime and the code cost. Most CISC instructions require extra hardware costs. For example, computing $1/x$ or $\sin(x)$ in one clock cycle is likely to be implemented using dedicated hardware. If fast computing is required, the trade-off between look-up tables and hardware modules should be investigated. Fast function computing based on look-up tables will be discussed later in this chapter.

7.4 ACCELERATED EXTENSIONS

7.4.1 Challenges

New functions and instructions should be easy to add to the processor core by the core users (instead of the core designers). The instruction set of a processor should be scalable. Adding accelerator instructions to a designed core should not induce redesign of the original assembly instruction set, the tools for assembly programming, and the RTL code of the processor core.

Accelerators can be designed as either loosely connected or tightly connected to the core. In Chapter 17, the accelerator design flow will be introduced and discussed in detail. In this chapter, you need to know only how to design the instructions for acceleration extensions.

When instructions are added to enhance the performance of a processor core, the current design of the processor core should be left unmodified as far as possible. This is extremely important. It may take a long time to reverify a processor if the accelerator design is not isolated.

An accelerator can be connected to a processor core via the I/O port connections. The core can control the accelerator by sending or receiving data to or from the I/O port. By connecting the interrupt interface of the accelerator to the interrupt controller of the core, the communication between the accelerator and the core can be efficiently managed by interrupts. In this way, data and control vectors can be transferred via the I/O port.

It seems that accelerator instructions are not necessary if an accelerator can be a general I/O device. The questions are, is it really necessary to design the instruction set and architecture for accelerator extension, and what benefits can we get from the so-called extendible design?

First, without special design for accelerated instruction extension, the behavior function of an accelerator cannot easily be integrated into the assembly simulator. The source codes of the assembly programming tools are usually not available to the core users. The behavior model of the accelerator may not be integrated easily into the assembly programming tools without touching the source code.

By connecting an accelerator via a general I/O port, the program running in the core has no direct link to the accelerator function and only the asynchronous communication is visible. Using the I/O port as the only connection media means that communication can be inefficient with long and unpredictable latency. If there is a lot of interaction between the core and the accelerator, the efficiency of this solution is low.

However, it is not trivial to add acceleration functions without changing the current design of a core. In order to design an extendible processor core with isolated accelerators, part of the hardware in the core should be predesigned for future accelerator support. For supporting tightly connected accelerators, the instruction decoder of the core should decode part of the accelerator instruction. To minimize the data access latency, an accelerator may need the register file in the processor core for operands, control vectors, and results.

7.4.2 Methodology

During the design of the processor core, the first step in the design for extension of tightly connected acceleration instructions is to specify which part of the instruction should be shared by the accelerator and the processor core. In general, interface functions should be decodable and executed by both the core and the accelerator.

The next step is to specify the instruction code format for accelerated instructions. Part of the code will be decoded by both the processor core and the accelerator, and part of the code will be decoded only by the accelerator. As an example, four instruction types are listed in Table 7.26.

As soon as the core decodes an accelerated instruction, type code = 11, the core will only decode and execute part of the instruction code. The rest of the code will be sent to the accelerator without further decoding. The shared control code carries messages to both the core and the accelerator. At the core side, the shared code tells if registers are needed or not by the accelerator. If the register file is used, further decoding will select operands and send them to the OPA and OPB buses. An example of shared coding is given in Table 7.27.

In Table 7.27, OP is the code indicating how to send operands from the general register file to the operand bus. RD is the code indicating how to get results from the accelerator to the general register file. The coding in Table 7.27 is suitable for acceleration of arithmetic computing, for example, C = A/B. In one instruction, the dividend and the divisor can be fetched to the accelerator from OPA and OPB buses in one clock cycle.

Table 7.26 Instruction Types.

Type	Code	Specification
1	00	load, store, move, or port access instructions for all kinds of data access
2	01	RISC and CISC instructions for single-step and iterative computing with all precisions
3	10	Program flow control instructions including jump, conditional jump, call, return, stack OP, etc.
4	11	Accelerate instructions for instruction extension

Table 7.27 Core-Accelerator Shared Codes.

Type	OP	RD	Core	Accelerator always gets local control codes from the core and...
11	0	0	Does nothing	...does nothing else
11	0	1	Receives result from I/O port	...keeps result data on the port
11	1	0	Sends OPA and OPB from register file	...gets data from OPA OPB
11	1	1	Sends OPA and OPB and gets result from I/O port	...gets data from OPA OPB and keeps result data on the port

7.5 INSTRUCTIONS FOR INSTRUCTION LEVEL PARALLEL (ILP) ARCHITECTURE

Typical ILP architectures are VLIW, superscalar, and SIMD. These architectures were discussed in Chapter 3.

7.5.1 Superscalar

The instruction format for a superscalar and a single datapath machine can be the same. This is an advantage for a superscalar because the same machine code can run on different machines. Code compatibility is kept and the code efficiency is high. The instruction level parallelization is managed by hardware, and there is no special binary code for parallelization; for example, the coding of || is not necessary.

7.5.2 VLIW Instructions

Parallelization of VLIW instructions is decided before execution by the compiler. Each instruction uses one bit to indicate that it may run in parallel with the next instruction. The assembly code may use || to denote that the following instruction will be executed in parallel, and each || consumes one bit of binary machine code. For example:

```
ADD R0, R1 ||  ADD R2, R3 || MOVE R4,DM0 || MOVE R5,DM1
       ADD R1, R2
       SUB R0, R4 || SUB R2, R5
```

In a four-way VLIW machine, the first four instructions will be executed in parallel. Then one instruction will be executed (three execution units will execute NOP). Finally, two instructions will be executed (two execution units will execute NOP).

The instruction flow controller is designed in such a way that the next instruction is always the default (PC <= PC + 1) if there is no jump or other program flow branches. Recall the design of conditional jump instructions: If a jump is taken in a processor with four pipeline stages as in Figure 7.14, all three instructions following the jump instruction will be flushed. In the worst case, $3 \times 4 + 3 = 15$ instructions are flushed in a four-way VLIW machine.

If jumps are taken often, for example while running coding algorithms, many cycles will be wasted. One way to minimize the overhead is to design the hardware without flushing while the jump is taken. The compiler for this architecture must then select instructions that are independent of the jump. The solution is illustrated in Figure 7.15.

If there are few independent instructions, NOP instructions can be used. In a VLIW processor, the concept given in Figure 7.15 is further developed in a way called "unbundling branch," meaning the condition test and the jump are not bounded in one instruction. A conditional jump instruction then is divided into two or three

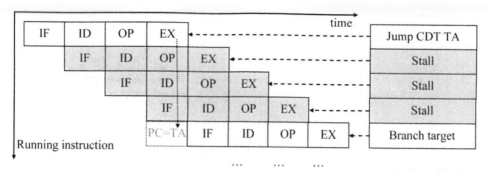

FIGURE 7.14

Pipeline stall when a conditional jump is taken.

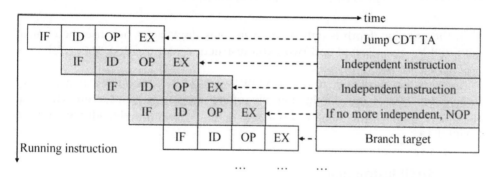

FIGURE 7.15

Code in a processor without pipeline stall flush.

independent instructions in order to make the jump control more explicit. One implementation of conditional jump instructions in a VLIW processor is given in reference [8]. It divides a jump instruction into the following:

- **Where:** Separate target address calculation of the jump.

- **Whether:** The condition test is separated from the jump.

- **When:** The jump instruction defines only the time to jump.

The jump instruction is divided into three pieces as instruction 1, 2, and 4 in Figure 7.16. Instruction 1 will test conditions and store the result of the test in the control path. To utilize the remaining clock cycles and to keep the pipelines full, calculating the target address 2 and another independent instruction 3 should be scheduled after the test condition instruction. The fourth instruction can therefore decide the jump (PC <= TA or PC <= PC + 1). After the jump instruction, instruction 5 will be executed if the jump is not taken or flushed if the jump is taken. Finally, instruction 6 is the fetched instruction with target address TA.

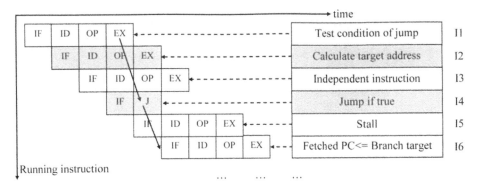

FIGURE 7.16

Explicit branch code with lowest cost of pipeline stall flush.

If the pipeline depth is only 4, there are only drawbacks, because the control path will be complicated and two extra instructions are needed. This method is useful only for deep pipelines.

To ease the hardware design of VLIW machine, it is preferred that a VLIW instruction set is designed following rules of data-stationary architecture. The data-stationary architecture will be discussed in the instruction decoder section of Chapter 14.

7.5.3 SIMD Instructions

The SIMD processor concept was introduced in Chapter 3. A SIMD processor is usually a slave processor manipulating vector data in parallel and controlled by a master processor [11]. For high hardware efficiency, complicated task flows including data quality control and threading control are not managed by the SIMD machine. Normally, tasks allocated to a SIMD processor are regular vector processing, with regular computing and regular memory addressing. The SIMD architecture is therefore simple and regular.

Typical SIMD instructions are:

```
Result vector <= (Vector operand A) operation (Vector
operand B)
     //transforms
Result <= Σ (Vector A) operation (Vector B)//accumulations
```

For example: 8 × 16b operand A from a vector register plus 8 × 16b operand B from another vector register; the 8 × 16b result is stored into a third vector register.

Another example is a convolution of an 8-tap FIR filter: 8 × 16b operand A from a vector register (eight data words) and 8 × 16b operand B from another vector register (eight coefficients).

Special features of SIMD processors are configurable data types and configurable operations on different data types (subword computing). This configurability can

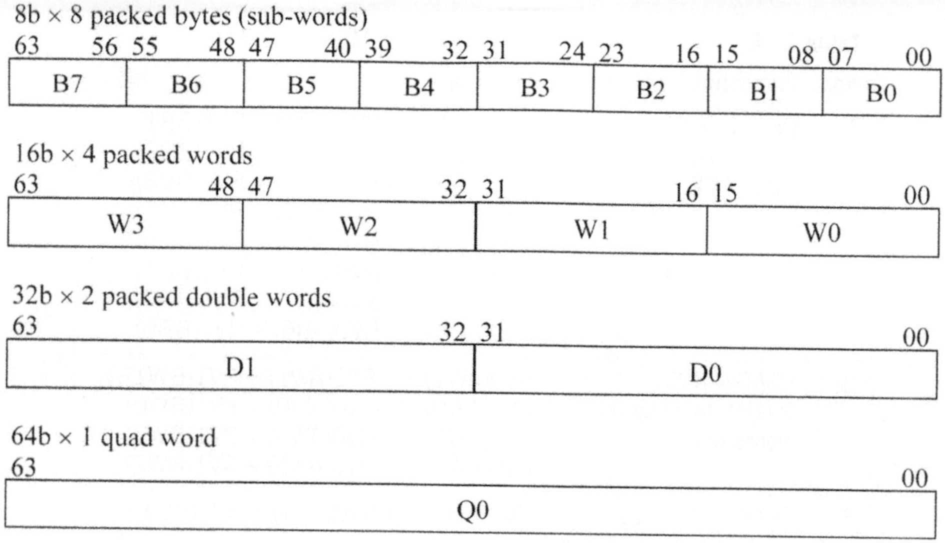

8b × 8 packed bytes (sub-words)

63	56 55	48 47	40 39	32 31	24 23	16 15	08 07	00
B7	B6	B5	B4	B3	B2	B1	B0	

16b × 4 packed words

63	48 47	32 31	16 15	00
W3	W2	W1	W0	

32b × 2 packed double words

63	32 31	00
D1	D0	

64b × 1 quad word

63	00
Q0	

FIGURE 7.17

Vector data types in a vector register (RV).

be conducted to SIMD instructions. A typical SIMD instruction carries configurations, including its data type, operands, destination, and arithmetic operations. For example, data types for a SIMD machine with 64-bit operands can be specified as shown in Figure 7.17.

Suppose there are eight vector registers RV0 to RV7. A SIMD vector instruction with configurable data types can be configured as shown in Table 7.28.

In Table 7.28, AB0 is the first byte in vector register A, AW0 is the first word in vector register A, and AD0 is the first double word in vector register A. A SIMD instruction set can include, for example, signed vector addition, signed vector minus, signed multiplication, unsigned vector addition, unsigned multiplication, vector logic operation, convolution, and so on. Unsigned vector computing can be used for video signal processing, because a pixel (luminance and chrominance) value can be only zero or positive. More SIMD instructions will be discussed in Chapter 20.

Example 7.5

Program eight multiplications with 16-bit operands and save the 32-bit results, using instructions available in Table 7.29.

Register allocation:

1. Load input data to RV0, vector register 0, and RV1, vector register 1.

2. Load coefficients to RV2, vector register 2, and RV3, vector register 3.

3. Results are stored in RV4, RV5, RV6, and RV7.

Table 7.28 An Example of a Configurable SIMD Instruction.

No.	Instruction	Behavior
1	PSADB RV0 RV1Packed signed bytes ADD	RV0(AB1) <= RV0(AB1) + RV1(BB1); RV0(AB0) <= RV0(AB0) + RV1(BB0) RV0(AB3) <= RV0(AB3) + RV1(BB3); RV0(AB2) <= RV0(AB2) + RV1(BB2) RV0(AB5) <= RV0(AB5) + RV1(BB5); RV0(AB4) <= RV0(AB4) + RV1(BB4) RV0(AB7) <= RV0(AB7) + RV1(BB7); RV0(AB6) <= RV0(AB6) + RV1(BB6)
2	PSADW RV0 RV1Packed signed words ADD	RV0(AW1) <= RV0(AW1) + RV1(BW1); RV0(AW0) <= RV0(AW0) + RV1(BW0) RV0(AW3) <= RV0(AW3) + RV1(BW3); RV0(AW2) <= RV0(AW2) + RV1(BW2)
3	PSADD RV0 RV1 Packed signed double word ADD	RV0(AD1) <= RV0(AD1) + RV1(BD1); RV0(AD0) <= RV0(AD0) + RV1(BD0)
4	PSADQ R0 R1 Packed signed quad word ADD	RV0(AQ) <= RV0(AQ) + RV1(BQ)

Table 7.29 Typical Vector Instructions.

No.	Instruction	Byte	Word	Double word	Quad word
1	Signed vector addition	PSADB	PSADW	PSADD	PSADQ
2	Unsigned vector addition	PUADB	PSADW	–	–
3	Signed vector subtraction	PSSBB	PSSBW	PSSBD	PSSBQ
4	Signed vector multiplication	PSMUB	PSMUW	PSMUD	–
5	Unsigned vector multiplication	PUMUB	PUMUW	–	–
6	Clean vector register (A XOR A)	PXORB	PXORW	PXORD	PXORQ

Computing:

```
1. PSMUW RV4D0 RV4D1, RV0W0 RV2W0, RV0W1 RV2W1
      /*(RV4D0 <= RV0W0 × RV2W0 and RV4D1 <= RV0W1 × RV2W1*/
2. PSMUW RV5D0 RV5D1, RV0W2 RV2W2, RV0W3 RV2W3
      /*(RV5D0 <= RV0W2 × RV2W2 and RV5D1 <= RV0W3 × RV2W3*/
3. PSMUW RV6D0 RV6D1, RV1W0 RV3W0, RV1W1 RV3W1
      /*(RV6D0 <= RV1W0 × RV3W0 and RV6D1 <= RV1W1 × RV3W1*/
```

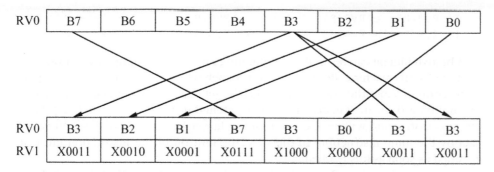

FIGURE 7.18

Data packing by vector shuffling.

```
4. PSMUW RV7D0 RV7D1, RV1W2 RV3W2, RV1W3 RV3W3
   /*(RV7D0 <= RV1W2 × RV3W2 and RV7D1 <= RV1W3 × RV3W3*/
```

In most cases vector computing algorithms are rather regular—for example, vector add and vector multiplication. However, data supplied for vector computing may not be so regular. The register file in a SIMD machine is prepared only for a vector operand of wide data access. The wide register file cannot be unbundled for individual data access in parallel. If data is irregular and computing is regular, data packing is required in order to organize the vector data for vector arithmetic. Data packing implies selecting data from different data vectors and merging them into a new data vector. Data packing can be time consuming. For example, vector data manipulation may follow a row or a column. If data was allocated and loaded following rows, vector operands must be repacked when computing on columns. Most SIMD instruction sets include permutation instructions that repack elements inside vectors, making them particularly useful for data processing and compression. RPACKB RV0 RV1, for example, is an instruction for shuffling eight bytes in RV0 based on shuffling control in RV1.

In this example, the lower four bits of each byte in RV1 is a control vector. When it is 1xxx, the corresponding byte in RV0 will not be moved. If, for example, RV1(B3) = xxxx1000, then the operation will be RV0(B3) <= RV0(B3). If a byte in RV1 is xxxx0xxx, then the corresponding byte in RV0 will be overwritten by another byte. If, for example, RV1(B0) = xxxx0011, then the operation will be RV0(B0) <= RV0(B3), see Figure 7.18.

Usually, a SIMD machine is a slave machine and does not manage branches. However, a branch function like "If T then A else B" can be handled as a conditional instruction by a SIMD machine. For example, the functional description of instruction PMAXB RV0 is:

```
If RV0(B7) ≥ RV0(B3) RV0(B7) <= RV0(B7) else RV0(B7)<=RV0(B3)
If RV0(B6) ≥ RV0(B2) RV0(B6) <= RV0(B6) else RV0(B6)<=RV0(B2)
```

```
If RV0(B5) ≥ RV0(B1) RV0(B5) <= RV0(B5) else RV0(B5)<=RV0(B1)
If RV0(B4) ≥ RV0(B0) RV0(B4) <= RV0(B4) else RV0(B4)<=RV0(B0)
```

A branch hint bit can be used for configuring a conditional jump instruction of an advanced SIMD machine. Because the control behavior of a streaming DSP program is rather predictable, the branch hint bit can be set during firmware development. If the hint bit is 0, the next instruction will be fetched from PC + 1. If the hint bit is 1, the next instruction will be fetched from TA (Target Address) carried as immediate data by the jump instruction. In this case, the hint bit of the instruction must be decoded as soon as it is fetched.

Complicated program flow control cannot be managed by a SIMD slave machine. This task is assigned to a RISC machine, operating as a master of the SIMD. In this case, tight coupling between the master machine and the SIMD machine is essential, and the communication latency between them must be minimized. The best way to minimize this latency is to design the SIMD as an autonomous datapath of the master machine. After the master machine has issued a task to the autonomous datapath, it will run this task stand-alone with fixed execution time, while the master can execute other instructions simultaneously. This architecture was introduced and denoted SIMT (Single Instruction Multiple Tasks) by the author's research team. SIMT will be discussed in more detail in Chapter 20.

7.6 MEMORY AND REGISTER ADDRESSING

The datapath executes arithmetic operations on operands from the register file and produces results to the register file. Load/Store instructions transfer data between the register file and the data memory. While executing a CISC instruction, the datapath might use data directly from the data memories. The operands can be classified as:

- **Register:** General and special registers addressed by a register name.

- **Memory:** A data word selected by pointing at its position in a physical memory.

- **Immediate:** A constant carried by an instruction to be used directly as an operand or to be stored in the register file.

- **Implied:** An operand or hardware operand that is implicitly executed by an instruction.

- **Vector:** A group of data stored in a data memory in a regular way and addressed in a regular way.

Addressing can be divided into register addressing, on-chip data memory addressing, and off-chip memory addressing. Addressing design will be discussed in detail in this section. An addressing algorithm should be sufficient, efficient, and reliable.

Sufficient means that an address shall reach all positions of the storage media, efficient means that the coding cost of addressing algorithms shall be acceptable, and reliable means that the addressing algorithm must produce the right address without exceptions.

7.6.1 Register Addressing

A register file is a first-level computing buffer. Each register in a register file has a unique name. Register addressing means pointing at a register using its name. There are three kinds of register files:

- **GRF:** A general register file.
- **SRF:** A special register file for memory addressing and control.
- **ACR:** Accumulation registers.

A register involved in ALU computing or load/store is typically in a general register file. I/O port registers are not allocated in the core. They are addressed in special ways and not classified as special registers.

General Register Addressing

In order to design a register file with high quality and low cost, the following questions related to general register addressing should be answered:

- How many data width formats (4-bit, 8-bit, 16-bit, 32-bit...) should be supported?
- What is the optimal register file size (8, 16, 32, 64, or more registers)?
- How many operands can be supplied in one clock cycle?
- How many results can be written back to the register file simultaneously?
- Should the accumulation registers be part of the register file?
- Should the special registers be available for general computing?

The size of a register file is the number of registers in this register file. A register stores a single data word with native precision. The number of registers in a register file is a trade-off between silicon cost and performance. The size of a computing buffer may not be enough if the size of the register file is too small. In this case, extra data swapping between the register file and the data memory will be required during the execution of a subroutine. In general, the size of a register file should be big enough to support subroutines without data swapping. However, silicon cost and power consumption will be high if a register file is too large. A large register file also needs longer binary code in order to address each register.

Benchmarking shows that at least 16 registers should be available in a general register file. Now the question is how large a register file should be (e.g., 64, 32, or

16 registers). There is no need to have 64 registers in a register file if a DSP processor does not run multiple tasks simultaneously. The silicon cost of a register file with 32 registers is usually acceptable for a single-issue DSP core.

Data types in a fixed-point DSP processor are typically 8, 12, 16, 24, or 32 bits. In most cases, the data width in a fixed-point DSP processor is 16 bits. For audio signal processing, a data type of 24 bits is popular. In baseband modems, the data width could be 14-bit real data, or 28-bit complex value.

There are two ways to support computing with different data types. The first way is to move the complexity to the computing device. An instruction always accesses a data word as an operand. This means that the design of the register file is easy. The ALU supports computing on different data types. For example, in a 16-bit fixed-point DSP processor, 8-bit computing can be based on 16-bit data, if a half word is processed and the other half word is masked and kept. After the computing, a complete result word is written back, including the half-processed word and the half-unchanged word. The datapath complexity will not be increased if half-word computing is only for unsigned arithmetic and logic operations.

Another way is to design a register file supporting accesses of different data types. The register file may supply half-word, word and double-word operands to the ALU. The register file will also accept results of different data types. Extra binary coding is required in order to distinguish between operands of different data types.

Figure 7.19 gives an example of registers with different data widths in a flexible register file. The $16 \times 16b$ register file acts as a computing buffer containing 16 operands of three data types: $16 \times 4\text{-bit}$ data, $16 \times 8\text{-bit}$ data, and $16 \times 16\text{-bit}$ data. By configuring RT (Register Type), the register file can be configured into three modes:

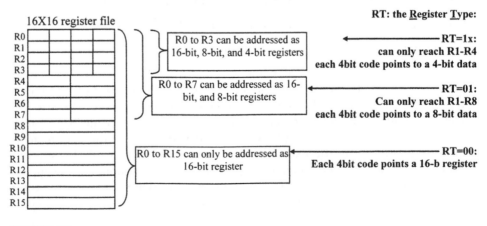

FIGURE 7.19

An example of flexible register file addressing.

Mode 1: 16-bit mode (RT = 00). This is a normal computing mode supporting arithmetic computing using 16-bit data. All registers from R0 to R15 can be accessed. Each register contains 16 bits.

Mode 2: 4-bit mode (RT = 1X). In this case, only the first four 16-bit registers are used. From R0 to R3, each register is divided into four sections, and these 4-bit registers are named R0-0, R0-1, R0-2, R0-3, R1-0.... R3-3 and R3-4.

Mode 3: 8-bit mode (RT = 01). In this case, only the first eight 16-bit registers can be accessed. From R0 to R7, each register is divided into two sections, and these 8-bit registers are named R0H, R0L, R1H, R1L,... R6H, R6L, R7H, and R7L.

How many operands can be addressed and supplied by a single instruction at the same time? Usually during the execution of one operation, two operands are read from, and one result is written to the general register file. When running several instructions in parallel, or running a SIMD instruction, more than two operands must be loaded and more than one result must be written back. A four-way datapath usually needs eight operands.

Any number of operands can be read from a register file, but it is not possible in general to write two or more results to one register file. This write dependence problem can be handled in two ways. One way is to fix the data dependence during compiling time. The other way is to split a register file into multiple datapath clusters. Each cluster (computing path) will have its own register file.

Register File and Accumulation Registers

Should accumulation registers be directly involved in ALU computing? In typical DSP processors, operands for ALU computing are from registers in the general register file. Operands for MAC computing can be both accumulator registers (ACR) and general registers. ACRs are allocated in the MAC and not directly connected to the ALU. The advantage of separating the ALU from the ACR is simplified hardware partitioning and design. The drawback of this separation is the extra cycle cost when moving data between the MAC and the general-purpose register file. More modern ASIP DSP processors use the general register file for both ALU computing and iterative computing. Two general registers are used as one accumulator register with eight not visible guard bits.

Special Register Addressing

Should special registers, SR, be part of the general register file? Should data in SR be operands of ALU computing? Data in special registers rarely are manipulated, and special registers are not usually ALU operands. If data in a special register must be manipulated, it can be moved to a general register. When the data has been processed, it is moved back to the special register. In this way the architecture can be kept simple, and special registers can be distributed everywhere in a DSP core.

The complete procedure of data manipulation for a special register normally consumes three to four clock cycles.

An example of data manipulation for a special register is given in Figure 7.20. Data manipulation for I/O port registers can be executed in the same way.

7.6.2 Data Memory Addressing

Data Memory Hierarchy

In most DSP processors, the register file is the first-level computing buffer for RISC instructions. The second-level computing buffers consist of scratchpad memories or cache memories. The main memory is the third-level computing buffer, see Figure 7.21. The main memory might be on-chip, but in most cases it is an off-chip component. Two questions need to be discussed here:

- What is the address space?

- How is the off-chip main memory accessed?

```
Data in SR cannot be involved in ALU computing
  MOVE R1,SR1
  MOVE R2,constant
  AND  R1,R2
  MOVE SR1,R1
// set control register
```

FIGURE 7.20

Manipulating data in a special register.

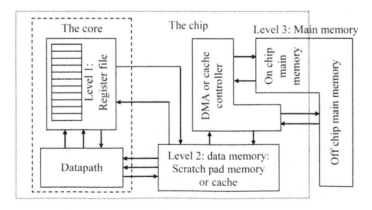

FIGURE 7.21

Memory hierarchy of a DSP processor.

The coverage of data memory addressing of an instruction set depends on the type of on-chip data memories. If cache is the on-chip memory, the address space of the instruction set might cover both on-chip and off-chip memories. Therefore, the complexity to access the off-chip memory can be hidden (the meaning of cache in France is to hide). In this case, there is a unified address space. If scratchpad memory is the on-chip memory, the main memory addressing has to be explicitly specified, and the address spaces of the on-chip scratchpad memory and the main memory are separated. There have to be two address spaces, one space for the on-chip scratchpad memory and one for the off-chip memory.

In most ASIP DSP, the on-chip data memories are scratchpad memories [12]. The address space of the assembly instruction set covers only the on-chip data memories. The size of this address space depends on what the instruction code can carry. For example, if the length of the address pointer is 16 bits, then the address space of the processor is $64\,K = 2^{16}$. The impact of the on-chip address space should be checked by running different applications. Off-chip memories can be accessed, for example, by DMA transfers. The address space of off-chip memories can be very large, from several megawords up to gigawords.

However, if the tasks running on an ASIP are very complicated and the data access to an off-chip memory cannot easily be predicted, then it will be difficult to know when and how to prepare a DMA task. DMA transactions requested by on-chip memory accesses will induce excessive or not acceptable data access latency. Therefore, when data accesses are not predictable, a scratchpad memory will not be the preferred solution.

Data Addressing Fundamentals

The main subject for this book is ASIP, application specific processors for streaming signal processing. Different types of addressing modes are required by advanced DSP algorithms. By checking the benchmark results for **Junior** using DSP kernel algorithms, we found out that there are too many clock cycles used for computing memory addresses.

Memory accesses can be considered as overhead for DSP algorithms. Thus, the cost from addressing and memory accesses should be minimized or hidden behind the computing of DSP algorithms. The best solution is to run all addressing and memory accesses in parallel with arithmetic operations.

Separate memory addressing hardware modules can be scheduled to work in parallel according to the data flow. For best performance the following requirements should be fulfilled:

- Datapath, addressing path, and memory subsystem are separate and work in parallel.

- The memory access to supply the operands should be executed just before the datapath needs the data.

- The memory access to store the result should be executed just after the result is available.

- The cycle cost of memory access shall be equal to or less than the cycle cost of arithmetic computing.

- Latency or address computing delays shall be minimized.

Addressing Modes Supporting Load and Store

Addressing modes are specifications of addressing algorithms for load, store, and CISC instructions. Load and store instructions supply data to the register file and collect results from it in order to supply operands for RISC instructions. CISC instructions may consume and produce vector data from and to data memories.

Usually, the data memory size of a DSP processor is at least several kilowords. A typical size of a data memory is 64 K words, and a word in the memory can be addressed using a 16-bit address. For the instruction code, we also need to specify at least one data register with which to associate the load and store operations. Therefore, we need at least 16 (address) + 5 (register name) = 21 bits to code the source and the destination addresses. Including the code cost of memory addressing modes (3 bits), the code cost of the memory pointer (1 bit), the code cost of the instruction-type specifier (4 bits), and the code cost of the operation (4 bits), the final size of the machine code will be 33 bits. This is a rather long instruction word for an instruction to load/store between a register and a memory.

By observing real DSP firmware, we will find that data and parameters used by a subroutine are rather local. This fact provides an opportunity to put all data and parameters of an algorithm into one segment. Based on data localization, the address can be separated into two parts: the offset and the segment (the block). The segment address can be shared by all data addresses of a subroutine, and the instructions in a subroutine will carry only the offset address. In this way, the code size can be reduced.

Also by observing real DSP firmware subroutines, regularities can be discovered in address computing. For example, post-increment on the address pointer often is executed as a preparation for addressing the next data in a data array. As discussed in the beginning of Chapter 5, the processor performance can be improved dramatically if we accelerate the addressing functions in hardware, for example, modulo addressing and bit-reversal addressing.

By utilizing the data storage regularity features in DSP applications just mentioned, addressing algorithms can be designed with both enough address space coverage and coding efficiency. Various addressing models can be found from DSP processors on the market. The most used addressing models are listed in Table 7.30.

In Table 7.30, "A" is the calculated address pointing at a data word in the data memory. Implied addressing is used only by specific instructions. SEG is the segment address register; OFFSET is an immediate data carried by the load/store instruction; GR is the general register; ADP is a special register, the address pointer; STEP is a special register for the step size; Top is the top register of the FIFO buffer in modulo addressing; and Bottom is the bottom register of the FIFO buffer in modulo addressing.

Table 7.30   Most Used Addressing Modes.

No.	Addressing	Algorithm Specification
1	Implied addressing	Address is implicitly specified by the operation code
2	Memory direct	A <= immediate data as address carried by instruction
3	Segment offset	A <= SEG + OFFSET
4	Register indirect	A <= Selected GR
5	Postincrement addressing	A <= ADP and INC (ADP) (ADP is an address pointer)
6	Predecrement addressing	DEC (ADP) and A <= ADP
7	Index addressing	A <= SEG + Index GR
8	Variable step size post increment addressing	A <= ADP and ADP <= ADP + STEP
9	Modulo addressing	If ADP != Top then A <= ADP and INC (ADP), else A <= Bottom

The data memory address is an immediate data carried by a load/store instruction while using memory direct addressing. Memory direct addressing also is called absolute addressing, which is used mostly for debugging functions during concept-level SW design. It is not very popular because it requires a long instruction code for the address. In many cases, an instruction word will be too long in order to carry an address code covering the entire address space (for example, 20-bit immediate data will be needed when the memory size is 1 megaword).

Segment offset addressing supplies the data memory address by adding the offset to the segment. The segment (block) register can be loaded with the segment address as part of the preprocessing of an algorithm. The offset is a short immediate data carried by an instruction. The size of the segment and the length of the offset are dependent on the application. Let us take a voice CODEC as an example: if the sampling rate is 8 kHz, then the 20 ms data frame contains 160 words. A segment of 256 words is enough for the input and computing buffer. By using the segment plus offset addressing model, the physical memory address is the sum of the segment and the offset. In the example depicted in Figure 7.22, 8 bits are used as the offset for addressing 256 words in a segment. The drawback of this addressing model is the risk of accessing data of other segments because segments can overlap each other. On another side, it gives experienced programmers more flexibility in using the data memories. In the example, R3 is the segment register.

While using register indirect addressing, see Figure 7.23, the data memory address is available in a pointed register when running a load/store instruction.

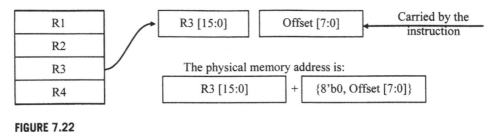

FIGURE 7.22

An example of segment plus offset addressing.

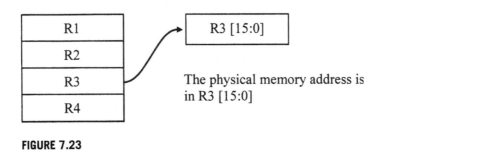

FIGURE 7.23

An example of register indirect addressing.

The register indirect addressing mode requires much shorter addressing code compared to memory direct addressing and segment offset addressing. Five bits are required to point to a register in a register file containing 32 registers. Eight bits are required for memory addressing, including three bits for addressing the model specifier. Based on register indirect memory addressing, advanced addressing modes such as increment addressing and index addressing can be developed.

Register increment addressing and register decrement addressing are extensions of register indirect addressing, which is especially useful for vector-based algorithms, for loading a variable table while initializing a program, and for pushing the stack. In modern digital communication and media signal processing applications, the unit for input or output data is usually a frame instead of a single sample.

Postincrement addressing can be used for both vector computing and stack push. Predecrement addressing is used mostly for stack pop (or stack push, depending on in which direction the stack grows). The addressing of stack pop is the opposite of the addressing of stack push. After pushing and postincrement computing on the stack pointer, it actually points at the next stack position for the next push. While executing pop, the stack pointer should move back one step and point to the latest pushed data. Predecrement addressing on the stack pointer is thus needed.

Note that, instead of using a general register, an address pointer is used for postincrement and predecrement addressing. The hardware implementation will be easier if the increment and decrement computing does not involve a general register file.

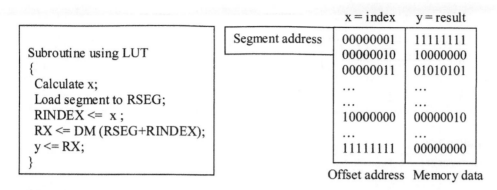

FIGURE 7.24

Index addressing for look-up tables.

Index addressing, illustrated in Figure 7.24, is a segment-offset addressing with variable offset. By using a variable offset, the computation of a complicated function such as $y = \sin(x)$ or $y = 1/x$ can be conducted in one clock cycle by using a look-up table (LUT). If $y = f(x)$, then x is the index value or memory address of the table. As the table content, y will be the precalculated results stored in a data memory. The function f is the segment address in the figure.

7.6.3 Hardware Accelerated Memory Addressing

An ASIP DSP is designed for a group of specific applications. Addressing acceleration for kernel algorithms, especially for some CISC instructions, can dramatically improve the performance or reduce the code cost. In this section, we will discuss accelerations such as modulo addressing, and addressing for FFT and other vector operations.

Recalling the FIR algorithm discussed in Chapter 1, the FIFO (first in, first out) buffer was implemented using a chain of registers. The FIFO register-chain consumes much power and is not very flexible. A FIFO register-chain-based FIR could be found in an ASIC, but it is not often implemented in a DSP processor. In reality, the FIFO buffer of a FIR filter is implemented using memories. The cycle cost for shifting data could be much higher to shift all data in a chain $x(n), x(n-1), x(n-2), x(n-3),$ $x(n-4), x(n-5)$ and $x(n-6)$, in a physical memory. For example, for shifting the data of the 7-tap FIR in Figure 7.25, at least $6 \times 3 = 18$ clock cycles are needed to complete a shift of all data samples in the computing buffer.

In reality, an address pointer is increased/decreased instead of shifting data. By decrementing the address pointer, it appears as if the data has moved one step forward in the buffer. When the bottom address of the buffer is reached, the pointer must be assigned to the top address of this buffer. Using the same way to store/load data into/from the buffer by moving the address pointer, a FIFO can be emulated. This addressing, called modulo addressing, is the way to address a FIFO buffer in a data memory.

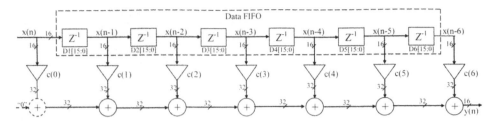

FIGURE 7.25

Behavior of a 7-tap convolution.

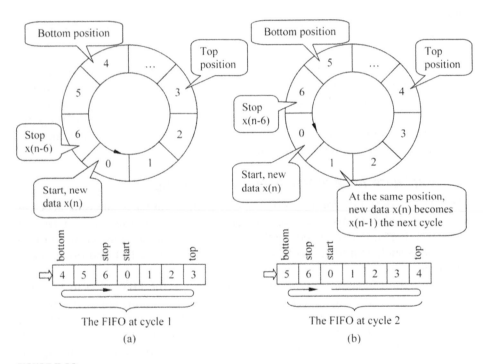

FIGURE 7.26

Illustration of a FIFO emulated in a data memory.

The principle of a FIFO buffer is illustrated in Figure 7.26. The depth of this FIFO buffer is seven. Figure 7.26(a) illustrates the data allocated in the FIFO during the first clock cycle, and Figure 7.26(b) illustrates the data in the FIFO during the second clock cycle. By comparing Figure 7.26(a) and (b), we recognize that top and bottom are physical boundaries of the FIFO and that the top and bottom positions do not change while shifting in data. However, the start point and the stop point of the data buffer in the FIFO change from Figure 7.26(a) to (b) after writing (pushing) in new data.

The FIFO supplies data from the start point, where the latest data is located. The oldest data is located at the stop point. While reaching the bottom of the FIFO, the address pointer jumps to the top of the FIFO by modulo addressing. The oldest data $x(n - 6)$ in Figure 7.26(a) is replaced by the latest data $x(n)$ in Figure 7.26(b) after writing (pushing) in new data to the FIFO. Both the start and the stop pointers of the FIFO change after writing in new data.

$$y(n) = \sum_{k=0}^{m-1} x(n - k)c(k) \tag{7.2}$$

An address pointer with modulo address function needs two associate registers: the FIFO top register (TOPR) and the FIFO bottom register (BTMR). A convolution is defined according to Equation 7.2, and the algorithm is shown in Figure 7.25. In the equation, $x(n - k)$ is the data vector, $c(k)$ is the coefficient vector, and $y(n)$ is the vector of the convolution result. A sample of $y(n)$ is computed by:

```
y(n) = x(n) × c(0) + x(n-1) × c(1) + x(n-2) × c(2)
     + x(n-3) × c(3) + x(n-4) × c(4)
     + x(n-5) × c(5) + x(n-6) × c(6)
```

To execute a tap of convolution in one clock cycle, the postincrement modulo addressing on data memory and postincrement addressing on coefficient memory must be executed implicitly behind the convolution algorithm. Data and coefficients can therefore be available when they are needed by the MAC unit. Specification of implied addressing for convolution includes:

1. Coefficient memory supplying coefficients without modulo addressing.

2. Data memory supplying data within a FIFO. The FIFO has:

   - BTMR to avoid underflow of the FIFO addressing.

   - TOPR to avoid overflow of the FIFO addressing.

   - DAR, data address pointer, pointing to data within the FIFO.

3. Postincrement (decrement) addressing when:

   - reaching the top of the FIFO, jump to the bottom of the FIFO.

   - reaching the bottom of the FIFO, jump to the top of the FIFO.

4. Two memories supply data and coefficient simultaneously.

5. All address operations should be executed in parallel with the data processing.

6. Data and coefficients must be available before computing.

The behavior model of the MAC instruction is therefore:

```
01    // CONV instruction: iteration from N to 1
02    OPB <= TM (CAR);            // load filter coefficient
03    OPA <= DM (DAR);                   // load sample
04    if DAR == TOPR then DAR <= BTMR   // Check FIFO bound
05    else INC (DAR);            // update sample pointer
06    INC (CAR);                 // update coefficient pointer
07    BFR <= OPA * OPB;                  // compute product
08    ACR <= ACR + BFR;                  // accumulate product
09    DEC (LCR);                         // DEC loop counter
10    if LCR != 0 then jump to 01;       // more coefficients?
11    else Y <= Truncate (round (ACR));     // final result
12    end.
```

Modulo addressing is performed by instructions 03−05. Implementation of modulo addressing in hardware will be discussed in Chapter 15. Supported by a FIFO in a data memory and modulo addressing hardware in an address generator, a convolution algorithm can be implemented using one instruction. One tap of convolution will consume only one clock cycle.

FIFOs are used in many applications. For example, a FIFO data buffer in an IP phone is essential for elimination of jitter introduced by channel latency. This FIFO needs modulo addressing for both load and store operations.

A memory FIFO usually is used instead of a register FIFO unless the FIFO size is very small such as the Biquad IIR filter in Chapter 2. For very small FIFO sizes, the overhead of initializing the bottom and top register might not be worth it.

Addressing Support for FFT

FFT is a frequently used basic algorithm in modern DSP applications. The signal flow chart of an 8-point Radix-2 FFT is given in Figure 7.27.

$$W_N^{nk} = e^{-j\frac{2\pi}{N}nk} \tag{7.3}$$

Two main approaches were discussed in Chapter 1: decimation in time (DIT) and decimation in frequency (DIF). Here W is defined according to Equation 7.3, where N is the number of FFT points; $x(n)$ is the time-domain sequence; $n = 0, 1, 2, \ldots N - 1$; $X(k)$ is the frequency-domain sequence; and $k = 0, 1, 2, \ldots N - 1$. The FFT algorithm is divided into three nested loops. The innermost loop is the computing kernel of a butterfly (The DIT butterfly kernel was discussed in Chapter 1). The middle loop is one layer of FFT including $N/2$ butterfly loops. The top loop includes $\log_2 N$ layers of the middle loop and bit-reversal addressing before or after FFT computing to disorder the inputs or reorder the final computing results.

Addressing for butterfly computing in a FFT should be postincrement register addressing with variable step size. Addressing mode 8 in Table 7.30 gives the addressing of A <= ADP and ADP <= ADP + STEP. Here STEP is the step size

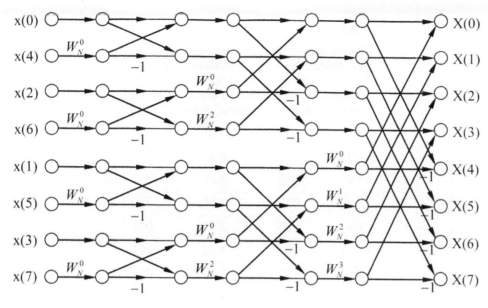

FIGURE 7.27

8-point FFT signal flowchart.

register. If it can be any register in a general register file, the addressing for FFT algorithms will be accelerated and flexible enough.

As a preparation for FFT computing based on the DIT algorithm, the inputs in Figure 7.27 should be reordered following bit-reversal addressing algorithm. In order to use the results from the FFT computing-based DIF algorithm, the output of the FFT computing should be reordered. Addresses before reordering and after reordering for an eight-point FFT are listed in Table 7.31.

Bit-reversal addressing algorithms can be formalized based on the behavior given in Table 7.31.

```
ADP <= {ADP [N-1; M], bit reversal (ADP [M-1:0])}
```

N is the width of the address code. $M = \log_2(\text{FFT length})$; for example, $M = 6$ for a 64-point FFT because $6 = \log_2 64$. {A, B} is the concatenation of a word by merging A and B. The operation of bit-reversal addressing on the bit-reversal part is to shuffle the MSB to the LSB, the bit next to the LSB to the bit next to the MSB, and so forth. The cost of software emulation of bit-reversal addressing is rather expensive when the FFT data size is large. However, the HW implementation is just a group of assignments at bit level, and its cost is negligible.

2D Data Addressing for Single Issue Computing

When processing 2D vector signals for image or video, pixel data in a DU (data unit) need to be accessed in a two-dimensional manner with minimum cost. Usually

Table 7.31 Bit-Reversal Addressing.

Before bit-reversal addressing		After bit-reversal addressing	
In-order vector	Binary codes	Vector with bit-reversal order	Binary codes
X[0]	000	X[0]	000
X[1]	001	X[4]	100
X[2]	010	X[2]	010
X[3]	011	X[6]	110
X[4]	100	X[1]	001
X[5]	101	X[5]	101
X[6]	110	X[3]	011
X[7]	111	X[7]	111

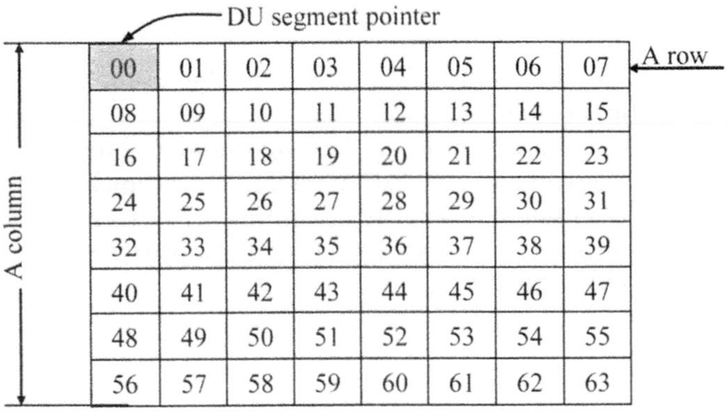

FIGURE 7.28

A data unit with 64 pixels.

it requires quite some effort to address 2D data because the addressing algorithm is complicated. To minimize the cycle cost for this, special acceleration may be required. A data unit is illustrated in Figure 7.28 as an example.

For a single-issue DSP processor there is no need to supply more than one data pixel at a time. A flexible addressing algorithm for one pixel is:

```
Address <= DU_segment + row_length*(column position-1)
        + row position-1
```

According to this example, a DU segment is the starting address of the DU. Row positions and column positions are both from 1 to 8. Using the row position and the column position together, a pixel in a DU can be addressed. For example, data 37 is on the fifth row and the sixth column, and the address is DU segment $+ 8 * (5 - 1) + 6 - 1 = 37$. 2D vector addressing is more complicated, and it will be discussed in Chapters 16 and 22.

7.7 CODING

Coding includes coding of assembly and coding of machine binary code. Assembly coding is to name assembly instructions and make sure that the assembly language is human-readable. Hardware machines can read only machine binary code, and the assembly code must be translated to machine binary code before execution.

7.7.1 Assembly Encoding

The purpose of assembly coding is to define a human-readable code, easy to use and remember without ambiguity, for firmware and assembler designers. In this section, assembly coding will be based on suggestions from the IEEE std. 694–1985 [13, 14]. As the lowest level hardware-dependent machine language, assembly code gives function descriptions based on microoperations. The assembly code also specifies the use of hardware such as the physical registers, the physical data memory addresses, and the physical target address for jump instructions.

The assembly language exposes microoperations in the datapath, addressing path and memory subsystems. Some microoperations in the control path, such as test of jump conditions and target addressing, also are exposed. Default operations carried by some instructions, such as PC <= PC + 1 and flag operations, will not be specified explicitly in the assembly language.

The specification of an assembly language is to describe the microoperations executed by the instructions and the constraints of the execution. Figure 7.29 depicts the partitioning and classification of microoperations. It suggests which microoperations need to be explicitly specified and coded into assembly code and binary machine code (explicit microoperation), what microoperations need to be specified in the assembly manual (implicit microoperations that programmers and compiler designers must know), and which microcode can be hidden from the assembly users (only for hardware designers).

You might think that the description of assembly code in the assembly manual should expose all microoperations. Actually, this is not necessary. Some microoperations are not exposed in the assembly manual because they are not directly used by assembly programmers, such as bus transactions and the details of instruction decoding.

All microoperations that the assembly users need to know will be exposed in the assembly language manual. However, to use the binary machine code efficiently,

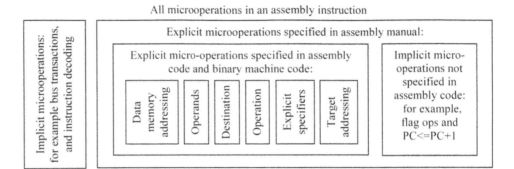

FIGURE 7.29

Assembly and binary coding convention.

some microoperations that are specified by the assembly language manual will not be coded into assembly and binary machine code. Examples of such operations are flag operations for ALU computing and PC <= PC + 1 after a normal instruction. These implicit functions will be decoded in hardware and in the assembly simulator from the operation code. When writing the RTL code for the control path and the datapath, all hidden microoperations must be exposed and implemented.

Except for the microoperations listed in Figure 7.29, more information should be specified for an instruction in order to make the assembly manual readable. Such information is:

Instruction name: Normally, the name of an instruction is a verb and adverbs describing the main operation; for example, "add with carry," "rotate right including carry," "rotate right without carry," and "jump if less than." Instruction names are used in the assembly manual, but do not show up in the actual assembly language.

Instruction type: An instruction shall be classified into a group or an instruction subset. In most cases, instructions are classified according to their operations, such as instructions for arithmetic operation or instructions for flow control. A good way to classify instruction types is according to the hardware partitioning. Matching of hardware partitioning to instruction classification will speed up the hardware development, especially the instruction decoder design and the hardware verification.

Instruction mnemonics: A mnemonic defines the main operation of an instruction. Normally, the first character of the mnemonic is the first letter of the verb describing the action. For example, "addc" is "add with carry," "rrc" is "right rotate including carry," and "jlt" is "jump if less than." Another way to define mnemonics is to separate the main operation and the pre- or postoperation to make the instruction more orthogonal. For example, we can configure the ADD instruction by separate specifiers CI (carry-in) and/or CO (carry-out).

Synonymous mnemonics: Sometimes, a synonymous mnemonic (extra name) is assigned to an instruction because this notation is preferred by some designers. ADD CI CO is an example of a synonymous mnemonic. Different notations based on the ADD instruction are:

```
ADD CI CO = ADC = Add with carry-in and generate carry out
   ADD CO = ADL = Add and generate carry out
   ADD CI = ADH = Add of higher part data using carry in
      ADD = ADD = Add without carry in, saturate if overflow
```

Operation: This is the main function of an instruction, such as arithmetic computing. Operations include addressing and fetching of operands, kernel arithmetic operations, storing the result, updating the PC, and updating flags. Finally, the cycle cost of these operations shall be specified.

Operand address: It points to the position where an operand is allocated.

Results address: It points to the position where the result should be stored.

Addressing algorithms: They specify the way to reach the final physical position in a data memory. The specification includes addressing modes (the specification of the addressing operation), the selection of data memory (which data memory), part of the address code, and finally the addressing cycle cost.

Configuration (Specifiers): Specifiers are used for configuring microoperations of an instruction. A specifier can be used for selecting operation features. For example, the specifier CI can be appended to the Add instruction to form "add with carry," which specifies whether the Add operation involves carry in or not.

Coding: Assembly coding and binary coding of an assembly instruction.

Constraint: Some instructions may not be used in some cases. Also an instruction may set a constraint for the following instruction. For example, a MAC instruction will set the constraint that the following single-cycle instruction cannot use the accumulator register (will be used by the MAC instruction).

Hardware information: Necessary information to expose hardware usage of some special instructions to programmers. For example, if any specific address pointers such as the boundary register in modulo addressing are used for the convolution instruction, then the way to use this register should be specified in the assembly language manual.

Dedicated functions: Special instructions may carry implicit functions, which should be specified in the assembly instruction set manual.

The best way to describe an instruction is to give good examples. This is especially important for application-specific instructions. These instructions might have to be executed together with other specific instructions. For example, a convolution instruction requires an initialization procedure, including setting the number of iterations and the boundary of the FIFO buffer. When a convolution instruction has been executed, dedicated instructions are required to move the result in the accumulation register to a general-purpose register or to the data memory. It is almost impossible to explain the convolution instruction without an example, including supporting prologue and epilogue code.

There are different ways of defining an assembly language. The most common way is to specify an instruction by operation and operands. The assembly language thus is close to the IR (intermediate representation) based on the TAC (three-address code) of compilers, and will be discussed more in Chapter 8. An example of this is:

```
ADD S DEST_REG, SOURCE1_REG, SOURCE2_REG
```

The behavior of the instruction is dest_reg <= saturation (source1_reg + source2_reg). This kind of notation exposes hardware microoperations to software engineers. Microoperations include the type of instruction (arithmetic), the computing (addition with saturation), the register addressing, the source storage (register names), and the destination storage (register name). Thus the instruction exposes the datapath operations in detail except for the flag operations.

Another kind of assembly notation looks more like a high-level language:

```
R0 <= SAT(R1 + R2 + C)
```

This notation is easier to understand and use.

7.7.2 Machine Code Coding

The purpose of binary coding is to define the binary image of the assembly instruction set. Assembly code cannot be executed directly on hardware. A necessary step is to code the instruction set from text format to machine code in binary format in order to run it on hardware.

Coding Fundamentals

Requirements on the coding of machine code include scalability, extendibility, low hardware decoding cost, and low program memory cost.

Scalability can be divided into processor scalability and coding scalability. In this chapter we will discuss only coding scalability, which stands for two things: the ability to accept new instructions (the scalability of the set) and the ability to accept new functions in one instruction (the scalability of an instruction). After benchmarking an instruction set, it might be necessary to add more instructions or to change microoperations for an instruction. To minimize the cost of design iterations, it is essential to:

1. Minimize recoding of assembly instruction and machine code.

2. Avoid redesign of available instructions.

3. Minimize changes in the assembler.

4. Minimize changes in the instruction set simulator.

Scalability depends on the availability of unused binary code. "Scalability of the instruction set" is a function of the available and orthogonal coding space for the operation codes. At the same time, there should be no need to recode other instructions. "Scalability of an instruction" is a function of the ability to code new microoperations using the available instruction code. Good scalability means that updating can be achieved with minimum effort.

Extendibility is a measure of the capability to add new functions after the core design has been released. It includes the ability to add new instructions to the processor without changing the processor core, the core ISA simulator, and the assembler. Extendibility is an essential feature of a reusable design. When a processor core is obtained from a third party, it does not always meet the system requirements. It might then be necessary, for example, to add a tightly connected accelerator. When adding and removing accelerators, we do not want to touch the core RTL code, the current simulator, assembler, and compiler. Design for extendibility starts from designing a set of extendible binary machine code. The instruction coding for extendibility will be introduced briefly later in this chapter. Details will be given in Chapter 17.

Orthogonal coding is to code the machine binary code into independent subsets. Two benefits can be achieved by orthogonal coding: coding scalability and structured hardware design of the instruction decoder. For example, by dividing the operation code into kernel operation code (add or sub) and feature operation code (carry-in, saturate on result) and coding it in an orthogonal way, features such as saturation can easily be added or removed.

An orthogonal instruction set is a different concept from orthogonal binary coding. An orthogonal instruction set gives the same right to each instruction. This means that any instruction can access any register in the entire register file. It also allows every instruction to access any data memory using any available addressing mode. A nonorthogonal instruction set has restrictions; for example, some operations can be performed based on only a limited number of registers, or some addressing modes are available for only some instructions.

With an orthogonal instruction set it is easier to design the assembler, the compiler, and the instruction decoder. This is also convenient for the programmer because there is no need to remember a lot of restrictions. A completely orthogonal instruction set requires a very deep code partitioning of the orthogonal binary coding, and this results in a very long machine code. In a DSP processor, in order to achieve efficient instruction coding, we have to sacrifice some orthogonal coding features.

Controllability is also a key issue. Controllability means that the code can reach and control every part of the hardware in a DSP core. A good coding of an instruction set gives a direct or indirect image of the hardware control in detail. Furthermore, the binary coding should expose enough information for the hardware instruction decoding. Hardware designers should get relevant information from coding and naming conventions in order to speed up the hardware design and simplify the hardware debugging. For programmers, regularity and explicitness are important. However, higher controllability, observability, and regularity of the instruction coding will increase the program memory size.

It is also important to keep the binary code without ambiguity, which is called decodability.

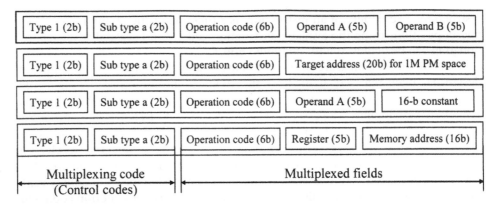

FIGURE 7.30

Code multiplexing.

Coding Technique

Different coding techniques are used in the industry, and the trade-off between the coding efficiency and flexibility has been investigated in many publications. Code multiplexing is used commonly in instruction set designs, which is to multiplex the main part of the binary code by the control of a short code field. An example of code multiplexing is given in Figure 7.30. One row represents one type of instruction, and one column represents one type of subcoding field. The length of the binary code depends on the longest (31 bits) code in the fourth row in Figure 7.30. If a 32-bit instruction word is acceptable, four bits (type and subtype) can be used for code multiplexing, and 28 bits can be used for coding each instruction when the length of binary code is 32 bits.

7.7.3 Examples

Figure 7.29 is a guide for designing an assembly instruction set manual. In order to gain more experience, you should study assembly instruction set manuals of commercial DSP processors. For designing an instruction, the following microoperations and related information should be specified:

- The assembly instruction name and type.
- The main arithmetic operation of the instruction, such as ADD.
- The full description and mnemonics of the instruction.
- The switch (specifier) for configurable operations.
- The operands.
- The memory addressing mode and addressing algorithms.

- The execution time (number of clock cycles).

- The result.

- The flags.

- The binary machine code.

- Exceptions (if any).

- One or more examples.

Example 7.6

A normal ALU operation coded using 24-bit binary machine code: Name ADD: Add 16-bit OPA to 16-bit OPB and result to OPA.

```
Assembly: ADDS OPA,OPB or ADDC OPA,OPB
Code:   Instruction multiplexer code is [23:22] = 2'b01
        ALU instruction multiplexer code is [21:20] = 2'b00
        Operation code [19:16] = 6'b 0001
        Switch S code [15] = 1'b1; add with saturation
        Switch C code [15] = 1'b0; add with carry out
        Unused code [14:10] = 3'bxxx
        Operand A and result address code [9:5] = 5b register
        address
        Operand B address code [4:0] = 5 bits register
        address
Operation:
        If S OPA<=SAT({OPA[15],OPA[15:0]} + {OPB[15],
        OPB[15:0]})
        If C {Carry,OPA[15:0]}<=
            {OPA[15],OPA[15:0]} + {OPB[15],OPB[15:0]}
Operands: OPA is from a register of the register file,
          on out bus 1
          OPB is from a register of the register file,
          on out bus 2
Run cycle cost: one clock cycle
Flags:  ALU Sign flag <=result[15]
        ALU Zero flag <= (result [16:0] = 17'b0)
        ALU Carry flag <= Carry
        ALU Saturation flag <= (result[16] != result[15])
Exception: no
Example:  ADDS R1,R2;
          // before execution R1 = 16'h7FF0 R2 = 16'h00FF
          // the internal result = 080EF [16]<>[15], overflow
          // after execution R1 = 16'h7FFF R2 = 16'h00FF
```

```
                        // ALU flags:
                        // Saturation flag = 1; Sign flag = 0;
                           Zero flag = 0
```

Example 7.7

A normal MAC instruction coded using 24-bit binary machine code Name MAC: Multiply 16-bit OPA and 16-bit OPB and accumulate result to ACRn.

```
    Assembly: MAC I ACRn,OPA,OPB or MAC F ACRn OPA OPB
    Code:   MAC instruction code type is [23:22] = 2'b01
            Sub type of Double cycle instruction is [21:20]
            = 2'b10
            Operation is code [19:16] = 6'b 1000
            When integer multiplication code [15] = I = 1'b1;
            When fractional multiplication code [15] = F = 1'b0;
            Unused code is code [13:12] = 2'bxx
            ACRn: code [11:10] is a select ACR0, ACR1, ACR2,
            and ACR3
            Operand A address is code [9:5] = 5 bits register
            address
            Operand B address is code [4:0] = 5 bits register
            address
    Operation:
        If I ACRn[39:0] <= ACRN[39:0] +
                Guard6({OPA[15],OPA[15:0]}*{OPB[15],OPB[15:0]})
        If F ACRn[39:0] <= ACRN[39:0] +
                {Guard5({OPA[15],OPA[15:0]}*{OPB[15],OPB[15:0]}),
                1'0}
    Operands:  OPA is a register from register file out bus 1
               OPB is a register from register file out bus 2
               ACRn is an accumulation register in MAC
    Result:    ACRn is an accumulation register in MAC
    Run cycle cost:
            two clock cycles,
            one for multiplication one for accumulation
    Flags:  MAC Sign flag <= ACRn[39]
            MAC Zero flag <= (ACRn [39:0] == 40'b0)
            MAC Saturation flag <= 1 when guards and sign are not
            equal
    Constraint:  ACRN cannot be used by a single cycle
                 instruction right after MAC instruction
    Example:  MAC I S ACR1 R1 R2;
        // before execution R1=16'h7FFF R2=16'h0002
           ACR1=40'h007FFFFF000
```

```
// after MAC before SAT ACR1=40'h008000EFFE
// finally after saturation ACR1=40'h007FFFFFFF
// MAC flags: Saturation flag = 1
// MAC flags: Sign flag = 0; Zero flag = 0
```

7.8 CONCLUSIONS

Assembly instruction set design of a DSP ASIP is a difficult task. This chapter gives only an introduction to instruction set design. More details are covered in Chapters 5, 6, 7, and 9.

Assembly instruction set design is summarized in Figure 7.31. The highlighted part in the figure is the focus of this chapter. Many design dimensions are possible

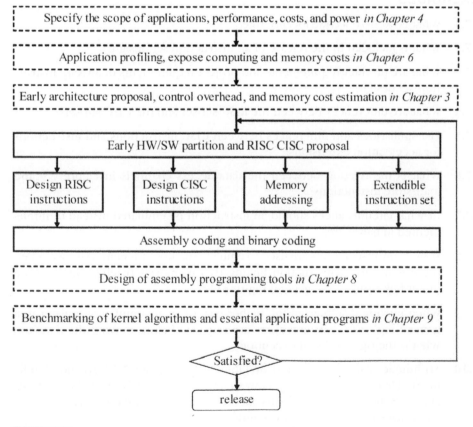

FIGURE 7.31

Summary of an assembly language design flow.

for an assembly instruction set, but it cannot be optimized for all at the same time. Enough function coverage should be supplied by RISC instructions. Performance and silicon efficiency as well as low power consumption can be achieved by introducing CISC instructions. These instructions will enhance both DSP computing and memory addressing.

After reading this chapter, you should know how to propose an assembly instruction set. However, in order to evaluate the proposed instruction set, knowledge in making assembly programming tools is necessary; this will be discussed in the next chapter. Signing off an instruction set is based on benchmarking, which will be discussed in Chapter 9. The quantitative sign-off of an instruction set also includes hardware cost estimation. This estimation skill will be acquired gradually while reading this book.

EXERCISES

7.1   What is 90%–10% code locality rule?

7.2   What is included in the release of an instruction set design?

7.3   What are product requirements and project requirements of an ASIP instruction set design?

7.4   What are differences between the RISC subset and the CISC subset?

7.5   List similarities and differences between CISC instructions and instructions for acceleration extensions?

7.6   What is the relation between the data access channels in Figure 7.9 and load/store instructions?

7.7   In what circumstances should an instruction be emulated instead of implemented in hardware?

7.8   How many instructions are used to emulate memory-to-memory-move when there is no direct instruction linking two memories (each memory has a direct link to/from the general register file)?

7.9   When moving results from the higher part of an accumulation register to a general register, what operations are required before moving the data and what is the right order of execution?

7.10   Arithmetic instructions listed in Table 7.7 and Table 7.12 are not implemented in hardware. They are emulated by using other available instructions and specified as assembly instructions for programmers. Discuss what advantages are behind the design decision.

7.11   Describe the behavior of REPEAT instruction and program Biquad IIR using REPEAT instruction based on pseudocode or the **Senior** instruction set.

7.12   When should we use conditional jump, and when should we use conditional ALU instruction?

7.13   Write bit-accurate behavior pseudocode to emulate 32bits*32bits integer multiplication using a 17bits*17bits signed multiplier, a 64-bit full adder, and a 64-bit accumulating register.

7.14   Try to make a short pseudo assembly code to support and describe the concept from Figure 7.15.

7.15   Try to make a short pseudo assembly code to support and describe the concept from Figure 7.16.

7.16   Design an 8-tap FIR based on Example 7.5 (using only instruction PSADD and PXORD in Table 7.29).

7.17   What are advantages and what are drawbacks when using special registers as operands of ALU computing?

7.18   Give an example of index addressing specified in Table 7.30.

7.19   What is modulo addressing, how many registers are required for modulo addressing, and how do you use these registers?

7.20   How many circulation buffers with modulo addressing are required for a FIR filter? How many circulation buffers with modulo addressing are required for autocorrelation?

7.21   Continue the drawing in Figure 7.26 (c) and (d) to further accept new data samples in the FIFO.

7.22   What are two special addressing modes that need to be accelerated for FFT computing? What are behavior models of these two addressing models?

7.23   Why does 2D addressing need to be accelerated? How do you accelerate 2D addressing for single-issue DSP processing?

7.24   What are microoperations? Discuss based on Figure 7.29 and classify what microoperations do not need to be specified in an assembly instruction set. What microoperations need to be specified in an assembly instruction set and do not need to be coded in binary machine code? Why?

7.25   What are differences between orthogonal assembly instruction set and orthogonal binary coding?

7.26   During the architecture design of a DSP core, the memory bus specification is based on all load/store/move instructions and "other" instructions. Find at least one instruction in the class of the "other" instructions.

7.27 Which of the following statement(s) is/are true? During instruction set design, 90%–10% locality means:

a. That 10% of instructions use 90% of the runtime.

b. That 90% of instructions use 10% of the runtime.

c. Both are not correct.

7.28 Which of the statement(s) is/are true? Instruction set design of a DSP core:

a. Is before application code profiling and after ASM benchmarking.

b. Is before ASM benchmarking and after application code profiling.

c. Also includes the specification of hardware module functions.

d. Gives decisions on how many operands can be carried by an instruction.

REFERENCES

[1] Fisher, J. A. (1999). Customized Instruction-Sets for Embedded Processors, DAC 99. New Orleans, Louisiana, USA.

[2] Muchnick, S. (1997). *Advanced Compiler Design and Implementation*. Morgan Kaufmann.

[3] Amdahl, G. (1967). Validity of the single processor approach to achieving large-scale computing capabilities, AFIPS. *Proceedings of the Spring Joint Computer Conference*, 483–485.

[4] www.openrisc.org.

[5] *Buyer's Guide to DSP Processors*. Berkeley Design Technology Incorporation (www.bdti.com).

[6] Salomon, D. (2006). *Data Compression, The Complete Reference, 3rd ed*. Springer.

[7] Hennessy, J. L., Patterson, D. A. (2002). *Computer Architecture, A Quantitative Approach, 3rd ed*. Morgan Kaufmann.

[8] Fisher, J. A., Faraboschi, P., Young, C. (2004). *Embedded Computing, A VLIW Approach to Architecture, Compilers, and Tools*. Morgan Kaufmann, Elsevier.

[9] www.tensilica.com, www.coware.com.

[10] www.ti.com.

[11] Rixner, S. (2002). *Streaming Processor Architecture*. KAP.

[12] Wehmeyer, L., Marwedel, P. (2006). *Predictable Scratch Pad Memory*. Springer.

[13] Barton, G. (1984). Towards an assembly language standard. *IEEE Micro*.

[14] IEEE standard for microprocessor assembly language. IEEE Std 694-1985.

[15] Kumar, M., Balakrishnan, J., Kumar, A. (2001). ASIP design methodologies: Survey and issues. *A. VLSI Design, Fourteenth International Conference*, 76–81.

[16] Franchetti, F., Kral, S., Lorenz, J., Ueberhuber, C. W.(2005). Efficient utilization of SIMD extensions. *Proceedings of the IEEE* 93(2), 409–425.

[17] Kitajima, A., Sasaki, T., Takeuchi, Y., Imai, M. (2002). Design of application specific CISC using PEAS-III, Rapid System Prototyping, 2002. *Proceedings 13th IEEE International Workshop*, 12–18.

[18] Clark, N. T., Zhong, H., Mahlke, S. A. (2005). Automated custom instruction generation for domain-specific processor acceleration. *IEEE Transactions on Computers*.

[19] Sun, F., Ravi, S., Raghunathan, A., Jha, N. K. (2003). A scalable application-specific processor synthesis methodology. *ICCAD*, San Jose, CA. USA.

[20] Goodwin, D., Petkov, D. (2003). Automatic generation of application specific processors. *Proc. of the 2003 International Conference on Compilers, Architecture, and Synthesis for Embedded Systems*, 137–147.

[21] Gschwind, M. (1999). Instruction set selection for ASIP design. *CODES99*, Rome, Italy.

[22] Khailany, B., Dally, W. J. et al. (2001). Imagine: Media processing with streams. *IEEE Micro*.

[23] Hinrichs, W., Wittenburg, J. P., Lieske, H., Kloos, H., Ohmacht, M., Pirsch, P. (2000). A 1.3-GOPS parallel DSP for high-performance image-processing applications. *IEEE JSSC* **35**(7).

[24] Eilert, J., Ehliar, A., Liu, D. (2004). Using low precision floating-point numbers to reduce memory cost for MP3 decoding. *MMSP*, Italy.

Software Development Toolchain

Processor firmware is developed using programming tools. In addition, programming tools can speed up hardware design and debugging. In this chapter, programming tools related to high-level languages (such as C) down to hardware-dependent low-level languages (such as machine code) will be reviewed. Since programming tools are not the main focus of this book, the presentation will be rather informal. A more formal and detailed description can be found in the compiler books in the reference list at the end of the chapter.

The discussion on generators for compilers and instruction set simulators is beyond the scope of this book.

8.1 WHAT IS TOOLCHAIN AND IDE?

The Integrated Development Environment (IDE) integrating tools in the toolchain cover all programming and debugging tools. In this context, "development" means program development or SW development for a processor; "environment" means tools for programming or SW design; and "integrated" means all tools in the toolchain are collected and integrated into one program, normally with a graphical user interface (GUI). The functions in an IDE can be divided into high-level language development tools and assembly language level development tools. In this chapter, only the assembly language level development tools, including compiler, assembler, linker, and simulator, will be introduced. Source code profiler will also be introduced briefly. Source code generation tools will be discussed later, in the chapter on firmware design and in the chapter on streaming signal processing. The following questions will also be discussed in this chapter:

1. What is an IDE/toolchain?

2. Why and how do you use a toolchain (from the toolchain user's view)?

3. What is the principle of each tool (from the toolchain designer's view)?

A compiler is a program that reads a program written in one language (usually a high-level language such as C), and translates it into an equivalent program in another language (usually assembly language). It reports errors in the source program. It is the primary tool used to design assembly code. A tool called assembler translates assembly language into an object file with relocatable binary machine code. Relocatable code is the code that can be loaded into any section in the main memory for execution. The linker finally combines several object files and library functions into a single and final executable program. Finally, an instruction set simulator (ISS) can load the program and execute it by simulating the behavior of the DSP processor for each instruction. This allows the software developer to run the program without the real hardware.

8.1.1 ASIP User's View on IDE

Generally, during the system-level design, application programs are developed based on high-level behavior languages, such as MATLAB or C. High-level language cannot be executed directly on processor hardware. The behavioral code must first be translated into machine code that the processor can understand. To qualify and translate the high-level behavioral language into the binary machine codes under hardware constraints, several steps are required. The first step is to translate the source code from behavioral language, for example C, into assembly code. Source code can also be translated into assembly code manually by assembly programmers if the source code is relatively simple. The assembly code quality depends on the experience of the assembly language programmers. High-quality code minimizes the execution time and the program size. However, a complex application requires an enormous amount of work to translate it by hand; thus a C compiler often is used to translate C code into assembly code. Before translating C code to assembly code, the C code may need to be modified according to the hardware constraints in order to help the compiler generate better code for the system. The modification may be guided by a C-HW-adapter program, which is a source code parser that identifies parts of the code where the C compiler will perform poorly.

During the next step, assembly language code from the compiler or the assembly language programmers is translated into binary machine code. All binary machine code of the application, including subroutines in libraries, need to be linked together and assigned to physical memory addresses. Finally, the code should be executed and debugged using the instruction set simulator before it is loaded to the hardware (program memory). A design flow from the programmer's perspective is given in Figure 8.1.

The source file editor can be any available text editor, but usually it has special functions to help the programmer as follows:

- Perform text creation and its modifications.

- Analyze the format of the program text.

- Hierarchically organize the source code.

- Print statements and comments in a decent and tidy way.

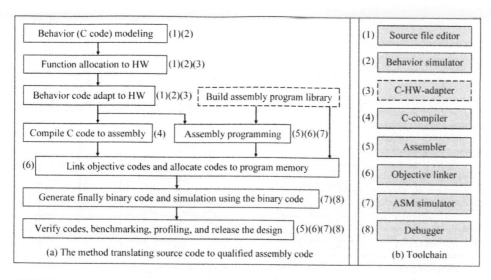

FIGURE 8.1

A DSP processor FW design flow based on a toolchain.

The system behavioral simulator is based on the behavioral language selected by the project. To speed up the behavioral modeling, we may use MATLAB or similar high-level languages with application-specific libraries. However, to compile behavioral language to assembly language, C may have to be used as the behavioral language. (The ways of compiling MATLAB code to assembly code were investigated by researchers [14, 15]).

To compile C source code to high-quality assembly code, ASIP hardware information should be adapted to the source code so that the distance between the source code and the assembly instruction set can be reduced and the compiler can further understand the source language. For this, certain coding templates may be proposed, and certain operation and operand libraries may be provided by the ASIP provider. The C library adaptation includes the hardware data types and the special operations (such as complex data operations in hardware). There may also be a need for special annotations in the C code to guide the compiler in translating parts of the source code to specific assembly language instructions.

However, extra features supplied by the C library for one ASIP in general cannot be recognized by other ASIPs. A program using hardware adaptation libraries cannot be used directly for firmware design on other hardware, meaning the code portability feature is low. Portability issues will be discussed in the chapter on firmware design.

8.1.2 ASIP Designer's View on IDE

Integrated design environment (IDE) is the integration of all tools in the toolchain of Figure 8.1 into a single program. To make an IDE for a processor, the knowledge

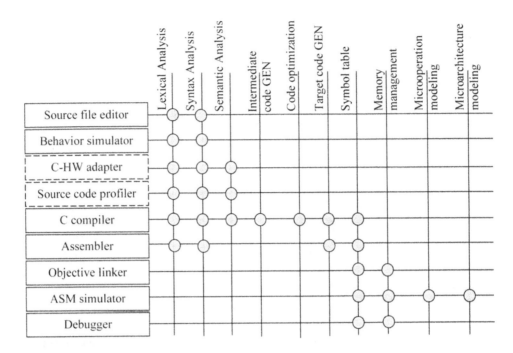

FIGURE 8.2

Basic knowledge behind a toolchain.

listed in Figure 8.2 is required. In the following sections, tools and their implementations will be discussed briefly.

From a developer's point of view, most tools in the toolchain can be divided into three parts: code analysis, code generation, and code modeling (simulation). Since the toolchain is not the focus of the book, the discussion of each part will be brief. However, in most processor design projects, an assembler and a simulator need to be designed for purposes such as debugging and benchmarking the hardware design.

The basic knowledge, from lexical analysis to microarchitecture modeling, behind a toolchain as shown in Figure 8.2 will be discussed in the following sections.

8.2   CODE ANALYSIS

Code analysis is also called code parsing [1, 2]. It reads code and arranges it into a form suitable for further processing by other parts of the tool. The code analysis is the first step of the profiler, compiler, and assembler. It can also be part of an intelligent editor. Code analysis usually is divided into two steps: lexical analysis and syntax analysis. At the input of the lexical analyzer, the program code is simply a sequence of characters. The analyzer reads the characters in sequence one by one and identifies each lexical token.

Tokens, the output from the lexical analyzer, are the input of a syntax analyzer. This is a program that analyzes the token stream and combines the tokens into larger units. For a C compiler, these units are declarations, expressions, and statements, which are combined further into functions and programs. Often, the syntax analyzer is the central engine in the tool so that the output of the syntax analyzer is the output of the tool, but internally the syntax analyzer generates trees and tables that represent the input.

8.2.1 Lexical Analysis

Lexical analysis also is called linear analysis or scan. Lexical analyzer is a scan on a program that reads characters of a line in the source program from the left to the right and groups these characters into tokens.

8.2.2 Syntax Analysis

Syntax analysis is to parse and analyze a language (C or assembly language would be the most interesting cases here) and to extract the structure according to the grammar of the language. Tokens are recognized and evaluated by the lexical analyzer as previously described. The syntax analyzer takes the sequence of tokens as the input and matches the token sequence with the rules of the grammar. Each grammar rule is a piece of program code. It is executed to take care of the input; usually it builds a parse tree that reflects the grammatical structure of the input, but it may do other things instead. For example, in an assembler, the code associated with the grammar rules for the various parts of an instruction, such as mnemonic and operand list, could remember the mnemonic and the operands in a global data structure so that this information is available for later use. The rule for a completed instruction later could generate the binary code for the instruction using this remembered information and write it to the object file.

A grammar is a set of syntactic rules of a language. It describes the structure, or syntax, of a language as a particular type of data. The grammar of arithmetic instructions will be taken as a simple example. It contains text files as a list of mnemonics, configuration of the operation, destination register, source operand A register, and optionally source operand B register:

```
ADD CIN R0 R1 R2
SUB saturation R31 R15 R30
ABS saturation R1 R1
MUL R16 R1 R2
```

The grammar of arithmetic instructions is a set of rules. Each rule contains the name of a symbol, followed by the production symbol ::=, definition of the symbol, and a semicolon (see the following code). An informal definition of a grammar starts with the description of the input, then incrementally defines each symbol in terms of other symbols, until the lowest level symbols are defined. When all symbols have been defined, the grammar will be used for parsing.

The following grammar describes the syntax of arithmetic instructions just listed. It is part of a simplified grammar of a part of an assembly instruction set.

```
ArithmeticInstruction ::= { InstructionLine };
InstructionLine ::= mnemonics Destination OPA [OPB];
//[] stands for optional term, meaning that OPB may not
exist.
  mnemonics::= Operation
                " "
                [Configuration] //optional term
                ;
 Destination ::= Register
 OPA ::= Register
[OPB ::= Register] //optional term
 Operation ::=
               "ADD"
             | "ABS"
             | "MUL"
             | "SUB";
 Configuration ::=
               "saturation" //carry out or saturation
             | "CIN"; //include or not include carry in
 Register ::=
               "R0"
             | "R1"
             | "R2"
             | ...
             | "R30"
             | "R31";
```

The parser for arithmetic instructions follows the grammar and performs syntax analysis by matching grammar symbols to elements in the input data. The result is a parse tree. Each node in the tree has a label, which is the name of a grammar symbol; and a value (if it exists), which is an element from the input data.

The Parse Tree

The execution of a parser implicitly describes a parse tree. Implicit means that although the parser does not generate a parse tree automatically, it does visit the nodes in the parse tree in depth-first order, and exposes the child information in each node. From this, it is rather straightforward to generate the parse tree if desired.

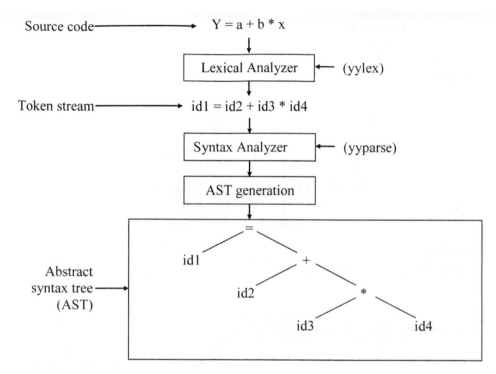

Source code ──────────────→ Y = a + b * x

Token stream ──────────────→ id1 = id2 + id3 * id4

FIGURE 8.3

A compiler front-end [6].

A parse tree consists of nodes and branches. Each node in a tree has zero or more child nodes. A node that has a child is called the child's parent node.

From Figure 8.3, yylex and yyparse will be discussed in the following section. Another example of a parse tree is given in Figure 8.4. It gives a detailed view of the tree structure.

Parser Generator

Lexical analyzers and parsers are essential for ASIP designers because they are basic components in the source code profiler, C-compiler, and assembler. To create a lexical analyzer and parser by hand is a very tedious job. Fortunately, a lexical analyzer and parser can be generated automatically based on a description of the patterns and the grammar to a parser generator. Lexical generators and parser generators were introduced in the mid-1970s, and systematically described and well used in the mid-1980s [1]. Well-accepted and free (from flex and GNU) lexical and parser generators are LEX and YACC [2]. There are several other similar free tools—for example, flex (The Fast Lexical Analyzer), available at GNU page, and rex, among others.

LEX generates a lexical analyzer (scanner). The generated lexical analyzer is a C program. LEX generates a scanner by adapting user-defined pattern matching rules to

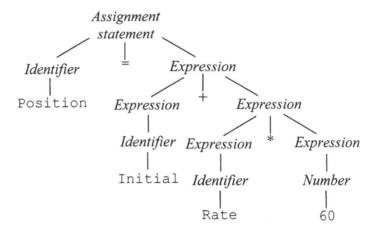

FIGURE 8.4

Parse tree for position = initial + rate * 60.

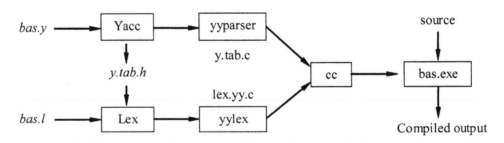

FIGURE 8.5

Parser generator.

the scanner generator. Inputs are matched using user-defined pattern matching rules and converted to tokens. Patterns are specified based on LEX regular expressions. Each pattern in LEX has an associated action that is executed when the pattern is found. An action usually returns a token of the matched string, for subsequent use by the parser.

The parser designer defines grammar rules as the input file to YACC. YACC generates C code of a syntax analyzer (parser) by adapting the user-defined language grammar. Rules of the grammar are used by YACC to analyze tokens generated from LEX. The generation of a lexical analyzer and a syntax analyzer using LEX and YACC is depicted in Figure 8.5.

In Figure 8.5, bas.y contains the grammar description, and the input bas.l contains the pattern descriptions. Ways of specifying rules were introduced in reference [2]. To generate a parser, the following scripts should be executed under UNIX (Linux) after loading and installing LEX/YACC on your computer:

```
lex bas.l # create lex.yy.c
yacc -d bas.y # create y.tab.h, y.tab.c
cc lex.yy.c y.tab.c -o bas.exe # compile/link
```

The generated executable code bas.exe is the generated parser. The lexical analyzer C code is generated from bas.l as yylex in file lex.yy.c. The parser C code is generated as y.tab.c. Finally, these codes are compiled, and the final executable code bas.exe is the parser with lexical analyzer.

8.2.3 Semantic Analysis

Semantics reflects the meaning of programs or functions. During semantic analysis, the semantic analyzer adds semantic information (computer language specification) to the parse tree, performs checking based on added information, and builds the symbol table. The semantic analysis checks if the parse tree (the combination of tokens) follows the language specification. The semantic analysis identifies semantic errors from the source program and gathers type information for further generation of target code.

During the semantic analysis phase, the source program is checked to find semantic errors, and the type information is collected for the code generation phase. The semantic analysis uses the hierarchical tree structure generated during the syntax analysis to identify operators and operands of expressions and statements. This checking procedure ensures that certain kinds of programming errors will be detected and reported. Examples of semantic checks can be:

Type checks. The analyzer should report an error if an operator is used in a wrong way or the type of an operand does not match the instruction (an incompatible operand)—for example, if an integer variable is added to a function. It can also check whether parameters of functions are correct both in types and numbers. For example, the expression "Hello" +1 is invalid because a string cannot be added to an integer in C. In an assembler, the semantic analyzer should report an error if the format of an instruction does not match its specification. For example, operands of ADD instruction can reside in only a general register file or be immediate data. If an ADD instruction carries its operand from a port register, there should be an error report.

Control flow checks. In a compiler, for example, a break statement in C causes the control flow to leave the enclosing while(), for(), or switch() statement. If brackets are used outside one of those constructs, an error is generated by the semantic analyzer. In the assembly language semantic analyzer, the limits of jump distances or memory sizes are specified by the assembly language or by a specific physical design.

Uniqueness checks. For example, an identifier cannot be specified twice with two meanings.

The type checks do not always have to result in an error. For example, a type mismatch can sometimes be resolved by converting the operand. If "*a*" in the statement "$a = a*2$" is a floating-point number, the integer "2" should be converted to a floating-point number before multiplying.

8.3 PROFILER AND WCET ANALYZER

Cost-free profilers are available, for example, gprof for gcc. However, most free profilers are not suitable for the profiling required when designing an assembly language. For example, many profilers sample the host PC during profiling (run) time, which is not useful for profiling when designing an assembly instruction set. The reason is simple; the assembly instruction set of any host machine (Pentium for example) will be much different from any assembly instruction set of an ASIP, and that is why we want to design an ASIP instead of using a Pentium.

Source code profiling and WCET (worst-case execution time) analyzing were discussed in Chapter 6. After introducing parsing in this chapter, the implementation of a dynamic and static profiler as an example will be explained further, based on the open source C compiler gcc [7, 8]. A profiler can be implemented using available resources, such as the compiler front-end from gcc [7].

The profiler was designed as an open source program available on this book's web page. Based on the front end of the GNU compiler, the parse tree (2) of the source code (1) in Figure 8.6 can be generated. The CFG (control flow graph (3)) can be derived from the parse tree, and all basic blocks can be identified and specified with annotations. The CFG exposes basic blocks. A basic block is a straight-line code sequence that is entered only at the beginning and has branches only at the end [3]. Computing costs of all basic blocks are accumulated and logged into the bb cost table (4). Accumulating total costs of each path of CFG based on the basic block cost table (4), the worst-case runtime costs can be predicted. (A good static WCET prediction for reliable computing might be harder, and is beyond the scope of the book.)

However, the worst-case runtime might be too long, and the system design following the worst case may induce too much silicon cost. It is important to expose the realistic runtime costs by dynamic profiling. The purpose of dynamic profiling is to give statistics on branch-taken probabilities of all branches based on typical data stimuli. Suppose the typical data stimuli are available (based on experiences); the statistics on branch-taken will help to remove many paths that are never or seldom executed. However, this method is not allowed for designing a system for reliable computing.

The runtime of a basic block (basic blocks were defined in Chapter 6) is the same from dynamic profiling and static profiling. The difference between dynamic

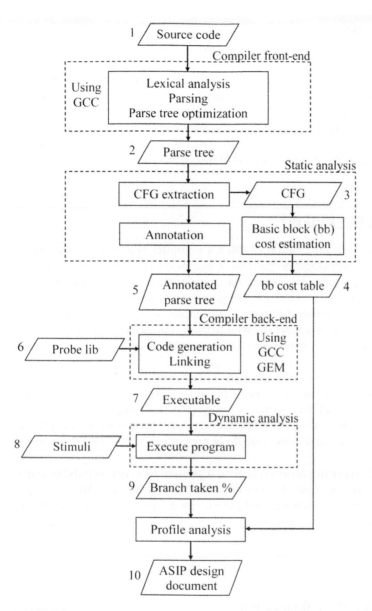

FIGURE 8.6

A profiling tool based on GCC [3].

profiling and static profiling appears on an upper level, induced by data-dependent branches. Therefore, the essential task of dynamic profiling is to analyze branches. All branch points thus are identified and annotated on the parse tree (5). Probes in the probe library (6) are further inserted to code (5) at all annotated branches.

After compiling and linking, code with annotated probes at all branches becomes the final executable file (7). This is actually the executable source code with branch probes.

Dynamic profile is executed by sending stimuli (8) to the executable file (7). After a long time execution using as much typical stimuli as possible, the statistics on all branches can be trustable and logged into (9). The log of branch taken (9) and early logged costs of basic blocks (4) finally are merged and analyzed. Realistic runtime costs will then be identified. The early HW/SW design proposal will be the final document to guide the design of the ASIP instruction set.

8.4 COMPILER OVERVIEW[1]

As mentioned in the beginning of the chapter, a compiler is a program that reads a program written in one language (usually a high-level language) and translates it into an equivalent program in another language (usually a low-level language) [6, 7]. Normally, for an ASIP firmware designer, the source language is C and the target language is assembly language executable on specific hardware, not on the host machine (the machine on which the compiler is executed). Therefore, in practice, all compilers generating ASIP assembly instruction sets are cross compilers.

Compilation consists of two phases: analysis and synthesis. The analysis phase is also called front-end. It breaks the source program into consecutive pieces and conducts lexical, syntactical, and semantic analysis. AST in Figure 8.7 stands for Abstract Syntax Tree. The output of the front-end of a compiler is an internal representation of the analyzed code, usually in tree form, along with the symbol table.

The synthesizer first creates an intermediate representation (IR) versus the parse tree of the source program. Code is then optimized on an IR level. After the optimization, the (source) code profiling can be performed based on optimized IR, and the profiling quality will be much higher [10]. Finally, assembly code is generated based on the optimized IR code.

The compiler front-end was discussed in the previous section; next, we discuss the compiler back-end.

8.4.1 Intermediate Code Generation

After semantic analysis, a compiler translates the source code to IR (intermediate representation), which is a machine-independent language. A typical IR is a so-called three-address code (TAC). Each line of TAC can be described as a quadruple (operator, operand1, operand2, result). The format is:

---

[1]We thank the compiler expert, Professor Christoph Kessler, for his contribution to this Section.

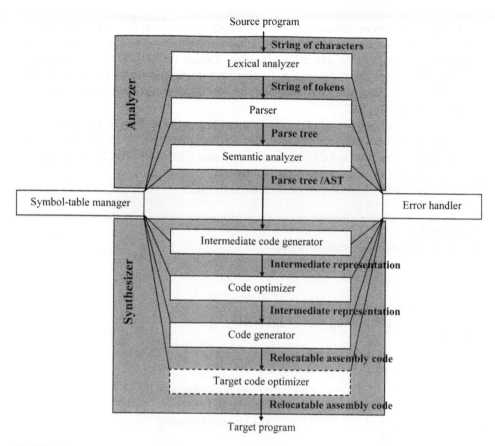

FIGURE 8.7

A compiler in general.

$$x := y \; OP \; z$$

where x, y, and z are variables, constants, or temporary variables that refer to the symbol table in the compiler front-end. OP is an operator representing an arithmetic or logic operation. Arbitrary arithmetic and logic computing can be translated into a series of TAC operations. For example:

$$a := b + c \times d$$

can be translated into TAC format if a new variable is introduced for the intermediate result:

```
x := c × d
a := b + x
```

Using TAC, every instruction (line) implements exactly one fundamental operation. The source and destination, in principle, can map directly to general registers. TAC is close to assembly language format (OP DEST OPA OPB). In an assembly language, OP is operation; DEST is the destination, equivalent to the left side identifier of TAC; and OPA and OPB are operands equivalent to identifiers on the right side of the TAC.

There are also statements for conditional jumps, procedure calls and returns, address and pointer assignments, and indexed assignment to be applied to arrays. TAC codes are specified in Table 8.1 and are just one variant of TAC. There are almost as many variants of TAC as there are compilers.

Using this table, C code can be translated to TAC.

Example 8.1

Translate the following C program to TAC.

```
main ( )
{
    int i; int b[8];
    for (i=0; i<8; i++) do { b[i]=i; }
}
```

The translation is:

```
i := 0                      // an assignment
L1: if i >= 8 goto L3       // a conditional jump
    b[i] := i               // an assignment
    i := i+1
    goto L1
L3 :                        // end.
```

Although the source program can be translated directly to the target language from the syntax tree, there are benefits of using an intermediate form:

- IR is the representation suitable for code-improving transformations. A machine-independent optimizer can be applied to the intermediate representation.

- Retargeting is easier. Creating a compiler for a different target machine can be done by replacing only the code generator. The front-end and the IR optimizer can be left intact or with only small changes.

8.4.2 Code Optimization

Because the major compiler research work is on optimizations, an enormous amount of algorithms and approaches are available for optimizations of code in a basic block up to interprocedure-level optimization. Implementation of optimizations can be

Table 8.1   TAC Code Table Example.

Assignment Statement	$x := y$ op z	x, y, z are names, constants, or compiler-generated temporaries; op is any binary operator
	$x :=$ op y	op is a unary operator
Indexed Assignment	$x := y[i]$ $x[i] := y$	x, y, i are data objects
Copy Statement	$x := y$	x, y are data objects
Unconditional jump	goto L	L is the label of the next statement to be executed
Conditional jump	if x relop y goto L	x, y are data objects. The relop is a binary operator that compares x to y and indicates true/false. L is the relocatable target address
Procedure calls	call p, n	p is the procedure, n is the number of parameters
Procedure returns	return y	y is an optional return value
Address and pointer assignments	$x := \&y$ $x := *y$	x is set to the location of y x is set to the value at the location y

divided into optimization before target code generation (machine-independent optimization) and optimization during or after code generation (machine-dependent optimization). Most optimizations are conducted on the IR level. Classical optimization after code generation usually is limited to peephole optimizations (replacing certain instruction combinations with faster or smaller alternatives), if performed at all. Modern post-pass optimizers can even reconstruct control flow from binary code.

The goal of code optimization could be to minimize the execution time cost, to shrink the size of the code, or to shrink the data memory cost. Recently, optimization may also target minimization of power consumption, which will not be discussed in this book. If it is not specially noticed, optimization by default is to minimize the runtime. Before applying an optimization scheme, the purpose of the scheme should be matched to the wish from designers.

There are different optimization algorithms conducting optimizations on different levels, from very low level of adjacent instructions to intraprocedure optimizations, up to interprocedure optimizations. Optimization algorithms can be found from compiler text books [5, 6]. The purpose of this chapter is to guide the selection of a compiler generator instead of designing a compiler generator. *What* to optimize will be introduced instead of *how* to optimize.

Data Flow Optimizations

Data flow exposes data dependence (detect hazard; see the glossary and Chapter 14), at the same time, independent computing is identified. If a piece of code is independent of surrounding codes, this piece of code can be moved. This principle gives opportunities for code movement. In the following pseudocode:

```
do {
a = 1
x = y + z
Until y > 23 }
```

the perceived constant expression $a = 1$ is independent of the do loop, and it can be moved without changing the meaning of the code. Therefore, this code can be changed to a more efficient way:

```
a = 1 {
Do x = y + z
Until y > 23 }
```

In this case the constant can be handled before runtime. This technique is called loop-invariant code motion. Expressions with constant operands with respect to the loop can be moved outside the loop.

Common subexpression elimination

A common subexpression (CSE) is created when two or more expressions compute the same value. The expression is calculated once to a temporary variable. The variable will be used instead of being computed once more. It will reduce both the cycle cost of execution and the memory cost. For example, the statement:

```
Array1 [i + 1] = array2 [i + 1];
```

will be transformed to:

```
temp1 = i + 1;
Array1 [temp1] = array2 [temp1];
```

Dead code, in other words code that is never executed or that does not affect the output of the program, can be eliminated. For example, this code fragment:

```
int global;
void foo(void) { int k = 1; global = 1; global = 2;}
```

will be transformed to

```
int global; void foo(void) { global = 2; }
```

Another example of dead code elimination is:

```
A=b(i)*c(i)
A=b(i+1)*c(i+1)
```

Most compilers will analyze this code and determine that the first line has no effect and will eliminate it.

Some expressions can be simplified by replacing them with a more efficient expression. For example, $i + 0$ will be replaced by i, and $i*0$ and $i - i$ will be replaced by 0, and so on.

Strength reduction is an optimization method that replaces expensive instructions with less expensive instructions. For instance, a popular strength reduction is to replace an integer multiplication by a constant power of two with a left or right shift.

An intuitive optimization is to minimize jumps. A jump induces pipeline stall and consumes extra clock cycles. A jump should be avoided if it is possible. For example, conditional execution was introduced in Chapter 7 to avoid jumps.

Code and data locality is preferred even if cache is usually not used in an ASIP. If all data is allocated within one segment in data memory, the value in the segment address register can be kept and costs of memory access can be reduced.

Loop Optimizations

Loop optimization modifies the code of a basic block inside a loop or adjacent basic blocks of neighbor loops to minimize the runtime or memory cost. Loop optimization gives the most significant impact on runtime of DSP firmware because most DSP algorithms are based on iterative computing.

An induction variable is increased or decreased by a fixed value during each iteration of a loop. Induction variables should be identified, and computing on it should be minimized in the loop. For example:

```
for (i=0; i < 10; ++i) {
    j = 10+i;
    ...;
}
```

should be changed to the equivalent code:

```
for (j=10; j < 20; ++j) {
    ...;
}
```

If a loop is very large, data in several data segments might be used by the loop, and changing segment value in the loop makes the loop irregular. Loop distribution technique is to break a loop into multiple loops. Data for each loop is allocated only one data segment. The loop program will be easier, and loop-level code vectorization can be supported.

Loop fusion technique reduces loop overheads by merging multiple loops into one. The condition of loop fusion is that the iterative result of one loop will not be the input of another loop. Each loop has its own loop overheads; by loop fusion, overheads of multiple loops can be minimized to the overhead of one loop. For example,

```
for (i = 0; i < 10; i++) {
    a[i] = 1;}
for (i = 0; i < 10; i++) {
    b[i] = 2;}
```

can be merged into one loop:

```
for (i = 0; i < 10; i++) {
    a[i] = 1;
    b[i] = 2;
}
```

Loop unrolling is used when the loop body (length of the inner-loop program) is small and the number of iterations is low. Loop unrolling duplicates the body of the loop multiple times, in order to decrease the loop overheads of the loop condition test and the jumps. Loop unrolling cannot be used if the number of iterations is unknown at the entry of the loop. Loop unrolling is not recommended if the code cost is critical. For example, when the size of the program memory is very small, loop unrolling shall be carefully used.

There are other loop optimization algorithms; some are for minimizing cache-misses, and some are not applicable for streaming signal processing. These algorithms will not be discussed because they are not the focus of the book.

8.4.3 Code Generation

The final phase in a compiler is the generation of target code. Code generation in a compiler translates the optimized IR to relocatable assembly code, the target code. Code generation is mixed with code translation and target hardware-dependent optimization. Code generation can be divided into instruction selection, instruction scheduling, and register allocation.

Target-Dependent Code Optimization

Target-dependent code optimization generates assembly code with hardware constraints. It is thus called machine-specific code optimization or back-end optimization. There are many techniques used by compiler companies to optimize code for a specific DSP processor. The following list gives frequently used methods for machine-specific code optimization:

1. Select best instructions based on the compiler-known function (CKF) annotation in the source code.

2. Select best subroutines based on the CKF annotation in the source code.

3. Select best addressing modes based on the CKF annotation in the source code.

Often, it is very difficult to make the compiler understand how and when to use the special features of the ASIP such as bit-reversed addressing or other hardware acceleration for special functions. The trade-off used in ASIP compilers is usually to use so-called compiler-known functions (also called intrinsics) that directly map to these features.

Instruction Selection

Instruction selection transforms the IR into a sequence of machine code or assembly language instructions. The principle of instruction selection is to match IR operations to templates of target codes with the same behavior. The target code templates are specified for a compiler generator following the assembly language manual of the ASIP. A template is a tree-like description of functions of an assembly instruction. The tree shall be specified similar to the tree of IR. Methods and algorithms are investigated to evaluate and decide the best match.

Register Allocation

Compared to memory, a register is much faster, but at the same time the register consumes 10 to 20 times more silicon area. Registers are RISC computing buffers because they are fast and can be accessed in parallel. However, an application program may require hundreds of thousands or even up to millions of variables, and it is clear that all of them cannot be stored in registers all the time.

A program that is executed sequentially usually requires only a few variables at a time. Only a few registers might be enough if variables can be loaded to the register file and prepared for the running program in time. Register allocation technique is therefore essential to keep as many useful variables as possible in registers for the application program, thereby reducing the number of memory accesses.

Register allocation is the process of choosing which variables should be stored in registers at any given time during the program execution.

The memory load and store of variables to and from registers is a meaningless operation because it does not exist in the source behavior code. The cost of memory load and store can be minimized by storing the least frequently used variables to data memory and giving the register space to most frequently used variables. Many algorithms are introduced, and optimization for register file allocation is the essential technology of compilers. The principle of most register allocation algorithms is the so-called graph coloring approach, which is discussed by practically every compiler book [6].

Instruction Scheduling

Instruction scheduling is the optimization at the backend stage to improve instruction-level parallelism via pipeline without changing the meaning of the source (IR) code. By reordering independent instructions, pipeline stalls can be minimized. The pipeline stalls, caused by structural hazards, data hazards, and control (branch) hazards, will be further discussed in Chapter 14.

Phase Ordering Issues in Code Generation

Code generation consists of three steps: the instruction selection, the register allocation, and the instruction scheduling. Classically, these three steps are conducted separately so that subsequent steps of code generation might be constrained by early ones. For example, instruction scheduling can be performed before register allocation or after register allocation. Early register allocation may introduce artificial data dependences that constrain the scheduler. On the other hand, the register allocation depends on the schedule. Any sequence of ordering the phases of register allocation, instruction scheduling, and instruction selection has advantages and drawbacks. An advanced approach is to integrate all subproblems in three steps of optimizations into a single and combined problem, the integrated code generation [10]. The integrated approach is time consuming. However, because the applications are fixed and known, the time consumed during the off-line optimization is not a critical design parameter.

8.4.4 Error Handler

It is important that a compiler or an assembler can detect errors in input codes and handle them. When an error occurs, the compiler or assembler emits an error message containing the location of the error in the source program and a message stating the type error, and then tries to continue with the compilation or assembling.

There are different kinds of errors—syntax errors, semantic errors (type errors, for example), or size errors (violating hardware constraints). Identification of semantic errors was handled by the semantic analyzer. The size errors should be specially discussed here. There are two sets of limits—the limits of the assembly instruction set and the limits of the hardware design. The limits from an instruction set are specified in the instruction manual. For example, the program address space is limited by the size of the program counter (PC). The data memory address space is limited by the size of the address pointer. Without special project specification, the error checker follows the size limit of the assembly instruction set manual. Based on one instruction set, several processors might be designed with different configurations. For example, one processor may contain just 16 kB program memory and 64 kB data memory, though the PC size is 16 bits (64 k instruction words) and the address pointer size is 18 bits (256 k data words). In this case, the error report follows the specification of the project instead of the assembly instruction set manual.

As soon as the error counter is not zero, an error flag is set, and the compiler or the assembler will stop execution after the semantic analyzer phase. There is no point in generating the target program when there are errors in the source program.

The compiler can also detect minor errors that will not stop the compilation, and will emit warnings about these errors instead.

8.4.5 Compiler Generator and Verification of a Generated Compiler

A compiler generator is a program. It generates a compiler according to a complete specification of the target language. Suppose that the compiler generator was 100% verified. Verification can be divided into the verification of the configuration of the compiler generator (templates) and the verification of the generated target code. Verification will be discussed in detail in Chapter 19.

8.5 ASSEMBLER

The assembler translates assembly instruction mnemonics, symbolic names for memory locations, and other entities into binary machine code. The input of an assembler is assembly code. The output file from the assembler is an object file, which contains relocatable binary codes and bookkeeping information (symbol table for the linker).

The translation consists of two steps. One is to calculate addresses of all symbolic address names. Another is to translate each assembly statement, mnemonic, register, configuration switch (specifier), and label into binary code.

An address label is local if it is available in an object file. It may also be public, which means that the label will be visible to others during the linking phase. An address label is external if it is not available in the object file but is expected to be resolved during linking. In most cases, an object file cannot be executed because some references, procedures, and data could be in other files. An assembler processes each file in a program individually. An assembler knows the addresses of only local labels. Another tool, the linker, will combine a collection of object files and libraries into an executable file by resolving external labels. The assembler assists the linker by providing lists of public labels and unresolved references (externals) in a linking table of the object code.

Because of dividing translation into two steps—finding labels and binary translations—an assembler's first pass reads each line of an assembly file and breaks it into its component pieces. These pieces, called lexemes, are individual words, numbers, and punctuation characters. For example, the line

```
L: ADD R1, R2
```

contains six lexemes: the label L, a colon, the opcode ADD, the register R1, a comma, and the register R2. If a line begins with a label, the label will be recorded in the symbol table, including the name of the label and the address of the instruction. This record prepares for linking of a target address for jumps and calls. During the first pass, the assembler as a parser generates a line for each instruction. When the assembler reaches the end of an assembly file, the symbol table records the location of each label defined in the file.

The assembler executes the second pass, producing machine codes, by using the information in the symbol table generated during the first pass. The assembler again examines each line in the file and concatenates the binary representations of its operation, operands, the address of the destination register, configuration switches (specifiers), and memory address into binary code of the instruction. If a symbol address is inside the table generated during the first pass, relative distance of jumps can be calculated for the program. Instructions and data words that reference an external symbol in another file cannot be completely assembled (they are unresolved) since the symbol's address is not in the symbol table.

Assemblers produce object files. An object file normally contains four or five sections:

1. The object file header describes the size and position of the other pieces of the file. (This part can be optional if the AIS system is simple.)

2. The text segment contains assembly language code for routines in the source file. (Routines cannot be executable because of unresolved references before linking.)

3. The data segment contains a binary representation of the data in the source file. (Data is incomplete because of unresolved references to labels in other files before linking.)

4. The relocation information identifies instructions and data based on absolute addresses. (Cannot be used before allocating the codes into physical memory.)

5. The symbol table associates addresses with external labels in the source file and lists unresolved references. (Will be merged into the global symbol table during linking procedure.)

Assemblers can provide other convenient features to make assembly programs shorter and easier to write. Macros are considered to be a pattern-matching and replacement facility of assembler. The macro function provides a simple mechanism to name a frequently used sequence of instructions. Instead of repeatedly typing the same instructions every time they are used, a programmer invokes the macro, and the assembler replaces the macro with the corresponding sequence of instructions. Similar to a subroutine, a macro allows a programmer to create and name a new abstraction for a common operation. The difference is that a macro call is replaced by the macro's body when the program is assembled, which means that the macro code generated by the assembler is not a subroutine. After this replacement, the resulting assembly is indistinguishable from the equivalent program written without macros. A macro does not cause a subroutine call.

Some instructions are not implemented in hardware, yet they are frequently used by assembly-level programmers. These instructions are emulated by other instructions supported by hardware. For example, if the instruction NEG R (negate the register value) is not implemented in hardware, it could be replaced by the real instruction SUB #0, R. The assembler could translate these pseudo instructions to the corresponding real instructions.

Figure 8.8 shows a simplified assembler for Senior Assembly Language. It is a simple assembler including part of linker functions. In the first pass in Figure 8.8, all mnemonics and tokens are stored into a token table. All binary codes of mnemonics are collected and matched to each mnemonic. Before the second pass, all binary codes are concatenated and grammars are checked. The code memory allocation and data memory allocation are performed before the third pass. If there are no grammar or linking errors (all code and data addresses can be allocated), the assembly is passed.

8.6 LINKER

A linker is a program that takes object modules with linking tables as inputs, combines them, and generates final executable binary machine codes.

To write assembly code by hand is time-consuming work. A firmware system usually is developed by a team of many designers, including library designers, subsystems designers, and firmware integration engineers. Therefore, a firmware system consists of many separate subsystems or subroutines. These subsystems and subroutines can be modified independently so that the cost of design iteration can be treated locally. An application firmware system finally is generated by linking involved subroutines and libraries together when all sublevel designs are ready.

Allowing separate assembly modules enables splitting a firmware design into pieces (in several files). Each file contains a logically related collection of subroutines and data structures, and each file is a module in the firmware design. A file can be compiled and assembled independently of other files, so that changes to one module do not require recompilation of the entire firmware design.

In some modules, there exist symbols, functions, and variables that need to be accessed by other modules. These will be declared as public. In the other modules where these symbols, functions, and variables are required, they are declared as external.

The linker finally combines all these modules into a single executable program. Machine code files can be from several compilations, some may be library files, and some may also be written manually by assembly designers. The linker resolves external references so that labels, data, and functions from the different files can be used by each other, see Figure 8.9.

When all the external references are resolved into the global symbol table, the linker of an embedded processor takes relocatable machine codes and alters relocatable addresses to real addresses. The linker will then place the code and the data into proper locations and create the output file. The linker performs the following tasks:

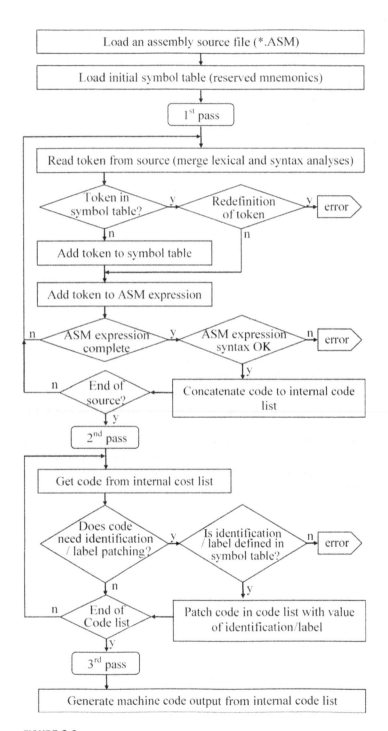

FIGURE 8.8

Senior Assembler flowchart.

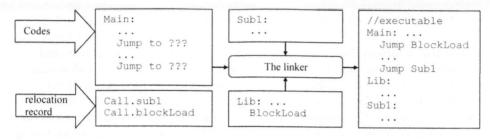

FIGURE 8.9

Linker.

- Searches the program libraries to find library routines used by the program by searching the library for still unresolved externals in the object files.

- Determines the memory locations of codes from all modules and relocates instructions by adjusting absolute references.

- Resolves references among files (match all labels to physical addresses and calculate all target addresses of jumps and calls).

The first task of a linker is to ensure that a program contains no undefined labels by cross-checking all symbol tables from all program modules and libraries. If there are unresolved external labels, the reason should be identified as being design bugs or missing subroutines. The linker matches the external symbols and unresolved references from other file modules to be linked. An external symbol in one file resolves to a public label from another file if both have the same name.

A program does not work and cannot be linked if an object file refers to an external label that does not exist in any object file or library. When the program uses a library routine, the linker extracts the routine's code from the library and incorporates it into the text segment of the program. This new routine may depend also on other library routines, so the linker continues to fetch other library routines until all external references are resolved.

At the next step, after resolving all external references, the linker determines the memory location for each module. Since the files were assembled separately, the assembler could not know where the instructions or data of a certain module will be placed. When the linker places a module in the memory, all absolute references must be relocated to reflect their true location. Finally, the linker produces an executable file that can run on the ASIP.

8.7 SIMULATOR AND DEBUGGER BASICS

An assembly language simulator can be an instruction set simulator (ISS) or a processor (computer) architecture simulator. The instruction set simulator exposes the bit and cycle accurate assembly language execution without linking to the actual hardware implementation. A processor architecture simulator is the executable hardware behavior model exposing implementation details and the pipeline and bus transactions accurate assembly language execution. ASIP simulator classification is shown in Table 8.2 and is discussed in the following sections.

Bit accurate means the outputs of the ISS to data memories and to registers is exactly the same as the outputs to data memories and registers in the hardware core. Cycle accurate means the clock cycle consumed by running instructions including running branch instructions, handling interrupts, and handling I/O ports is the true number of clock cycles. Pin accurate means that input and output to and from each pin of the simulator is the same as the input and output to and from each pin of the processor RTL code. Finally, pipeline accurate means that the execution of architecture simulator and the execution of RTL code are exactly synchronous on the machine clock. That is, all registers and memories get data and send data at exactly the same time (clock cycle) in both the architecture simulator and RTL code. However, all accuracies are compared and required on architecture level instead of microarchitecture level, meaning operations inside modules are not exposed and compared.

Table 8.2 Simulator Classification.

	ISS	Architecture simulator
Core	1. Bit and cycle accurate simulation of the assembly instruction set.	1. Bit accurate and pipeline accurate simulation of the core architecture.
	2. The model of the assembly instruction set manual.	2. Simulator is synchronized by a machine clock.
	3. Used as the original ISS of the core, not used for a design.	3. Used for hardware design of the core. It is not a mandatory tool if the processor is simple.
Processor or system	1. Bit accurate, cycle accurate, and pin accurate simulator plus peripheral modules.	1. Bit accurate, pipeline accurate, and pin accurate of the processor architecture including architecture model of peripheral modules.
	2. Configured for a design with physical memory size and I/O specifications.	2. Configured for a design with physical memory size and I/O specifications.
	3. Integrated with accelerator behavior models.	3. Integrated with architecture model of accelerators.
	4. Used as the behavior model of a design.	4. For HW design of the processor and SoC design.

8.7.1 **Instruction Set Simulator (ISS)**

We refer to a simulator as the assembly instruction set simulator. It is the behavioral model of a DSP core. The simulator mimics the behavior of the processor, reading instructions, executing instructions, and maintaining internal variables. A simulator is a program that:

- Supports hardware debugging and firmware design.

- Includes modeling of each assembly instruction as a subroutine.

- Includes the behavioral model of all addressable registers and memories.

- Runs binary machine code stored in program memory.

- Accesses data stored in data memories or registers in the simulator.

- Gives the same behavior as the spec of the assembly instruction set.

- Enables early development of application software in parallel with hardware development.

An ISS may only expose the execution behavior of assembly code regardless the cycle cost. It is called a simulator without cycle accuracy. An ISS report both execution of function and cycle cost is called cycle-accurate simulator. An ISS is called pipeline accurate simulator when it exposes the cycle cost and the execution in each pipeline in detail.

If it is not specified, a simulator is the ISS of the core but without peripheral components. By adding all behavior models of all peripherals, a core ISS can be promoted to be a processor ISS. Definition of an ISS is given in Figure 8.10.

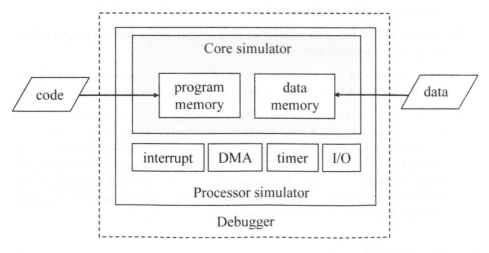

FIGURE 8.10

Simulator definitions.

Peripheral components include interrupt controller, timer, DMA controller, I/O ports, and accelerators of a design.

Principle of an Assembly ISS

Inputs of an assembly ISS include:

1. A program as executable binary code, which is loaded into the program memory. Any necessary subroutines shared by multiple programs should also be included and loaded.

2. Data loaded into the data memories, which will be processed by the executed program. Data include inputs, outputs, coefficients, and program parameters.

3. Configuration of a design. A specific processor design may not use all memories covered by the address space of the instruction set. For example, in a project, the size of the program memory may be just 32 k instruction words even though the PC can potentially cover the jump distance of one megaword according to the specification of the instruction set manual.

4. Configuration and assignment of peripheral devices of the design. For example, port 00 is assigned to the interrupt controller, port 01 is assigned to timer, port 02 is assigned to DMA controller, and port 03 is assigned to the input port of external data.

5. Debugging information, for example, run mode (stepwise, run N cycles, stop at a position, etc.).

The outputs of an ISS are:

1. Data stored in the register file and other addressable registers at the time when the simulation is stopped.

2. Data stored in the data memories at the time when the execution of the program is suspended or terminated.

3. Current status of the processor (PC, flags, processor status, control registers, stack pointer, stack data) and status of the finite state machines of peripherals.

4. Error report if there are errors. Hardware execution and pipeline should be exposed in detail during hardware debugging so that the architecture simulator for hardware debugging is useful. At the same time, fast simulation is needed for firmware design. The ISS for firmware designers should be fast. In an advanced processor design project, *both* the fast and bit accurate but less cycle and pipeline accurate simulator *and* the slower but more accurate simulator might be required to cover different needs.

Core Instruction Set Simulator

A core ISS simulates executions of all assembly instructions, not including peripheral functions. It simulates execution of a single program without interrupt handling and I/O access. The inputs of the simulator include the binary code as the program to simulate and the data as the stimuli to process (see Figure 8.10).

The top-level ISS of a simple DSP core is given in Figure 8.12. The first part of the core ISS is the code (data) loader. It loads the executable binary machine code to the program memory of the ISS. Usually, the loading start address is PC = 0. There should be memory overflow check during loading. After loading, the start PC and execution modes of the ISS will be configured by the debugger.

The first step running ISS is to check if there is an accepted interrupt. If there is an accepted interrupt, the current PC is stacked and the interrupt entry is loaded to PC. If there is no interrupt, a normal instruction is loaded.

As soon as an instruction is fetched, the code type of the fetched instruction will be extracted and checked. Instructions with different types will be switched to different subroutines in an ISS. If there is a branch instruction, the branch conditions will be checked, and if the condition is not true, there will be no execution. If the condition is true, the target address of the jump will be calculated. If there is an overflow, the ISS will switch to debugging mode.

A load/store instruction transfers data between the general register file and a data memory. The memory address is calculated first. Overflow and data dependence of the calculated address should be checked. For example, there could be a write-after-write dependence when loading data. It happens at a store instruction when the pointed register of the source register is locked by the destination of a multiple cycle instruction just before storing (see Figure 8.11). Overflow checking on a calculated memory address could be based on the limit of the instruction set specification or the specification of a project.

The CISC simple loop is an example of running a special iterative instruction in ISS. The example executes a single iterative instruction using multiple clock cycles. While running the instruction, the ISS continually checks the overflow of the calculated memory address. When the loop is finished, the ISS jumps to the last step (PC++) of the ISS. There could be other CISC instructions. The dashed-line part will check other CISC instructions and lead the ISS to simulate specific functions.

Instructions of register-to-register operations include ALU operations and move data between registers. The dependence check of reading operands or writing a result is required. If the previously executed instruction is not finished yet, the operand is dependent on the result of the previous instruction. If the previous instruction is

	Clock cycle n	Clock cycle n+1	Clock cycle n+2
Double-cycle instruction	Pipeline EXE1	Pipeline EXE2	Result available
Store instruction		Load data when it is not available	Data can be loaded

FIGURE 8.11

Data dependency check for load/store instructions.

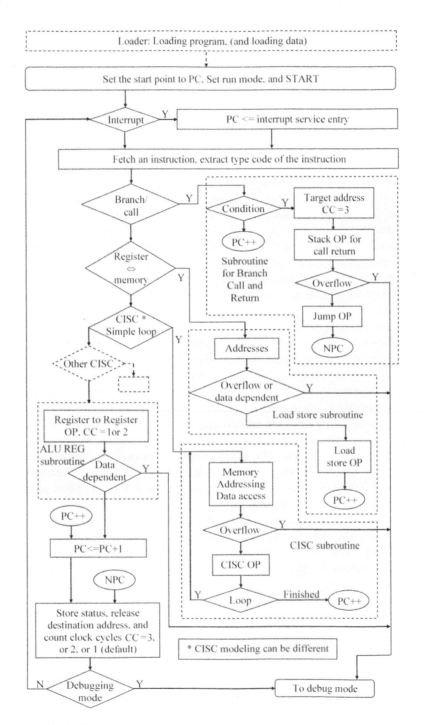

FIGURE 8.12

A simplified instruction set simulator.

a multicycle instruction, the previous instruction and the current instruction may write results to the same register, a confusing and dependent write. These cases of dependence should be reported.

The last subroutine in ISS performs the computation of the next PC and writing status, and releasing the destination register. The destination register is locked when the instruction starts running and when the destination register is released after writing the result to it. Finally, the request of entering debug mode should be checked before fetching the next instruction. The ISS should stop running and give resources to the debugger if the user wants to stop running the code and debugging the execution. An interrupt might be accepted in a processor ISS while fetching an instruction (see the top part in Figure 8.12).

The default overflow checker checks if the calculated address is beyond the limit of the assembly instruction set or negative. The default overflow check is used when running machine codes on a processor core. When designing a processor using the processor core, the physical size of memories and I/O usage might be configured according to the hardware configuration. For example, the size of the physical program memory might be smaller than the size limit of the PC (program counter). In this case, the processor hardware configuration should be used in the checker.

The data-dependent checker checks if an operand address is locked by a running instruction or if the destination address is locked by another running instruction. When running an instruction, the destination address of the running instruction is locked in the dependence checker. After writing results, the instruction releases the locked destination address.

Modeling an Instruction

The implementation of an assembly instruction is to model all microoperations specified for the instruction and all default microoperations for all instructions. The modeling language is usually C. In this section, we will discuss how to implement microoperations in each instruction.

One or more subroutines are used to model an instruction. An assembly instruction can be classified into four categories: the branch instruction, the CISC instruction, the load/store instruction, and the normal instructions including arithmetic and move operations. An instruction of any kind can be divided into four operations: the address (data memory or register addresses) computing, including check of overflow; the datapath computing, with checking of register dependences; the result store, with checking of write dependences; and the computing of the next PC, including storing status and releasing the destination address from the dependence checker.

An ISS has two working modes: running mode and debugging mode. Running mode was described in Figure 8.12. In debugging mode, the simulator is paused, keeps all information about the running program, exposes all addressable memories and registers, and gives access right to the debugger. After running each instruction, the simulator checks if there is a requirement of debugging or if there is a bug to be debugged.

Microoperations carried by an instruction were specified by the first figure in the coding section in Chapter 7. Microoperations are distributed into several subroutines specified in Table 8.3. Several subroutines are classified and described.

Modeling of an assembly instruction is the collection and description of involved microoperations in the right order. For example, in the subroutine of an arithmetic or load/store instruction, the microoperations include the following:

1. The subroutine of a fetched instruction is called when a fetched instruction is decoded.

2. Computing of operands and result addresses are explicitly specified in an instruction. The implementation will be simple and direct. The dependence checks are implicitly microoperations.

3. Operands will be loaded from (to) the calculated addresses.

4. Arithmetic or load/store operations will be executed, and the result will be stored to the place allocated during the address computing. Write dependence check should be executed implicitly.

5. Flag computing of the arithmetic instruction is executed, and flags are stored in flag registers.

6. The last step is to compute the next PC. The default microoperation of the next PC is $PC <= PC + 1$.

A subroutine of an instruction (see Figure 8.12) carries other subroutines. For example, the branch subroutine carries another subroutine checking overflow of the program memory. Special instructions are modeled in special subroutines. Modeling of an instruction could consist of a group of subroutines or a dedicated subroutine.

Another example is the call and return instructions using hardware stack operation. When executing call or return instructions (the first subroutine), the first microoperation is to check if the further operation is a push or a pop. Push execution (the second subroutine) is always associated with the checking of overflow (the third subroutine), and pop execution is always associated with the checking of underflow.

Another example is a loop instruction for accelerating the most frequently appearing iteration, the convolution. When a simulator executes a simple loop instruction, PC is kept $PC <= PC$ and a loop counter down-counts once after each step of iteration. In each iteration step, all iterative microoperations are executed. When the down-counting of the loop counter reaches 0, the loop is finished and the last step of the simple instruction is $PC <= PC + 1$ instead of $PC <= PC$.

Cycle-accuracy Issues

Cycle accuracy means the counting of the execution cycles in the simulator is the same as in the real hardware. A simple cycle accurate simulator as in Figure 8.12 gives cycle counts by accumulating execution cycles specified by the instruction

TABLE 8.3   Description of Subroutines for Modeling an Instruction.

	Address computing	Datapath	Result	PC and status
Branch	1. Condition test 2. Target address computing 3. Check overflow	Register supply target address or no operation	No	PC <= Target address or PC <= PC+1 Store status Stack OP for Call/Return
LoadStore	1. Memory address computing 2. Check register dependence	1. Register access	Release destination register	PC <= PC+1 Store status
Simple CISC loop	1. Memory address computing 2. Check register dependence	CISC datapath operation	1. Lock result register during CISC execution 2. Release destination register after CISC exec	PC <= PC during CISC execution PC <= PC+1 after CISC, Store status
Register to register	1. Check register dependence	1. Register access 2. Datapath operation	Release destination register	PC <= PC+1 Store status
Repeat instruction	Custom subroutine			PC <= PC+1 Store status
Accelerate instruction	1. Check register dependence if read RF	No	Depends	PC <= PC+1 Store status
Others	Custom subroutine			

set manual. Cycle cost to set up the pipeline after reset shall be taken into account. Dynamic pipeline behavior such as pipeline stall should be counted also based on the cycle counts of jump taken and not taken. Cycle costs of asynchronous events such as interrupts or I/O stall can also be counted from predefined cycle costs. Because firmware designers are not interested in execution details in each pipeline, the simple simulator in Figure 8.12 gives sufficient accuracy on cycle cost accumulation to firmware designers.

Maintaining Internal Variables

An ISS needs to log enough runtime data for debugging. Data and variables that should at least be kept are listed:

- Executed instructions in the order of execution associated with PC values.
- Data in data memories.
- Registers in general register file.
- Special addressable registers (addressing registers, stack pointers, status registers, loop counters, flag register, and PC).
- I/O port registers.

Error Checking and Reports

Checking functions were discussed in Figure 8.12 and Table 8.3. Overflow, data dependence, control exception, and data exception need to be checked. Overflow check is based on two parameter tables: the limitation table of the instruction set and the limitation table of a design. Parameters to check include at least:

- Program memory size and target address limit.
- Data memory size and data memory address limit.
- Register file size and register addressing.
- Stack size, the lower bound, and the higher bound.

The dependences are at least:

- Read prohibit: A result register address is locked when the result is not yet produced. The register cannot supply operand.
- Write prohibit: When a result address is locked by a vector instruction and another instruction tries to write to it.

A control exception of a design happens when the execution does not follow the design configuration. It could be the illegal use of hardware in a design. For example, the firmware tries to access a not yet specified port register. The port register exists in the ISS; however, it is not used by the design. Control exception can be recognized by checking the configuration of the design.

The data exception is checked in each instruction. There might be more guard bits in an ISS than the number of guard bits implemented in hardware. More guard bits in ISS are used for checking exceptions during iterative arithmetic computing. ISS therefore is used for reliability checking of firmware. An ISS can find errors that may not be found in hardware.

8.7.2　Processor Simulator

A processor simulator as depicted in Figure 8.10 is the core simulator plus function models of peripheral modules. A processor is always a design using an assembly instruction set. The design has its own configuration of data memories and the program memory. The design also contains peripheral modules, typically interrupt handler, timer, and DMA controller. The design also includes I/O port specifications and accelerators. By integrating the core ISS with a set of configurations, the peripheral modules, I/O, and accelerators, a processor simulator should give bit accurate, cycle accurate, and pin accurate simulation of the processor.

Modeling of Peripheral Modules and Accelerators

The function of each peripheral module will be discussed in Chapter 16. Modeling of peripheral behavior is to describe its finite state machine and custom functions behind the finite state machine. Only fundamental issues will be discussed in this subsection since peripheral modules will be covered in detail in Chapter 16. In this section, protocols connecting and using peripheral modules, accelerators, and other I/O devices will be discussed.

A peripheral module is connected to the core via I/O port registers. At least three or four port registers are used by one peripheral module. These registers are:

- Control register: Via it, the core sends control vectors to peripheral module.

- Status register: Via it, the peripheral module exposes its status to the core.

- Data in (core to port) register: Core sends data to the peripheral module.

- Data out (port to core) register: Peripheral module gives data to the core.

Each register has its port address. A peripheral module therefore holds at least three to four port addresses. The core can reach a peripheral register by port-read or port-write instructions. By writing in control vectors, a peripheral module is configured. By writing in data to data register, input of a task is prepared and a task in the peripheral module can be executed. After the task execution, the statuses are available in the status register, and the result is available in the data out register. Communications between the core and peripheral modules usually are initialized by interrupts. The result from a peripheral module includes also cycle cost. The cycle cost of a task executed in a peripheral module should be exposed to ISS for debugging of scheduling.

Error Checking and Reports

Except for the overflow check and the dependence check discussed earlier, more checking on synchronization and dependence between the core and peripheral modules and accelerators are required.

8.7.3 Architecture Simulator

A processor architecture simulator is a program modeling processor devices (or components) to expose pipeline and hardware usage and to simulate outputs and performance on a given input. In this book, a processor architecture simulator is a microarchitecture simulator. It gives behavior of each hardware module, such as the behavior of ALU, or the behavior of MAC.

Different from an ISS, a processor architecture simulator uses a hardware description and performs function simulation beyond RTL codes. The architecture level simulator is used for:

- Fast architecture level simulation on system level (SoC HW simulator), because the SoC hardware simulator must be fast and expose pipeline accurate information between processor cores and SoC level modules.

- Debugging superscalar architecture and processors with cache. This is because the cycle costs in a superscalar or processor with cache cannot be reported by the traditional ISS.

- Design entry for system-level hardware synthesis. A good architecture simulator might be from an architecture description and can be synthesized to RTL code in the future.

- Bridging the gap of formal verification between RTL and system level in the future.

An important decision is on the selection of system-level language. The language should be on a higher level than HDL. It should be fast and supply modeling of clock and pipeline, modeling of hardware multiplexing, and different custom data types. SystemC has been a choice.

8.8 DEBUGGER AND GUI

8.8.1 Debugger

A debugger is a program to test and debug assembly programs. The assembly code to be debugged is loaded in an ISS and executed by the ISS. A debugger should support at least the functions described in the following sections.

The first function is to find the reason why the running assembly code crashed. When the program crashes, the debugger points out the crash position in the original code, which is a line in the binary code. The executed binary code is translated from binary code to readable assembly code so that the debugging engineer can read it. There are different reasons of crashing; for example, the program tried to reach an address beyond the limit, the program tried to use an instruction not available on the current version of the processor, or the program accessed a protected register and stopped by the data dependence checker in ISS.

A debugger interacts with ISS and offers different running modes, such as step-by-step mode (running single-stepping), N-instruction mode (running N instructions), and breakpoints mode (pausing the program at breakpoints). When the ISS pauses, the debugger can track the values of variables. An advanced debugger gives guidance of debugging together with the error message reports.

The GNU Debugger (GDB) is a typical open source GNU software system running on UNIX and Linux systems [7]. GDB allows the programmer to see what is going on inside another program (for example, the assembly code) while it executes (in ISS). It exposes what the program was doing at the moment it crashed. GDB can do four main things to catch bugs:

- Start a program in the simulator and interact with the simulator. Specify anything (data or control status) that might affect its behavior.

- Stop the program on specified conditions (the program might crash by itself).

- Examine what happened when the program crashed or what the results were when the program stopped.

- Change codes to experiment the correction of a bug or change data to stimulate the occurrence of a bug, and continue running the code.

The GDB user interface is a CLI (command line interface), but there are GUI front-ends such as DDD (Data Display Debugger), which includes a command window (Input command and get output from GDB), the source window (to observe the source file to simulate and debug), the assembly window (to show the executed program), and the register window.

8.8.2 SW Debugging

Once the software is written, the functionality must be proven by running a simulator via a debugger. Instead of using a simulator and debugger, firmware debug can also be conducted using a hardware development board. It is a board designed for real-time firmware development containing the processor and debugging interfaces. The debugging interface includes the interface for data (input and output) accesses and the interface for debugging (no damage reading data in memories and registers).

In any case a debug process includes loading code, loading data, running code, and observing output. Signal processing applications tend to require large amounts of data streaming in and out of the processor, so it is helpful if tools have an easy way to deal with I/O. For large data sets, running the software on a development board may be the easiest way to debug the firmware. Some development boards are equipped with specialized I/O ports tailored for applications such as audio or video processing. These ports can ease the testing of these applications.

8.8.3 GUI

User interface (UI) is the interface between a program running on the host computer and the programmer. The host computer is the machine running ISS and debugger. In particular, UI in this book is the interface between the assembly programmer and the assembly toolchain, including a compiler, an assembler, a linker, an ISS, and a debugger. The UI supports programmer interaction with the toolchain running on the host computer. The UI provides an interface between the programmer and the program running behind UI, sending inputs to and getting outputs from the program behind UI running on the host machine.

There are three kinds of UI: CLI (command line interface), TUI (text user interface), and GUI (graphical user interface). A GUI supplies a graphical image and text to represent the information and actions available to a user. GUI design enhances the usability of programs carried by GUI. The GUI exposes the execution of a program to the GUI user by emphasizing important messages, supplying hyperlinks pointing to bugs, and sending control to the program behind GUI. They include graphical elements such as windows, buttons, menus, and scroll bars.

Because UNIX (Linux) and Windows are the most used host operating systems, a GUI for an ASIP should be executed on both UNIX and Windows. A Tcl/Tk-based GUI for "Senior DSP processor Toolchain" is available on this book's web site. Tcl is a reasonable platform-independent scripting language, and Tk is a GUI ToolKit for Tcl. Functions of a GUI interface for assembly code debugging include:

- Interfacing the programmer to ISS, text file editor, assembler, and linker.

- Interfacing the programmer to the configuration of an ASIP chip design.

Detailed GUI interface functions are:

1. Code load: Load source assembly code to the assembler.

2. Data load: Load data memories. (Data load is optional. Data might be loaded by running DMA if data is available in the main memory of the ASIP.)

3. Edit: Edit the source code by calling a source code editor in the host machine.

4. Project settings: Including
   a. Setting of I/O ports and peripheral modules.
   b. Connecting accelerators used by the design (see Chapter 17).

5. Configuring sizes of program and data memories.

6. Setting of execution modes:
   a. The start PC.
   b. The run mode as "Run," running the infinite loop.
   c. The run mode as "N-instructions."

    d. The run mode as "Step."

    e. The run mode as "Stop at a break point."

    f. Multiple break points.

7. Run: Including two steps:

    a. Run assembler and linker to get executable binary machine code.

    b. Run simulator; run the binary code according to the setting.

8. Display:

    a. Display data stored in register file and memories at the stop point of the code running.

    b. Highlight and display the break point and possible reasons of the crash.

    c. Display the executed binary codes.

    d. Display source codes.

9. Modify:

    a. Modify register values and break points.

    b. Reset all registers.

8.9 EVALUATION OF PROGRAMMING TOOLS

Usually it is difficult to know the quality or the weakness of programming tools (especially high-level design tools and compilers) until they have been used for a long time. The demonstrators of most tools show excellent features; however, they may not achieve the declared performance in your project when your application is different from the selected benchmark subroutines. A tool that is good for one application might not be good enough for other applications.

Many ASIP programming tools are developed in-house. Rich application and hardware knowledge can be advantages of the team, but poor tool design experiences could prove their drawbacks. To speed up the ASIP design, using third-party tools, tool generators, or tool platforms might be a good choice. Therefore, complete evaluation of a toolchain before making a decision is extremely important. The evaluation may include the following topics:

1. How were the tools verified? Are the functional design and verification of the simulator, assembler, and the compiler really trustworthy? Do you also trust the compiler generator?

2. The way to check verification quality of compilers and simulators might be different. Fully toggled does not mean verified.

3. Opinions and evaluations from reference customers are essential. Avoid using tool suppliers without rich reference lists. Remember, the extra cost of a good toolchain is usually lower than the cost of the project delay induced by a poor toolchain.

4. Check the optimizer in the compiler carefully. Be sure that algorithms are suitable for your applications. Many algorithms are used for minimizing the caches mismatch, and most ASIPs do not have a cache.

5. The speed of the ISS simulation shall be checked carefully. The pipeline accurate simulation is not necessary for firmware debugging. The following points are useful for simulator decision:
   a. Speed of a simulator depends on the host machine.
   b. Support cycle and bit-accurate simulation.
   c. Support special check of exceptions that do not exist in hardware.
   d. Support scalability for adaptation of hardware accelerators.

8.10 CONCLUSIONS

A processor cannot be used without enough support from a firmware programming toolchain. In this chapter, the assembly-level programming toolchain was introduced briefly, including the principles of compiler, assembler, linker, simulator, and debugger. The introduction was prepared for toolchain users instead of toolchain platform developers. The purpose is to enhance the knowledge of firmware design tools for hardware designers and toolchain evaluators; designing a toolchain is beyond the scope of this book.

EXERCISES

8.1 Try to collect microoperations to be simulated in an assembly instruction set simulator. What microoperations are not simulated there?

8.2 What are errors from the first error report "port" in Figure 8.8? Give an example.

8.3 What are errors from the second error report "port" in Figure 8.8? Give an example.

8.4 What are errors from the third error report "port" in Figure 8.8? Give an example.

8.5 Which of the following statements are right? The DSP processor behavioral model is:
   a. The assembly instruction set simulator.

   b. A golden model that supplies bit-accurate simulation.

   c. The behavior description of an algorithm.

8.6 How many kinds of errors can an instruction set simulator in Figure 8.12 report? In what way does the simulator in Figure 8.12 report errors?

REFERENCES

[1] Lesk, M. E., Schmidt, E. (1975). Lex—A lexical analyzer generator. *Computing Science Technical Report No. 39*. Bell Laboratories, Murray Hill, NJ. A PDF version is available at ePaperPress.

[2] Johnson, S. C. (1975). Yacc: Yet another compiler compiler. *Computing Science Technical Report No. 32*. Bell Laboratories, Murray Hill, NJ.

[3] Skglund, B. (2007). Code profiling as a design tool for application specific instruction sets. Master thesis, LiTH-ISY-EX-07/3987-SE, Linköping University, Sweden (www.ep.liu.se).

[4] Leupers, R. (1997). *Retargetable Code Generation for Digital Signal Processors*. KAP.

[5] Aho, A. V., Lam, M. S., Sethi, R., Ullman, J. D. *Compilers: Principles, Techniques, and Tools*, 2nd ed. (2006). Addison-Wesley.

[6] Muchnick, S. (1997). *Advanced Compiler Design and Implementation*. Morgan Kaufmann.

[7] www.gnu.org.

[8] http://www.ecsl.cs.sunysb.edu/gem/. GCC Extension Modules.

[9] Karuri, K., Al Faruque, M. A., Kraemer, S., Leupers, R., Ascheid, G., Meyr, H. (2005). Fine-grained application source code profiling for ASIP design. DAC 2005, Anaheim, CA, USA.

[10] Kessler, C., Bednarski, A. (2006). *Optimal Integrated Code Generation for VLIW Architectures, Concurrency and Computation: Practice and Experience*, 18, 1356-1390. John Wiley & Sons, Ltd.

[11] Larus, J. R. *Assemblers, Linkers, and the SPIM Simulator*. Microsoft Research, Microsoft. https://pages.cs.wisc.edu/~larus/.

[12] Conte, T. M., Gimarc, C. E. (1995). *Fast Simulation of Computer Architectures*. KAP.

[13] Rowen, C. (2004). *Engineering the Complex SOC*, Chapter 7. Prentice Hall.

[14] Cichon, G., Fettweis, G. (2003). MOUSE: A Shortcut From Matlab Source to SIMD DSP Assembly Code. *International Workshop on Systems, Architectures, Modeling, and Simulation (SAMOS'03)*, page(s) 126, 21.–23.07.03.

[15] Robelly, P., Cichon, G., Seidel, H., Fettweis, G., (2005). A HW/SW Design Methodology for Embedded SIMD Vector Signal Processors. *International Journal on Embedded Systems (IJES)*, January 2005.

Evaluation of an Instruction Set

The evaluation of an assembly instruction set usually includes the benchmarking, profiling, and coverage analysis of the assembly instruction set.

Benchmark is a type of program designed to measure the performance of a digital signal processor in a particular application. *Benchmarking* is the execution of such a program in a way that allows processor users to track the number of machine clock cycles consumed by a specific section of code. Assembly code profiling in this chapter is different from the source code profiling discussed in Chapter 6. The assembly code profiling is to evaluate the appearance (usage) of each assembly instruction in the machine code of an application. Coverage analysis is to evaluate the functional coverage of the instruction set. It measures the flexibility of the assembly instruction set.

A proposed instruction set must be carefully evaluated and optimized before releasing it for further implementation of the microarchitecture of a processor core. On the other hand, while selecting a programmable COTS component, the instruction sets of several programmable devices should be checked carefully and compared based on the requirements of the system. In this chapter, fundamental knowledge and examples of ASIP evaluation will be given.

9.1 BENCHMARKING

Benchmarking by definition is some kind of measurement of performance. In this book, we define benchmarking as measuring the performance of an instruction set of a DSP processor. Benchmarking can be conducted in many ways. However, one common way of benchmarking is to measure the number of clock cycles required to accomplish a defined task. Other measurements can also be important, such as memory usage and power consumption. Memory usage can be further divided into the cost of program memory and data memory required by a benchmark. DSP benchmarking usually is conducted by running the benchmarking assembly code on the ISS of the DSP processor.

Benchmarking of power consumption measures the dynamic power consumed while running a sequence of code associated with a sequence of data pattern instead of only one line of code. The power consumption depends on the instruction sequence and data patterns, which is specified as the "power consumed during the transaction from one instruction to another associated with the change of stimuli data." Power measurement exposes the consumption during the transaction. Due to the complexity of transaction combinations, power consumption can be measured based only on long-time statistics. Benchmarking of power consumption must be published together with test stimuli (code and data pattern). Measuring of DSP power consumption should be on a specific physical design, specified by its silicon technology, supply voltage, data memory size, program memory size, and included peripheral modules. The measuring data will not be valid while any of these specifications are missing. The power measure of a DSP core (without memory and peripheral) is meaningless because the power consumed by memories is usually the dominant power. The accurate power measure should be based on running codes on real silicon with sufficient data. The measuring of power consumption without available silicon is very difficult and not accurate. The early power estimation of a processor is based on power calibration of the RTL component for a silicon process, the estimated silicon cost, and predicted wire capacitance. Measuring power consumption has not been within the scope of the book.

For DSP processor designers, benchmarking measures the quality of a proposed instruction set and checks and improves the performance of such an instruction set. For the users of DSP processors, benchmarking evaluates the performance of a certain processor or makes comparison of several different processors targeting certain applications.

DSP benchmarking can be divided further into the benchmarking of DSP algorithm kernels and the benchmarking of DSP applications. By benchmarking DSP kernels, essential performance and cost can be estimated. Benchmarking the kernel algorithms runs the kernel assembly code such as filters FIR, IIR, LMS (Least Mean Square) adaptive filter, and transforms (FFT and DCT) on the ISS of the processor.

The second kind of benchmarking, the benchmarking of applications, runs assembly application code on ISS of the processor. Overheads can be identified by benchmarking applications. However, the design cost for application code can be very high. The benchmarking of an application is usually on the cost extensive part instead of the complete applications. The cycle cost of DSP applications can be categorized and partitioned as seen in Table 9.1.

The distribution of cycle costs will vary significantly from application to application. The ratio listed in Table 9.1 is accumulated from research and consulting experience. If you are working in an application domain, you should be able to make similar tables of cycle costs based on frequently used instruction sets in their application. Cycle cost estimation and minimization will also be discussed in Chapter 18.

Traditional performance measurement is based on MIPS (millions of instructions per second). Another way is to measure the performance in MOPS (millions of operations per second) or MFLOPS (millions of floating-point operations per second). However, they are not necessarily good metrics because one single DSP instruction might be an N-step iterative operation, and is equivalent to five to ten

Table 9.1 Classification and Estimated Time Costs of General DSP Tasks.

Class	Description	Costs
Kernel	Iterative computing kernels, such as FIR, IIR, correlations, FFT, DCT, adaptive filter, etc.	40–90%
Memory	Vector move, data swap, task configuration, inputs, and outputs.	0–50%
Control	Program flow control, data quality control for fixed-point HW, and interprocessor control.	0–5%
Asynchronous event handling	Thread control, Interrupt control, Context saving and restore, and waiting for streaming data.	0–5%

times the instructions of a RISC processor. An efficient DSP CISC instruction might be equivalent to several RISC instructions.

To benchmark a DSP instruction set, two kinds of cycle cost measurement frequently are used. One is the cycle cost per algorithm per sample data. For example, 30 clock cycles are used to process a data sample using a 16-tap FIR filter. Another kind of measurement is the data throughput of the firmware of an application per megahertz. For example, 500 voice samples are processed in an acoustic echo canceller, with 2048 coefficients using one million clock cycles (in one MHz). If the voice sampling rate is 8 kHz, the computing cost of 16 MHz will be required.

The benchmarks of basic DSP algorithms usually are written in assembly. The first reason is because the purpose of benchmarking is to measure the quality of the assembly instruction set; by nature, the benchmarking should be in assembly language. The architecture of a DSP processor is dedicated to certain applications. The quality of assembly code somewhat reflects the quality of hardware utilization by the code. The second reason is that most DSP assembly programs are relatively simple and can be managed by programmers. The third reason is that the effectiveness of programs written in high-level language is very much dependent on the compiler. The assembly code generated by a compiler sometimes doesn't really expose the efficiency of the processor; instead it shows the quality of the compiler optimization. For a given architecture, benchmark programs written in assembly are usually highly optimized, in the sense that all dedicated hardware features toward certain algorithms can be maximally utilized.

However, if the compiler is essential and the firmware design time is very short, benchmarks written in high-level language and compiled by the compiler will become practical and even necessary. Instead of checking the quality of ISA, it examines the mixed qualities of ISA and compiler. BDTI (Berkeley Design Technology Incorporation [1]) always supplies benchmarks based on hand-written assembly code. Meanwhile, EEMBC (EDN Embedded Microprocessor Benchmark Consortium [2]) allows two scoring methods: Out-of-the-box benchmarking and Full-Fury benchmarking. Out-of-the-box (not requiring any extra effort) benchmarking is based on the assembly code directly generated by the compiler.

Full-Fury (also called optimized) benchmarking is based on assembly code generated and fine-tuned by experienced assembly language programmers.

It is not easy to make an ideally fair comparison by benchmarking algorithm kernels running on the target processors. Each processor has dedicated features and is optimized for some algorithms, but not optimized for some other algorithms. A processor holding a poor benchmarking record of an application might have a very good benchmarking record of another application. A typical case is that a radio baseband processor will never be used as a video decoder processor. For fair comparison, processors not from the same category should not be compared.

The benchmark code is not necessarily relocatable. It means that immediate addressing mode might be used in the benchmark, or the prologue of the benchmark code might be too easy. The benchmark code cannot be loaded to any position in the program memory. That is one reason that the benchmark score of a processor cannot be reached in applications. The consequence is that a benchmark code shall not be used directly in applications.

The silicon cost in a modern DSP processor is dominated by the memory cost. Therefore, the silicon cost optimization is usually the optimization of memory usage, which is currently the hottest area in DSP-related research. It is a challenge to measure and evaluate the efficiency of memory usage. One accepted way is to measure the improvement that can be achieved in comparison to a reference design. The memory efficiency includes both the data memory and program memory efficiencies. Benchmarking of memory usage will be included in this chapter.

Application-level benchmarking checks the performance of a processor for a specific real application, like MPEG4 for video decoding. When benchmarking an application, more specifications or descriptions shall be applied to the benchmarking result. For example, a specific benchmarking performance of MPEG-4 could be based on a test sequence called "coast guard," which has the resolution of CIF (352×288 pixels per frame), 30 frames per second, and the coding scheme conforms to MPEG-4 part 10 baseline profile and level 1. Here profile and level represent different coding parameter sets defined in the MPEG-4 standard. Clock frequency often is used as the measuring metric of application-level benchmarking. For the sake of simplicity, in this book we will discuss only the benchmarking of algorithm kernels.

9.1.1 Benchmarking DSP Kernel Algorithms

A DSP kernel algorithm is a small portion of code (assembly program) that forms the heart of an algorithm. When describing basic DSP algorithms at assembly level for benchmarking purposes, only the kernel parts of the algorithms are considered. This is relevant because DSP kernel algorithms will take the majority of the execution time in most DSP applications. The optimization and benchmarking of kernel code are usually easy because the size of a kernel code is typically short.

In order to carry out comprehensive comparison on different COTS processors, sets of benchmark programs are selected to capture all necessary aspects of the performance of DSP processors. A good reference is the *Buyer's Guide to DSP*

Processors [1]. It exposes performance of COTS DSP processors by running a set of benchmarks on most COTS DSP processors available in the market. Benchmarking DSP kernels is also called BDTI Benchmarking because benchmarking BDTI on popular general-purpose DSP processors has been well accepted by DSP users and the semiconductor industry.

There are some guidelines to the benchmark code introduced by BDTI and accepted by DSP users and processor vendors [1]:

- Benchmark code is not relocatable code.

- For processors using caches, the cycle cost of cache mismatch is not included.

- All data are available in the processor's data memories.

- Cycle cost of cleaning vector buffers should not be included in benchmark.

- Necessary prologue (loading data and program configuration) is included.

- Necessary epilogue (rounding, saturation, saving result, and saving data buffer pointer) is included.

- Excessive loop unrolling and table look-up are not allowed.

Well-accepted DSP kernels are listed in Table 9.2.

Table 9.2 Kernel DSP Algorithms.

Algorithm kernel	Descriptions or specifications
Block transfer	To transfer a data block from one memory to another memory
256p complex FFT	256-point FFT including all computing, addressing, and memory access
Single FIR	A N-tap FIR filter running one sample
Frame FIR	A N-tap FIR filter running K samples
Complex FIR	A N-tap FIR filter running one sample complex data
Simple IIR	A Biquadrate IIR (2nd order IIR) running one sample
LMS Adaptive FIR	Least significant square adaptive filter including convergence control and coefficient adaptation
16-bit/16-bit division	A positive 16 bits divided by 16 bits positive data
Vector add	$C[i] <= A[i] + B[i]$ (i is from 0 to N-1)
Vector window	$C[i] <= A[i] * B[i]$ (i is from 0 to N-1)
Vector max	$R <= MAX \{A[i]\}$ (i is from 0 to N-1)
FSM	Finite state machine (not yet standardized)
DCT	8X8 Discrete cosine transform

The EEMBC (EDN Embedded Microprocessor Benchmark Consortium) is another source of knowledge about DSP benchmark. EEMBC is more application-oriented. Its benchmarks are classified into five application groups: automotive/industrial, consumer, networking, office automation, and telecommunications. EEMBC gives source code for benchmarking and benchmark flowcharts [2].

The following example is the benchmark of an FIR filter using the **Senior** instruction set. The algorithm can be represented using the formula $y(n) = \Sigma x(n-k)h(k)$; $x(n-k)$ is the input, $y(n)$ is the output, and $h(k)$ is the coefficient of the filter. The behavioral level pseudocode (single-sample FIR) is shown in Listing 9.1.

LISTING 9.1   Pseudocode for single-sample FIR on **Senior**.

```
Reset ACR /* ACR is an accumulator register */
DM0(DP) <= The latest Sample /* Data FIFO buffer allocated in
DM0*/
DP <= DP + 1 /* The oldest sample is replaced by the latest
sample*/
For i=0 to 15 do
  (
   ACR =< ACR + DM0(AP0)*DM1(AP1) /* 16-tap convolution of a
   sample*/
   AP0 <= AP0 + 1, AP1 <= AP1 + 1 /*Implied modulo addressing
   on AP0*/
  )
Round and Sat ACR
Output result to data memory
Store the data pointer DP
```

The pseudocode is the behavioral model of convolution for one data sample. Following the pseudocode, parameters are defined for the benchmarking. The first parameter is N, the number of taps of the filter ($N=16$ in this example). The benchmark will be implemented using the **Senior** instruction set.

LISTING 9.2   **Senior** assembly code for single-sample FIR.

```
        ;; FIR filter:
        ;; Inputs: r1 contains a new sample value
        ;; Outputs: Filtered sample into resultptr array
        ld0     r15,(resultptr) ; Result pointer
        ld0     r0,(dataptr)    ; Load dataptr to reg
        set     bot0,bottom     ; FIFO bottom pointer
        set     top0,top        ; FIFO top pointer
        move    ar0,r0          ; Set address reg to dataptr
        set     ar2,coeffptr    ; Coefficient pointer
```

```
        st0        (ar0++%),r1        ; A sample data from r1 to DM0
                                      ; (ar0++%) signifies modulo
addressing
        clr        acr0
; ─────────────── The prologue consumes 8 cycles ───────────────
;
; ─────────────── The second part of the program ───────────────
        repeat endloop,16
        convss ap acr0,(ar0++%),(ar2++) ; Convolution loop
        endloop
; The iteration uses 16+1 = 17 clock cycles
;
; ─────────────── The third part of the program ───────────────
        add        r16,r15,1          ; Prepare the next value for the
                                        result ptr
        move       r0,ar0             ; Save the next value for the
                                        data ptr
        move       r1,sat rnd acr0 ;
        st0        (resultptr),r16
        st0        (dataptr),r0
        nop                           ; This must be here since r1 is
                                        not
                                      ; yet ready in the pipeline
        st0        (r15),r1
; ─────────────── The epilogue consumes 7 cycles ───────────────
```

The **Senior** assembly code of the single-sample FIR is shown in Listing 9.2. The total cycle cost of processing one sample by the 16-tap FIR filter is $8 + 17 + 7 = 32$ clock cycles. Comparing the cycle cost of **Junior**, using the same pseudocode as the reference of the assembly benchmarking, the cycle cost could be reduced to 32 cycles from 143 cycles. Comparing the state-of-the-art DSP processors with a single MAC unit (single-issue DSP processor), the best score from BDTI benchmarking is also about 31 clock cycles.

Notice that this benchmark is not relocatable code because the initial parameters are immediate data instead of variables in data memory. The code that is not relocatable cannot be used for real applications. The relocatable code should be as depicted in Listing 9.3.

LISTING 9.3 Senior relocatable assembly code for single-sample FIR.

```
─────────── The re-locatable single-sample FIR ───────────
    ;; FIR filter:
    ;; Inputs: r1 contains a new sample value
    ;; Outputs: Filtered sample into resultptr array
```

```
        move      ar3 , r31           ; R31 contains segment address
        nop                           ; Must wait due to pipeline
                                        effects

        nop
        ld0       r15 , ( ar3 , 0)    ; Load result pointer
        ld0       r0 , ( ar3 , 1)     ; Load dataptr to reg
        ld0       r14 , ( ar3 , 2)    ; Load FIFO bottom pointer
        ld0       r13 , ( ar3 , 3)    ; Load FIFO bottom pointer
        ld0       r12 , ( ar3 , 4)    ; Load coefficient pointer
        move      ar0 , r0            ; Set address reg to dataptr
        move      bot0 , r14
        move      top0 , r13
        move      ar2 , r12

        st0       ( ar0++%) , r1      ; A sample data from r1 to DM0
                                      ; ( ar0++%) signifies modulo
addressing
        clr       acr0
;--------------- The prolog consumes 14 cycles ---------------
;--------------- The second part of the program ---------------
        repeat    endloop , 16
        convss    ap acr0 , ( ar0++%) , ( ar2++) ; Convolution loop
        endloop
;----------- Convolution iteration consumes 17 cycles -----------
;--------------- The third part of the program ---------------
        add       r16 , r15 , 1       ; Prepare the next value for the
                                        result ptr
        move      r0 , ar0            ; Save the next value for the
                                        data ptr

        move      r1 , sat rnd acr0 ;
        st0       ( ar3 , 0) , r16
        st0       ( ar3 , 1) , r0
        nop                           ; This must be here since the
                                        value
                                      ; in r1 is not ready yet in the
                                        pipeline
        st0       ( r15) , r1
;--------------- The epilog consumes 7 cycles ---------------
```

The real cycle cost of running relocatable code is actually $14 + 17 + 7 = 38$ clock cycles instead of 31 clock cycles from the benchmarking. The classic way of DSP kernel benchmarking does not seem to be 100% accurate; however, this is the traditional way to demonstrate DSP benchmark by several DSP semiconductor

Table 9.3   Benchmark of Selected DSP Kernel Algorithms.

Algorithm kernel	Specification	Junior benchmark	Senior benchmark	BDTI average score (2001)	Improvement from Junior to Senior
Block transfer	N = 40 from DM1 to DM2	242	45	47	5.3 times
Single FIR	16 coefficient single sample	173	32	31	5.4 times
Frame FIR	16 coefficient 40 sample	7492	893	973	8.3 times
16-bit/16-bit division	Case sensitive e.g., 41/7	199	54	52	3.6 times
Vector window	40 window factors	363	167	165	2.2 times

companies. It is the truly executable code with minimum cycle cost. This example tells us that the cycle cost obtained from benchmarking might be slightly different from the real cycle cost running relocatable code. Nevertheless, the extra cycle cost to reach relocatability is relatively low if the benchmarking is based on the execution of many data samples.

9.1.2 Some Benchmarking Examples

Benchmark programs based on the **Junior** instruction set were introduced and discussed in Chapter 5. The evaluation of **Junior** was performed, and its weaknesses were identified. The **Senior** instruction set therefore was proposed in Chapter 7. In the following sections, benchmarking of kernels on **Senior** ISA is given in Table 9.3.

Table 9.3 shows a significant improvement of performance by comparing a traditional RISC plus a MAC (**Junior**) to a professional single-issue DSP processor **Senior**.

9.2 INSTRUCTION USE PROFILING

Instruction use profiling is another evaluation of the instruction set. Instead of measuring the performance, instruction use is measured by counting the number of times that each instruction is executed while running an application. For example, after running a voice CODEC (coder/decoder) on a DSP processor, statistics of the use of all instructions can be counted. Instruction use profiling should be conducted dynamically by identifying and accumulating the appearances of each instruction by using an instruction set simulator (ISS).

The profiling use tells which instruction is used the most in an application. It also tells which instruction is the least used by an application. If several instructions are not used by a class of applications and the processor is designed only for this class of application, these instructions can be removed to simplify the hardware design and reduce the silicon costs. However, sometimes the saving of silicon cost and design cost is almost negligible by removing an instruction. The cost reduction could be significant only when many instructions are removed. On the other hand, since the most used instructions can be identified, optimizing them can significantly improve the performance or reduce the power consumption.

9.3 COVERAGE ANALYSIS

A complete application code classified in Table 9.1 includes both kernel subroutines and miscellaneous control, glue, and other functional code. These miscellaneous functions also need to be implemented, and the assembly instruction set should have enough coverage to support the efficient implementation of the miscellaneous functions. Coverage analysis is to evaluate if the designed assembly instruction set is sufficient for efficiently implementing all applications in the application domain. Coverage analysis uses algorithms more close to applications and not collected by BDTI. Coverage analysis also collects miscellaneous functions including:

- Supporting special addressing algorithms

- Control functions such as program flow control, which are instructions to collect and select subroutines; and data quality control for fixed-point HW, including data condition test and data analysis

- Thread control and interrupt control

- Interprocessor control

- Instructions for debugging

9.4 CONCLUSIONS

Methods of evaluating an assembly instruction set were discussed. The evaluation usually includes ASM benchmarking, profiling the instruction appearances, and coverage analysis. Benchmarking methods and benchmarks are available from

BDTI [1] and EEMBC [2]. The profiling of instruction usage could be implemented as a function in an ISS (Instruction Set Simulator). Coverage analysis is important to widen the coverage of the processor and as a result to reduce the design and manufacturing cost.

REFERENCES

[1] www.bdti.com

[2] www.eembc.org

Design of DSP Microarchitecture

10.1 INTRODUCTION TO MICROARCHITECTURE

Design in general is to divide and conquer. From the top layer to the bottom layer, each lower layer is the refinement of its upper layer, and all layers follow the same specification. The step size between layers should be chosen in such a way that refinement is feasible and design time can be minimized. From top to bottom, hardware design usually is partitioned into architecture level, microarchitecture level, and RTL (Register Transfer Level).

Microarchitecture design is the activity to refine the architecture-level functional description into descriptions of each hardware module and to connect all modules together. It is the step planning for RTL coding of a hardware module. Design of processor microarchitecture is one of the two focuses of this book. The other focus is the design of the assembly instruction set. In this chapter, the microarchitecture design in general and the DSP ASIP microarchitecture will both be discussed.

10.1.1 Microarchitecture versus Architecture

Architecture design of a system gives specifications of the hardware system structure including behavior of the system in general, high-level description of hardware modules in the system, and interconnection and interworking between these hardware modules in the system.

Architecture design does not involve the implementation of hardware modules in detail. Instead, the architecture design specifications are used as inputs for the microarchitecture design task.

In Chapter 3, the design of processor architectures was discussed, which is to specify a hardware platform for an assembly language instruction set. The processor specification includes the processor functions in general, the architecture of its high-level hardware modules (the datapath including RF, ALU, and MAC; the control path; bus system; and memory subsystem), and interconnections between top-level hardware modules. The processor architecture also is called the assembly instruction set architecture (ISA). In this book, the ISA is the specification of what a processor can do by executing each instruction. Implementation details and performance issues are not specified by the ISA.

10.1.2 Microarchitecture Design

Microarchitecture design in general is to specify the implementation of functional modules including intermodule functions in detail. It includes function partition, allocation, connection, scheduling, and integration. The microarchitecture design specifications are used as inputs for the RTL coding.

As a special case, processor microarchitecture design is the implementation of each functional module in a processor core based on its assembly instruction set with physical implementation constraints, meeting performance and cost targets. The inputs of the processor microarchitecture design are the assembly language manual, the processor architecture description, and the physical constraints. The output of the processor microarchitecture design is the description of the implementation of the hardware modules (such as ALU, RF, MAC, and control path) and the bit-accurate intermodule connections for further RTL coding of the processor.

Processor microarchitecture design is the detailed hardware implementation of each assembly instruction into involved hardware modules. It includes the partitioning of each instruction into several microoperations, the allocation of each microoperation to specific hardware modules, hardware reuse and minimization, and the scheduling of each microoperation into different pipeline stages.

The microarchitecture design input includes the architecture design, the assembly language manual, and hardware requirements. The output of the microarchitecture design is a kind of hardware implementation specification on paper, which basically consists of documents to be used as inputs for RTL coding. There has been some research in describing the microarchitecture of processors using ADL (architecture description language) in order to make the microarchitecture design executable and eventually synthesize it down to RTL code.

Actually, most assembly instruction sets are relatively independent of their microarchitecture. This means that the same assembly instruction set can be implemented into different architectures. For example, both a simple single-issue processor and a multi-issue superscalar processor can be implemented using the same assembly instruction set. A typical case is to promote the performance of an old processor by an advanced superscalar using the same instruction set. The advanced processor executes the same instruction set, keeping assembly level compatibility for legacy firmware.

10.2 MICROARCHITECTURE-LEVEL COMPONENTS

A chip consists of system-level components like processor cores, bus subsystems, memory subsystems, peripheral modules, and I/O subsystems. A DSP core contains modules like ALU, RF, MAC, and control path. These modules also are called module-level or microarchitecture-level components. Inside each module exists RTL components. Microarchitecture design is thus equivalent to planning for RTL coding. Essential RTL components of the microarchitecture will be reviewed in this section. The details of arithmetic component design (adder, multiplier) can be found in references [1,2].

10.2.1 Basic Logic Components

Register

A register can be used either as a storage component or a component to isolate functions between pipeline stages. A register consists of one or several register cells, usually D flip-flops. The two basic functions of a register cell are: (1) to store the input value when the rising (or falling) clock edge arrives; and (2) to keep the output value until the next rising (or falling) clock edge arrives. A reset function is essential for a register in order to reset it into a defined state independent of input data. A synchronous reset means that the register output will be zero when the rising (or falling) clock edge arrives and the reset control is activated (usually low). An asynchronous reset will force the register output to be zero immediately after the reset signal is activated.

In order to support product tests of integrated circuits, all register cells in a digital integrated circuit shall be reconnected in series (scan chain) during test mode. Test patterns can then be scanned into all registers and results can be scanned out. A complete one-bit register as a RTL component is illustrated in Figure 10.1.

Multiplexer

A multiplexer is a circuit for selecting one of multiple inputs as the output according to a selection control scheme. In Figure 10.2(a), when the control = 00, input 1 is selected as the output; when control = 10, input 3 is selected as the output. Multiplexers are used widely to supply different inputs for a component so that the component can be used in a flexible way. This is called hardware multiplexing.

Verilog code of a 4-to-1 multiplexer is given in Listing 10.1.

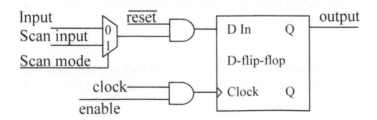

FIGURE 10.1

One bit register cell with synchronous reset, clock gating, and scan chain connections.

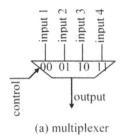 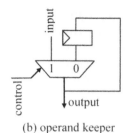

(a) multiplexer   (b) operand keeper

FIGURE 10.2

Multiplexer and operand keeper.

LISTING 10.1 Verilog code for a 4-to-1 multiplexer.

```
case(control)
        2'b00: output   <= input1;
        2'b01: output   <= input2;
        2'b10: output   <= input3;
        default: output <= input4;
endcase
```

Operand Keeper

An operand keeper consists of a two-way multiplexer and a register (see Figure 10.2(b)). When the control = 0, the multiplexer selects the value stored in the register as the output so that the old output value is kept. When the control = 1, the input is connected to the output. Operand keeper can be replaced by the enable function of a flip-flop. The enable signal equates to control = 1. In RTL design, the keeper must be supplied to all registers in a processor core. The reason is simple—there are usually several registers connected to one data source (for example, a bus). While updating a register, other registers connected to the physical data source shall not accept the data. In SoC platforms, many devices are connected to a bus. However, only the selected components accept the current data on the bus, whereas other devices keep their old data regardless of the transitions on the bus. Operand keepers will block unnecessary propagations of transitions, and this will relatively reduce the power consumption of the processor.

Decoder

A decoder has the reversed function of a multiplexer, and usually has n-bits input and 2^n-bits output. Verilog code of a 2^2 decoder is given in the example in Listing 10.2.

LISTING 10.2 Verilog code for a 2^2 decoder.

```
case(input)
        2'b00: output <= 4'b0001;
```

```
2'b01: output <= 4'b0010;
2'b10: output <= 4'b0100;
default: output <= 4'b1000;
```

10.2.2 **Arithmetic Components**

Full Adder

An N-bit full adder is usually a combinational circuit performing addition based on two's complement numerical representation. If the result is R $[N : 0]$ = A $[N - 1:0]$ + B $[N - 1:0]$, then the carry-out of the addition is given by the MSB pin R $[N]$. If the result is R $[N - 1:0]$, the carry-out will be represented separately. The RTL code also should support the case when the full adder supplied by a synthesizer does not have a carry-in. If we need the carry-in to be involved in the computing, a full adder containing $N + 1$ bits is needed to perform the N-bit arithmetic. The RTL model of a full adder is described by the following Verilog code in Listing 10.3.

LISTING 10.3 Verilog code for a full adder.

```
module adder #(parameter N=8)(
        output wire [N-1:0] sum,
        input wire [N-1:0] A,B,
        input wire carryin ,
        output wire carryout);

        wire [N+1:0] result;

        assign result = {1'b0,A,1'b1} + {1'b0,B,carryin };
        assign carryout = result[N+1]; // result[0] is not useful
        assign sum = result[N:1];

endmodule
```

Multiplier (Multiplier Primitive)

A two's complement signed multiplier usually has two N-bit inputs and one $2N$-bit output. Listing 10.4 is an example of a multiplier coded in Verilog. Although users of Verilog 2001 compliance tools have the luxury of signed datatypes, old versions of Verilog had only unsigned datatypes. Therefore, signed multiplication is somewhat tricky in plain Verilog, and Listing 10.4 shows an example of how it can be done.

Try to find out what will happen if there are no guard bits for operand A and operand B (opa and opb are both 16-bit) when both operands hold the maximum negative value.

LISTING 10.4 16b x 16b two's complement multiplication in Verilog using only unsigned multiplication.

```
// --------------------------------------------------------
// Signed Multiplier
// --------------------------------------------------------
always @(opa or opb)
begin : signed_multiply
  // ------------------------------------------------------
  // OPA and B: 17-bits including a guard
  // ------------------------------------------------------
  // If negative, change it to positive by 2's complement
  // ------------------------------------------------------
  if (opa[16] == 1'b1)
    internal_opa_pos = ~opa + 1'b1;
  else internal_opa_pos = opa;
  if (opb[16] == 1'b1)
    internal_opb_pos = ~opb + 1'b1;
  else internal_opb_pos = opb;
  // ------------------------------------------------------
  // Unsigned Multiply : 17-bit x 17-bit = 34-bit result
  // ------------------------------------------------------
  unsigned_mul_result = internal_opa_pos * internal_opb_pos;
  // ------------------------------------------------------
  // Determine sign of result, 2's complement if negative result
  // ------------------------------------------------------
  if (opa[16] != opb[16])
    mul_result = ~ unsigned_mul_result +1'b1;
  else mul_result = unsigned_mul_result;
end // signed_multiply
// --------------------------------------------------------
```

10.3 HARDWARE DESIGN FUNDAMENTALS

The general hardware design flow, illustrated in Figure 10.3, can be found in most embedded system design books. Each step of this design flow will be discussed in the following sections.

10.3.1 Function Partitioning

In microarchitecture design, functions shall be partitioned according to the available hardware, classification, and complexity. When the available hardware is a multiple-processor platform, or a processor with multiple accelerators, the function

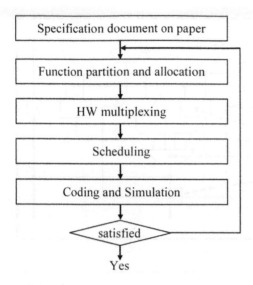

FIGURE 10.3

Hardware design flow.

partitioning and allocation will be a complicated design challenge. We will discuss functional partitioning and allocation using a multiprocessor for streaming signal processing later in this book. In this chapter, the function partitioning and allocation for a DSP processor core is relatively easy and can be merged into one step.

10.3.2 **Function Allocation**

Partitioned functions are connected sequentially in one or several task flowcharts. In microarchitecture design, each function in a task flowchart should be allocated to the corresponding hardware. The procedure is given in Figure 10.4.

In Figure 10.4, the flowchart on the left side shows the sequential connections of the partitioned function building blocks; the right side shows the architecture of a hardware unit executing functions listed in the flowchart.

In case the hardware required by the flowchart is not available, the functional partitioning should be modified, or different algorithms should be selected to adapt to the available hardware.

In Figure 10.5, a hardware multiplier is initially not available for mapping the functional flowchart to the hardware. If we do not change the hardware, algorithms should be changed, and the multiplication will be replaced by iterative computing using full adders. Another choice is to add a hardware multiplier. Using advanced silicon technologies, we find that the cost of an added multiplier is not so significant and may be justified.

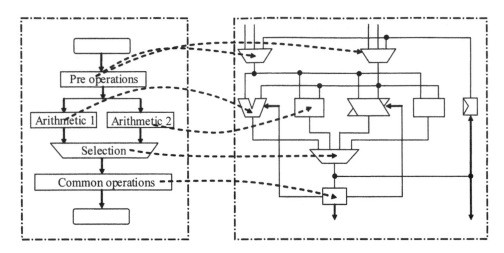

FIGURE 10.4

Function allocation in microarchitecture design.

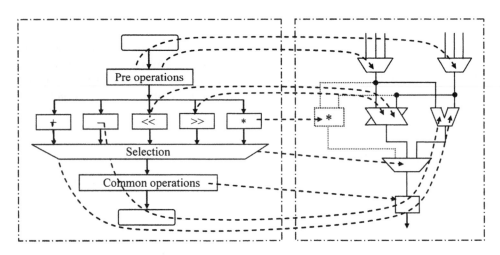

FIGURE 10.5

Hardware modification during function allocation.

10.3.3 HW Multiplexing

Hardware multiplexing is a technique to share a hardware device or a functional unit (a full adder, a multiplier, a register file, etc.) for different purposes at different times. Hardware multiplexing can be implemented either by software or by configuring the hardware.

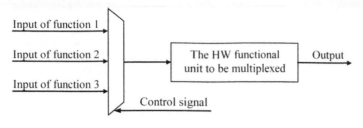

FIGURE 10.6

Concept of hardware multiplexing.

By selecting different inputs by control signals, a hardware module can offer computing hardware to different inputs at different times. This is called time division-based hardware multiplexing. The concept of hardware multiplexing is illustrated in Figure 10.6.

A typical design of hardware multiplexing for a datapath device (a building block or a module) is to divide its functions into three groups: the preprocessing circuits, the kernel processing circuits, and the postprocessing circuits. The preprocessing circuit collects and preprocesses operands and feeds preprocessed operands to the kernel component. A kernel component is usually an arithmetic or logic device consuming a significant number of gates. By feeding different preprocessed operands to the kernel component, different output values can be generated through the same kernel processing circuit. The postprocessing component collects and selects results from kernel components, and conducts postprocessing.

The example in Figure 10.7 gives detailed descriptions of hardware multiplexing, including the preprocessing, the kernel processing, and the postprocessing. The preprocessing in Figure 10.7 is for hardware multiplexing. The postprocessing-1 in Figure 10.7 is not for hardware multiplexing. However, the postprocessing-2 contains functions of hardware multiplexing. Hardware above MP2 is multiplexed for both saturation on results or not saturation on results.

In Figure 10.7 MA and MB are multiplexers to select the left and right operands for the adder (ADD) or the multiplier (MUL). MP is the multiplexer selecting outputs from the ALU or MUL. By using a hardware multiplexing technique, the four additions can be allocated into one adder and the four multiplications can be allocated into one multiplier. Suppose the gate count of a 16-bit full adder is 200, the gate count of a 16-bit multiplier is 5000, and the gate count of a 16-bit 2-to-1 multiplexer is 50—the hardware cost without hardware multiplexing will be $(2 \times 4 \times 5000 + 4 \times 200) = 41600$ gates. With hardware multiplexing as depicted in Figure 10.7, the gate count is $5000 + 200 + 4 \times 50 = 5400$ gates, saving gate counts up to eight times. Therefore, when the kernel device is multiplexed, the hardware cost is significantly reduced. In the pseudocode in Listing 10.5, the behavior of the hardware shown in Figure 10.7 is specified.

Eight arithmetic components are used in the behavior code in Listing 10.5. We might expect the synthesis tool to minimize the hardware cost and to keep the

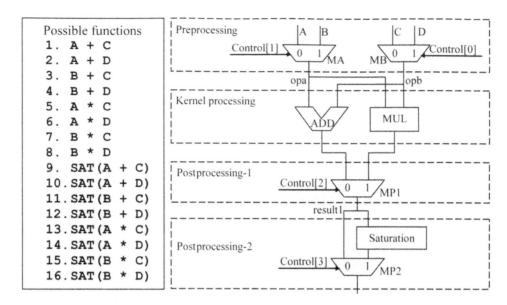

FIGURE 10.7

An example of hardware multiplexing.

LISTING 10.5 Verilog code for Figure 10.7.

```
always @* begin
        case(control[2:0])
                3'b000: result1 = a+c;
                3'b001: result1 = a+d;
                3'b010: result1 = b+c;
                3'b011: result1 = b+d;
                3'b100: result1 = a*c;
                3'b101: result1 = a*d;
                3'b110: result1 = b*c;
                default: result1 = b*d;
        endcase

        case(control[3])
                1'b0: result = result1;
                default: result = saturation(result1);
        endcase
end
```

behavior of the code. For small circuit blocks, the RTL code is short, and the synthesis tool might realize that the four adders in the code can be merged into one. However, in reality, the size of the behavior code could be large and a synthesis tool may not be able to optimize the behavior HDL code.

One difference between the RTL (Register Transfer Level) code and behavioral code is that the hardware multiplexing can be specified and exposed in the RTL code. The RTL code in Listing 10.6 shows an example of exposing the hardware multiplexing details.

LISTING 10.6 Verilog code for Figure 10.7 with hardware multiplexing.

```verilog
always @* begin
    case(control[1])
        1'b0: opa = a;
        default: opa = b;
    endcase

    case(control[0])
        1'b0: opb = c;
        default: opb = d;
    endcase

    mul = opa * opb;
    add = opa + opb;

    case(control[2])
        1'b0: result1 = add;
        default: result1 = mul;
    endcase

    case(control[3])
        1'b0: result = result1;
        default: result = saturation(result1);
    endcase
end
```

In this case, preprocessing is conducted by MA and MB. OPA = A when control [1] = 0. OPB = C when control [0] = 0. Postprocessing is conducted by MP, and control [2] to MP is used to select the result; result = ADD when control [2] = 0.

10.3.4 Scheduling of Hardware Execution

Scheduling decides the execution sequence, the runtime, and the relationship between the inputs and outputs of hardware modules. Scheduling as a design step can be found in different design hierarchies, but here the focus will be on pipelining of microoperations.

A pipeline diagram is a good method for describing the pipeline inside a processor. Figure 10.8 is a simplified example where a pipeline diagram is used to specify the pipeline of a DSP core.

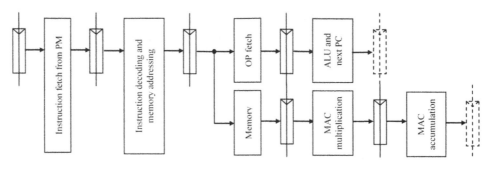

FIGURE 10.8

Pipeline diagram for a DSP processor.

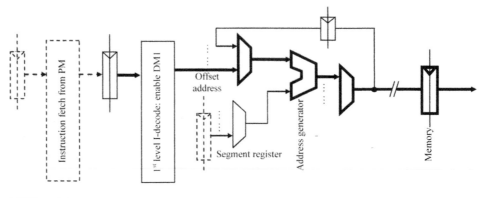

FIGURE 10.9

Pipeline diagram describing the critical path of memory addressing.

In this diagram, the pipeline is described, including activities both in the datapath and the control path. The details of processor pipelining will be discussed in later chapters. In Figure 10.8 we want to show only that pipeline diagrams could be used to analyze and describe the pipeline and the data flow.

Pipeline diagrams can also be used to describe a critical path in a circuit block. In Figure 10.9 one example of a computing circuit for memory addressing is described. The critical path can often be predicted before the quantitative analysis of its RTL code. Following the figure, we can assume that the critical path starts from the offset constant decoding in the instruction decoder, goes through the multiplexer selecting operands for address calculation, enters the address calculation logic and the multiplexer after it, and via a long interconnection, ends up at the flip-flop. Based on design experiences, this path (highlighted in Figure 10.9) could be critical, because it is the longest path in the pipeline diagram.

Pipeline diagrams are suitable for describing pipeline scheduling when the pipeline depth is short. In order to describe the timing or scheduling of hardware

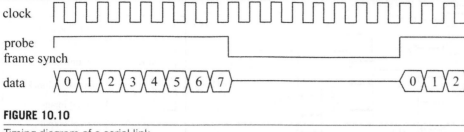

FIGURE 10.10

Timing diagram of a serial link.

with long pipeline depth (e.g., the finite state machine of a timer) or the timing protocol of a serial peripheral port, a normal cycle accurate timing diagram is preferred over a pipeline diagram. Figure 10.10 is an example of a cycle accurate timing diagram.

The advantage of using a timing diagram as in Figure 10.10 is its large observation window. Thus the figure can describe the scheduling/timing involving many clock cycles. If asynchronous events are involved, for example if a "handshaking finite state machine" is used for packet data processing, the timing diagram in Figure 10.10 is very useful.

10.3.5 Modeling and Simulation

There is no accepted hardware description language at microarchitecture level, but SystemC and C sometimes are used as modeling languages. However, it is designed for higher-level modeling, and the use of SystemC for microarchitecture modeling is done mostly within academia.

10.4 FUNCTIONAL SPECIFICATION AT MICROARCHITECTURE LEVEL

In this section, hardware design fundamentals will be used to specify the hardware microarchitecture. Different methods for microarchitecture functional specification will be introduced following a top-down design methodology. In the following chapters block diagrams, schematics, flowcharts, finite state machines, and truth tables will be presented.

10.4.1 Intermodule Block Diagram

Block diagrams are used to describe the design on a high level. The block diagrams partition the high-level functions and allocate partitioned function into hardware modules. Interconnects give relations between hardware modules. Block diagrams contain neither timing information nor data/control flow information. They are used as a high-level schematic entry of the hardware hierarchies and relations.

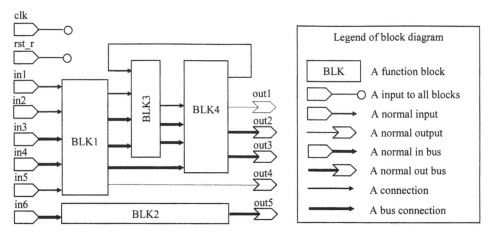

FIGURE 10.11

An example of a block diagram.

Figure 10.11 shows a high-level view of a design. A description of legends is given on the right side of the figure. Remember that a complete input-output description should be made if the block diagram is to be the reference of the sensitivity list of the RTL code. Normally, thin wires are used for single-bit signals and thick wires for buses. Inputs such as clock and reset are global signals connected to all modules and sublevel modules.

10.4.2 Microarchitecture Schematic

Schematics or circuit schematics give detailed descriptions of the circuit topology. The difference between a block diagram and a schematic is that the schematic specifies both connections and functions, whereas the block diagram specifies only connections between black boxes. A formalized schematic of a data processing module, including preprocessing, kernel processing, and postprocessing, is shown in Figure 10.12.

Control flow and pipeline details might not be included or exposed enough by the circuit schematic. In order to specify hardware, flowcharts or state machines are sometimes needed in addition to schematics. An example of an ALU schematic is shown in Figure 10.13.

This schematic describes a bit-accurate datapath by annotating connection wires. Information supplied in the schematic is detailed enough for RTL coding, though the control information is not included yet.

10.4.3 Module Functional Flowchart

Flowcharts frequently are used in software design and engineering. In software, a flowchart is called a task graph. In hardware design, flowcharts also are widely used.

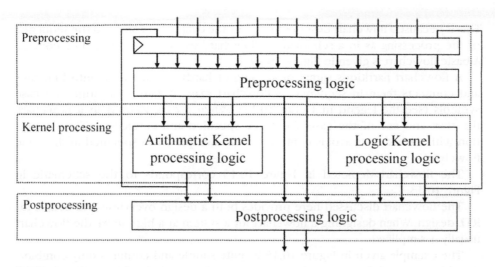

FIGURE 10.12

A formalized circuit schematic.

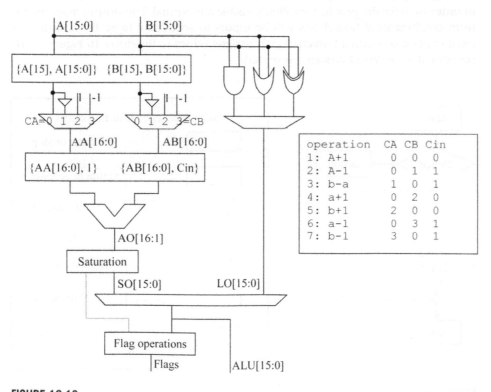

FIGURE 10.13

A schematic-based design of an ALU.

This is because a flowchart describes not only the data processing and the sequence of the processing as in a schematic; a flowchart also describes the flow control. A basic flowchart is given in Figure 10.14.

A flowchart partitions signal processing or hardware functions into branches and connects them in sequence. A flowchart may contain subroutine and case symbols (see the legend in Figure 10.14). When a case is used in a flowchart, the start and end of the case should be specified and connected. The connection with dashed lines avoids confusion with multiple cases specified in the same flowchart.

The flowchart depicted in Figure 10.15 corresponds to the schematic in Figure 10.13.

The flowchart discussed here provides both a design overview and details for RTL design. When designing a flowchart for a system at a high level, the flowchart usually is simplified.

The example given in Figure 10.15 is quite simple and contains only combinational logic. A real design may have more complicated design hierarchies, and the flowchart may contain both sequential (clocked) and combinational logic functions. It is a good practice to separate the combinational logic from the sequential logic in order to make the pipeline explicitly visible and formal. The output assignments from combinational flowcharts will be inputs to sequential logic flowcharts. The most explicit sequential flowchart only assigns values to flip-flops. In Figure 10.16, sequential logic flowcharts are presented.

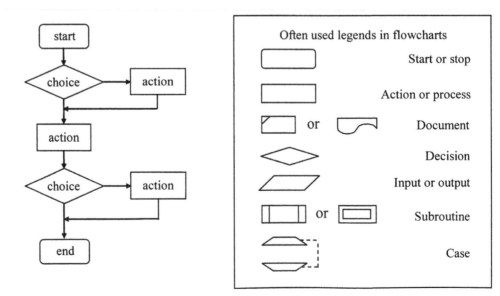

FIGURE 10.14

A simple flowchart and common legends in flowcharts.

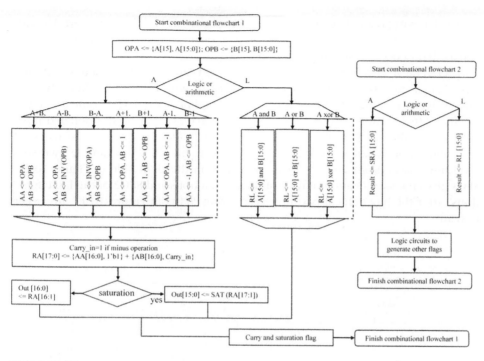

FIGURE 10.15

A flowchart-based design of ALU combinational logic.

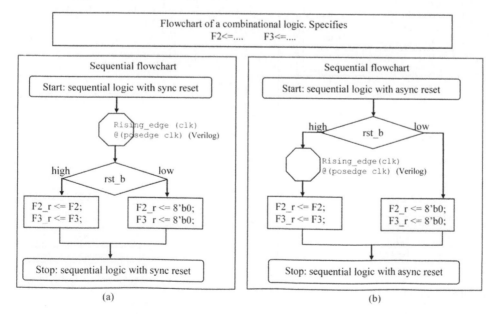

FIGURE 10.16

Flowchart of sequential logic.

The left part (a) is a flowchart with synchronous reset and the right part (b), with asynchronous reset. The clock is described as rising_edge(clk) when using VHDL and as @(posedge clk) when using Verilog.

Hierarchical flowcharts are necessary if a design cannot be described by only one flowchart. Figure 10.15 is redrawn in Figure 10.17 with hierarchy, where the top-level flowchart on the left side of the figure is simplified compared to the flowchart in Figure 10.15. Sublevel flowcharts (SFC1 and SFC2) describe the detailed functions. Here, the entry of the sublevel flowchart is identified using the name of the sublevel flowchart surrounded by a double frame. In this example, two sublevel flowcharts (SFC1 and SFC2) are depicted on the right side in Figure 10.17. SFC1 gives the specification of the arithmetic functions and SFC2 of the logic functions. After the hierarchical partitioning, the flowcharts in Figure 10.17 are easier to read.

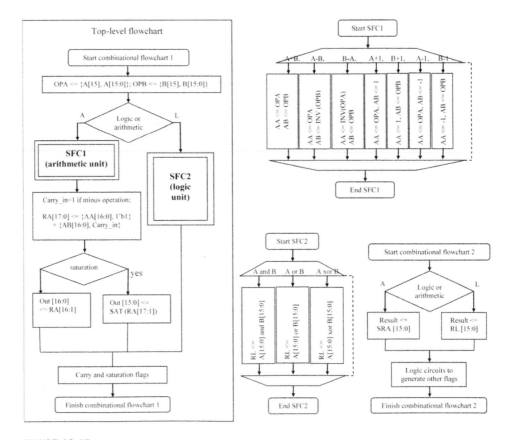

FIGURE 10.17

Hierarchical flowchart of an ALU.

10.4.4 Finite State Machine

A finite state machine (FSM) is useful for describing sequential behaviors, interwork models, and protocols, especially when asynchronous events are involved or the number of clock cycles of an event cannot be predictable. A processor core usually is designed using flowcharts, except for the instruction flow control (the PC circuit), where a finite state machine is preferred.

Finite state machines can be of three different types: Medvedev, Moore, or Mealy (see Figure 10.18). The input of the finite state machine is vector X and the output is Y. The STATE of the finite state machine is the output register value, which is also used as input.

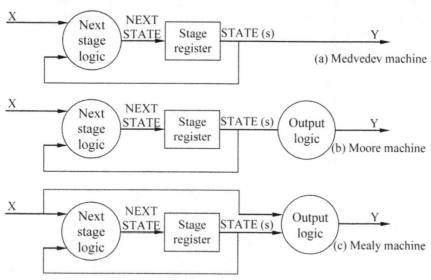

FIGURE 10.18

Finite state machines.

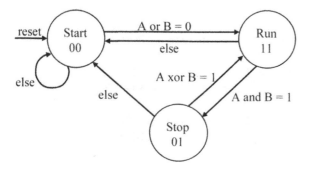

FIGURE 10.19

Medvedev state machine.

The Medvedev state machine is simple and commonly used for example counters. The output vector (Y) is taken from the state vector (s). If further decoding of the state vector is required, a Moore machine is useful, especially when the finite state machine is based on a counter. A Mealy machine is not recommended because it is not as easy to specify the microarchitecture of a processor using this type of machine.

The RTL code of a Medvedev state machine is given in Listing 10.7 and Figure 10.19.

LISTING 10.7 RTL code for a Medvedev state machine.

```verilog
module medvedev_fsm(
        input wire rst, clk,
        input wire A,B,
        output wire X,Y);

        localparam START = 2'b00,
                   RUN   = 2'b11,
                   STOP  = 2'b01;

        reg [1:0] state;
        reg [1:0] nextstate;

        // Flip-flops for next state (async reset)
        always @(posedge clk or negedge rst)
            if(!rst) state <= START;
            else     state <= nextstate;

        // Combinational generation of nextstate
        always @* begin
            nextstate = state;

            case(state)
                START:
                    if((A | B) == 0) nextstate = RUN;
                RUN:
                    if((A & B) == 1) nextstate = STOP;
                    else             nextstate = START;
                STOP:
                    if((A ^ B) == 1) nextstate = RUN;
                    else             nextstate = START;
            endcase
        end

        assign {X,Y} = state; // Output == inner state
endmodule
```

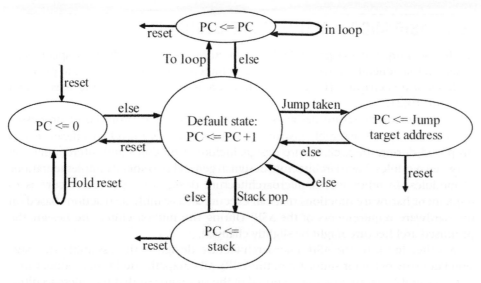

FIGURE 10.20

PC FSM.

To improve readability, it is preferable to separate the combinational logic from the sequential logic. The combinational logic describes what the next state will be, depending on external conditions X and the current state. An example of a FSM for a Program Counter (PC) is given in Figure 10.20. This FSM provides the next program counter value according to the program flow. We will describe the principle of PC FSM in later chapters.

Reset always has the highest priority, and the PC content will be 0 if the reset is activated. Except for branch and other special cases, the PC will be incremented each clock cycle. When running a single-step loop, the PC will keep its value until the loop is finished. When running a conditional branch instruction and the branch is taken, the PC is loaded with the target address of the branch. When executing stack pop, the stack value is assigned to the PC.

Hierarchical design of a FSM can be specified in a similar way, as described in the flowchart section.

10.4.5 Truth Table for Coding and Decoding

Truth tables frequently are used for designing combinational logic with multiple inputs and outputs. The typical case is to use a truth table as a description or specification of a decoder, such as an instruction decoder. A design example using a truth table can be found in Chapter 12, where the output of a shifter is specified based on the number of bits to shift.

10.5 ASIP MICROARCHITECTURE DESIGN FLOW

In this section, the design of ASIP architectures will be addressed specifically based on the general knowledge of microarchitecture design discussed in previous sections of this chapter. The general ASIP microarchitecture design flow is shown in Figure 10.21.

ASIP microarchitecture design starts from an available proposal of processor architecture and its assembly instruction set. An ASIP core level specification was proposed during the architecture design, including modules and interconnection between modules. Microarchitecture design in general is to specify implementations of modules. In other words, microarchitecture design of a processor core is to implement hardware functions that can execute all assembly instructions based on the hardware requirements of the ASIP. During the microarchitecture design, the proposed architecture might be slightly changed.

Another input to the ASIP microarchitecture design is the assembly language simulator (the behavior simulator of the ASIP). All datapath and PC microoperations of the assembly instructions are exposed in the simulator so that the microarchitecture designer knows what to allocate and how to do the scheduling according to the exposure from the simulator. If the assembly language simulator is not yet available, the assembly instruction set manual can also be the design input. However, hidden microoperations might not be found from this manual.

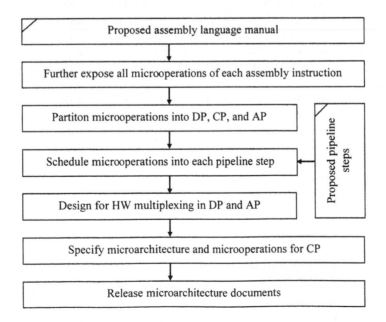

FIGURE 10.21

ASIP microarchitecture design flow.

The output of the ASIP microarchitecture design is a design document for RTL coding.

10.5.1 Exposing Microoperations

The first step in the ASIP microarchitecture design is to expose all microoperations of each assembly instruction. Microoperation refers to the basic computing step of arithmetic or logic operations specified by an assembly instruction. It can be explicitly specified by the assembly instruction or implicitly required by the instruction.

How to specify and expose hidden microoperations is not a trivial task. There are two types of hidden microoperations. The first type cannot be identified from the machine code, yet it can be found from the assembly instruction set manual. Examples are flag operations and modulo addressing supporting a convolution. Most instruction-level accelerations targeting specific algorithms implicitly include operations, and these operations can be found from the assembly instruction set manual. The second type of hidden microoperations is not specified by the assembly instruction set manual. They can be disclosed only based on design experience. Typical hidden operations are bus transactions. Fortunately, bus transactions can be identified from the architecture specification, which was discussed in Chapters 3 and 7.

10.5.2 Allocation and Partitioning of Microoperations

The second step in ASIP microarchitecture design is to partition and allocate microoperations into the datapath (DP), the addressing path (AP), and the control path (CP). Actually, the most typical microoperations—arithmetic, addressing, and control operations—were partitioned and allocated to DP, AP, and CP in Example 10.1.

Example 10.1

Identify all microoperations.

Identify all microoperations included in the convolution instruction specified as:

```
Name CONV: Convolution, to run MAC N times
Assembly: CONV N ACR1 DM1, DM2
Binary Machine Code: skipped in this example
Operation: Execute N times on
        ACR1 [39:0] <= ACR1 [39:0] + DM1(p1)[15:0] * DM2(p2)[15:0]
        DM1 address pointer is p1 with post incremental step size
        1
        DM2 address pointer is p2 with post incremental step size
        1
        DM1 modulo addressing top/bottom registers are MTOPR MBTMR
Operands:
        16b for multiplication, 40b for accumulation
```

```
Run cycle cost: two clock cycles
Flags: MAC Sign flag <= ACR1 [39]
       MAC Zero flag <= 1 when result ACR1 [39:0] = 40'b0
Exception: no
```

Identified microoperations in datapath are:

No.	Name	Microoperation
1	Signed OPA	Get data from DM1, copy sign bit [15] to [16]
2	Signed OPB	Get data from DM2, copy sign bit [15] to [16]
3	Multiplication	Signed multiplication OPA MUL OPB
4	Fractional result	Left-shift one bit of MUL result and LSB = 0
5	Guard for accumulation	Copy result [33] to [39:34]
6	Accumulation	ACR1 <= ACR1 + guarded result of MUL
7	Flag operation	Sign and Zero flag operations

Identified microoperations in addressing path are:

No.	Name	Microoperation
1	Pointer 1 postincrement	p1 <= p1 + 1
2	Pointer 1 modulo	If p1 = MTOPR the p1 <= MBTMR else no-op
3	Pointer 2 postincrement	p2 <= p2 + 1

Identified microoperations in control path are:

No.	Name	Microoperation
1	Loop counter	LC <= LC − 1
2	PC operation	If LC = 1 (loop finish) PC <= PC + 1 else PC <= PC

Bus transactions are:

No.	Name	Microoperation
1	DM0 bus	Get address from address bus; Send data to data bus
2	DM1 bus	Get address from address bus; Send data to data bus

10.5.3 Pipeline Scheduling Microoperations

The processor pipeline structure is specified according to the processor architecture and profiling of applications. The physical speed (working frequency) of the processor can be higher when the pipeline is deeper. At the same time, a processor with more pipeline stages will use more clock cycles when a branch is taken. More pipeline stages also require more hardware running in parallel. Functions in different pipeline stages that are executing in parallel are illustrated in Figure 10.22.

IF, ID, OF, EX, and WR stand for instruction fetch, instruction decoding, operand fetch, execution, and write result, respectively. In some cases a second cycle of execution could take the place of the WR stage. There are five pipeline steps in the processor requiring five groups of hardware running in parallel. In Figure 10.22, the gray part indicates five groups of hardware running five pipeline steps in parallel: the fifth instruction is fetched, the fourth instruction is decoded, operands of the third instruction are fetched, the execution unit is running for the second instruction, and finally, the result of the first instruction is written back.

The functions specified for each pipeline step are given in Table 10.1.

Table 10.1 exposed requirements of design for parallelization. Functions listed in the table shall be allocated to different hardware circuits and be executed in parallel. The next example is convolution (CONV), taken from Example 10.1. Table 10.2 shows the detailed pipeline scheduling.

Hidden physical critical paths need to be taken into account based on design experience. For example, the data memory might be placed physically far away from the logic circuits (the datapath and the addressing path). Thus long delays due to interconnects could be expected between the logic circuits and the memory. To balance the execution time of different pipeline stages is always an implementation challenge.

10.5.4 HW Multiplexing of Microoperations

The fourth step in ASIP microarchitecture design is to minimize the hardware cost by using hardware multiplexing techniques. A formal way to design for microoperation hardware multiplexing is given in Figure 10.23.

IF1	ID1	OF1	EX1	WR1	IF6	ID6	OF6	EX6
	IF2	ID2	OF2	EX2	WR2	IF7	ID7	OF7
		IF3	ID3	OF3	EX3	WR3	IF8	ID8
			IF4	ID4	OF4	EX4	WR4	IF9
				IF5	ID5	OF5	EX5	WR5

FIGURE 10.22

An example of pipelined execution.

Table 10.1　Microoperation Scheduling of the Pipeline in Figure 10.22.

Pipeline	Microoperations in DP	Microoperations in CP	Microoperations in AP
1		Fetch an instruction from program memory	
2		Instruction decoding; Register address calculation;	Memory address calculation;
3	Read register file;		Send data to memory;
4	Execution clock 1; Load to register; ALU; Logic; Shift; Multiplication; Flag of ALU; Flag of Accumulation		Memory accept data;
5	Execution clock 2; Accumulation Write result to register; Flag of MAC;	Calculate the next PC	Write result to memory

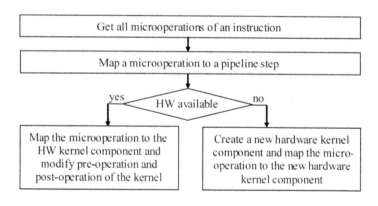

FIGURE 10.23

Hardware multiplexing design flow.

The technique using preprocessing and postprocessing circuits for hardware multiplexing was given earlier in this chapter.

10.5.5　Microoperations Integration

Finally, the fifth step in ASIP microarchitecture design is to coordinate and integrate the DP, AP, and CP. Before the integration phase, we first need to check that the

Table 10.2   An Example of Microoperation Scheduling for CONV.

Pipeline	N	Microoperations in DP	Microoperations in CP	Microoperations in AP	BUS transactions
1			Fetch CONV		
2			Decode	DM0 DM1 addressing(1)	
3	1			DM0 DM1 addressing(2)	Data (1) to MAC
4	2	MUL data (1)		DM0 DM1 addressing(3)	Data (2) to MAC
5	3	MUL data (2); Accumulate data (1)		DM0 DM1 addressing(4)	Data (3) to MAC
6	4	MUL data (3); Accumulate data (2)		DM0 DM1 addressing(5)	Data (4) to MAC
		. . .		. . .	. . .
N + 2	N			DM0 DM1 addressing(N)	Data (N − 1) to MAC
N + 3		MUL data (N − 1); Accumulate (N − 2)	IF PC + 1	NOP	Data (N) to MAC
N + 4		MUL data (N); Accumulate (N − 1)	ID PC + 1	Addressing for PC + 1	NOP
N + 5		Accumulate data (N)	OPF PC + 1	. . .	OPF PC + 1

complete set of microoperations of all assembly instructions has been mapped to hardware.

The hardware multiplexing technique may impact the performance and introduce a longer critical path. When a hardware module is over multiplexed, the preprocessing and postprocessing circuits will add delay and might generate critical paths. Trade-off between cost and performance is therefore necessary during the integration phase. A way to remove the critical path introduced by the preprocessing and postprocessing circuits is to add redundant kernel components.

We normally expect the critical path to be in the kernel device when we design a DSP core. If we examine the datapath of a DSP processor, we usually find that the critical path goes through the multiplier and added preprocessing and postprocessing logic functions. Pipeline modifications of preprocessing and post-processing circuits are therefore essential design techniques. We will discuss this technique more in Chapter 18.

During the integration phase, we also need to check the consistency of the microarchitecture with the architecture proposed during the instruction set design. It should be noted that modifications of an early proposed ASIP architecture are usually unavoidable.

10.6 CONCLUSIONS

Driven by requirements for performance, cost, and power consumption, DSP architectures are diverging. One type of DSP architecture may only be suitable or only required for a specific application. We cannot say that one type of DSP processor architecture is good or not. Instead, we may say only whether this type of DSP processor architecture is suitable or not for a certain application. Because of the vast number of applications, different types of ASIPs are needed with different microarchitecture.

Microarchitecture and ASIP microarchitecture design have been introduced in this chapter. However, principles and general guidance addressed here are not enough. Experience also is needed, and can be accumulated only through real design projects. You should read this book again after going through a hardware design project.

EXERCISES

10.1 Discuss the differences between architecture design and microarchitecture design.

10.2 Design signed and unsigned multipliers with integer and fractional outputs using signed integer multipliers.

10.3 Write the RTL code of Figure 10.13 and write the RTL code of Figure 10.15. Compare and discuss the two code implementations.

10.4 Code Figure 10.16 using both Verilog and VHDL.

10.5 Code Figure 10.20 using both Verilog and VHDL.

10.6 Describe the process of mapping an instruction set to hardware.

10.7 Check Table 10.1 and estimate how many full adders are required if all functions are allocated in parallel to all pipeline stages in Table 10.1.

10.8 Are the following statements correct? Microarchitecture design of a DSP core:

a. Is to design or specify the core instruction set.

b. Is to specify functions and the implementation of hardware modules in the core according to its assembly instructions set.

c. Is the description of an application.

d. Is to decide how many operands can be carried by an instruction.

10.9 Which of the following statements are correct? Hardware multiplexing technique:

a. Can decrease the critical path of logic circuits.

b. Can decrease the silicon cost.

c. Can supply multiple functions using a minimum number of hardware kernel devices.

10.10 Following the method used in Example 10.1 and Table 10.2, expose, partition, allocate, and schedule all microoperations for autocorrelation instruction.

REFERENCES

[1] Dave, N., Ng, M. C., Arvind. (2005). Automatic synthesis of cache-coherence protocol processors using Bluespec, Formal Methods and Models for Co-Design. MEMOCODE '05. Proceedings, Third ACM and IEEE International Conference, 25–34.

[2] http://www.da.isy.liu.se.

Design of Register File and Register Bus

The datapath in a DSP processor usually consists of ALU, MAC, RF (register file), and instruction-level acceleration units. In this chapter, we will discuss the design of register files, including function design and physical designs of the register file and register bus. It is assumed that the register file size and data type have already been decided. These issues were discussed in Chapter 7.

11.1 DATAPATH

The datapath is the execution unit in a DSP processor consuming input data and producing results. The datapath in a DSP processor consists of MAC (arithmetic module for multiplication and accumulation), ALU (arithmetic, logic, and data manipulation unit), RF (register file), bus subsystem, and some specific functional units (see Figure 11.1). The data processing in the datapath is controlled by control signals decoded by the instruction decoder in the control path. To prepare for datapath computing, operands should be available in the RF. Load and store should also be executed between RF and data memories.

In Figure 11.1, a processor core is represented by gray blocks, consisting of datapath (DP), control path (CP), and addressing path (addressing generator, AGU). The design of the datapath microarchitecture is the procedure of mapping data processing and load/store instructions to the datapath hardware, with minimized hardware cost using hardware multiplexing techniques. The inputs of the datapath function design are part of the assembly manual, including ALU instructions, MAC instructions, and Load/store instructions. The physical requirements on hardware, such as power or area requirements, are also datapath design inputs. The outputs of a datapath design consist of the schematic description of datapath modules and RTL coding guidelines.

Recall the microarchitecture design methodology discussed in Chapter 10. Operations in an instruction usually is mapped into three parts of hardware in the datapath: the preprocessing hardware, the kernel hardware, and the postprocessing hardware. Preprocessing operations are applied to the input operands before they enter the kernel component. Preprocessing components include multiplexers,

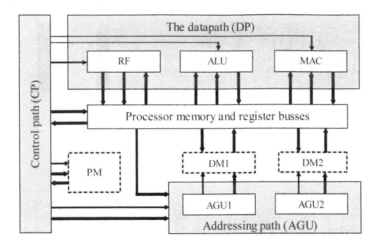

FIGURE 11.1

A datapath in a DSP processor core.

inverters, combiners and splitters of buses. Kernel components are full adders, multipliers, and other large and silicon-consuming components. A kernel component is multiplexed by several instructions at different times. Postprocessing is the operations applied to the output of the kernel component before it can be output as the final result. Postprocessing includes output selection, miscellaneous operations on the output, and flag computing.

Part three of this book (Chapters 10–15) describes the microarchitecture design of DSP processors. Microarchitecture design involves the kernel component selections and the design of preprocessing and postprocessing of hardware modules. Based on the same instruction set, microarchitecture or hardware circuits could be specified much differently by different designers. Therefore, all discussion in Part three of this book can be taken as examples or case studies.

11.2 DESIGN OF REGISTER FILES

11.2.1 General Register File

A general register file (RF) consists of a group of registers used as the first level of computing storage buffers. A physical data memory (typically with a single read/write port) can access one data at a time, and read and write cannot be executed simultaneously. A register file, on the other hand, supports simultaneous reads and writes in one clock cycle, and an unlimited number of operands can be supplied by a RF at the same time. A RF in a DSP supplies at least two operands simultaneously. The RF gets data from data memories by running load instructions while preparing for an execution of a subroutine. While running the subroutine based on RISC instructions, registers in the register file are used as computing buffers. After running the subroutine, results in the RF will be stored into data memories by running store instructions. Reading data from a register in a RF or writing data to a register in a RF is a microoperation. A RF access is addressed and controlled by decoded control signals from the control path.

The size of a register file is specified during the instruction set design. A RF can be very small, including only eight registers. It can also be very large in some cases; for example, the RF in Imagine [1], a streaming DSP processor, has several thousands of addressable registers. The size of a RF in a normal DSP processor is about 16 to 64 registers. The number of registers in a register file can be decided by benchmarking the firmware. Too many registers in a RF requires unaffordable silicon cost (both logic cost and program memory cost because of the extra code size required by coding of a register address). Too small a RF will introduce too much data swapping between the registers and the data memories.

11.2.2 Design of a Simple Register File

Let us start from a simple RF depicted in Figure 11.2. The RF includes n addressable registers and uses $\log_2 n$ bits for register addressing. For example, four bits are required to point to a register if the size of the RF is 16. The word width of each addressable register is one bit. When designing a 16-bit processor, we simply copy and paste this circuit 15 times to build a 16-bit RF. Every register in the RF can be addressed to accept new data, and the output of every register can be selected by operand selection multiplexers. In this circuit, only one write per clock cycle to the RF is allowed, and the RF can supply multiple (two in this example) operands simultaneously.

There is no extra clock cycle cost when reading from and writing into the register file. The processor pipeline can be simplified when using such a register file. Two control signals are required to write one data value to a register in the RF; one is the data selection control, and another is the register selection control (r_select), both coming from the instruction decoder in the control path. The data selection control selects which port (from the ALU, the MAC, the memories, or the external ports) can write data into the register file. The register selection control (r_select) selects which register can update data. Not selected registers (r_select = 0) will keep their old values. In the design of Figure 11.2, only one of 16 r_select pins can be high at a time. That is, only one of the registers can be selected to accept new data during one clock cycle, and the rest of the registers will keep their old data because other r_select signals let their registers "keep" their values.

The output of each register is connected to all operand selection multiplexers. Each operand bus is connected to an operand selection multiplexer. In Figure 11.2, two operands can be supplied in each clock cycle. If Ctrl_opa = 0000, the first register R0 will be connected to operandA bus. In the same way, if Ctrl_opb = 1111, the last register R15 will be connected to operand B bus.

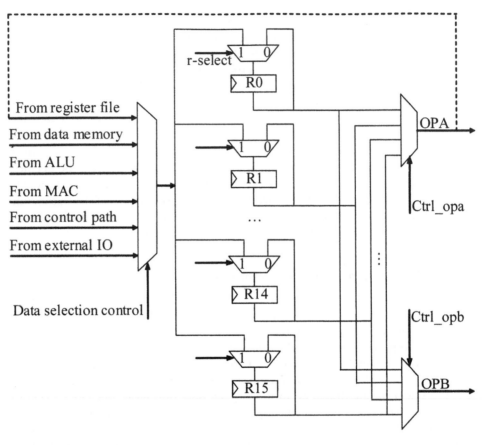

FIGURE 11.2

A simple 1-bit register file.

Note that the fan-out of the operand selection control pin from CP to RF is very high. A register file with 32 16-bit registers will be used as an example in the following text. The silicon costs will be estimated. The clock load and the load of the control signal for operand selection will be calculated in Figure 11.3.

A register file might consume significant gate count in a DSP core. For example, a register file with 32 general registers in a 16-bit processor contains $32 \times 16 = 512$ flip-flops. The gate count of 512 flip-flops is about 5.12k gates. The gate count for all $32 \times 16b$ keepers is $32 \times 16 \times 6 = 3.1k$ gates. The gate count for operand selection including OPA and OPB is $2 \times 16 \times (32 + 16 + 8 + 4 + 2 + 1) = 2.1k$ gates. The gate count of the data selection control is at least $16 \times (8 + 4 + 2 + 1) = 0.24k$ gates. Therefore, the total gate count of the register file without driving buffers is about 10.6k gates. Including extra driving buffers, the gate count of this register file will be around 15k. It is more than the gate count of a simple 16-bit MAC.

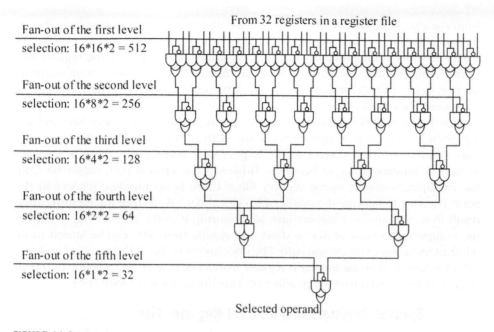

From 32 registers in a register file

Fan-out of the first level
selection: 16*16*2 = 512

Fan-out of the second level
selection: 16*8*2 = 256

Fan-out of the third level
selection: 16*4*2 = 128

Fan-out of the fourth level
selection: 16*2*2 = 64

Fan-out of the fifth level
selection: 16*1*2 = 32

Selected operand

FIGURE 11.3

Fan-outs of control signals for a 16-bit 32-to-1 multiplexer.

The fan-in of the clock pin in a flip-flop is at least 2. The fan-out of the clock driver to the RF will be at least $512 \times 2 = 1024$. It means the clock driver to the register file will drive at least 1024 gates.

Figure 11.3 is the circuit schematic of the 32-to-1 multiplexer for an operand selection in a RF with $32 \times 16b$ registers. There are 16 multiplexers, each selecting a bit of an operand. There are five control bits ($\log_2 32 = 5$) to drive a 32-to-1 multiplexer, each controlling a level of hardware multiplexing in the multiplexer. The fan-outs of five control levels are shown in Figure 11.3.

Fan-out of 512 is very heavy, and this is a typical hidden critical path between control path and datapath. It might not be recognized until reaching the logic synthesis of the RTL codes because the critical path cannot be identified from the datapath schematic. It is a typical mistake of an inexperienced design team.

11.2.3 Pipeline around Register File

Pipeline around register file includes the pipeline of result store (toward RF) and the pipeline of operand fetch (from RF).

The operand cannot arrive in time to the computing device because of the hidden critical path induced by the operand multiplexer in Figure 11.3. That is why a pipeline stage is prepared for operand fetching. From observing a data supplying path from the register out-port to a kernel computing device, the path is relatively short, consisting of only one or two small multiplexers. However, the hidden delay from the excessive fan-out of control signals might require a pipeline stage. Let us further check where an operand goes. An operand might be data to be stored to a data memory. A data memory might be allocated far away from the register file. Considering the delay from the multiplexer in Figure 11.3 plus the long interconnection delay from the register file to a data memory, a pipeline stage of supplying an operand becomes essential.

Computed results need to be stored to the register file. Is it necessary to have a pipeline stage to store results? Since the path is relatively short from the out-port of most arithmetic units to the in-port of a register file, there is no need to have a pipeline stage in between. However, the critical path might be from the multiplier out-port to the register file if there is no pipeline register in the path. The way to manage the predictable critical path is to store the multiplication result in an accumulator register instead of storing it to the register file. Normally, the computing latency of ALU is short, and results from ALU can be stored to RF without inducing extra critical path. The conclusion is that if the result of multiplication is stored in an accumulator register instead of in a RF, the pipeline of "result store" can be merged into the pipeline of "execution" for most ASIP DSPs.

11.2.4 Special Registers in a General Register File

Registers in a register file can be either general registers or special registers. The circuit around a general register is only a multiplexer. Special registers could be allocated inside a general register file. In case a register is a general register carrying special register functions, extra logic circuits will be added around the register. This is the typical case in an ASIP when a special register frequently is involved in ALU computing. For example, if a register in the register file is both a general register and an address pointer, addressing logic will be added around this register.

A register file consisting of general and special registers is called a multiple-function register file. The circuit design for a multiple-function register file starts from the circuit of a general register file. Figure 11.4 is an example of a register file including an address pointer as a special register. Address pointer design will be discussed in Chapter 15. In this chapter, before discussing the principle of the addressing logic, the extra circuit for addressing is simply copied and merged into a general register circuit. The purpose is to show how a circuit is added to a general register.

The original general register circuit ((a) in Figure 11.4) is the copy of one register circuit in Figure 11.2. The modified circuit ((b) in Figure 11.4) consists of a circuit block drawn using thin lines and another block of circuit drawn using thick lines. The block of circuit with a thin line is a copy of Circuit (a), which functions as a general register. The block of circuit with thick lines in the dash line block is the addressing logic, which supplies extra function so that the register can be used as both a general register and a special register. Functions inside the dashed line will be discussed in detail in Chapter 15.

Based on the case in Figure 11.4, the way to add special register functions into a general register can be depicted in Figure 11.5.

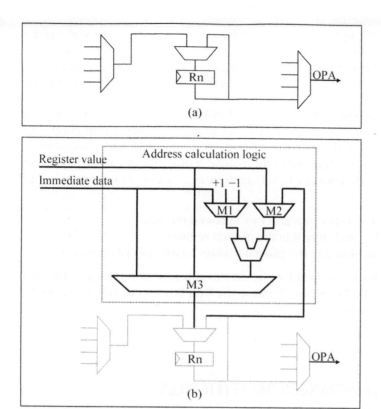

FIGURE 11.4

Integration of addressing logic in a general register file.

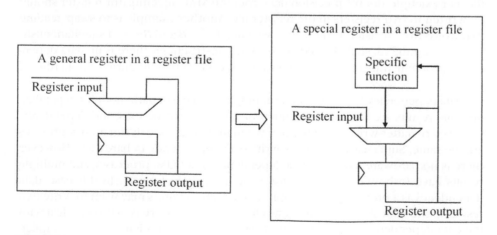

FIGURE 11.5

Design of a special register in a general register file.

The advantage is the opportunity to directly manipulate data in a special register when it is integrated in a general register file. Two drawbacks can be found. The first is that there are fewer general-purpose registers available when using the register as a special register. The other drawback is the extra complexity added to the compiler.

Not all special registers can be integrated inside the general register file. For example, special registers of peripheral devices usually are allocated in peripheral modules. Loop counter and stack registers usually are allocated in the control path. If a special register is not integrated in the general register file, it cannot be used as an operand supplying data directly to ALU. In this case, the data process for a special register is:

1. Move the data in a special register to a general register.
2. Process the data and store it in the general register.
3. Move the result stored in the general register to the special register.

Three clock cycles are needed for one data manipulation in this case. This is a typical solution for a DSP processor because the data in a special register is usually not needed very frequently.

11.3 DESIGN OF ADVANCED REGISTER FILES

11.3.1 Register File for Cluster Datapath

There are two cases where multiple data can be written to a register file simultaneously. In the first case, a special instruction writes two or more data to a register file. For example, double-precision data from the MAC accumulator register should be written to two registers in the register file. Another example is to swap, reading two words from register A and B and assigning $A <= B$ and $B <= A$ simultaneously. When executing these dedicated double-write instructions, there is no data dependency problem; thus these double writes are easy to handle during the hardware design.

Data dependency happens when running multiple datapath modules in parallel. Multiple results are from the execution of multiple instructions, and dependency such as write-after-write might happen when more than one destination addressees are the same. Superscalar managers write data dependency in hardware. However, there is no hardware handling data dependency in a VLIW processor, and multiple results have to be written to a register file in one clock cycle. In this case, data dependency is a problem. For example, two execution units may want to write two results to the same register in the same clock cycle. There are two ways to deal with the data dependency. One way is to manage all data dependency in the compiler. Another way is to design hardware to take care of data dependency problems in runtime and to avoid resource contention.

The hardware avoiding data dependency during writing results is called a cluster datapath. In a cluster datapath, a register file is divided into several subregister files. Each result bus of a datapath cluster is connected to only one subregister file. A result of each datapath cluster can be written to only one subregister file, and the data dependency problem can be naturally avoided. Each subregister file sends data to all datapath clusters, so that any datapath cluster can get data from any subregister file; enough flexibility is available.

Figure 11.6 gives an example of a two-way datapath in a VLIW processor. The datapath is divided into two clusters. The general register file is divided into two isolated register files (RF-I and RF-II) and assigns them to two clusters (datapath-I and datapath-II). Therefore, datapath I can write its result to only the left register file (RF-I), and the datapath II can write data to only the register file on the right side (RF-II). Using this architecture, cross reading (datapath I reading data from register file in datapath II) operands from another cluster is allowed.

In Figure 11.6 , both A and B are double-word buses supplying two operands or a long operand. The drawback might be the imbalanced usage of each subregister file. For example, sometimes results to RF-I could be much more than the results to RF-II. The number of registers sometimes in RF-I might not be enough, and the number of registers in RF-II might be more than enough at the same time.

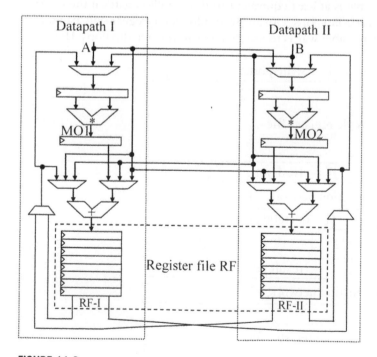

FIGURE 11.6

Write separation using two register files.

11.3.2 **Ultra Large Register File**

A more flexible register file structure is depicted in Figure 11.7. The hardware circuit gives all arithmetic and logic modules (ALU1 to ALUn) permission to read data from any register and to write data to any register. To avoid data dependency, the write of an ALU result from each ALU module must be regulated so that write dependency can be avoided during the coding phase (the way used for VLIW). If predefined associations between ALU modules are fixed during instruction specification, the datapath becomes SIMD (Single Instruction Multiple Data).

There are two connection networks performing permutation operations while fetching operands and writing results. The connection network-I gives flexibility through permutation of operands. For example, registers 1 to 8 can be selected through port "a" to the connection network-I. The connection network can switch "a" to any operand port of ALU1 to ALUn. In this way, any register can be assigned to any operand port of any datapath. The connection network-II gives flexibilities to writing results. For example, through connection network-II, all result ports of ALU1 to ALUn can be connected to any write port of the wide register file. Thus any result port can write data to any register.

The silicon cost of an ultra large register file is huge. A large register file in a video processor could be several thousands words. The silicon cost of a 16k 16b word general register file is at least equivalent to three million gates if the register file is implemented using flip-flops. Obviously, an ultra large register file cannot be based on flip-flops. A possible way is to use memory to replace flip-flops. However,

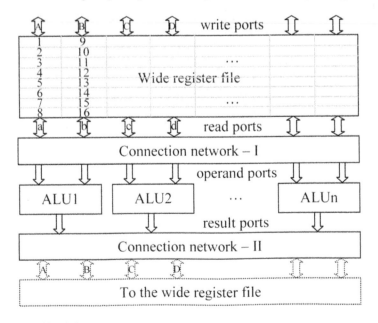

FIGURE 11.7

Parallel datapath using a wide register file.

memories cannot be accessed multiple times in one clock cycle. To do that, multiple memories have to be used to supply multiple accesses. A one-write multiread register file can be implemented using memory blocks in Figure 11.8. When writing a data word to the register file, the data comes to and is written into all SRAM blocks from 1 to n. When reading data, each SRAM block can be addressed separately so that each SRAM can give one random operand word. In Figure 11.8, n operand words can be supplied from the register file.

When the size of an SRAM is rather large, the gate count of the SRAM could be estimated as the number of bits divided by two. The equivalent gate count of an SRAM with 16k × 16b = 256k bits is roughly 128k gates. It is only 5-10% of the gate counts of the 16k ×16b register file based on flip-flops. Considering routing efficiency, the relative silicon costs of the ultra large register file using SRAM will be even lower. Here the routing efficiency is the total silicon cost without interconnections over the total silicon cost with interconnections. It is less than 0.6 for a register file and more than 0.9 for a 16k ×16b SRAM. Including the cost of interconnections, the silicon cost of a single operand SRAM register file is about 1/15 of the silicon costs of a 16k ×16b register file using flip-flops. Obviously, using SRAM, the silicon costs of an 8-port 16k ×16b SRAM register file shown in Figure 11.8 is about the half of the same register file using flip-flops shown in Figure 11.7.

In the same way, multiple read and multiple write register files using SRAM can be simply implemented in Figure 11.9.

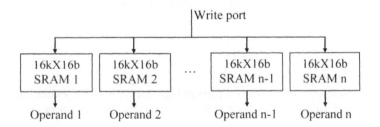

FIGURE 11.8

Ultra large register file using SRAM.

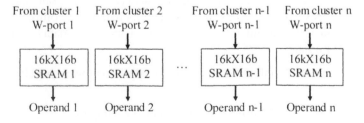

FIGURE 11.9

Multi-in multi-out register file using SRAM.

To reduce the silicon costs of an ultra large register file, latches can be used to replace flip-flops. Readers can find design examples from most VLSI books. Be careful when introducing latches in a design. Existing latches in a design will be warned by synthesis tools because a latch is not really a reliable storage device. Latches can be introduced to a register file with a special BIST (built-in self test). However, latches shall never be used in a normal synchronous logic circuit.

11.4 CONCLUSIONS

The design of a datapath is based on experience. Hardware multiplexing is an essential technology to design a datapath. The register file consumes both significant silicon area and power. In embedded systems, a low-power design technique is essential so that a register file within an ASIP must be designed with high efficiency and low costs. Extra latency from fan-out of logic circuits around register in/out ports should be predicted during the microarchitecture design phase.

When writing multiple data to a general register file, the data dependency problem must also be taken into account because supporting reads and writes in parallel is essential.

A register file may give the major impact on power consumption. During the functional microarchitecture design, a low power design techniques should be used. Synthesis for clock gating should be performed during logic synthesis.

EXERCISES

11.1   Find the maximum fan-out of all control pins in Figure 11.2 to drive a general register file with $8 \times 16b$ registers, $16 \times 16b$ registers, $32 \times 16b$ registers, and $64 \times 16b$ registers.

11.2   Design a register file ($32b \times 16$) for a baseband signal processor. The register file can supply either two 16-bit integer words or two 32b ($2 \times 16b$) complex words. The register file accepts either one complex data ($2 \times 16b$) or one real data (16b) in one clock cycle. The register file accepts data from complex-MAC (16-bit real and 16-bit imaginary data), real ALU (one 16-bit data), and data memories (32-bit data).

11.3   Is the circuit in Figure 11.7 only for VLIW, only for SIMD, or for both VLIW and SIMD?

11.4   Can the circuit in Figure 11.7 be used for multicluster datapaths? If it can be used for multicluster datapaths, how is it used? If it cannot be used for multicluster datapaths, how can it be changed so that it can eventually be used for multicluster datapaths?

11.5 Draw a schematic circuit of a 24b (processor data bus width) register file. The register file has four registers (R0, R1, R2, and R3). The register file gets inputs from register file port A [23:0] (OPA), the result bus of the ALU, the result bus of the MAC, M [23:0], data from the external port P [15:0] with left alignment, data from data memory D [23:0], and data from coefficient memory T [23:0]. The register file supplies two operands at the same time; the operands' names are A [23:0] and B[23:0].

11.6 Based on the gate count estimation of the SRAM register file in Figure 11.8, estimate the silicon costs when the register file size is 8k × 16b.

REFERENCES

[1] Khailany, B., Dally, W. J., Kapasi, U. J., Mattson, P., Namkoong, J., Owens, J. D., Towles, B., Chang, A., Rixner, S. (2001). Imagine: Media processing with streams. *IEEE Micro*, 35–46.

[2] Pirsch, P. (1997). *Architecture for Digital Signal Processing*. John Wiley & Sons.

[3] Lapsley, P., Bier, J., Shoham, A., Lee, E. A. (1997). *DSP Processor Fundamentals, Architectures and Features*. IEEE Press.

[4] Weste, N. H. E., Eshraghian, K. (1997). *Principles of CMOS VLSI Design, 2nd edition*. Addison-Wesley.

ALU HW Implementation

A datapath in a DSP processor typically consists of ALU, MAC, RF, and instruction-level acceleration units. In this chapter, implementation of the ALU hardware will be discussed based on examples.

12.1 ARITHMETIC AND LOGIC UNIT (ALU)

In a real DSP processor, an ALU could cover more functions such as shift-rotation functions, bit manipulation functions, and some application-specific functions. ALU gets operands from a RF and sends results to a RF. Usually an ALU executes only RISC instructions, meaning that all operands to ALUs are from RFs, and the execution cost of each ALU instruction is one clock cycle. Iterative instructions are assigned to a MAC unit. In most cases, ALU operands and results are with native data width (the data width of register and memory bus). Double-precision computing usually is executed in a MAC unit.

Figure 12.1 gives a top view of an ALU. Two 16-bit inputs are selected for ALU. Inputs of an ALU can be from the register file (RF) and as immediate data from the control path and carried by an instruction. The preprocessing of operand A and operand B includes guard insertion, operand inversion, and preparing for the carry-in. After preprocessing, operands will be shared by all the processing units, including the arithmetic unit, the shift unit, the logic unit, and the special function unit. Each processing unit gives computing on the same operands after preprocessing. One of the outputs from processing units will be selected for postprocessing. The postprocessing includes saturation on the result and the flag computing.

An ALU could be part of the MAC in an early DSP processor in the 1980s when gate count was critical. Today, the silicon cost is not so expensive when using modern silicon technology. For example, by using a 65-nanometer digital CMOS process, more than 200 k gates or 400 k bits memory cells can be allocated within 1 mm^2. The silicon cost of an ALU of a 16-bit DSP processor is less than 5 k gates. Therefore ALU hardware has been separated from the MAC hardware in modern DSP processors. This separation can bring advantages such as parallel execution and shorter silicon verification time.

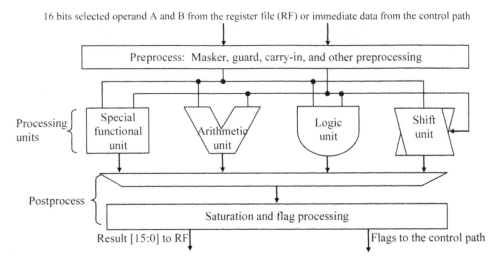

FIGURE 12.1

ALU schematic.

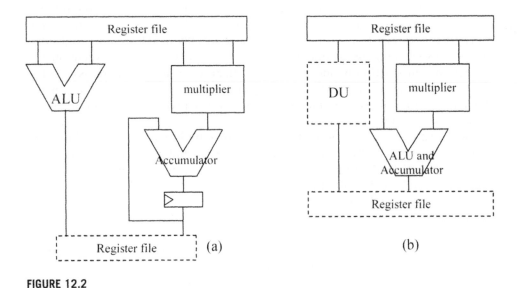

FIGURE 12.2

(a) ALU separated from MAC, (b) ALU inside MAC.

Figure 12.2 (a) shows the most commonly used architecture of DSP processors. In Figure 12.2 (b), the full adder after the multiplier is used both for accumulation and ALU functions. Other data manipulations such as logic and shift operations might be implemented in the accumulator or in another functional unit DU (data manipulation unit) beside the MAC. An ALU usually supports single-precision arithmetic operations, and the MAC usually supports iterative operations with double-precision results because the result of a single-precision multiplier supplies double precision.

Other than running regular arithmetic and logic operations, the ALU could be designed to support irregular functions, such as special acceleration for coding or decoding based on bit manipulations, data quality monitoring and enhancement, and packaging and depackaging data packets.

Compared to the critical path of a MAC unit and the memory access path, path delay in an ALU is much shorter. Path delay induced by operand fetching to the ALU will not be timing critical, and preprocessing of operands can therefore be scheduled together with the ALU operation within one clock cycle.

12.2 DESIGN OF ARITHMETIC UNIT (AU)

An arithmetic unit for arithmetic ($+$, $-$, $++$, $--$, average, absolute, negate, etc.) consists of one or multiple full adders with surrounding logic circuits for hardware multiplexing and operand preprocessing.

12.2.1 Implementation Methodology

Recall microarchitecture design fundamentals discussed in Chapter 10. The basic steps of designing hardware module microarchitecture in an ASIP are:

1. Collect instructions running in the module (AU in this section).

2. Collect all microoperations of these instructions (arithmetic ops).

3. Remove microoperations not allocated to the module.

4. Decide kernel operations and select the proper kernel component.

5. Plan for hardware multiplexing.

6. Design preoperations following the instruction manual.

7. Design postoperations following the instruction manual.

8. Verification.

The procedure of microarchitecture implementation of a data processing module is illustrated in Figure 12.3.

Each instruction contains multiple microoperations. Only part of them will be implemented in one module; the remaining microoperations will be implemented in other modules. By analyzing microoperations allocated in the module, kernel operation will be decided. Preoperations and postoperation functions will then be implemented around the kernel components.

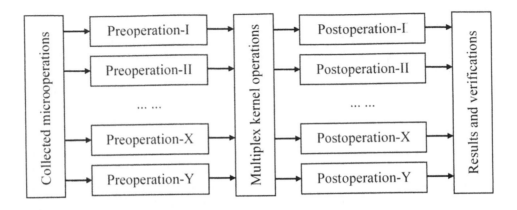

FIGURE 12.3

Mapping functions to a hardware module.

12.2.2 Select Kernel Components

The Precision of the Kernel Component

A kernel component is a RTL primitive that performs the essential function of a module, consumes the majority of the silicon cost of the module, and is multiplexed to minimize the silicon cost. It is usually a primitive of arithmetic operation, such as a parameterized adder (or full adder) or a multiplier. The kernel component selection includes three steps:

1. Components selection.

2. Determination of precision.

3. Determination of the number of kernel components required.

Most arithmetic operations are based on additions; obviously, the kernel components in AU are full adders. What we need to decide now is the precision of each full adder and how many full adders are needed.

As discussed in Chapter 2, one guard bit is required for single-step arithmetic computing, and multiple guard bits are required for iterative computing. To be able to get carry-out and be able to run saturation on the results of the AU, a guard bit is required in each operand before the arithmetic operations. The precision of the full adder is at least 17 bits assuming 16-bit operands. Normally, a full adder as a primitive may or may not have a carry-in bit (depending on the cell library). Thus it is good to design RTL code adapting to all possible cell libraries by using the LSB of the full adder as the carry-in bit. To conclude, the internal operand data widths of a 16-bit AU is 18-bit.

```
IKA [17:0] <= {A [15], A [15:0],"1"}// Internal kernel OPA = IKA
IKB [17:0] <= {B [15], B [15:0],CIN}// Internal kernel OPB = IKB
```

IKA and IKB are Internal Kernel operands A and B. A and B are inputs of the arithmetic unit. After computing, the LSB of the result from the 18-bit adder is truncated because it was only used to support the carry-in on the input side, with no meaning on the output. The useful 17-bit output from the full adder is FAO [17:1]. The principle of using an 18-bit full adder as the computing primitive for 16-bit arithmetic computing is depicted in Figure 12.4.

The data width of inputs A and B to the adder primitive is 16 bits. The fixed-point note $Q_D(M, F)$ is $Q_2(16,0)$ for integer representation and $Q_2(1,15)$ for fractional representation. After adding one guard bit and one carry-in bit, fixed-point note $Q_D(M,F)$ becomes $Q_2(17,1)$ for integer representation and $Q_2(2,16)$ for fractional representation. The result [16:0] including a guard bit can be represented as $Q_2(17,0)$ or $Q_2(2,15)$. The Q notation for data representation was introduced in Chapter 2. The guard bit of the result [16] can be used as the carry-out bit, and Result [16:15] can be used for saturation operation.

The Number of Required Kernel Components

Special AU instructions listed in Table 12.1 are good examples to illustrate the way to decide how many kernel components should be used.

To implement MAX and MIN, one full adder will be enough because the MAX or MIN function is operand selection according to the comparison of the two operands. The full adder will be used as a comparator.

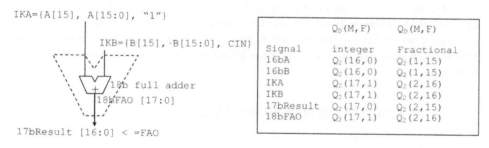

FIGURE 12.4

18-bit full adder is the primitive in a 16-bit arithmetic unit.

Table 12.1 Special Arithmetic Instructions.

Mnem	Description	Operation
MAX	Select larger value	GR ← max {GR, GR}
MIN	Select smaller value	GR ← min {GR, GR}
DTA	Difference of two absolute values	GR ← |GR| − |GR|
ATD	Absolute of the difference	GR ← |GR − GR|

It seems that three full adders are needed for DTA function ($|A| - |B|$) because one adder is for $|A|$, one is for $|B|$, and the other is for the subtraction. When $|A| - |B|$ is decomposed, it will be identified that four cases can be defined according to the sign bits of A and B: A15B15 $<=$ {A [15], B [15]}, which can be specified using the code in Listing 12.1.

LISTING 12.1 Verilog code for $|A| - |B|$.

```
case(A15B15)
     // Both positive: Plain subtraction, A-B
     2'b00: result = A + INV(B) + 1;

     // A positive, B negative: Plain addition, A-(-B)=A+B
     2'b01: result = A + B;

     // A negative, B positive: (-A) + (-B) = INV(A)+1+INV(B)+1
     2'b10: result = INV(A) + INV(B) + 2;

     // Both negative: Plain subtraction, B-A
     2'b11: result = INV(A) + B + 1;
endcase
end process;
```

There are two "+" operators: one is for normal addition and the other is for carry-in. If the carry-in is only 1, we need only one adder. However, the carry-in could be 2, so we have to use two full adders. In a similar way, instructions DTA and ATD are also analyzed, and the conclusion is that we do not need more than two full adders.

12.2.3 Implementing Simple AU Instructions

In this section, the implementation will be discussed for an arithmetic unit using a kernel computing device as a full adder given in Figure 12.4. Some simple arithmetic instructions in Table 12.1 will be the design input.

In Table 12.2, OP is the operation code, CINC is the control code selecting carry-in, and SAT is the code of saturation. If each instruction is implemented without hardware multiplexing, 11 instructions are implemented into the 11 circuits in Figure 12.5.

In Figure 12.6, the 11 circuits in Figure 12.5 are integrated, and each instruction can be executed by selecting the right control signal. Each S block represents saturation. For example, if we want to execute A $-$ B with saturation, we can simply select the control signal $= e$.

The solution in Figure 12.6 is obviously not a good solution because it consumes too much hardware, including 11 18-bit full adders, four inverters, and

Table 12.2 Simple Arithmetic Instructions.

	Instructions	Function	OP	CINC	SAT
1	ADD SAT	A + B with saturation	000	00	1
2	ADD COUT	A + B without saturation	000	00	0
3	ADD CIN SAT	A + B + C in with saturation	000	1x	1
4	ADD CIN COUT	A + B + C in without saturation	000	1x	0
5	SUB SAT, CMP	A − B with saturation and compare	100	01	1
6	SUB COUT	A − B without saturation	100	01	1
7	ABS(A)	ABS(A) Absolute operation, saturation	111	00	1
8	NEG(A)	NEG(A) Negate operation, saturation	101	00	1
9	INC(A)	Increment and saturation	001	00	1
10	DEC(A)	Decrement and saturation	011	00	1
11	AVG	(A+B)/2 Average operation, saturation	010	00	1

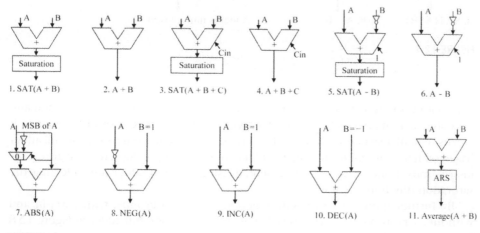

FIGURE 12.5

Direct mapping of simple arithmetic instructions to hardware.

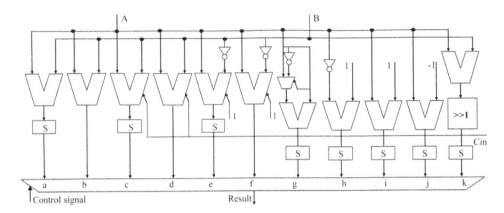

FIGURE 12.6

Simple AU implementation without hardware multiplexing.

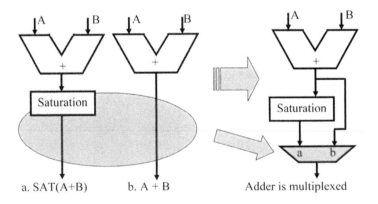

a. SAT(A+B)   b. A + B   Adder is multiplexed

FIGURE 12.7

Merge of SAT (A + B) and A + B instructions.

eight saturation function blocks. Actually, only one instruction can be selected and executed at a time, so that one full adder will be enough, instead of 11 full adders. It means that all instructions in Table 12.2 could be implemented using one full adder based on hardware multiplexing technique. In the example depicted in Figure 12.6, instructions 1 and 2 can be implemented using one full adder and selecting the saturation function in Figure 12.7.

By further grouping arithmetic functions into two groups, with carry-in and without carry-in, more instructions can be mapped to one full adder in Figure 12.8 by selecting different values as the carry-in signal.

So far, instructions 1 to 4 were implemented using only one full adder. Looking into instruction 5 and 6, the differences between the first four instructions can be

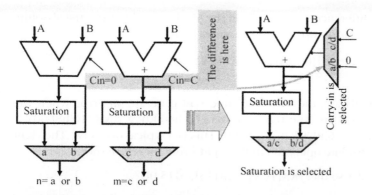

FIGURE 12.8

Merging instructions into one hardware unit by selecting carry-in.

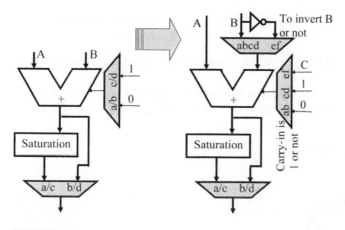

FIGURE 12.9

Merging the add and subtract instructions into one hardware unit.

identified—that is, to either invert or not invert operand B, and to either supply 1 to carry in or not. The implementation of the subtract instruction is given in Figure 12.9.

So far on the right in Figure 12.9, six instructions are mapped to the circuit with only one 18-bit full adder. The implementation of carry-in was discussed in Figure 12.4; the carry-in is actually the concatenation of IKA = {A [15], A [15:0], 1'b1} and the concatenation of IKB = {B [15], B [15:0], carry_in}.

The next instruction, compare A to B, actually has the same operation as the previous instruction, A − B with saturation. The difference is that the write to destination is disabled, but this difference is not visible in the AU. Instruction (g) is the absolute operation. Intuitive absolute operation on operand A is:

```
if(A[15]) result = INV (A[15:0]) + 1;
else result = A[15:0];
```

In fixed-point hardware, the behavior code will generate an error when the input is A [15:0] = 16h8000. The result of ABS (16'h8000) = INV (16'h8000) + 16'h0001 = 16h7FFF + 16h0001 = 16h8000. This is a typical overflow. The right implementation of absolute operation in the right order therefore should be:

1. Guard operand A so that IKA[17:1] = {A[15], A[15:0]}.

2. ABS operation on IKA[17:1] instead of A[15:0].

3. Saturation operation after ABS operation.

The right hardware behavior code of absolute operation on A should be:

```
if(A[15]) result = saturate (INV({A[15], A[15:0]}) + A[15]);
else result = A[15:0] + A[15] /* A[15]=0 in this case */.
```

The hardware implementation of the absolute operation on A is shown in Figure 12.10.

The Average operation in Table 12.2 is somewhat simplified because it does not round the result. The operation result is $2^{-1}* (A + B) = (A + B) >> 1$. Here 2^{-1}

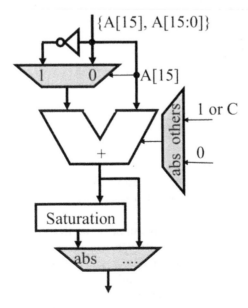

FIGURE 12.10

Implementation of an absolute operation.

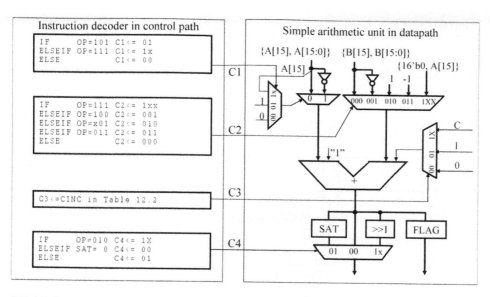

FIGURE 12.11

Implementation of simple arithmetic instructions with HW multiplexing.

is one bit arithmetic right-shift. If the shift-out bit is one, the result will be slightly less than the value it should be. A more complete Average operation includes an implied round operation, and it will be discussed later in this section. By integrating all functions with hardware multiplexing, the arithmetic unit is implemented, which supports 11 instructions in Table 12.2 with minimum hardware cost.

The instruction decoder in the control path supplies correct control signals according to the instruction codes. By implementing those four if-else-then blocks on the left side in Figure 12.11 at RTL level, control signals will be supplied according to the running instruction.

Extra logic delay will be introduced after hardware multiplexing. Normally, the extra logic delay around ALU in Figure 12.11 will not be the critical path of the DSP processor.

12.2.4 Implementing Special AU Instructions

A few special AU instructions were listed in Table 12.2. Let us start from implementing the most complicated instruction, DTA, differences of two absolute values. An intermediate control variable, the sign bits of A and B:$A15B15 <= \{A2[15], B2[15]\}$, is defined first. The instruction implementation according to four cases of $A15B15$ was discussed and shown in 12.1.

By using two adders, $|A| - |B|$ is implemented by using the code $A15B15$ written as just shown. The logic function of $|A| - |B|$ is implemented in Figure 12.12 following the description.

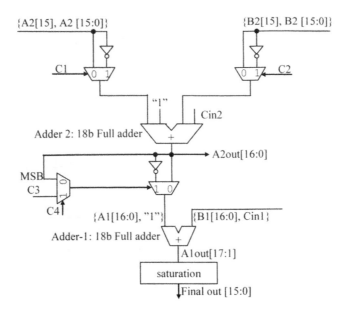

FIGURE 12.12

Implementing |A| − |B| into hardware.

Table 12.3 Implementing |A| − |B|.							
A1[15]	**B1[15]**	**Operation**	**C1**	**C2**	**C3**	**C4***	**Cin1**	**Cin2**
0	0	A + INV(B) + "1"	0	1	0	1	1	0
0	1	A + B	0	0	0	1	0	0
1	0	INV(A) + INV(B) + "2"	1	1	0	1	2	0
1	1	INV(A) + B + "1"	1	0	0	1	1	0

Control signals are decoded according to {A1[15],B1[15]}, and the decoding is given in Table 12.3. Notice that C4 is marked with "*". It is not a control variable for |A| − |B| in Table 12.3. It will be a control variable for the instruction |A − B| in Table 12.4.

C4 in Table 12.3 will be used by other functions. It is not efficient to use the complicated circuit depicted in Figure 12.12 only for the calculation of |A| − |B|. The hardware in Figure 12.12 can be multiplexed to execute other instructions. For example, |A − B| is implemented by assigning C1 = 0 and C2 = 1 and Cin2 = 1 to get A2out <= A − B. Control signals for instruction |A − B| are specified in Table 12.4. The execution of | A − B| in Figure 12.12 becomes:

```
A2out = A + INV (B) + 1;
if(A2out[MSB]) A1out = INV (A2out) + 1;
else A1out = A2out;
```

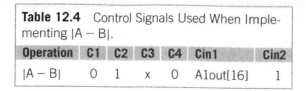

Table 12.4 Control Signals Used When Implementing |A − B|.

Operation	C1	C2	C3	C4	Cin1	Cin2
|A − B|	0	1	x	0	A1out[16]	1

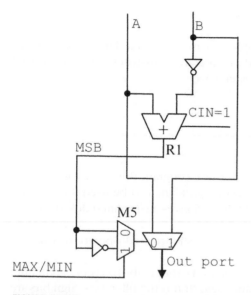

FIGURE 12.13

Implementation of MAX/MIN function in an arithmetic unit.

The MAX (A, B) and MIN (A, B) functions are implemented in Figure 12.13. The MSB (guard bit) of the result from comparing A and B is the control signal to select A or B as the output. For example, when executing the MAX function, the M5 is controlled by the guard bit and the selection of an operand is on the out port. When A >= B, A is selected as the output because the guard of the full adder is 0.

The Average function can be implemented by performing an arithmetic right-shift on the output of the full adder. If the LSB (shift out) simply is truncated, the truncation may introduce an offset error (discussed in Chapter 2). The rounding operation using the shift-out bit can eliminate the truncation error. To eliminate the truncation offset error, an extra full adder is needed for rounding. Fortunately, we have two full adders in the AU because the |A − B| and |A| − |B| functions require two full adders. Therefore, the Average function is implemented as depicted in Figure 12.14.

FIGURE 12.14

Implementation of average operation with round.

Based on discussions above and the implementations depicted from Figure 12.11 to Figure 12.14, the readers are suggested to integrate all instructions in Table 12.2 to one circuit using two full adders as kernel devices.

12.3 SHIFT AND ROTATION

Shift and rotation can be used for scaling, bit test, program flow control, protocol package processing, and different kinds of coding. It can also be used for software division and other arithmetic operation. Table 12.5 gives a simplified description of shift and rotation functions.

FAC here stands for the carry flag of ALU. The shift-related operations can also be represented using Figure 12.15.

N in Table 12.5 is the number of shifts or rotations; the symbol {} is concatenation. The difference between arithmetic shift and logic shift is the fill-in bits. Sign bits are kept after the execution of arithmetic shift. Zero will be filled in after the execution of logic shift. Arithmetic left-shift seldom is used, yet can be found in some other DSP processors (not in **Senior**). It is used for "up-scaling." Since this kind of scaling may introduce overflow errors, firmware designers should be careful when using it. The result of arithmetic left shift should be saturated if the significant bits are shifted out. When executing rotation includes carry, the carry flag is inserted during the rotation.

Shift is controlled by a shift control vector. Usually, the shift control vector specifies the number of bits to shift, and it is part of operand B. Four control bits are required for a 16-bit barrel shifter. A shift control vector (operand B) can be an immediate data if the number of shifts is known during programming. A shift control vector can be a variable in a register in RF.

There are basically two ways to implement the shift and rotation hardware. One is to implement the shifter and rotator using a shifter primitive. Another way is to implement the shifter and rotator based on a truth table. If the synthesis tool is advanced, we get roughly the same gate cost. If the synthesis quality cannot be guaranteed, a design based on the primitive consumes less silicon.

Table 12.5 Shift and Rotation Functions.

	Function	Result Specification
1	Arithmetic right shift	{ $N'A[15]$, $A[15 - N:N]$ }; $FAC <= A[N - 1]$
2	Logic right shift	{$N'0$, $A[15 - N:N]$}; $FAC <= A[N - 1]$
3	Logic left shift	{$A[15 - N:0]$, $N'0$}; $FAC <= A[15 - N + 1]$
4	Right rotation with FAC	{FAC, $A[15 - N:N]$} if $N = 1$; {$A[N - 2:0]$, FAC, $A[15 - N:N]$} if $N > 1$; $FAC <= A[N - 1]$
5	Right rotation without FAC	{$A[15 - N:0]$, $A[15:15 - N + 1]$}
6	Left rotation with FAC	{$A[15 - N:0]$, FAC} if $N = 1$; {$A[15 - N:0]$, FAC, $A[15:15 - N + 2]$} if $N > 1$; $FAC <= A[15 - N + 1]$
7	Left rotation without FAC	{$A[N - 1:0]$, $A[15 - N:N]$}

12.3.1 Design a Shifter Using a Shifter Primitive

A shifter primitive is a RTL component consisting of $\log_2 n$ stages of multiplexers. Here "n" is the data width to shift. A typical shifter primitive is illustrated in Figure 12.16.

A shifter primitive is a right shifter. A 16-bit shift requires four control bits. The CTRL [LSB] selects either the same bit or the left neighboring bit (with 2^0 bits offset of the same bit). The control signal of LSB (CTRL [LSB]) determines whether to shift 1 bit (2^0 bit) or not. The CTRL [1] selects either the aligned bit or the bit next to the neighboring bit on the left, which is the bit with 2^1 offset on the left, shifting either 0 bit or 2^1 bit. In the same way, the CTRL [2] selects either the aligned bit or the $2^2 = 4^{th}$ bit on the left side, shifting either 0 bit or 2^2 bit. Finally, the CTRL [MSB] selects either the aligned bit or the $2^3 = 8^{th}$ bit on the left.

Using different combinations of control bits, all possible shifts can be obtained from 0 bit to 15 bits. For example, to shift seven bits, control signal 0111 should be used. Thus the first level MUX will shift one bit, the second level MUX will shift two bits, and the third level MUX will shift four bits. Since there is no shift on the last level MUX, the final number of shifts is $1 + 2 + 4 + 0 = 7 = 0111 \, b$.

Except for the control signals and the data input signals, the third group of inputs are the fill-in bits, which are supplied to fill-in the empty positions after the shift

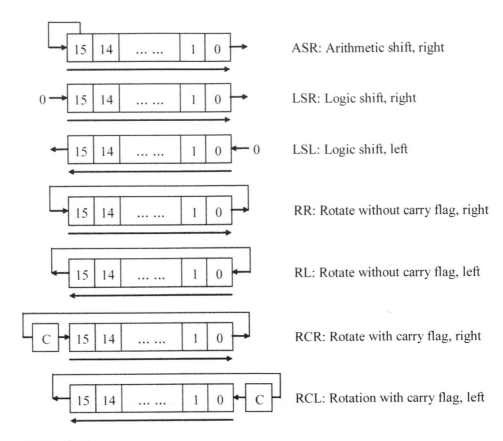

FIGURE 12.15

Shift and rotation operations.

operation. By supplying different fill-in bits, different kinds of shift or rotation can be implemented. For example, for the arithmetic right-shift, all fill-in bits should be the MSB, sign bit, of the input. If all fill-in bits are connected to the ground, the shifter becomes a logic right shifter.

The shifter primitive itself is a right shifter. To implement a logic left shift, a swap on input bits is required; swaps include LSB to MSB, the next LSB to the next MSB, and so on. A barrel shifter is depicted in Figure 12.17.

The circuit in Figure 12.17 can be used for right-shift when Left_right is R and left-shift when Left_right is L. Arithmetic and logic shift can be controlled using Fill_in_control. When it is A, the sign bit is the fill-in for arithmetic shift; when it is L, zero is the fill-in for logic shift. To execute rotation, a fill-in table is required.

On a behavioral level, the rotation operation is to assign the shift-out bits from one side of the shifter as the inputs to another side of the shifter (see Figure 12.15). It is not allowed in RTL code because it forms a combinational loop. A combinational loop is an illegal design and cannot be compiled by any HDL compiler. The right way

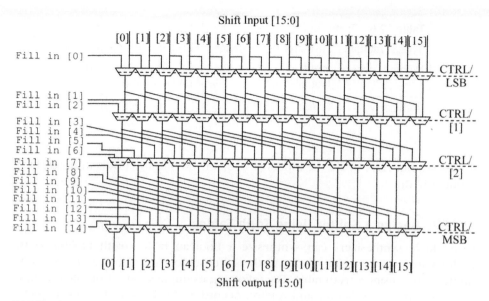

FIGURE 12.16

A 16-bit shifter primitive.

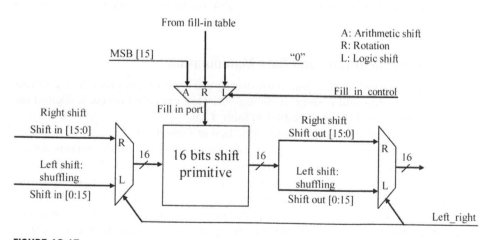

FIGURE 12.17

Right-shift and left-shift based on one shifter primitive.

is to design a fill-in table for rotation by using corresponding input bits. Table 12.6 gives an example of right rotation and left rotation.

For example, the fill-in bits for right-rotate 2 bits without carry-in are the inputs [1:0], or the output [15:0] = {input [1:0], input [15:2]}. The fill-in bits for left-rotate

Table 12.6 Truth Table of Fill-in Bits for Rotation without Carry-in.

Type of rotation	Fill-in for right rotation	Fill-in for left rotation
Without carry	input[15:0], [0] is the first fill-in bit	input[15:0], [15] is the first fill-in bit
With carry	{input[14:0], FAC}	{FAC, input[15:1]}

4 bits without carry-in are the inputs [15:12], or the output [15:0] = {input [11:0], input [15:12]}.

12.3.2 Design a Shifter Using Truth Tables

Good logic synthesizers can synthesize a Boolean table (truth tables) to the netlist with minimum gate count. All the shift-rotation functions can therefore be implemented using a truth table. The following example shows a simplified implementation of a 4-bit shifter and rotator. According to the specification depicted in Figure 12.15, the implementation of a shifter using a truth table is given in Table 12.7.

FAC is the Carry flag of ALU. There should be another table to generate the new FAC of the shift result; try to design a table of the new carry flag after shift or rotation.

12.3.3 Logic Operation and Data Manipulation

Logic operations and bit manipulations usually are used for program flow controls, protocol processing, and coding/decoding. They are executed on each aligned bit-pair of operand A and B and specified in Table 12.8.

Functions of bit manipulations are specified in Table 12.9.

Two operands are required for bit manipulation. One operand contains data to be manipulated. Another operand is the mask selecting the bit to be manipulated. At least four bits are needed to point to one bit to process. Field manipulation functions are similar to the bit manipulation specified in Table 12.9. The difference is the pointer. Only one bit should be pointed when coding a bit manipulation instruction. For field manipulation both the starting point and the finish point (or the length) of the field should be provided. Field manipulations are specified in Table 12.10.

Again, two operands are required for field manipulation; one carries data to be manipulated, and the other carries the position and length of the field to be processed. The control word will be used to select a field from m to n to be manipulated.

The implementation of the logic part depicted in Figure 12.18 is very simple. Since no saturation operation on result is required, guard bits are not needed.

Table 12.7 Table Used to Implement a 4-bit Shifter/Rotator.

N of shift or rotation	Result bit	ASR	LSR	LSL	RR	RL	RCR	RCL
0	R[0]=	A[0]	A[0]	A[0]	A[0]	A[0]	A[0]	A[0]
0	R[1]=	A[1]	A[1]	A[1]	A[1]	A[1]	A[1]	A[1]
0	R[2]=	A[2]	A[2]	A[2]	A[2]	A[2]	A[2]	A[2]
0	R[3]=	A[3]	A[3]	A[3]	A[3]	A[3]	A[3]	A[3]
1	R[0]=	A[1]	A[1]	0	A[1]	A[3]	A[1]	FAC
1	R[1]=	A[2]	A[2]	A[0]	A[2]	A[0]	A[2]	A[0]
1	R[2]=	A[3]	A[3]	A[1]	A[3]	A[1]	A[3]	A[1]
1	R[3]=	A[3]	0	A[2]	A[0]	A[2]	FAC	A[2]
2	R[0]=	A[2]	A[2]	0	A[2]	A[2]	A[2]	A[3]
2	R[1]=	A[3]	A[3]	0	A[3]	A[3]	A[3]	FAC
2	R[2]=	A[3]	0	A[0]	A[0]	A[0]	FAC	A[0]
2	R[3]=	A[3]	0	A[1]	A[1]	A[1]	A[0]	A[1]
3	R[0]=	Fill in this row as your home work						
3	R[1]=	Fill in this row as your home work						
3	R[2]=	Fill in this row as your home work						
3	R[3]=	Fill in this row as your home work						

Table 12.8 Logic Operations.

No.	Function	Result specification
1	Invert	Result <= INV (A [15:0])
2	AND	Result <= A[15:0] and B [15:0]
3	OR	Result <= A[15:0] or B [15:0]
4	XOR	Result <= A[15:0] xor B [15:0]

Table 12.9   Bit Manipulations.

No.	Function	Result specification
1	Invert bit n	Invert one bit at position n and pass the rest of the bits
2	Clean bit n	Clean one bit at position n and pass the rest of the bits
3	Set bit n	Set one bit at position n and pass the rest of the bits
4	Check bit n	Copy bit n to FAC (the carry flag of ALU)

Table 12.10   Field Manipulations.

No.	Function	Result specification
1	Invert a field from bit m to bit n	$\{A[15:m+1], \text{INV A } [m:n], A[n-1:0]\}$
2	Clean a field from bit m to bit n	$\{A[15:m+1], A[m:n]=(m-n)\text{'b0}, A[n-1:0]\}$
3	Set a field from bit m to bit n	$\{A[15:m+1], A[m:n]=(m-n)\text{'b1}, A[n-1:0]\}$

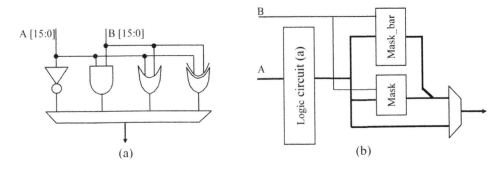

(a)   (b)

FIGURE 12.18

Logic circuits in ALU.

12.4 **ALU INTEGRATION**

12.4.1 **Preprocessing and Postprocessing**

Preprocessing includes operands selection and operations on operands from ALU in-port to a kernel component, including copying guard bits, inverting, selecting carry-in, and generating immediate data. Operands could come from either the register file or from the control path.

Postprocessing includes result selection, saturation, and flag operation. The ALU result is selected from arithmetic unit, shift unit, logic unit, or special function unit. Flag operations can be divided to flag operation and flag prediction. Flag operation is based on the result of ALU, which generally includes the following arithmetic:

1. Zero flag $<=$ 1 if all result bits are zero.

2. Sign flag $<=$ MSB of the result.

3. Carry-out flag $<=$ the lowest guard bit closest to the sign bit.

4. Overflow flag $<=$ when guards and the sign bit of the result are not equal.

Flag prediction is to calculate flags based on operands instead of on the result. It is necessary when the result from the execution unit is time critical or when the flags are required for other operations within the same clock cycle. Some flags can be predicted using input operands together with the control signals of the operation. For example, AU zero flag can be predicted when both operands are zero and the operation code is not INC or DEC in AU. Sign and Carry-out flags are always predicted by the full adder primitive. Therefore, flags can be predicted if necessary.

Carry flag FAC can be generated from both the arithmetic unit and the shifter. During the shift and rotation, FAC is used as the link bit. Saturation is a special flag. FAC flag and saturation flag are not used at the same time, which means that if FAC is enabled the saturation is disabled. The computing of ALU saturation is:

```
if(result[16] != result[15])
  if(sat_result[16])sat_result[15:0] <= 16'h8000; //MAXNEG
  else              sat_result[15:0] <= 16'h7fff; //MAXPOS
else sat_result[15:0] <= result[15:0]; // no saturation.
```

12.4.2 **ALU Integration**

Usually the critical path of a DSP processor is not inside the ALU. The integration of the ALU is relatively easy. Figure 12.19 illustrates an integrated ALU as an example. We would like to emphasize one thing: all material discussed in the ALU section are example-based. For different applications, the ALU could be rather different.

If there are extra available binary machine codes spaces in an ALU instruction, an instruction can carry extra ALU functions on one instruction. Typically, conditions

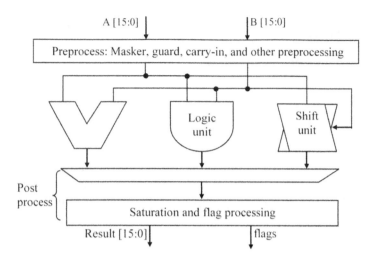

FIGURE 12.19

An example of ALU integration.

can be carried by an ALU instruction, and conditional execution is allowed to minimize the control overhead.

More functions (or instructions) could be merged into one instruction if several ALU operations/instructions are frequently executed together in a fixed sequence. For example, a conditional shift operation frequently is executed after a subtraction (positive integer division). In this case, arithmetic and shift should be conducted sequentially by one instruction, and the hardware of the ALU therefore is integrated and illustrated in Figure 12.20. In this ALU, logic – arithmetic – shift can be executed using one instruction, or individually by control of multiplexers.

However, the execution order of AU, SU, and LU must be prefixed during the instruction set design phase. It is not possible to change the order of execution of arithmetic and logic/shift when they are in the same pipeline stage. If the order of executions of combinational modules within one pipeline is changed by reconnecting inputs and outputs signals, the synthesis tool will think that the design contains combinational loops.

12.5 CONCLUSIONS

Datapath design is to implement microoperations, carried by instructions, to datapath hardware modules. Hardware multiplexing technique is used to minimize the hardware costs. Gate count can be reduced dramatically when advanced hardware multiplexing techniques are used properly.

Microoperations in an ALU, including AU, LU, and SU, are implemented in this chapter based on several examples. The purpose of this chapter is to let you gain

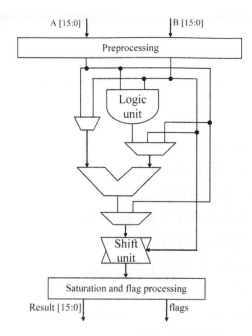

FIGURE 12.20

Another example of ALU integration.

design experience. More homework and lab work are mandatory for students with no real design experience.

EXERCISES

12.1 In which case is the saturation operation needed in AU and in which case not? In which case is the saturation operation needed during shifting and in which case not? Is saturation needed for logic unit? Why?

12.2 Implement $|A[3:0]| - |B[3:0]|$ in one clock cycle using the minimum number of components.

12.3 Design an AU using one full adder and implement the functions and flags "sign," "zero," and "saturated":

a. $A[3:0] + B[3:0] + $ Carry-in

b. Saturate $(A[3:0] + B[3:0])$

c. Saturate $(A[3:0] - B[3:0])$

d. ABS $(A[3:0])$

e. Negate $(A[3:0])$

f. INC (B[3:0])

g. If (A[3:0] >= B[3:0]) result <= A else result <= B.

12.4 Design a 4-bit Barrel shifter primitive circuit (VHDL and schematics) using multiplexers.

12.5 Specify and implement the following functions:

a. Result [3:0] = Arithmetic right shift A [3:0] by B [1:0].

b. Result [3:0] = Logic right shift A [3:0] by B [1:0].

c. Result [3:0] = Logic left shift A [3:0] by B [1:0].

d. Result [3:0] = Right rotation A [3:0] by B [1:0] without carry.

e. Result [3:0] = Right rotation A [3:0] by B [1:0] with carry.

f. Result [3:0] = Left rotation A [3:0] by B [1:0] without carry.

g. Result [3:0] = Left rotation A [3:0] by B [1:0] with carry.

12.6 Design a circuit for the instruction: The result <= max{|A|, |B|}, using circuits in Figure 12.12 and Figure 12.13 as references.

12.7 When executing Result [3:0] <= Absolute (operand A[3:0]) on signed two's complement data, do you need the carry-out flag? Why or why not?

12.8 Design a complete circuit of absolute operation ABS (A [3:0]) on RTL.

12.9 Find errors in Figure 12.21.

Instruction	Function Specification	OP
1	A + B with saturation	0000
2	A + B without saturation	0001
3	A + B + C in with saturation	0010
4	A + B + C in without saturation	0011
5	A − B with saturation	0100
6	A − B without saturation	0101
7	A compare to B with saturation	0110
8	ABS (A) absolute operation	0111
9	NEG (A) Negate operation	1000
10	(A + B)/2 average operation	1001

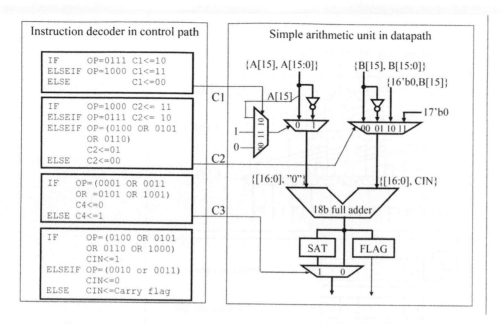

FIGURE 12.21

Question 12.9.

12.10 Design an 8-bit shifter. The shifter shall be designed based on the Shift primitive, on the upper side in Figure 12.22. The design shall include the block diagram using only one BSL as a black box.

   a. Specify the way of left-shift and right-shift using hardware multiplexers.

   b. Specify filling-in Fillin[7] to Fillin[1] for arithmetic right-shift, logic left, logic right-shift, rotate left without carry flag, and rotate right without carry flag using filling-in tables.

12.11 Integrate the following instructions into one circuit based on Figure 12.12 and Figure 12.13.

   a. result $<=$ max {| A |, | B |}

   b. result $<=$ | A | $-$ | B |

   c. result $<=$ | A $-$ B |

   d. result $<=$ max {A, B}

   e. result $<=$ average {| A |, | B |} with rounding 0.

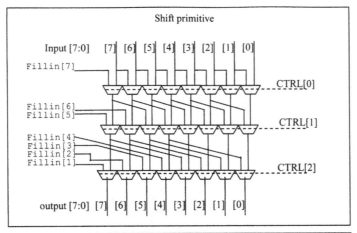

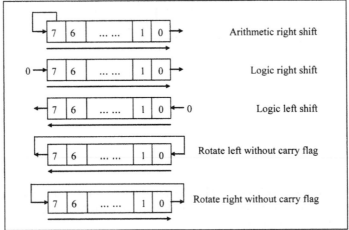

FIGURE 12.22

Question 12.10.

REFERENCES

[1] Omondi, A. R. (1994). *Computer Arithmetic Systems Algorithms, Architecture and Implementation*. Prentice Hall.

[2] Pirsch, P. (1997). *Architecture for Digital Signal Processing*. John Wiley & Sons.

[3] Lapsley, P., Bier, J., Shoham, A., Lee, E. A. (1997). *DSP Processor Fundamentals, Architectures and Features*. IEEE Press.

[4] Wanhammar, L. (1999). *DSP Integrated Circuits*. Academic Press.

[5] Weste, N. H. E., Eshraghian, K. (1997). *Principles of CMOS VLSI Design, 2nd edition*. Addison-Wesley.

MAC Hardware Implementation

The acronym MAC has two meanings: an instruction or a hardware module. MAC instruction stands for multiplication and accumulation operation. In this chapter, MAC as hardware module stands for multiplication and accumulation unit. A MAC unit is probably the most important hardware module in any datapath of DSP processors. It supports algorithms including filtering, auto/cross correlations, transforms, and double-precision arithmetic operations. This chapter deals with the microarchitecture design of MAC unit. Tables illustrating instructions were discussed in Chapter 7.

13.1 INTRODUCTION

13.1.1 Review of Convolution

Among DSP kernel algorithms, the most frequently used one is convolution. All linear time-invariant (LTI) systems can be modeled using the corresponding transfer function $H(z) = Y(z)/X(z)$ in the Z (frequency) domain. In the time-domain, the output is the convolution of the input signal with the system coefficient set; thus we have $y(n) = \Sigma x(n - i) * c(i)$, with i from 0 to $(k - 1)$; k is the length of the iteration. The detailed view of a convolution is given in Figure 13.1.

The mathematical presentation is given in the following equation:

$$y(n) = \sum_{i=0}^{k-1} x(n - i)c(i) \qquad (13.1)$$

In the equation, $x(n - i)$ is the data vector, $c(i)$ is the coefficient vector, and $y(n)$ is the vector of the convolution result. The example in Figure 13.1 is a 5-tap convolution $k = 5$. Directly mapping functions into hardware, the computing will be executed in the following way:

1. Data $x(n - i)$ is shifted through a four-register FIFO buffer, so that $x(n)$ becomes $x(n - 1)$, $x(n - 1)$ becomes $x(n - 2)$, and so on to the next clock cycle.

439

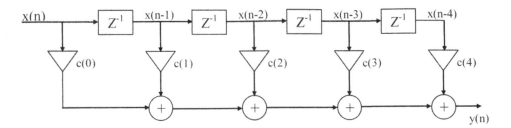

FIGURE 13.1

A 5-tap convolution behavior diagram.

2. All arithmetic executions are mapped to hardware in parallel.

3. There are five multipliers and four full adders.

4. A sample $y(n)$ is computed within a clock cycle according to the following equation:

$$y(n) = x[n]^*c[0] + x[n-1]^*c[1] + x[n-2]^*c[2] + x[n-3]^*c[3] + x[n-4]^*c[4].$$

When executing a convolution algorithm in a general-purpose processor, only one MAC unit is usually available, and one step of the convolution can be executed in each clock cycle assuming that the memory subsystem can supply the right data and the right coefficients each clock cycle (suitable addressing modes, post incremental, and modulo addressing were discussed in Chapter 7).

13.1.2 MAC Fundamentals

For our purposes, a MAC unit is a hardware module consisting of at least a multiplier and an accumulator. The multiplier in a MAC unit is normally a signed two's complement general multiplier. The accumulator in a MAC unit is a full adder supporting two's complement arithmetic computing with double-precision. The accumulator register keeps the result from the accumulator supporting iterative computing. By specifying pre- and postoperations around the multiplier and the accumulator, a MAC unit can support general arithmetic and logic functions with high precision.

The multiplier usually can supply a double-precision ($2N$) result of two single-precision (N) operands within one clock cycle. The accumulator can execute one step of arithmetic computing with double-precision inputs ($2N$) and double-precision output ($2N$) in one clock cycle. To execute both multiplication and accumulation operations, two clock cycles usually are required. Basic MAC circuits are described in Figure 13.2.

In Figure 13.2, (a) is the MAC circuit using one pipeline stage. It performs ACR <= ACR + MOA × MOB, whereas (b) is the MAC using two pipeline stages. It performs

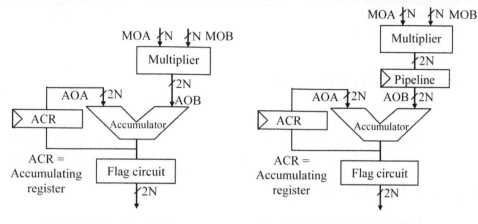

(a) MAC using one pipeline stage   (b) MAC using two pipeline stages

FIGURE 13.2

Basic MAC circuit.

```
always @(posedge clk) begin
Pipeline_REG <= MOA * MOB;
ACR <= Pipeline_REG + ACR;
End
```

Using circuit (a), the design of instruction decoding is simple because this is a typical data-stationary architecture (See more discussions in the section of instruction decoding in Chapter 14). The pipeline costs of the MAC instruction is the same as the cost of other instructions, and the drawback is the low clock rate. The circuit (b) requires two clock cycles while executing a MAC instruction. Most RISC instructions of a DSP processor consume only one execution cycle so that the complexity of instruction decoding is relatively easy. However, as MAC instruction is using two or more execution cycles, the DSP instruction decoding is more complicated. The management for single- and double-execution cycles will be discussed in detail in Chapter 14.

Please notice that the pipeline in Figure 13.2(b) is not perfect and the delays in two pipeline steps are not balanced. To reach a perfect pipeline partition, the inserted pipeline should be inside the multiplier. However, pipeline inside a multiplier limits the portability of the design.

It has been mentioned in Chapter 2 that guard bits are needed for iterative computing. It was also discussed that the number of guard bits in a MAC should not be less than $\log_2(\max \Sigma |c(n)|)$ to avoid overflow ($c(n)$ is a set of coefficients). When coefficients are adaptive variables, such as the LMS (least mean square) algorithm, the value of $\max \Sigma |c(n)|$ should be decided when each $|c(n)|$ is possible maximum value.

In reality, once the number of guard bits is decided, the decision is fixed in the hardware. Firmware designers can either divide long iteration into a group of smaller loops or scale down the accumulator inputs during the iteration. In most general-purpose DSP processors, eight guard bits are enough.

13.2 MAC IMPLEMENTATION

MAC units in different DSP processors could be quite different. Different designers can even design a MAC in different circuit styles based on the same instruction set. There is neither a theory nor absolute rules to follow when designing a MAC unit. Therefore, the design can be discussed based only on examples. The case study here is a design of a MAC unit in a normal single MAC DSP processor with a separated ALU and MAC unit.

13.2.1 MAC Instructions

In the remainder of this chapter, essential MAC instructions of **Senior** are used as references in examples, including circuits for long precision arithmetic computing, moves between general registers and accumulator registers (ACR), and iterative computing. Examples will demonstrate implementations of single MAC hardware with 16-bit operand A and operand B as inputs. Operands could come from the memories (iterative computing) or the registers (single-step computing). Hardware includes one 17×17 two's complement multiplier, 8 guard bits for accumulation, one 41-bit two's complement accumulator and several 40-bit accumulation registers. Accumulation registers (ACR1, ACR2, . . . ACRn) support double-precision arithmetic and iterative computing. Flags will be generated from the result from either the multiplier (for multiplication instructions) or the accumulator (for double-precision and iterative computing). MAC instructions usually are divided into four groups:

- Multiplication and MAC arithmetic operations.

- Move and data-type conversion.

- Double-precision arithmetic and logic operations.

- Iterative MAC instructions.

13.2.2 Implementing Multiplications

The implementation in this section is based on the signed integer multiplier primitive with $N + 1$ bit precision (a single-precision multiplier). By using the $N + 1$ bit component, multiplications with higher precision can be emulated.

In order to emulate multiplications of double-precision operands using a single-precision multiplier, long operands must be divided into multiple single-precision operands. A higher precision multiplication can then be emulated by accumulating several signed and unsigned multiplications. A signed multiplication is achieved directly from the signed two's complement multiplier. An unsigned multiplication is achieved from a signed two's complement multiplier by modifying the signed inputs to unsigned inputs; that is, adding a 0 as the sign bit on the left side of MSB. Signed multiplications will be used to emulate the multiplication of the higher part data. Unsigned multiplication will be used to emulate the multiplication of the lower part data. Different types of multiplications can be classified as shown in Figure 13.3.

Seven multiplications, MU1 to MU7, are specified in the following example based on Table 13.1, following the discussion in Chapter 7.

The 16-bit signed and 16-bit unsigned multiplication can be implemented based on a 17b × 17b signed multiplier. In general, a $(N + 1) \times (N + 1)$ bits signed multiplier can give N bits signed and unsigned multiplication. As we know, a signed

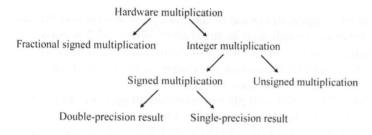

FIGURE 13.3

Examples of multiplications.

Table 13.1	Behavior Operations of Different Multiplications.
Instructions	**Behavior Specifications**
MU1	Signed integer multiplication, double-precision result ACR [31:0] <= {A[15], A[15:0]} × {B[15], B[15:0]}
MU2	Signed-unsigned integer multiplication, double-precision result ACR [31:0] <= {A[15], A[15:0]} × { "0", B[15:0]}
MU3	Unsigned-signed integer multiplication, double-precision result ACR [31:0] <= { "0", A[15:0]} × {B[15], B[15:0]}
MU4	Unsigned-unsigned integer multiplication, double-precision result ACR [31:0] <= {"0", A[15:0]} × {"0", B[15:0]}
MU5	Signed integer multiplication, single-precision result no round ACR [31:16] <= SAT(2^{16} × ({A[15], A[15:0]} × {B[15], B[15:0]}))
MU6	Signed fractional multiplication, double-precision result ACR [31:0] <= SAT (2 × ({A[15], A[15:0]} × {B[15], B[15:0]}))
MU7	Signed fractional multiplication, single-precision rounded result ACR [31:16] <= SAT(Round(2 × ({A[15], A[15:0]} × {B[15], B[15:0]})))

and unsigned positive number has the same value. If we add a 0 as the sign bit of any number (either positive or negative), this number will be a positive number, or an unsigned number. If 0 is added as the sign bits for both N-bit operands of the multiplier, the $N + 1$ bits signed multiplier will be an N bits unsigned multiplier. If we assign the MSB of the 17-bit operand with sign bit extension, a 16b × 16b signed multiplication can be supplied by a 17b × 17b signed multiplier.

The 16b × 16b signed/unsigned hardware multiplication therefore is implemented using a 17b × 17b signed (two's complement) multiplier, and the width of the result will be 34 bits. Although 34 bits are available on the physical output port of the multiplier, only the lower 32 bits are significant after saturation since the input operands were actually only 16 bits. It means that the output [31:0] is the right output instead of the output [33:0]. The output [33:32] is produced by the sign extension bits.

Instruction MU1 is the original operation of the multiplier. Instruction MU7 is used mostly when executing a single-step multiplication. When MU7 is used, the result precision is higher than that of the precision from instruction MU1. (Read Chapter 2 again to understand why the precision is higher on the result of fractional multiplication.) Both MU1 and MU7 are used frequently.

In Table 13.1 a 17b × 17b signed multiplier is used to manage both signed and unsigned multiplication in one hardware multiplier. One example is to emulate the long multiplication ACR <= X[31:0] × Y[15:0]. If ACR is a 48-bit register, the emulation will be:

```
ACR[47:0] =
$signed({X[31],X[31:16]})*$signed({Y[15],Y[15:0]}) << 16) +
        $signed({1'b0,X[15:0]})*$signed({Y[15],Y[15:0]});
```

Two result bits, [49:48], are redundant bits and are deleted. Unsigned multiplications are emulated using a signed multiplier by forcing the sign-bit to be 0. In a 16-bit processor, ACR has 32-bit visible resolution. The lower 16 bits of the 48 bits result will be truncated because of the hardware limitation.

Another example is to have ACR <= X[31:0] × Y[31:0]. If ACR here is a 64-bit register, the emulation will be:

```
ACR[63:0] =
($signed({X[31],X[31:16]}) * $signed({Y[31],Y[31:16]}) << 32)+
        ($signed({1'b0,X[15:0] }) * $signed({Y[31],Y[31:16]})
        << 16) +
        ($signed({X[31],X[31:16]}) * $signed({1'b0,Y[15:0]})
        << 16) +
        ($signed({1'b0,X[15:0]})    * $signed({1'b0,Y[15:0]})
        << 0);
```

Two result bits, [65:64], are redundant bits and are deleted. If the ACR has only 32-bit visible resolution, the last step of multiplication may not be necessary if slightly reduced accuracy is acceptable. The simplified emulation code (the code is not for logic synthesis) will be:

```
assign tmp2[47:0] =
 ($signed({X[31],X[31:16]})* $signed({Y[31],Y[31:16]}) << 16)+
 ($signed({1'b0,X[15:0] })* $signed({Y[31],Y[31:16]}) << 0)+
 ($signed({X[31],X[31:16]})* $signed({1'b0,Y[15:0]}) << 0);
assign ACR[31:0] = tmp2[47:16];
```

Thus the lower 16 bits of the accumulation will not exist in hardware because of the limited precision of the accumulator and the accumulator register.

Instruction MU5 produces a single-precision integer result. Principally, the result of integer multiplication with single precision sits in the lower 16 bits of the 32-bit result. However, flag computing becomes easier if moving the result from the lower 16 bits to the upper 16 bits. The data width of the result of instruction MU5 and MU7 are 16 bits after the truncation of 32-bit data. During the truncation, truncation errors might be introduced. Sometimes, truncation errors are negligible, other times, significant. Because round operation may consume one extra clock cycle, round arithmetic is optionally needed before the truncation. Round operation was discussed in Chapter 2. A round operation for a 16-bit result is

```
ACR[33:0] = {ACR[33:16], 16'b0} + {17'b0, ACR[15], 16'b0};
```

The round operation is implemented using the accumulator (also called accumulator adder), with at least double-precision 32 bits. An accumulator in this book has 40-bit resolution, including eight guard bits. The saturation arithmetic is required for a special case of fractional multiplication, $(-1) \times (-1)$. We accept (-1) as an operand, and we do not accept the result of $(-1) \times (-1)$ without saturation because

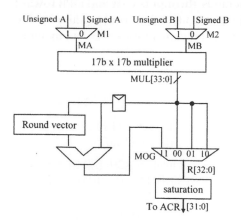

SA when M1=0: MA = {A[15], A[15:0]}
UA when M1=1; MA = { "0", A[15:0]}
SB when M2=0; MB = {B[15], B[15:0]}
UB when M2=1; MB = { "0", B[15:0]}

```
When MOG=00 // Integer MUL
  R[32:0] = MUL [32:0]
When MOG=01 // Fractional MUL
  R[32:0] = {MUL[31:0],"0"}
When MOG=11 //Round short fractional
  R[32:0]<= {MUL[31:0],"0"} +
            {16'b0, MUL[14],16'b0}
            // use result R[32:16]
When MOG=10 //Short integer
  R[32:0] = {MUL [32:16], 16b'0}
            // use result R[32:16]
```

FIGURE 13.4

Implementation of multiplications.

+1 is not a valid data representation in the fractional system. The saturation and round hardware will be discussed later in this section. Figure 13.4 illustrates the hardware implementation of multiplication operations in Table 13.1.

In Figure 13.4, SA/SB stands for signed operand A/B, and UA/UB stands for unsigned operand A/B. According to this example, multiplication with rounding takes two clock cycles and other multiplications take one clock cycle each.

It is important to mention that signed multiplication may not be available from some of the Verilog language simulator. In this case, an unsigned multiplier must be used as the multiplier primitive for the functional simulation, and the signed result can be achieved from the example given in Chapter 10.

13.2.3 Implementing MAC Instructions

MAC instructions can be divided into single-step MAC and iterative convolution. However, step-control is implemented in the control path. In MAC module, MAC operation and convolution operation are the same. MAC instructions listed in Table 13.2, including multiplication and accumulation, were discussed in detail in Chapter 7.

A 17b × 17b signed multiplier is an available component for MAC and convolution instruction. According to the instruction specifications of **Senior**, eight guard bits are required for MAC and convolution instructions in Table 13.2. However, there are already two guard bits produced by the 17b × 17b multiplier. Eight guard bits therefore are implemented using two MSB from the result of the multiplier and six extra guard bits. To implement fractional MAC and convolution instructions, five extra guard bits are enough because three extra bits are available after left-shifting one bit on the result of the multiplier. Guard bit specification and insertion are specified in Table 13.3.

In Table 13.3, the MUL is the output of the 17b × 17b multiplier. When running a single-step MAC instruction, MAC operands, the inputs OPA and OPB toward the MAC unit are fetched from the general register file. When running an iterative convolution, data are loaded directly from two parallel memories, the data memory

Table 13.2	Single-Step MAC and Convolution Instructions.
Instructions	**Behavior Specifications**
I1	signed integer multiplication and accumulation ACRn[39:0] <= ACRn[39:0] ± 6Guard(A{[15],A[15:0]} × {B[15],[15:0]})
I2	signed fractional multiplication and accumulation ACRn[39:0] <= ACRn[39:0] ± {5Guard({A[15],A[15:0]} × {B[15],[15:0]}),0}
I3	signed integer convolution running MAC N times N steps Loop: ACRn <= ACRn ± 6Guard({A[15],A[15:0]} × {B[15],[15:0]})
I4	signed fractional convolution running MAC N times N steps Loop: ACRn <= ACRn ± {5Guard({A[15],A[15:0]} × {B[15],[15:0]}),0}

Table 13.3 Guard Operations in MAC.

Specification: Add guard bits to...	Guards	Data to be guarded
the input from RF to ACR	{8R[15], R[15:0]}	R[15:0]
the integer MUL output from 17 b× 17b multiplier	{6MUL[33], MUL[33:0]}	MUL[33:0]
the Fractional MUL output from 17 b× 17b multiplier	{5MUL[33], MUL[33:0], 1'b0}	{MUL[33:0], 1'b0}

and the coefficient/data memory. Data typically is supplied with postincremental modulo addressing, and coefficients are supplied with postincremental addressing. The behavior of a convolution is illustrated in the following code:

```
01 OPA <= DM (DAR); OPB <= TM (TAR); /* One pipeline step */
02 BFR <= OPA * OPB; ACR <= ACR + BFR; /* Two pipeline steps */
03 if DAR == TOPR then DAR <= BTMR else INC (DAR);
04 INC (TAR); DEC (LCR);
05 if LCR != 0 then jump to 01
```

DAR is the data address pointer register; TAR is the coefficient address pointer register; TOPR is the top pointer register of the FIFO buffer; BTMR is the bottom pointer register of the FIFO buffer; and LCR is the loop counter. The FIFO buffer has exactly the size of the filter order and works as the data buffer for the convolution. The initial value in LCR is $N - 1$ ($N - 1$ down to 0), and N is the length of iteration. In hardware, running jobs listed above are pipelined in parallel. Therefore, without counting the overhead of the pipeline setup, N steps of convolution in a DSP processor will take $N + 1$ clock cycles.

Normally, a fractional single-step MAC is preferred because the precision of the result is higher. While running a convolution as an iteration loop, the selection of integer multiplication or fractional multiplication is the trade-off of the precision and the reliability of the result. By using integer convolution, an overflow on a result will be further avoided. By using fractional multiplication, the highest precision is kept.

An implementation of MAC circuit is illustrated in Figure 13.5. Pipeline register is inserted at the output of the multiplier in our example. The implementation depicted in Figure 13.5 is based on the previous design in Figure 13.4.

The accumulation (plus) is executed when M4 = 0, M5 = 00, M6 = 0, and CIN = 0. The result of accumulation is stored in an accumulator register, ACR. ACR is a register in MAC storing accumulative operands and intermediate result during iterative computing. The data width of an ACR is $2N + G$. N is the significant data width of operands (16 bits in this case), and G is the number of guard bits (8 bits in this case). Subtraction (minus) is executed when M4 = 0, M5 = 00, M6 = 1, and CIN = 1. A round vector is generated from the round vector module. When M5 = 10 and M6 = 0, this circuit carries out round operation for the multiplication-with-round

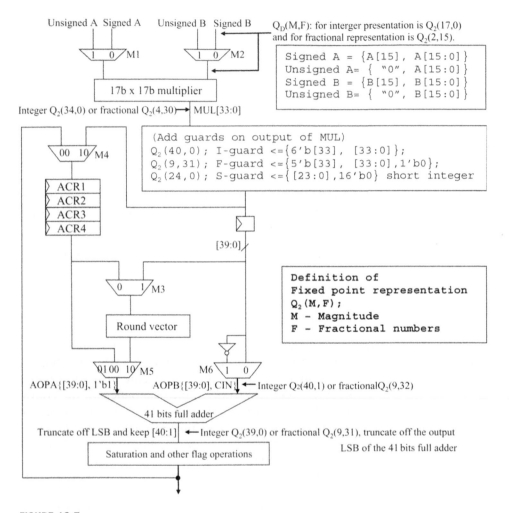

FIGURE 13.5

Implementation of a MAC unit.

instruction. When M5 = 00, M4 = 1, and M5 = 1, this circuit rounds data from an accumulation register ACR.

To implement instructions with minus or absolute operation, the subtraction function (OPA-OPB) should be OPA + INV(OPB) + 1. A carry-in bit is needed. Adders supplied by logic synthesis tools may or may not have carry-in as a separate input, but a 100% portable implementation of carry-in is preferred. A 41-bit full adder is used instead, and the LSB of the 41-bit full adder can be the carry-in bit:

```
AOPA [40:0] = {M5_out[39:0], "1"};        //operand A
AOPB [40:0] = {M6_out [39:0], Carry_in};  //operand B
```

Table 13.4 Control Signals for Multiplication, MAC, and Convolution.

Instruction	M1	M2	M3	M4	M5	M6	Guard mode	CIN
M1	0	0	x	10	xx	0	I	x
M2	0	1	x	10	xx	0	I	x
M3	1	0	x	10	xx	0	I	x
M4	1	1	x	10	xx	0	I	x
M5	0	0	x	10	xx	0	I	x
M6	0	0	x	10	xx	0	F	x
M7	0	0	0	00	10	0	F	0
I1/3+	0	0	x	00	00	0	I	0
I1/3−	0	0	x	00	00	1	I	1
I2/4+	0	0	x	00	00	0	F	0
I2/4−	0	0	x	00	00	1	F	1

The LSB of the 41-bit full adder is used only to implement the carry-in, which is not significant and should not show up in the result. Therefore, the alignment of result from the accumulator is:

```
to_saturation[39:0] <= from_adder[40:1];
```

M6 and CIN are used for implementing minus computing for decumulation, which is to subtract the result of multiplication instead of adding it.

```
ACR[39:0] = ACR[39:0] - A*B // decumulation
```

Here the multiplication can be signed or unsigned, integer or fractional.

I1/3+ stands for instruction I1 or I3 with accumulation mode. I1/3− stands for instruction I1 or I3 with decumulation mode. Table 13.4 specifies control signals for multiplication, MAC, and convolution instructions.

13.2.4 Implementing Double-Precision Instructions

A double-precision instruction conducts arithmetic and logic operations with double-precision operation, using double-precision operands, producing double-precision results. In the case of this chapter, the precision is 16 bits and the double-precision is 32 bits. It is essential to keep the double-precision of miscellaneous computing between two iterative operations using double-precision accumulations. When requirements on higher dynamic range and precision are exposed

Table 13.5   Arithmetic Computing with Double-Precision.

Instructions	Behavior specification
D1	Double-precision data add/sub double-precision data Saturate (ACRx[39:0] ± ACRy[39:0])
D2	Double-precision data add/sub single-precision data align to LSB Saturate (ACRx[39:0] ± {24'b OPB [15],OPB[15:0]})
D3	Double-precision data add/sub single-precision data align to MSB Saturate (ACRx[39:0] ± {8'b OPB [15],OPB[15:0], 16'b0})
D4	Double-precision data plus/sub single precision immediate Saturate (ACRx[39:0] ± 24'b immediate[15], immediate[15:0])
D5	Absolute operation on a double-precision data if ACRx[39] Saturate (INV(ACRx[39:0]) + "1") else ACRx
D6	Compare two double-precision data and set flags set flag: Saturate (ACRx [39:0] − ACRy [39:0])
D7	Simple scale by MUX instead of by shift logic

during behavior modeling, the algorithm should be translated to assembly code using double-precision instructions. Frequently used double-precision instructions are listed in Table 13.5.

Instructions listed in Table 13.5 are rather similar to ALU instructions except that the data width here is 32 bits (40 bits with guard bits). Following Table 13.5, all instructions have implied saturation operation, which means that 64-bit arithmetic is not supported. Remember that all results of the arithmetic operation are actually stored in a 40-bit register; thus a saturation operation is needed to generate the final result of 32 bits.

Arithmetic operations in the MAC could be real double-precision based on two accumulation registers (ACRx and ACRy), or they could be pseudo double-precision based on one long operand (ACRx) and one operand with single-precision from a register in RF.

Data scaling can be used to improve the capacity of guards when running long iteration loops. Eight guard bits may not be enough when the iteration length is more than 256. During iterative computing, arithmetic right-shift on the multiplier output equals more guard bits to prevent overflow. One bit right-shift equals an extra guard bit. To simplify the hardware implementation, a simple scaling could be implemented using offset connection. For example, concatenation {[39], [39:1]} means being divided by 2 or arithmetic right-shift one bit. Because frequently used scaling factors are 2.0ACR, 0.75ACR, 0.5ACR, and 0.25ACR, the simple scaling can be implemented as shown in Figure 13.6. The corresponding behavior for the various scaling parameters is shown in the following RTL code:

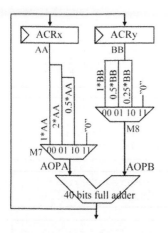

```
While scaling,
  AA=BB or ACRx=ACRy
2.0ACRx: M7=01 M8=11
1.0ACRx: M7=00 M8=11
0.5ACRx: M7=10 M8=11
.25ACRx: M7=11 M8=10
.75ACRx: M7=10 M8=10
```

FIGURE 13.6

Implementation of simple scaling in a MAC unit.

```
ACRn2_0 ={ACRx [38:0], "0"};              // ACR*2.0
ACRn0_5 ={ACRx [39], ACRx [39:1]};        // ACR*0.5
ACRn0_25={{2{ACRx[39]}}, ACRx[39:2]};     // ACR*0.25
ACRn0_75={ACRx[39],ACRx[39:1]}+{{2{ACRx[39]}},ACRx[39:2]};//ACR*0.75.
```

Finally, by adding M7 and M8 into Figure 13.7, implementing absolute computing using M10, and minus computing using M6 and CIN in Figure 13.7, the hardware implementation of basic arithmetic functions is depicted in Figure 13.7; control signals are listed in Table 13.6.

The absolute instruction is implemented exactly the same way as described in Chapter 12. The MSB of M7 output is used as the control bit. When MSB = 1, the output of M7 is negative; invert operation on output of M7 is required.

Each instruction in Table 13.6 was specified in Table 13.5. For example, D1+ means the instruction D1 is configured for plus operation. The selection of immediate data or data in a register file is not visible in the MAC hardware module.

13.2.5 Accessing ACR Context

The accessing ACR context is usually between register(s) in the general register file and ACR in the MAC unit. In the example throughout the chapter, the native data type is 16-bit, the internal visible data type in the MAC is 32-bit, and the physical data type inside the MAC is 40-bit including guards. Data types must be converted during moving data from or to the MAC. The operations of moving data between an ACR in MAC and a register in register file (RF), accessing ACR contexts, are:

- Move data into ACR, concatenate the data to double-precision, and guard it.

- Save the 16-bit result from an ACR to a register in general register file.

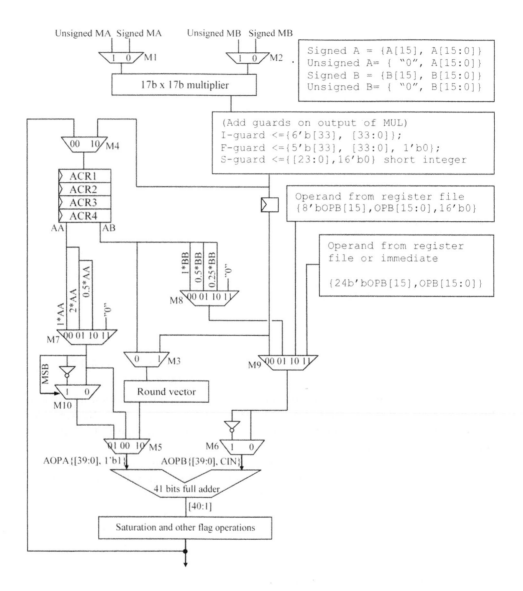

FIGURE 13.7

Implementation of arithmetic functions with double-precision.

- Store the 40-bit data from an ACR to a register in register file (ACR context switch).

- Load 40-bit data from three general registers to ACR (ACR context switch).

The following examples expose different implied concatenations when moving data into ACR in the MAC (see Table 13.7).

Table 13.6 Control Signal for Double-Precision Arithmetic Instructions.

Instruction	Configuration	M5	M6	M7	M8	M9	M10	CIN	Notes/control path
D1+	Plus	00	0	00	00	01	X	0	
D1−	Minus	00	1	00	00	01	X	1	
D2+	Plus	00	0	00	00	11	X	0	
D2−	Minus	00	1	00	00	11	X	1	
D3+	Plus	00	0	00	00	10	X	0	
D3−	Minus	00	1	00	00	10	X	1	
D4+	Plus	00	0	00	00	11	X	0	Use Immediate
D4−	Minus	00	1	00	00	11	X	1	Use Immediate
D5		01	0	00	11	01	MSB	MSB	
D6		00	1	00	00	01	X	1	
D7:2.0	Scale to 2.0	00	0	01	11	01	X	0	
D7:1.0	Scale to 1.0	00	0	00	11	01	X	0	
D7:0.5	Scale to 0.5	00	0	10	11	01	X	0	
D7:0.25	Scale to 0.25	00	0	11	10	01	X	0	
D7:0.75	Scale to 0.75	00	0	10	10	01	X	0	

Table 13.7 Loading and Preparing Data into an ACR.

Instructions	Functions	ACR are loaded with
L1	Keep lower part	ACRx <= {8'bR[15], R[15:0], ACRx[15:0]}
L2	Clean lower part	ACRx <= {8'bR[15], R[15:0], 16'H0000}
L3	keep higher part	ACRx <= {ACRx [39:16], R[15:0]}
L4	Long sign extension	ACRx <= {24'bR[15], R[15:0]}
L5	Load A and B from RF	ACRx <= {8'bRa[15],A[15:0], Rb[15:0]}
L6	Restore guards	ACRx <= {R[7:0], ACRx[31:0]} //After executing instruction L5

The data A[15:0] and B[15:0] are operands from the general register file. When loading data to the higher part of the accumulation register ACRx, guard bits are needed. Sometimes, the higher part needs to be loaded and the lower part should be kept, while sometimes, the higher part needs to be loaded and the lower part should be cleaned. The instruction L5 loads both the higher and lower part of data from two registers. Instructions L5 and L6 are used to restore ACR value from stack. The hardware implementation of these instructions can be found in Figure 13.8.

The implementation of the assignment box will be an 8-to-1 40-bit multiplexer accepting six concatenated inputs listed in Figure 13.8.

Store instructions move data from ACRx in the MAC to registers in the register file. When moving data from ACRx to registers, the programmer needs to know the purpose of the move. The purpose could be context saving or result saving. Contexts in ACRx must be temporally stored to stack via the general register file when the processor hardware resource is temporally given to an interrupt. After the interrupt service, contexts in stack must be restored to ACRx via the general register file using instruction L5 and L6. The internal data width is 40 bits, including eight guard bits. Remember that the bus width is usually 16 bits. It therefore takes three clock cycles to save the data in an ACR to three general registers: one cycle for ACR[15:0], one cycle for ACR[31:16], and one cycle for ACR[39:32]. If the context switching is frequently invoked, a special instruction may be required. In this case, special connections between the MAC and the RF are required.

Instruction L7, L8, and L9 are general move instructions, see Table 13.8. Instruction L10 is used to store results from an ACR to a general register. Round and saturations were specified in Chapters 2 and 7. Instructions L11 and L12 are used

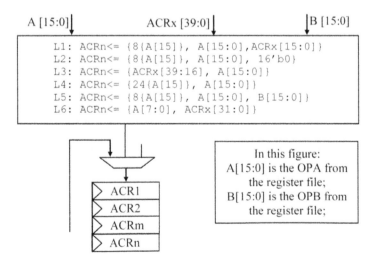

FIGURE 13.8

Hardware for loading an ACR.

Table 13.8 Typical Moves from MAC to Register File.

Instructions	Specifications on the result
L7	Ra <= ACRx[31:16]; //Ra is a register in RF
L8	Ra <= ACRx[15:0];
L9	Ra <= ACRx[31:16]; Rx <= ACRx[15:0]; // Both loads happen simultaneously, (x is register number a+1)
L10	Ra <= Saturation (Round (ACRx[39:0])); // After round and saturation, move ACR[31:16] to Ra
L11	Ra <= ACRx[31:16] I Ra+**1** <= ACRx[15:0] I Ra+**2** <= {8'b0, ACRx[39:32]}
L12	Ra <= {8'h00, ACRx[39:32]}; // guard to a general register

for context saving. Hardware implementation of instructions L7 to L12 is on the register file bus level and in the general register file except L10.

Round hardware was discussed in the previous section. Saturation operation is based on the availability of guard bits and the correctness of the left-most guard bit. When performing saturation, the MSB guard will be used as the reference. The saturation function is given in the following pseudocode when eight guard bits are used:

```
if((ACR[39:31] != 9'h0) || (ACR[39:31] != 9'h1ff))
    if(ACR[39]) result[31:0] = 32'h80000000; // a negative
overflow
    else result[31:0] = 32h'7FFFFFFF; // a positive overflow
else result[31:0] = ACR [31:0]; // no overflow.
```

13.2.6 Flag Operations and Other Postoperations

Because control code usually is implemented using ALU instructions, flags in MAC module are not used much. However, the MAC module flags can be used when conditional execution is not available or for an exception. Flags in the MAC include the saturation flag (FMO), the sign flag (FMS), and the zero flag (FMZ). The definition of flags in the MAC is similar to that of an ALU. Because 64-bit precision is seldom used in a DSP processor, the carry-out flag will not be discussed in this example of the chapter.

Flags may sense either the result of the accumulation adder or the result of the multiplier. An instruction decoder will select the flags. When running multiplication instructions, flags are from the output of the multiplier. Otherwise, flags are based on the result at the output port of the accumulation adder. The MAC flags typically do not change when the MAC involves load/store executions.

A physical critical path might be identified when flags are from the output of a multiplier. Flag prediction techniques can be used. Flag prediction is executed using operands of the multiplier, and functions are:

```
//Sign prediction
  Sign <= 1 if sign of operand A and sign of operand B are the
same
  Zero <= 1 if operand A = 0 or operand B = 0
```

A saturation flag is required only when executing fractional multiplication and both operand values are maximum negative.

To store a result from an ACR to a register, the processing on an output data must be in the following order.

1. Guarding operands.

2. Kernel operation (iteration) and scaling.

3. Round after iteration.

4. Saturation and removing guards.

5. Truncation and output.

The list is the correct execution order to implement the logic hardware. If the order is not correct, errors will be introduced. For example, if the round operation is after the saturation operation and if the result is "saturated to the MAX positive," the round will damage the saturated result. Finally, the implementation of the MAC module in the example throughout the chapter is summarized in Figure 13.9.

13.3 A MAC DESIGN CASE

The case shown in this section will be a design of a single MAC module.

Example 13.1

Design a MAC hardware module for a 16-bit fixed-point DSP processor. The internal resolution for iteration is 40 bits, including 8 guard bits. Two accumulation registers, ACR1 and ACR2, are needed. Designing a two pipeline-step MAC, the multiplier output is registered. The following instructions must be implemented:

1. Load ACR1H (ACR1[31:16]) from general register port A/B and fill in guards;

2. Load ACR2H (ACR2[31:16]) from general register port A/B and fill in guards;

3. Load ACR1L (ACR1[15:0]) from register port A/B and keep the higher part;

4. Load ACR2L (ACR2[15:0]) from register port A/B and keep the higher part;

5. ACR1(or ACR2) = RFP1 [15:0]×RFP2[15:0] SS integer result in one clock cycle;

6. ACR1(or ACR2) = RFP1 [15:0]×RFP2[15:0] SS fractional result in one clock cycle;

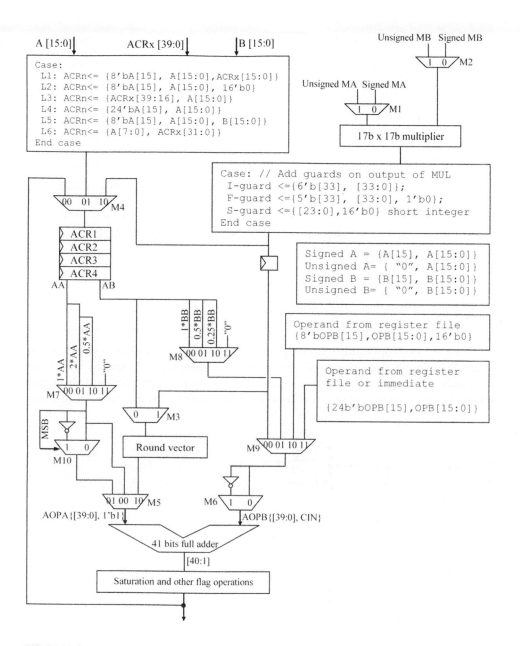

FIGURE 13.9

The final complete MAC circuit example.

7. ACR1(or ACR2) = RFP1 [15:0]×RFP2[15:0] SU integer result in one clock cycle;

8. ACR1(or ACR2) = RFP1 [15:0]×RFP2[15:0] US integer result in one clock cycle;

9. ACR1(or ACR2) = RFP1 [15:0]×RFP2[15:0] UU integer result in one clock cycle;

10. ACR1(or ACR2) = ACR1(ACR2) + RFP1 [15:0] × RFP2[15:0]; SS integer;

11. ACR1(or ACR2) = ACR1(ACR2) + RFP1 [15:0] × RFP2[15:0]; SS fractional;

12. ACR1(or ACR2) = ACR1(ACR2) + Mem1 [15:0] × Mem2[15:0]; SS integer; (conv)

13. ACR1(or ACR2) = scale (ACR1(ACR2)) using factors 2.0, 1.5, 0.75, 0.5, 0.25;

14. ACR1(or ACR2) = ACR1 + ACR2;

15. ACR1(or ACR2) = ACR1 − ACR2 or ACR1(or ACR2) = ACR2 − ACR1;

16. ABS ACR1(or ACR2);

17. Send ACR1H, Saturation (round(ACR1));

18. Send ACR2H, Saturation (round(ACR2)).

SS means both operands for multiplication are signed. US and SU mean that one multiplication operand is signed and another is unsigned. UU means that both operands are unsigned. The first four instructions are implemented and shown in Figure 13.10. The function of guard8 is to copy the sign bit [15] to guard bits [23:16].

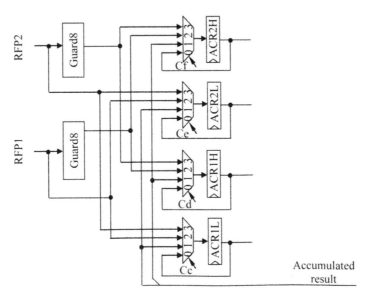

FIGURE 13.10

Implementation of the MAC instructions 1–4.

ACR1 is the concatenation of ACR1[39:0] = {ACR1H[23:0], ACRC1L[15:0]}. In the same way, ACR2 is the concatenation of ACR2[39:0] = {ACR2H[23:0], ACRC2L [15:0]}. Multiplexers, Cc, Cd, Ce, and Cf, are used to select inputs: keeping its own ($Cx = 0$), from MAC result ($Cx = 1$), from RFP1 ($Cx = 2$), and from RFP2 ($Cx = 3$). Here x can be c, d, e, or f. Control signals are specified in Table 13.9.

The implementation of multiplication instructions (instructions 5–9) is shown in Figure 13.11. When Cg(or Ch) = 0, the operand of the multiplication will be signed. The inputs to the multiplier are {A[15], A[15:0]} and {B[15], B[15:0]}. When Cg(or Ch) = 1, the operand of the multiplication is unsigned. The inputs to the multiplier

Table 13.9 Implementing the Control Signals for Load Instructions 1–4.

Instruction	Cc	Cd	Ce	Cf	Instruction	Cc	Cd	Ce	Cf
ACR1H<=RFP1	0	2	0	0	ACR1H<=RFP2	0	3	0	0
ACR2H<=RFP1	0	0	0	2	ACR2H<=RFP2	0	0	0	3
ACR1L<=RFP1	2	0	0	0	ACR1L<=RFP2	3	0	0	0
ACR2L<=RFP1	0	0	2	0	ACR2L<=RFP2	0	0	3	0

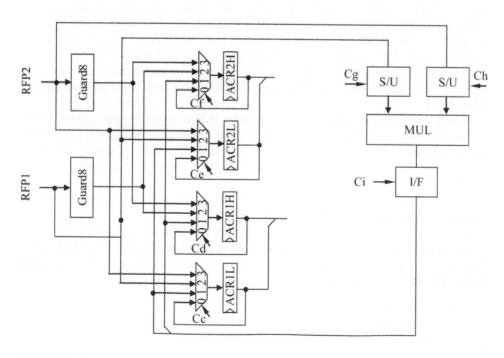

FIGURE 13.11

The implementation of multiplication instructions in Example 3.1.

Table 13.10 Implementing the Control Signals for Multiplication Instructions.

Instruction	Cc	Cd	Ce	Cf	Cg	Ch	Ci
ACR1 S × S integer	1	1	0	0	0	0	0
ACR2 S × S integer	0	0	1	1	0	0	0
ACR1 S × S fractional	1	1	0	0	0	0	1
ACR2 S × S fractional	0	0	1	1	0	0	1
ACR1 S × U integer	1	1	0	0	0	1	0
ACR2 S × U integer	0	0	1	1	0	1	0
ACR1 U × S integer	1	1	0	0	1	0	0
ACR2 U × S integer	0	0	1	1	1	0	0
ACR1 U × U integer	1	1	0	0	1	1	0
ACR2 U × U integer	0	0	1	1	1	1	0

are {"0", A[15:0]} and {"0", B[15:0]}. A 17b × 17b signed integer multiplier is used with output MUL[33:0]. After guarding, the 40 bits output of I/F will be {6'bMUL[33], MUL[33:0]}. The fractional output is {5'bMUL[33], MUL[33:0], "0"} when Ci = 1. Control Signals for Multiplication instructions are specified in Table 13.10.

The implementations of the MAC and the convolution instructions are shown in Figure 13.12. Control Signals for MAC and convolution instructions are specified in Table 13.11.

When executing a scaling instruction, one ACR is selected and connected to the scaling box. The selected ACR will be used as operand A and B. Operands A and B will be added together by the 40-bit full adder after the operations specified in Table 13.12 on both operands A and B. The box is controlled by the control signal C_m.

The plus and minus instructions (instructions 14–15) are implemented by specifying control signals in Table 13.13.

The instruction of absolute ACR1 or ACR2 is implemented in Table 13.14.

In Table 13.14, A1S is the sign bit of ACR1; it is ACR1[39], and A2S = ACR2[39]. When Cj = 1, operand B is controlled by the ABSC block in Figure 13.13 instead of by Ck. The function of ABSC is:

```
if(Cc==1) begin
  if(ACR1[39]==1) ABSC=2; else ABSC=0;
  end else
if(Ce==1) begin
  if(ACR2[39]==1) ABSC=3; else ABSC=1;
```

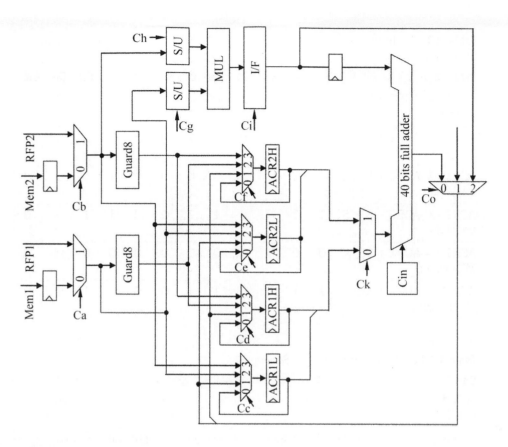

FIGURE 13.12

The MAC unit after the MAC and convolution instructions are added.

```
    end else begin
  ABSC=0; //default
  End
```

Operand A will be 0 (Cn = 2), and the output of the full adder will be

```
FullAdder = OperandB + 0 + Cin;
```

The result of the full adder will be saturated to keep all corner cases correct. The saturation operation on the result is:

```
if((FullAdder[39:31]!= 9'h1ff) || (FullAdder[39:31] != 9'h0)
    if(FullAdder[39]) result = 40'hff80000000;
    else              result = 40'h007fffffff;
else result = FullAdder[39:0];
```

Table 13.11 Implementing the Control Signals for the MAC and Convolution Instructions.

Instruction	Ca	Cb	Cc	Cd	Ce	Cf	Cg	Ch	Ci	Ck	Co	Cin
ACR1 <= ACR1 + S × S integer	1	1	1	1	0	0	0	0	0	0	0	0
ACR2 <= ACR2 + S × S integer	1	1	0	0	1	1	0	0	0	1	0	0
ACR1 <= ACR1 + S × S fractional	1	1	1	1	0	0	0	0	1	0	0	0
ACR2 <= ACR2 + S × S fractional	1	1	0	0	1	1	0	0	1	1	0	0
ACR1 <= ACR1 + M1 × M2 integer	0	0	1	1	0	0	0	0	0	0	0	0
ACR2 <= ACR2 + M1 × M2 integer	1	1	0	0	1	1	0	0	0	1	0	0

Table 13.12 Function Inside the Scaling Box.

Cm	1	2	3	4	5
Scaling factors	2.0	1.5	0.75	0.5	0.25
Operand A	Pass no OP	Right shift one bit	Right shift one bit	Right shift one bit	= 0
Operand B	Pass no OP	Pass no OP	Right shift two bits	= 0	Right shift two bits
Result	A + B	0.5A + B	0.5A + 0.25B	0.5A + 0	0 + 0.25B

Table 13.13 Implementing the Control Signals for the Add and Subtract Instructions.

Instruction	Ca	Cb	Cc	Cd	Ce	Cf	Cj	Ck	Cn	Co	Cin
ACR1 <= ACR1 + ACR2	1	1	1	1	0	0	0	1	0	0	0
ACR2 <= ACR1 + ACR2	1	1	0	0	1	1	0	1	0	0	0
ACR1 <= ACR1 − ACR2	1	1	1	1	0	0	0	3	0	0	1
ACR2 <= ACR1 − ACR2	1	1	0	0	1	1	0	3	0	0	1
ACR1 <= ACR2 − ACR1	1	1	1	1	0	0	0	2	1	0	1
ACR2 <= ACR2 − ACR1	1	1	0	0	1	1	0	2	1	0	1

Table 13.14   Implementing Control Signals for the Absolute Operation.

Instruction	Ca	Cb	Cc	Cd	Ce	Cf	Cj	Cn	Co	Cin
ABS (ACR1)	1	1	1	1	0	0	1	2	1	A1S
ABS (ACR2)	1	1	0	0	1	1	1	2	1	A2S

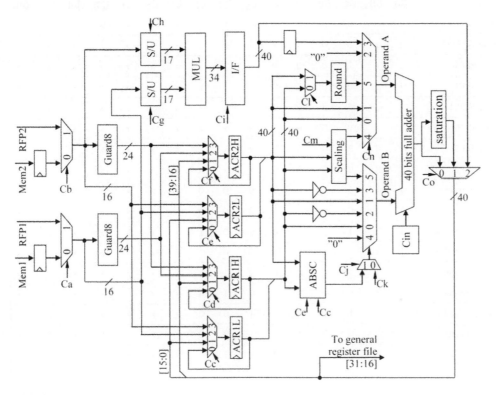

FIGURE 13.13

The final circuit of Example 13.1.

The round operation will be:

```
FullAdder = OperandB + RoundVector;
```

Operand B and the round vector must be from the same accumulator register. The round vector is:

```
RoundVector = {23'b0, ACR[15], 16'b0};
```

After rounding, the saturation is conducted by selecting Co = 1. After integration of the entire designs, the final MAC circuit is shown in Figure 13.13 and the control signals of the MAC unit are specified in Table 13.15.

Table 13.15 Implementing Control Signals for Instructions.

Instruction	Ca	Cb	Cc	Cd	Ce	Cf	Cg	Ch	Ci	Cj	Ck	Cl	Cm	Cn	Co	Cin
ACR1H <= RFP1	1	x	0	2	0	0	x	x	x	x	x	x	x	x	x	x
ACR2H <= RFP1	1	x	0	0	0	2	x	x	x	x	x	x	x	x	x	x
ACR1L <= RFP1	1	x	2	0	0	0	x	x	x	x	x	x	x	x	x	x
ACR2L <= RFP1	1	x	0	0	2	0	x	x	x	x	x	x	x	x	x	x
ACR1H <= RFP2	x	1	0	3	0	0	x	x	x	x	x	x	x	x	x	x
ACR2H <= RFP2	x	1	0	0	0	3	x	x	x	x	x	x	x	x	x	x
ACR1L <= RFP2	x	1	3	0	0	0	x	x	x	x	x	x	x	x	x	x
ACR2L <= RFP2	x	1	0	0	3	0	x	x	x	x	x	x	x	x	x	x
ACR1 S × S integer	1	1	1	1	0	0	0	0	0	x	x	x	x	x	2	x
ACR2 S × S integer	1	1	0	0	1	1	0	0	0	x	x	x	x	x	2	x
ACR1 S × S fractional	1	1	1	1	0	0	0	0	1	x	x	x	x	x	2	x
ACR2 S × S fractional	1	1	0	0	1	1	0	0	1	x	x	x	x	x	2	x
ACR1 S × U integer	1	1	1	1	0	0	0	1	0	x	x	x	x	x	2	x
ACR2 S × U integer	1	1	0	0	1	1	0	1	0	x	x	x	x	x	2	x
ACR1 U × S integer	1	1	1	1	0	0	1	0	0	x	x	x	x	x	2	x
ACR2 U × S integer	1	1	0	0	1	1	1	0	0	x	x	x	x	x	2	x
ACR1 U × U integer	1	1	1	1	0	0	1	1	0	x	x	x	x	x	2	x
ACR2 U × U integer	1	1	0	0	1	1	1	1	0	x	x	x	x	x	2	x
ACR1 + S × S integer	1	1	1	1	0	0	0	0	0	0	0	x	x	3	0	0
ACR2 + S × S integer	1	1	0	0	1	1	0	0	0	0	1	x	x	3	0	0
ACR1 + S × S fractional	1	1	1	1	0	0	0	0	1	0	0	x	x	3	0	0
ACR2 + S × S fractional	1	1	0	0	1	1	0	0	1	0	1	x	x	3	0	0
ACR1 + M1 × M2 integer	0	0	1	1	0	0	0	0	0	0	0	x	x	3	0	0
ACR2 + M1 × M2 integer	1	1	0	0	1	1	0	0	0	0	1	x	x	3	0	0
Scaling ACR1	1	1	1	1	0	0	x	x	x	0	5	x	*	4	0	0
Scaling ACR2	1	1	0	0	1	1	x	x	x	0	5	x	*	4	0	0

Continued

Table 13.15 Continued

Instruction	Ca	Cb	Cc	Cd	Ce	Cf	Cg	Ch	Ci	Cj	Ck	Cl	Cm	Cn	Co	Cin
ACR1 <= ACR1 + ACR2	1	1	1	1	0	0	x	x	x	0	1	x	x	0	0	0
ACR2 <= ACR1 + ACR2	1	1	0	0	1	1	x	x	x	0	1	x	x	0	0	0
ACR1 <= ACR1 − ACR2	1	1	1	1	0	0	x	x	x	0	3	x	x	0	0	1
ACR2 <= ACR1 − ACR2	1	1	0	0	1	1	x	x	x	0	3	x	x	0	0	1
ACR1 <= ACR2 − ACR1	1	1	1	1	0	0	x	x	x	0	2	x	x	1	0	1
ACR2 <= ACR2 − ACR1	1	1	0	0	1	1	x	x	x	0	2	x	x	1	0	1
ABS ACR1	1	1	1	1	0	0	x	x	x	1	x	x	x	2	1	A1S
ABS ACR2	1	1	0	0	1	1	x	x	x	1	x	x	x	2	1	A2S
SAT (Round ACR1)	1	1	1	1	0	0	x	x	x	0	0	0	x	5	1	0
SAT (Round ACR2)	1	1	0	0	1	1	x	x	x	0	1	1	x	5	1	0

13.4 MAC INTEGRATIONS

13.4.1 Physical Critical-Path

A physical critical-path means a long path (possibly the longest path) in a netlist circuit after synthesis of the RTL code. It is a path from one output of a flip-flop to an input of a flip-flop, making the longest delay while running the circuit. A complete execution path of the MAC instruction in a 16-bit DSP processor includes one 17b × 17b multiplication, one 41-bit full adder operation, and pre- or postprocessing circuits. The physical critical path in a DSP core is most likely in the MAC hardware. As illustrated in Figure 13.14, in this DSP core an operand has to go from the register file, through various operand multiplexes, and generate a result in the multiplier before it ends up in a pipeline register. This very long path is likely to be a critical path.

When specifying the hardware functions before the RTL coding, it is important to collect the timing parameters of RTL components from the cell libraries as early as possible. Possible critical paths should be predicted before the RTL coding. The critical paths should be predicted according to the worst-case (worst temperature and worst process margin). It is important to predict the critical path by using the correct methodology. Static timing analysis is the correct way to predict this, whereas dynamic timing analysis should not be used. Here, static timing analysis is the measurement of the path delay by accumulating the delays of each component in a logic tree. This is the right way to find the critical path. Dynamic timing analysis is the measure of circuit speed by running a circuit simulator. In this case the appearance of critical paths depends on the stimuli.

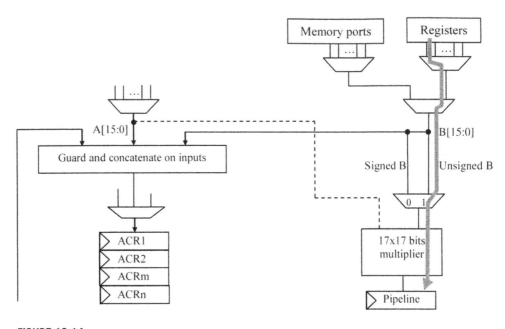

FIGURE 13.14

A possible physical critical path in a MAC unit.

After identifying a critical path, the path can be eliminated by inserting or modifying pipeline not only inside the MAC circuit, but also between the MAC and the register file as well as the memory bus.

As shown in Figure 13.14, there are two operands A[15:0] and B[15:0] that might come either from the general register file or from different memories. In a modern DSP processor, multiple data memories are available on the chip. When designing the instruction set and MAC hardware module, all possible data sources should be supplied to the MAC module to achieve flexibility. Operands of the MAC module include registers and memories. The heavy input selection logic and the fan-out of control signals are thus introduced, which makes the critical path longer. Figure 13.15 gives an example showing how many inputs could be introduced to the MAC module.

The output selection circuit of a register file causes a long delay because of the large fan-out of control signals, which was discussed in Chapter 11. In many high-end DSP processors, a separate pipeline stage (pipeline-op) is used between operand selection circuit and the input of the multiplier in Figure 13.15.

13.4.2 Pipeline in a MAC

The simplest way to implement a MAC is to design a so-called one-cycle-MAC, see Figure 13.16(a), the MAC as a combinational logic without extra pipeline steps

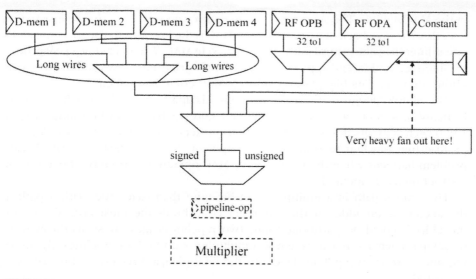

FIGURE 13.15

Multiple input selection logic in MAC.

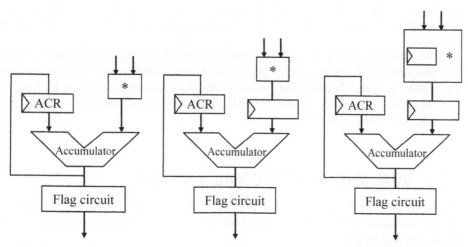

(a) MAC in one clock cycle   (b) MAC using two clocks   (c) MAC using three clocks

FIGURE 13.16

Pipeline in a MAC.

in the MAC module. The advantage of using one-cycle-MAC is its simple instruction decoder. The drawback of using one-cycle-MAC is the low speed. The machine clock rate is limited by the critical path of the one-cycle-MAC. When making a DSP processor intended for low-end applications such as a hearing aid, a low-price GSM mobile phone, or a DECT mobile phone, one-cycle-MAC could be enough.

The simple design of the one-cycle-MAC datapath limits the machine speed. In many processors, a two-cycle-MAC see Figure 13.16(b), can be found instead. A two-cycle-MAC divides the multiplication and accumulation into two clock cycles by inserting pipeline registers at the output of the multiplier. The data hazard problem happens when the result from a MAC operation is used by the following instruction as an operand.

The critical path in a multiplier can be more than twice the critical path in the accumulation adder. If the multiplier contributes the most critical path on the chip, it could be partitioned into two pipeline stages. Most synthesis tools or datapath generators can generate a multiplier with built-in pipelines. By inserting one more extra pipeline stage into the multiplier, a three-cycle-MAC datapath is created.

A three-cycle-MAC, see Figure 13.16(c), can support higher machine clock rate and higher performance. However, it will be more difficult to design the control path for a three-cycle-MAC. While executing a hardware loop convolution, a three-cycle-MAC requires two cycles to set up the pipeline for iteration. A result data dependence check shall be kept for two clock cycles. The loop controller should prepare the flag of loop finish in advance to compensate for the extra pipeline requirement. When a single MAC instruction is executed, three clock cycles are required.

The multiple-cycle-MAC is useful and used mostly for dedicated DSP hardware circuits instead of a general DSP processor. Actually three-cycle-MAC or multiple-cycle-MAC may not be necessary when using deep submicron silicon technology. The critical path of an on-chip SRAM might be longer than the critical path introduced by MAC hardware.

13.5 DUAL MAC, MULTIPLE MAC, AND VLIW

Multi-MAC and VLIW technology will be introduced briefly in this section. Details will be discussed in Chapter 20. In this chapter, a short introduction of multi-MAC concepts will be given.

A dual MAC datapath is simplified and depicted in Figure 13.17. Comparing the dual MAC with a single MAC module, the dual MAC is almost the duplication of two single MACs except for the sharing of accumulator registers.

There are two MAC units, MAC-C1 (cluster 1) and MAC-C2 (cluster 2). To avoid writing two data to one accumulator, the accumulator registers also are divided into two groups (ACR-C1 and ACR-C2), each for a MAC circuit. MAC-C1 can write

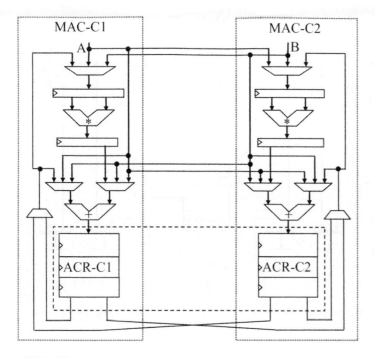

FIGURE 13.17

A dual MAC datapath.

data only to ACR-C1, and MAC-C2 can write data only to ACR-C2. However, MAC-C1 and MAC-C2 can read any data from both ACR-C1 and ACR-C2. Write dependences therefore are avoided and all operands from all accumulator registers in both ACR-C1 and ACR-C2 can be available for both MAC clusters.

A complex MAC (with four multipliers) is simplified and shown in Figure 13.18. Operands are complex data from two complex data sources (complex data registers or complex data memories). Operands are:

```
OPA = AR + jAI;
OPB = BR + jBI;
```

Here AR is the real part of operand A (OPA) and AI is the imaginary part of the OPA. The output of the complex MAC is from ACRR and ACRI, and the result of a complex data MAC is:

```
CMACResult = ACRR + AR*BR - AI*BI + j*(ACRI + AR*BI + AI*×BR);
```

To adapt a CMAC to a butterfly computing unit for FFT, the enhanced CMAC becomes the one depicted in Figure 13.19.

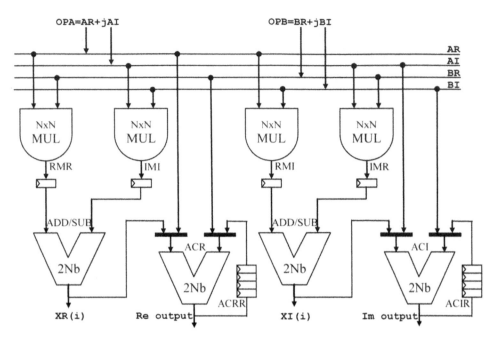

FIGURE 13.18

A simplified MAC for complex valued computing.

In Figure 13.19, the butterfly of the decimation in time (DIT) algorithm is implemented following the specification given in Chapter 7. Result of a butterfly executed by the circuit in Figure 13.19 is:

```
Output1Real=(BR + AR*CR - AI*CI); Output1Imag=BI + AR*CI + AI*CR;
Output2Real=(BR - AR*CR + AI x CI); Output2Imag=BI - AR*CI - AI*CR;
```

13.6 CONCLUSIONS

A MAC module is a basic hardware circuit in the datapath of a DSP processor. The design of the MAC module is essential in DSP hardware design. In this chapter, the hardware implementation of the MAC module was introduced based on case studies. The discussion of a MAC instruction set was given in Chapter 7. Based on the MAC instruction subset specification, the MAC module implementation is divided into:

- Multiplication.
- MAC and convolution.

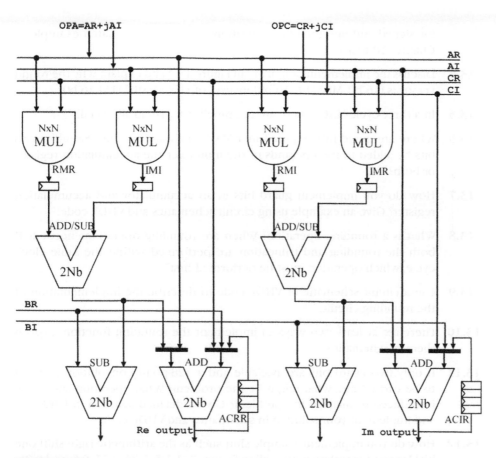

FIGURE 13.19

Modified complex valued MAC supporting FFT.

- Arithmetic operations with double precision.

- Data load/store between MAC and general register file.

EXERCISES

13.1 Based on the implementations of MU1–MU7, I1–I4, D1–D7, and I1–I6 in this chapter, try to implement all **Senior** MAC instructions.

13.2 While executing integer and fractional multiplications on two's complement data, when should saturation be used after the multiplication, and why?

13.3 Examples in this chapter are based on signed multipliers. Use an unsigned multiplier to implement the instructions from M1 to M4 in Table 13.1

for signed and unsigned multiplications. (Hint: read a related example in Chapter 10 first.)

13.4 Discuss the implementation of M7 in Figure 13.4 and explain why the round vector is {16'b0, MUL[14],16'b0} instead of {16'b0, MUL[15],16'b0}.

13.5 In a three-cycle-MAC, where are the pipeline registers allocated?

13.6 Where are guard bits required in a MAC hardware module? Should guard bits be added to the operands of the multiplier, the accumulation register, or both?

13.7 How do you implement guard bits in an accumulator and accumulation register? Give an example using circuit schematics and VHDL code.

13.8 What is a rounding operation? When are rounding operations needed? If both the rounding and truncation are performed within the same clock cycle, which operation must be performed first?

13.9 Use a circuit schematic or VHDL code to describe the implementation of the rounding circuit.

13.10 There are at least two ways to implement the rounding function; explain them in schematics.

13.11 If only two guard bits are available and a 128-point convolution is going to be carried out, what might be the problem? What operation should be introduced to manage the problem? Give a detailed solution in hardware and implement your solution in schematics and VHDL code.

13.12 How do you implement a simple shift such as the arithmetic right-shift one bit? Design for scaling with scaling factors 3, 2, 1.5, 1.25, 0.75, 0.5, and 0.25.

13.13 Annotate fixed-point representation $Q_2(M,F)$ at all input and output ports of all components (except for multiplexers) in Figure 13.7 and Figure 13.9 following examples given in Figure 13.5.

13.14 Why is a saturation operation always necessary after iteration?

13.15 Why should saturation be avoided during iteration?

13.16 In which case is the unsigned multiplication needed?

13.17 Give an example of using signed and unsigned multiplication to execute $16 \times 32b$ multiplication using a $17b \times 17b$ multiplier and a 48-bit accumulator.

13.18 When giving a 16-bit output to memory from a 40-bit accumulation register with eight guard bits, saturation, truncation, and rounding operations are needed. Only one order is correct: which executes the first, which the second, and which the third?

13.19 Design a MAC hardware module for a 12-bit fixed-point DSP processor. The internal resolution for iteration is 32 bits, including eight guard bits. Four accumulation registers ACR0, ACR1, ACR2, and ACR3 are needed. Designing a two-pipeline-step MAC, the multiplier output is registered. The following instructions must be implemented (x or y can be either 0, 1, 2, or 3):

    a. Load ACRxH (ACRx[23:12]) from register port A[11:0] and fill in guards.

    b. Load ACRxL (ACRx[11:0]) from register port A[11:0].

    c. Send result from ACRxH to general-purpose register.

    d. ACRx = Port A [15:0] $\times$ Port B[15:0].

    e. ACRx = ACRx + Port A [15:0] $\times$ Port B[15:0].

    f. ACRx = scale (ACRx) using factors 2.0, 1, 0.5, and 0.25.

    g. ACRx = ACRx + ACRy.

    h. ACRx = ACRx − ACRy.

    i. ACRx[23:12] = Saturation(Round ACRx)).

    j. ABS ACRx[23:0].

13.20 There is only one 5b $\times$ 5b signed integer multiplier, one 16-bit full adder, two 16-bit registers, and some multiplexers. Design instructions and a datapath to calculate 8b $\times$ 8b two's complement integer multiplication. A pipeline register is required between the multiplier and the adder.

    a. Describe the bit-accurate computing procedure R [15:0] = A [7:0] * B [7:0] using HDL or pseudocode.

    b. Define instructions and use these instructions to program the 8b×8b multiplication.

    c. Design the datapath circuit.

    d. Design a table of all control signals for each instruction.

13.21 The data width on a processor memory bus is 4 bits. Each general register in a file holds 4 bits. The data within an accumulator register is 12 bits ACR [11:0], including four guard bits. Please design the output circuit that passes the final computing result with native data precision 4 bits from the ACR to the general register file with minimum truncation error. Describe the circuit functions in detail using either HDL code or circuit.

13.22 While using a 2-cycle MAC module (one cycle for multiplication), which of the following pseudocode is/are correct?

```
{1:  ACR1 <= ACR1 + R1*R2;
 2:  R3 <= round(saturation(ACR1)); }
     /*code1*/


{1:  ACR1 <= ACR1 + R1*R2;
 2:  R3 <= saturation(round(ACR1)); }
     /*code2*/


{1:  ACR1 <= ACR1 + R1*R2;
 2:  NOP;
 3:  R3 <= round(saturation(ACR1)); }
     /*code3*/


{1:  ACR1 <= ACR1 + R1*R2;
 2:  NOP;
 3:  R3 <= saturation(round(ACR1)); }
     /*code4*/
```

13.23 By using pseudo instructions and using a single MAC datapath, make a short firmware to execute ACR1 = Real $[(a + jb) \times (c + jd)]$ and ACR2 = Imaginary $[(a + jb) \times (c + jd)]$.

REFERENCES

[1] Omondi, A. R. (1994). *Computer Arithmetic Systems Algorithms, Architecture and Implementation*. Prentice Hall.

[2] Pirsch, P. (1997). *Architecture for Digital Signal Processing*. John Wiley & Sons.

[3] Lapsley, P., Bier, J., Shoham, A., Lee, E. A. (1997). *DSP Processor Fundamentals, Architectures and Features*. IEEE Press.

[4] http://www.xilinx.com/support/documentation/user_guides/ug073.pdf.

[5] Fettweis, G., et al (1996) Strategies in a Cost-Effective Implementation of the PDC Half-Rate Code for Wireless Communications. *46th IEEE Vehicular Technology Conference (VTC'96)*, pages 203–207, 28.04-01.05.96.

Control Path Design

The organization and hardware microarchitecture implementation of the control path will be discussed in this chapter. Microarchitecture implementation of PC FSM and instruction decoder will be discussed in detail.

14.1 CONTROL PATHS

The control path is a hardware module in a processor core. It handles both synchronous and asynchronous tasks. A synchronous task is the running of the program, and an asynchronous task is the handling of events not in the program flow, such as handling an interrupt. Two parts of hardware can be found in a control path: the instruction flow controller and the instruction decoder. The peripheral modules and I/O modules are not functionally part of the control path.

The control path in a basic DSP processor is illustrated in the left part of Figure 14.1, with gray background. Inputs of a control path include flags from the datapath, target address, configuration vectors, and interrupt vectors. Outputs of a control path include control signals to the datapath and address generator, as well as the immediate data decoded from the decoded instruction.

The program memory address is generated from the PC finite state machine (FSM). The PC FSM manages asynchronous jobs before fetching a new instruction. If there is an accepted interrupt, the PC FSM stacks the PC pointing to the next instruction to be fetched. At the same time, the PC FSM fetches an instruction from the interrupt service entry, pointing to an instruction in program memory. If there is no accepted interrupt, a normal instruction is fetched by the PC FSM according to the program flow.

The fetched instruction is decoded and becomes control signals. Some of the control signals go to the address generator controlling address calculation and memory enable signals. Some of the control signals go to the datapath, controlling register file access and data manipulation. Some of the control signals stay in the control path for the decision of the next instruction. Finally, some tightly coupled peripherals might be controlled by the instruction decoder. In most cases, to isolate the design of the DSP core, the instruction decoder does not supply control signals to

475

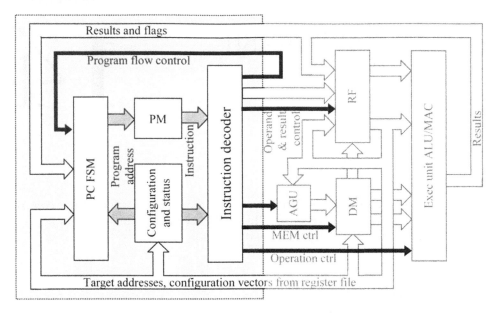

FIGURE 14.1

Control path in a DSP processor.

peripherals. Peripherals instead are controlled via reading and writing peripheral registers (special registers). A typical and simplified control path is depicted in Figure 14.2.

Before going into the details of the control path, you need to know how an instruction is executed in a processor. Most DSP processor architectures are based on Harvard architecture. In Figure 14.3, the instruction execution procedure is illustrated by a flowchart.

In this flowchart, an instruction is fetched from the address pointed by the PC (program counter). After being decoded by the instruction decoder, the instruction will be executed either in the path of data computing or move or in the path of branch execution. Here a branch instruction could be a normal jump or a function call and return. In Figure 14.3, pipelining and instruction-level parallelization have not been mentioned; these will be discussed in detail later in this chapter.

14.2 CONTROL PATH ORGANIZATION

The design of a control path consists of two steps: the organization of the control path and the microarchitecture hardware implementation of the control path. The organization includes functional specification, partition, allocation, and scheduling of pipeline.

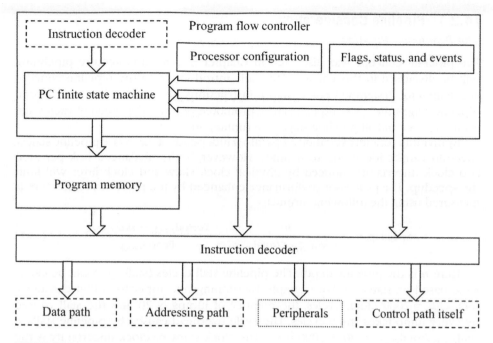

FIGURE 14.2

Simplified function description of control path.

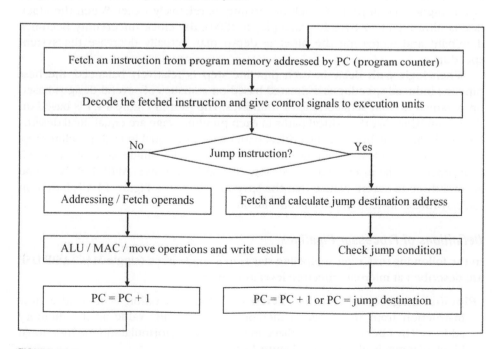

FIGURE 14.3

Procedure of executing an instruction.

14.2.1 Pipeline Consideration

The Processor Pipeline

The pipeline concept was discussed in Chapter 3. Microarchitecture pipelining will be discussed in this chapter. The early introduction to the pipeline concept is essential for designing branch (jump) instructions. To partition jobs into several pipeline stages, the system speed can be promoted because the speed of each stage is much faster and all pipeline stages are running in parallel.

By dividing parallel execution steps into n independent hardware pipeline stages, speedup can be ideally up to n times. However, both imbalanced task partition and clock uncertainty induced by physical clock skew and clock jitter will limit the speedup. The processor performance enhanced by the pipeline architecture is measured using the following formula:

$$\text{Speedup} = \frac{n}{1 + \text{stall}_{\text{cycles}}/\text{total}_{\text{cycles}}} \times \frac{\text{Period}_{\text{clock}} - \text{skew}_{\text{clock}}}{\text{Period}_{\text{clock}}} \qquad (14.1)$$

Here n is the pipeline depth. The pipeline stall cycles ($\text{stall}_{\text{cycles}}$) can be calculated based on statistics. For example, by running one application that consumes $\text{total}_{\text{cycles}} = 2000$ clock cycles, the number of jumps is 20 and the stall caused by every jump takes three cycles. The cycle overhead caused by pipeline stalls is $\text{stall}_{\text{cycles}}/\text{total}_{\text{cycles}} = 20*3/2000 = 3\%$. The clock skew or clock uncertainty is the deviation of clock arrival time from the ideal arrival time. The deviation increases when design scale (chip size) is larger and silicon feature size is smaller. When clock rate is higher, the impact of clock uncertainty is relatively larger. When the clock rate is ultra high (>2 GHz, for example, in 2006), the clock uncertainty becomes dominant and increasing the pipeline depth may actually decrease the system speedup.

When logic path delay in each pipeline step is relatively balanced, the best pipeline-induced speedup is estimated based on statistics of several designs based on parameters from semiconductor foundries. The plot in Figure 14.4 is based on the assumption that the critical paths in each pipeline stage are equal. Real designs are much different, because the critical path cannot be equal in each pipeline step; the real speedup is lower than the curve in Figure 14.4. For single issue (single MAC) DSP processor, the pipeline depth for low-cost and low-power ASIP DSP should be around three to seven. Notice that the speedup shown in Figure 14.4 is based on logic synthesis instead of full custom silicon design.

Definitions of Processor Pipelines

In the following text, typical pipeline stages in a single issue (single MAC) ASIP DSP are described at microarchitecture level as examples.

Pipeline stage 1: Instruction Fetch (IF). The operation is to read a new instruction from the program memory using the PC value as the memory address. The operation is formalized in the following formula, and the functional block diagram is illustrated in Figure 14.5.

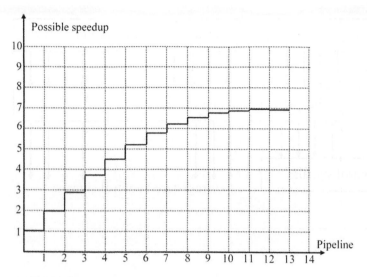

FIGURE 14.4

Speedup versus the number of pipeline stages.

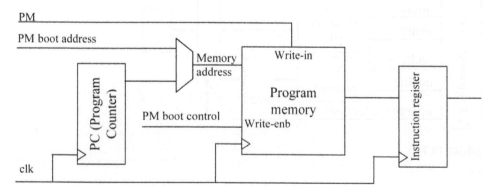

FIGURE 14.5

Instruction fetch hardware and pipeline.

```
instruction_register <= program_memory[program_counter];
```

Pipeline stage 2: Instruction Decoding (ID). The fetched instruction is stored in the instruction register. In this clock cycle the fetched instruction will be decoded. Decoded control signals might be clocked or not. The function of the instruction decoder is described by the following formula and the functional block diagram is illustrated in Figure 14.6:

```
control_signal
  <= decoding_logic (instruction, processor_configuration,
status);
```

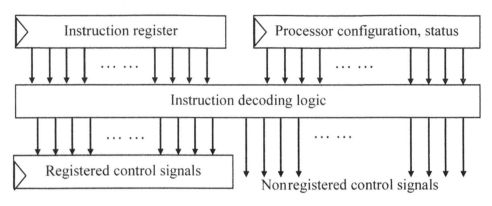

FIGURE 14.6

Instruction decoding.

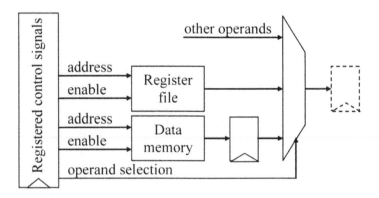

FIGURE 14.7

Operand fetch.

Pipeline stage 3: Operand fetch (OP). Operands can be fetched either from the register file or from the data memories. The memory address or register address should be provided during the instruction decoding. During the OP pipeline stage, the operand is supplied to the input of the computing devices in the datapath. The operation of OP is given in the following formula and the functional block diagram is illustrated in Figure 14.7:

```
operand <= operand_source (operand address);
```

Pipeline stage 4: Execution (EX). The execution of the instruction may take one clock cycle or multiple clock cycles. The ALU, register file, or memory access operation normally takes one clock cycle, and the MAC operation usually takes two or more cycles. Normally, the execution of the MAC instruction in the first clock cycle is multiplication, and the execution in the second clock cycle is accumulation. The operation of EX is given in the following formula and the functional block diagram is illustrated in Figure 14.8:

```
result <= OP (operand A; operand B; control_signals);
```

Pipeline stage 5: Write back or store result (WB). Storing results may take one clock cycle or part of EX without using a complete clock cycle. In some processors, when the write (input) port of the register file is complicated, one cycle is necessary for writing back the result. The function is described by the following formula and the functional block diagram of WB is illustrated in Figure 14.9:

```
Register_file [address] <= result;
```

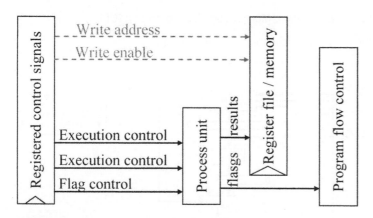

FIGURE 14.8

Pipeline stage for execution.

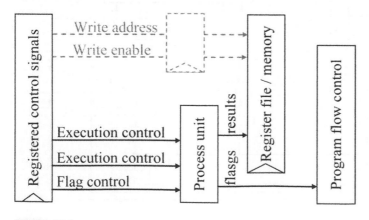

FIGURE 14.9

Pipeline stage for result store.

Example 14.1

Design pipeline architecture for **Senior** processor according to the specification given in the appendix.

The appendix gives a simplified pipeline structure. Refinement of the pipeline is shown in Figure 14.10.

The pipeline design is based on six stages:

- Instruction fetch.

- Instruction decoding.

- Operand Fetch and Memory address computing.

- Execution cycle one: ALU, MUL (for MAC), Register access, Memory access.

- Execution cycle two: Accumulation, Multiplication for convolution (CISC).

- Execution cycle three: Accumulation for convolution (CISC).

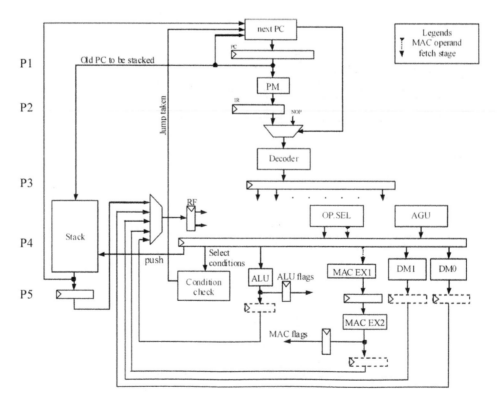

FIGURE 14.10

The pipeline design of the senior DSP core.

The pipeline for convolution is special because the operand is from the output of data memories.

Analysis

The pipeline structure is simple and efficient. There might be a hidden critical path: that is the added delay to the pipeline 5, the multiplication for the convolution. The added delay is from the (possible long) interconnection between the memory output port to the multiplier in port.

14.2.2 Interrupt Management

Why Do We Need Interrupts?

A processor has only one PC FSM, and it is designed to do one thing at a time. However, users want the DSP processor to execute multiple tasks, including I/O handling. It is possible to run tasks not existing in the main program using an interrupt calling the task to run. In this case, the DSP processor appears to do many things at the same time. The processor cannot run all tasks simultaneously. The processor shares its time among the various running programs. It appears that the processor is doing many things at once to the user (for example, audio is decoded when decoding the video).

Many peripheral modules need to send information to and from a DSP core, and they expect to be able to get the processor's attention when they need to do this. The core has to balance the information access with various parts of the machine and make sure they are handled and organized. There are two basic ways that a DSP core could do this:

Polling: The DSP core points to each peripheral module and asks if they want to do something by checking the status register of the peripheral. This is called polling the devices. Polling could be used when the system is simple and the computing load is low. It is not popular for advance applications. Much time is wasted by polling all devices one by one. The service latency of each device could be high.

Interrupting: Another way is to let a device request a service when necessary. This is the basis for the use of interrupts. When a device has data to transfer, it generates an interrupt that says, "I need your attention now." The processor then stops the running task and serves the request. Multiple requests from multiple peripheral modules can be handled based on the priority level of each peripheral module.

Interrupt Concepts

An interrupt can be hardware interrupt or software interrupt, both indicating the need for attention and requiring the core to do something for the interrupt. A hardware interrupt is an asynchronous signal from hardware outside the core. A software interrupt is a synchronous event in software running in the core. In most cases, an interrupt is implicitly a hardware interrupt.

A hardware interrupt causes the processor to save its state of execution via a context switch, and begins execution of an interrupt handler, the software handling the interrupt service. Software interrupts usually are implemented as instructions in the instruction set, which causes a context switch to an interrupt handler similar to a hardware interrupt. Interrupts are a commonly used technique for multitasking— for example, the main task and the I/O task.

A request of a hardware interrupt is collected by an interrupt controller, which is a peripheral hardware module outside the DSP core. The interrupt controller accepts a request of an interrupt and asks the DSP core to give a service. The DSP core stops executing the current task, gives the hardware resource to the interrupt, and serves the interrupt. After serving an interrupt, the DSP core continues the task interrupted.

A complete handling process of an interrupt is distributed to the interrupt controller, the core hardware, and the software. In this section, the complete interrupt process will be listed. However, only the hardware in the core for interrupt handling will be discussed. The interrupt controller will be discussed in Chapter 16, and the interrupt handling software will be discussed in Chapter 18.

Interrupt Process

An interrupt handling is depicted in Figure 14.11. An interrupt is initialized by an interrupt request (a hardware signal requests an interrupt service) from a hardware pin (step 1) of interrupt source (a hardware device requests service from another device). The interrupt priority (the order of importance of interrupt services) is checked and compared with the priority of the current running program. If the interrupt request has the higher priority, the interrupt will be accepted; otherwise, the interrupt will be rejected (step 2). Either an interrupt is accepted or rejected; an interrupt acknowledgment will be sent back to the hardware requiring the interrupt (also step 2).

As soon as the interrupt is accepted, context saving will be performed. Contexts (typically the next PC and the current flag values) of the interrupted program will be stored (step 3) on the stack so that the program can be resumed later. Other state does not need to be stored if the interrupt service program is very easy and the register usage is restricted and predictable. Normally, before serving a normal interrupt, context saving will be performed (step 4). After performing context saving, the interrupt service will be executed (step 5). The interrupt service routine is the function requested by the interrupt. After executing the interrupt service routine, the stored context will be restored (step 6). Finally, the PC and the flag of the interrupted program will be restored (context restore), the End Of Interrupt (EOI) signal is sent to the interrupt controller, and the interrupted program will continue (step 7).

Contexts of DSP applications include only register data, and special and general register values. A context switch is the process for context storing and context restoring of the interrupted task. To temporally store the context of a task, the context of the interrupted program will not be damaged during the execution for

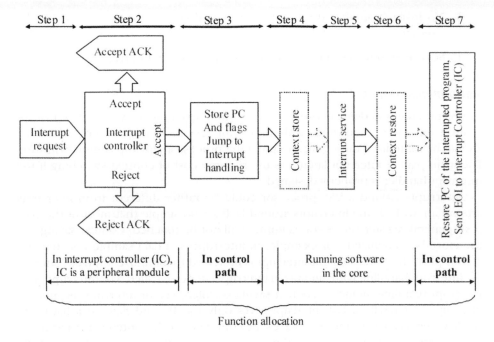

FIGURE 14.11

Interrupt handling process.

the interrupt service, and the task can be continued after restoring the context. The processor hardware resources can thus be shared by multiple tasks. A context switch in a DSP processor is usually a register context switch for a thread. Memory segments used by different threads will be isolated instead of shared. A thread is an interrupt service task. It can be an application program, an I/O subroutine, or a system service routine.

Handling of interrupt requests are conducted in the Interrupt Controller (IC). IC will be discussed in detail in Chapter 16. Context switching and interrupt service will be implemented as the assembly code and be executed by running the code in the core. Software will be discussed in Chapter 18. In this chapter, we will discuss only the reaction of an interrupt and the execution of "return from an interrupt service" in the control path hardware, which are steps 3 and 7 in Figure 14.11.

The interrupt controller as a peripheral module will be discussed and designed in Chapter 16. In this chapter, the discussion will be limited to functions allocated in the control path. To accept an interrupt, the control path must leave the machine in a well-defined state (precise interrupt). It includes:

1. The PC + 1 and the current flag are saved in a known place, normally pushed on the stack.

2. All instructions fetched before accepting the interrupt must be fully executed.

3. No instruction after the interrupt accepted (PC + 1 or PC + 2...) shall be executed.

4. Send EOI to IC when executing the "return from an interrupt."

Performance issues

Interrupt latency is defined as the period of time from the arrival of the interrupt request until start time running the interrupt service routing. Interrupt latency should be minimized during the design of the processor and application firmware. The main part of interrupt latency is the runtime cost of context switching if the interrupt handler is properly designed.

Interrupts around a DSP processor could be rather different from interrupts around an MCU. Some interrupts around DSP are so simple that most of the context (general register values, for example) will not be touched. Context saving can therefore be minimized. For example, an interrupt is "a data sample updated at a sensor port." In this case, a fast interrupt is preferred to minimize interrupt latency.

Interrupt handling with minimized interrupt latency is called "fast interrupt", and interrupt handling including context saving is called "vector interrupt" or "normal interrupt." Generally, a fast interrupt saves only the PC and flags to a hardware stack, whereas a vector interrupt saves the context (usually contents in register file, accumulator registers, PC, flags, and processor configuration) into a software stack. Steps 4 and 6 in Figure 14.11 will not be executed when running a fast interrupt.

In some cases, minimum interrupt latency is required also for some vector interrupts. A shadow register file is required to handle at least one interrupt with minimum latency. A shadow register file is another identical register file including all registers that application program can access (all general registers, accumulator registers, and special registers). As soon as an interrupt is accepted, the shadow register file replaces the main register file as the computing buffer for interrupt service. By supporting of the shadow register file, the context in the main register file can be kept and one vector interrupt can be served with minimum latency.

14.3 CONTROL PATH HARDWARE DESIGN

14.3.1 Top-level Structure

A simplified block diagram of control path architecture in a DSP processor is illustrated in Figure 14.12.

The program memory stores the programs in binary format (code) to be executed, and the code is loaded either by a booting FSM or by the MCU via DMA during the system initialization phase. The program memory could be booted either during the reset (power-on) process or during the initialization of a new application. The booting FSM is suggested to be independent to the instruction flow controller because the instruction flow controller does not work until the code is loaded.

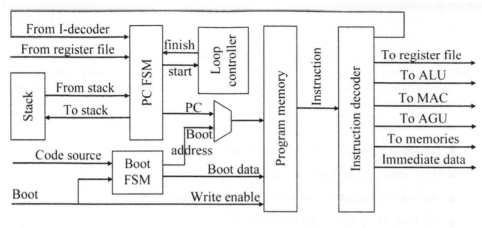

FIGURE 14.12

Simplified control path diagram.

The instruction flow controller (also called the PC FSM) consists of the PC and its FSM. It gives the next PC as the address to the program memory. Conditions for the decision of the next PC include asynchronous conditions and synchronous conditions. Asynchronous conditions are exceptions and interrupts. Synchronous conditions include the current PC, the running instruction, the processor configuration, and the flags from the latest previous execution.

Processor configuration vectors are stored in a group of control and status registers. The running program can configure the DSP processor to get certain features for specific applications. For example, a processor can be configured with integer mode, saturation can be enabled after certain instructions, certain peripheral ports can be enabled, and so on. Configurations are performed by assigning configuration vectors to the configuration registers. A set of configurations should be relatively stable while running a thread.

The concept of configurability and programmability is going to be defined here. A programmable device is driven by a running program. One or several instructions are executed within an execution step (a clock cycle) sequentially in a programmable device. A configurable device is configured, and the configuration vector does not change frequently (not in every clock cycle). An ASIP DSP processor is usually a programmable device with configuration features, which means the processor is both programmable and configurable. It is programmable because it executes instructions during each execution step. It is also configurable because some relatively stable functions or features can be "configured" for a relatively long period of time. The goal to make functions programmable is to get more flexibility. The cost of program memory can be reduced by changing programs to configuration vectors, which do not change in every clock cycle. Configuring can be suitable for relatively stable functions or features. The trade-off between configurability and programmability is a popular research task, and it will still be a hotspot in the research of ASIP in the long run.

The instruction decoder essentially decompresses the fetched instructions and generates all control signals. The inputs of the instruction decoder include the fetched instruction, flags, processor status and configurations, and asynchronous inputs. Outputs of the instruction decoder include at least:

- The control information and branch information to PC FSM.

- The control signals to no-overlap loop controller.

- The control of the hardware stack (if there is a HW stack).

- The control signals to address calculation circuits.

- The control signals to data memories.

- The control signals to ALU.

- The control signals to register file.

- The control signals to MAC.

- Immediate data.

An immediate data can be carried as is or in some kind of compressed form by an instruction. The loop control circuit generates the control signals to support a nonoverlapped hardware loop. The loop control circuit could be either a simple loop counter supporting a single instruction loop or a relatively complicated circuit supporting a loop with multiple instructions. A loop controller could be even a small slave PC finite state machine supporting a loop running in parallel with the main program. In this case, the loop controller may require a small local program memory, a local PC, and a local instruction decoder.

The HW stack is a hardware stack buffer inside the control path. The hardware stack is easy to use by programmers, and its hardware cost is high. It stores the PC value and flags as the return information of a fast interrupt. When executing CALL-RETURN, only PC instead of flags will be stored. The stack in the control path is only for the program flow control.

14.3.2 **Design of Program Memory and Peripherals**

In most cases, the program memory is SRAM, which stores the binary code of the firmware and is booted from the main memory outside the DSP chip during the power-on procedure or an initial phase for a new application. In special cases, the program memory can also be a hard-coded ROM if the application is fixed. When the program memory is a ROM, programs are fixed in hardware during the chip implementation phase.

High silicon cost may be the problem when using SRAM. Low flexibility will be the problem when using hard-coded ROM as the program memory. A trade-off between using ROM and SRAM is to use nonvolatile memory technology (EEPROM, for example). The area cost is much lower, and the flexibility is as high as SRAM, at least if the program rarely needs to be changed. Drawbacks are the extra costs from the special silicon process and relatively low access speed.

The PC width, the number of bits in a PC, defines the maximum program memory size. For example, if the program memory size is 256 k words, the length of PC should be 18 bits ($2^{18} = 256$ k). It is good if the PC width is the same or less than the native data width. For example, the native data width is 16 bits and the width of PC is also 16 bits. If the PC width is larger than the native data width, a register value in the general register file will not cover all addresses of program memory. A general register will not be directly used as a target address of a global jump. That usually makes the firmware design more complicated.

A compromised solution is to collect all programs (including the subroutine library) of one application into one page of the program memory. If the page size is not larger than 2^N (N is the native data width), a native width data will be enough to cover all program addresses of the application.

If the PC width is too long, the code size of direct jump on immediate data will be too long. To use page-based program memory, all required subroutines must be collected in one memory page. The consequence is that a subroutine may not be shared by several applications and subroutines must be duplicated and allocated in each page.

14.3.3 Loading Code

The booting FSM loads code (program in binary format) to the PM in the DSP. The booting FSM could be a peripheral device or a module in the control path. Since the boot FSM could be tightly connected to PM, it will be discussed here instead.

Normally, the application program is stored in the off-chip memory, a ROM (read-only memory) or a RAM (random-access memory). Three phases of booting should be managed by the booting FSM:

1. Initialize ports and devices involved in booting.

2. Load initial program into program memory.

3. Finish a booting by releasing booting resources and start the PC FSM.

During the initialization phase, the booting FSM first initializes a peripheral port in the DSP chip as the channel to connect the port to the off-chip memory and load the binary code to PM. Before loading the codes, the booting FSM initializes the off-chip memory, its port, and configures its access mode. After assigning the initial address to the PC, the booting procedure starts.

Most ROMs supply data in byte format, and the length of an instruction word could be much longer, 16, 24, 32 bits, or longer. During the program booting phase,

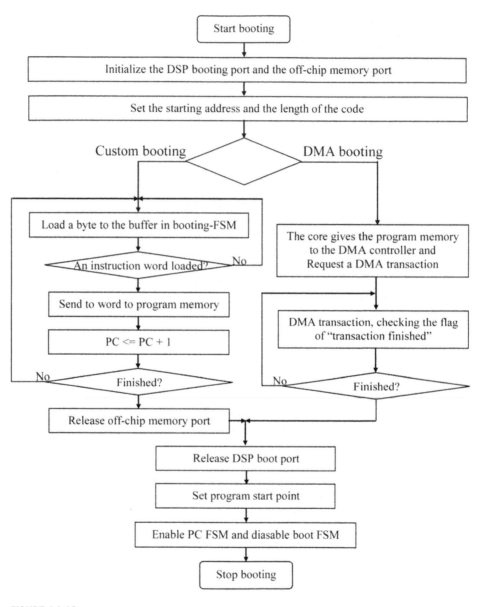

FIGURE 14.13

Example of program load at boot time.

the data type will be reformatted to adapt to the instruction word width. Therefore, a buffer for data-type adaptation is required inside the booting FSM.

When the booting is finished, the ROM and the port for booting first should be disconnected. Before shutting down, the booting-FSM initializes the PC-FSM by assigning the PC starting address. The PC starting address is the starting point of the binary code loaded to the PC. The default PC starting address can be zero.

DMA will be discussed in Chapter 16. If there is no booting FSM in a DSP processor, the microcontroller (MCU) can boot programs for it. The program can also be booted by scanning in the binary code to the program memory using a JTAG machine (see more in Chapter 19). The booting procedure is roughly the same as the flow in Figure 14.13, except that the master is MCU instead of the booting FSM. While booting, the MCU holds the DSP processor by locking or disconnecting the PC. After the booting, the MCU will write the initial address to the PC of the DSP processor and release the PC or connect the PC to the DSP processor (see the PM addressing port in Figure 14.15).

The program memory could contain a small bootloader in ROM pointed by POR (Power on Reset), which loads the main program into the SRAM part of the PM.

14.3.4 Instruction Flow Controller

The instruction flow controller (program flow controller) updates the program counter (PC) for fetching the next instruction. A basic instruction flow controller is a finite state machine (PC FSM) with the state diagram illustrated in Figure 14.14.

The PC FSM points to the next PC as the address of the next instruction to be fetched from the program memory. The default state is PC <= PC + 1. The initial state of the PC FSM is from the reset and holding of the reset. Reset may happen at any time (see the dashed lines). One special state is PC <= PC, which is used to support a single-instruction nonoverlapped hardware loop. While running a hardware loop instruction, for example, the convolution instructions, the PC FSM holds the current loop instruction for N cycles. Another special state is

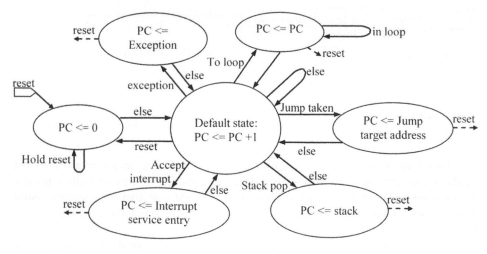

FIGURE 14.14

Example of a PC FSM.

Table 14.1   Decision of the Next PC.

Next PC	Decision priority	Conditions
PC <= 0	Highest	Reset and hold on reset
PC <= exception handling or interrupt service	2nd	Exception in hardware or Accept interrupt by interrupt handler
PC <= jump target address	3rd	CALL or Jump taken and jump on a constant
PC <= Stack pop	3rd	Return from a CALL or an interrupt
PC <= PC	3rd	To a loop and In a loop
PC <= PC + 1	Lowest	Default

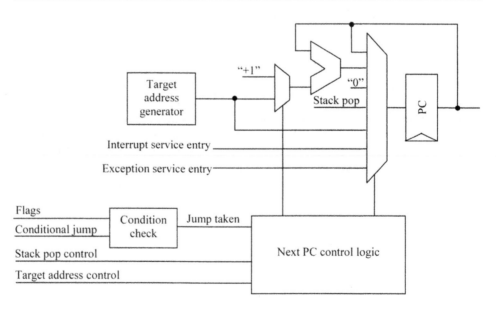

FIGURE 14.15

PC FSM circuit.

the stack pop, which is the return function when finishing an interrupt. The rest of the states are used for supporting different kinds of jumps or calls with different ways to reach the target address. The PC FSM can also be specified using a state transfer table. Conditions for the next state in the table are depicted in Table 14.1.

Conditions of conditional jump are flags from the results of the latest previously executed arithmetic instructions. "Default" in Figure 14.14 means "if there is no other condition." The condition "In a loop" means a simple loop is executed and the loop finish flag is not yet true. The logic design should follow the decision

Table 14.2 Testing Flags to get Conditions of Jumps.

CDT	Description/specification	Flag tested
–	Unconditionally true	none
EQ	ALU equal/zero	$AZ = 1$
NE	ALU not equal/not zero	$AZ = 0$
UGT	ALU unsigned greater than	$AC = 0$ and $AZ = 0$
UGE/CC	ALU unsigned greater than or equal	$AC = 0$
ULE	ALU unsigned less than or equal	$AC = 1$ or $AZ = 1$
ULT/CS	ALU unsigned less than	$AC = 1$
SGT	ALU signed greater than	$AN = AV$ and $AZ = 0$
SGE	ALU signed greater than or equal	$AN = AV$
SLE	ALU signed less than or equal	$AZ = 1$ or $AN <> AV$
SLT	ALU signed less than	$AN <> AV$
MI	ALU minus/less than	$AN = 1$
PL	ALU positive/greater or equal	$AN = 0$
VS	ALU has overflowed	$AV = 1$
VC	ALU has not overflowed	$AV = 0$
MEQ	MAC or MUL Equal	$MZ = 1$
MNE	MAC or MUL Not equal	$MZ = 0$
MGT	MAC or MUL greater than	$MN = 0$ and $MZ = 0$
MGE/MPL	MAC or MUL positive or zero	$MN = 0$
MLE	MAC or MUL less than or equal	$MN = 1$ or $MZ = 0$
MLT/MMI	MAC or MUL negative or less than	$MN = 1$
MVS	MAC was saturated	$MS = 1$
MVC	MAC was not saturated	$MS = 0$

order as depicted in Table 14.1. "First" means the highest decision priority, and "Lowest" means the lowest decision priority. Jump taken decisions are from the computing of current flags in Table 14.2.

Flags used in Table 14.2 are listed in Table 14.3.

Jump taken is a STATE of the PC FSM. The conditions of "Jump taken" include: (1) "It is a jump instruction" and (2) "the computing result on the selected flags is true." Jump taken means to load the PC with the target address in the program memory

Table 14.3   Flag Specifications.

Flag	For signed computing, flag is 1 when	For unsigned computing, flag is 1 when
AZ	ALU result is zero	ALU result is zero
AN	ALU result is negative	(ALU result has MSB set)
AC	ALU saturated (or carry out)	ALU carry out/borrow
AV	ALU result overflowed	no meaning
MZ	MAC result is zero	No unsigned computing in MAC
MN	MAC result is negative	
MS	MAC result was saturated	
MV	MAC result overflow (>40bits), sticky	

when there is a jump and the jump condition is satisfied. The simplified PC FSM circuit is illustrated in Figure 14.15.

14.3.5 Loop Controller

A loop controller is a hardware module in the control path of a DSP processor supporting iterative computing with minimized cycle cost overhead. It counts the number of iterations, checks the finish of a running loop, executes loop without extra cycle cost, and sends flags when the loop execution is finished.

When running an iteration loop, the loop controller will decide how to fetch the next instruction for the loop execution instead of running a jump instruction; extra cycle cost induced by pipeline stall can therefore be avoided. While running the last step of the iteration loop, the loop controller will decide to leave from the loop and to fetch the next instruction after the loop.

If the hardware loop feature is not used, more instructions and more execution cycles will be used. For example, if the pipelining of a processor is IF-ID-OP-EXE..., at least $3N$ extra clock cycles will be consumed by running the loop control program and pipeline stalls. Here N is the number of iteration. Wasted clock cycles and codes can be observed from codes given in Chapter 5.

The loop controller is a submodule inside the program flow controller. There are two kinds of hardware loops. The simple hardware loop supports loops consisting of single instruction. A simple loop controller is depicted in Figure 14.16. A simple loop controller consists of a loop counter and the loop flag circuit. The simple loop controller counts the loop execution and sends a flag when a loop is finished.

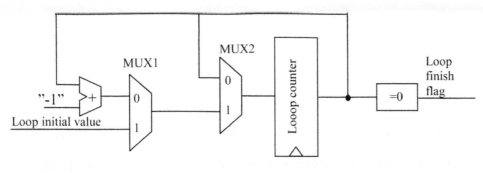

FIGURE 14.16

Single instruction loop controller.

In Figure 14.16, the simple loop controller supports a single-instruction loop such as convolution instruction. This loop controller can be configured by writing the length of a loop to the loop counter. A loop controller keeps down-counting while running the loop instruction. When reaching zero, a loop finish flag will be set.

In Figure 14.16, the initial value to the loop counter is loaded by setting MUX1 = 1 and MUX2 = 1. The loop counter keeps its value before starting the loop by having MUX2 = 0. During the execution of the loop instruction, the loop counter keeps counting downward until it reaches zero. To match the pipeline of the control logic and the finishing operations of the loop, flags might be set to true when the value in the loop counter is another value (e.g., one instead of zero).

Another loop controller, REPEAT loop controller, supports both single-instruction loop and loops carried by the REPEAT instruction, which is REPEAT M N. It carries M instructions in a loop and runs the loop N times. In this case, two counters are required; one is MC (the inner loop program counter, with initial value M from the REPEAT instruction), which counts the steps of the inner loop; and the other is LC (the loop counter, the number of iteration of the loop, with initial value N from the REPEAT instruction). The behavior of the REPEAT loop controller is described by the following pseudocode:

```
RL = M; LC = N;
/* M is the code length register, N is the loop length */
/* RL points the remaining lines to be executed in the loop */
/* LC is the loop counter */
LoopStart = PC; /* Keep the start point of the REPEAT */
IL:
if (RL != 0) {
    RL = RL - 1;
    PC = PC + 1;
    goto IL;
} else if (LC != 0) {
```

```
          RC = RC – 1;
          RL = M;
          PC = LoopStart;
          goto IL;
     }
     else LoopFlag = 1;
```

Following the behavior description, the PC FSM should be modified and redrawn in Figure 14.17.

The circuit schematic of REPEAT loop controller is shown in Figure 14.18.

M is loaded when MUX1 = 1. MC down-counts while running REPEAT. When MC = 0 and LC <> 0, the M will be loaded to MC, and MC starts down counting again. N is loaded when LoadN = 1. NC down-counts when ZeroFlag is 1. When LC is 0, the LoopFlag is set.

14.3.6 PC Stack

There are two kinds of stack: the PC stack or hardware stack, and the software stack. A PC stack is used only to stack PC and flags. It is used to support interrupt and procedure calls. When an interrupt is accepted, both the PC and flag values should be pushed onto the stack. When there is a procedure call, only the PC should be pushed onto the stack. To return from an interrupt, both the PC and the flag values should be popped. To return from a procedure call, only the PC should be popped from the stack.

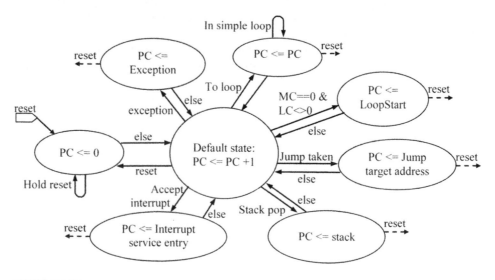

FIGURE 14.17

PC FSM that supports REPEAT.

The block diagram depicted in Figure 14.19 is an example of a low-power stack with four stack registers. A low-power stack performs push and pop by changing the stack pointer instead of pushing or popping through the register chain one by one. To simplify the circuit, only one bit data is presented in Figure 14.19. If PC and

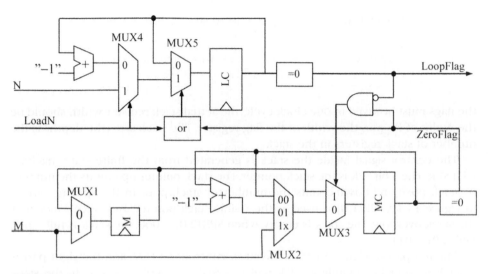

FIGURE 14.18

Loop controller as an example for advanced hardware loop.

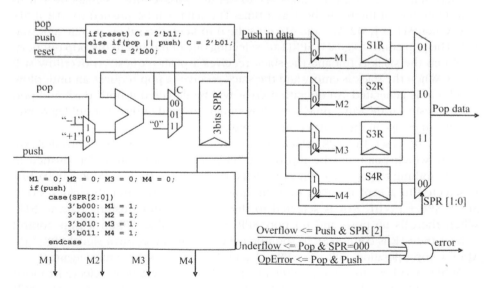

FIGURE 14.19

An example of "push-down" PC stack for fast interrupts.

Table 14.4 Possible Stack Operation Errors.

Error Type	Error Conditions
Illegal operation	The push and pop control signals are true at the same time.
Underflow	A pop occurs while all SxRs are empty (SCR = 000).
Overflow	A push occurs while all SxRs are used (SCR = 011).

the flags must be saved in one clock cycle, the actual stack register width should be the width of PC plus the width of the flag register. The stack size (the depth) is the number of stack registers in the stack.

The control signal inside the stack is generated from the finite state machine FSM-SPR. The 3-bit SPR is the stack pointer. The stack pointer up-counts the number of stack push and down-counts the number of stack pop. In the specific circuit in Figure 14.19, SPR[1:0] is used as the visible stack pointer, and SPR[2] indicates the stack overflow. The stack is empty when SPR[2:0] = 000; the stack is full when SPR[2:0] = 011.

The left part of Figure 14.19 is the stack controller, a FSM, and the right part is a stack register file including stack registers S1R, S2R, S3R, and S4R. In the stack controller, decoding control signals is based on the input signals, push or pop, and the old SPR value. Control signals M1 to M4 are push control signals of S1R to S4R. Only one of them can be 1 at a time. The data will be pushed to S1R if M1 is 1. When M1 is 1, M2, M3, and M4 must be 0 to keep the values in S2R, S3R, and S4R. The pop control signal is SPR[1:0], selecting one of the SxR to the pop out port.

When the stack is full and the stack receives a push request, an overflow will occur. When the stack is empty and the stack receives a pop request, an underflow condition is generated. An operation error will be detected if both push and pop are requested at the same time. Overflow, underflow, and OpError should generate an error message in the instruction set simulator.

After the reset, the SPR (stack pointer register) in the stack controller will be 000, which points to the top of the stack waiting for data to be pushed in (while nothing to pop out). When a push comes, M1 <= Push & (SPR = 000), S1R is selected to accept the data from the push-in data port. After this clock cycle, SPR = SPR + 1 = 01. S1R is selected to the pop port after pushing data to S1R. When there is no push and no pop, SPR keeps its value. When a pop control comes, the S1R is connected to the pop-data port by the control signal SPR[1:0]. After a pop operation, SPR <= SPR − 1 and the stack will be empty again.

When another push comes after pushing data to S1R, S2R is selected to load because M2 <= Push & (SCR = 001). After pushing the value in S2R, the SCR becomes 10 and will point to S2R as the pop register. The SCR is updated (plus one or minus one) after a push or pop command. A stack access error occurs in the cases shown in Table 14.4.

14.3.7 **Senior PC FSM Example**

Example 14.2

Design the PC FSM part dealing with conditional jumps and calls. The processor shall handle the insertion of NOPs to ensure correct execution.

The processor has the following pipeline:

1. Next PC(NP).

2. Instruction Fetch(IF).

3. Instruction Decode(ID).

4. Operand Fetch(OF).

5. Execute(EX) [Flags are set here].

6. Write back(WB).

The first thing to note is that flags will be set in the EX phase, which means that the instruction before the jump has to complete the EX phase so flags are set correctly for the jump.

Listing 14.1 lists a small program to be used in this Example. The program implements a small loop starting at PC = 0. To investigate this program closer, a pipeline table is shown in Table 14.5, which depicts how the various instructions propagate through the pipeline. The WB phase is omitted since it does not matter for our discussion.

LISTING 14.1   Simple Jump Example.

0	Label	add	r3 , r4
1		sub	r17 ,1
2		jump.eq	label
3		**or**	r3 , r7
4		set	r5 ,13
5		**xor**	r5 , r5
6		ror	r13 ,7

The rising edge of the clock (i.e., when registers are written) comes between the rows; this means that the flag register for **sub** will be written when **sub** exits the EX phase (i.e., between the **sub** and **jump** instruction in the EX column). For this reason the earliest time we can decide the jump, to take the jump or not, is when the **jump** instruction is in the EX phase.

During the time we have waited for the flags to be set, more instructions have been fetched and executed if nothing is done about it. In Table 14.5 we can clearly see how three more instructions are injected into the pipeline after the jump instruction (they are marked gray in the table). Since it is up to the processor to handle the program flow correctly, we can first ensure that the PC isn't incremented more than necessary, leaving us with the pipeline table shown in Table 14.6. Observe that the earliest we can affect the PC is in the ID phase, when the processor can identify the instruction and thus take action upon it.

To stall the PC, however, did not solve the problem that erroneous fetched instructions will be executed. We can solve this problem by inserting nop:s, leaving us with the pipeline table

Table 14.5 Pipeline Table for Listing 14.1.

NP PC+	IF PC(instr)	ID	OF	EX
1	0(add)	–	–	–
2	1(sub)	add	–	–
3	2(jump.eq)	sub	add	–
4	3(or)	jump.eq	sub	add
5	4(set)	or	jump.eq	sub
0	5(xor)	set	or	jump.eq
1	0(add)	xor	set	or
2	1(sub)	add	xor	set
3	2(jump.eq)	sub	add	xor
4	3(or)	jump.eq	sub	add
5	4(set)	or	jump.eq	sub

Table 14.6 Revised Pipeline Table for Listing 14.1.

NP PC+	IF PC(instr)	ID	OF	EX
1	0(add)	–	–	–
2	1(sub)	add	–	–
3	2(jump.eq)	sub	add	–
3	3(or)	jump.eq	sub	add
3	3(or)	or	jump.eq	sub
0	3(or)	or	or	jump.eq
1	0(add)	or	or	or
2	1(sub)	add	or	or
3	2(jump.eq)	sub	add	or
3	3(or)	jump.eq	sub	add
3	3(or)	or	jump.eq	sub

Table 14.7   Revised Pipeline Table with nop:s for Listing 14.1.

NP PC+	IF PC(instr)	ID	OF	EX	FSM state
1	0(add)	–	–	–	N
2	1(sub)	add	–	–	N
3	2(jump.eq)	sub	add	–	N
3	3(or)	jump.eq	sub	add	N
3	3(or)	or	jump.eq	sub	J1
0	3(or)	or	nop	jump.eq	J2
1	0(add)	or	nop	nop	J3
2	1(sub)	add	nop	nop	N
3	2(jump.eq)	sub	add	nop	N
3	3(or)	jump.eq	sub	add	N
3	3(or)	or	jump.eq	sub	J1
3	3(or)	or	nop	jump.eq	J2
4	3(or)	or	nop	nop	J3
5	4(set)	or	nop	nop	N
6	5(xor)	set	or	nop	N
7	6(ror)	xor	set	or	N

shown in Table 14.7. You might wonder why we opted to stall the PC. This can be answered by looking at Table 14.7; when the jump is not taken, stalling the PC enables us just to continue from the stalled PC. This also has benefits for the call instruction since we shall push the PC of the instruction to be executed so that execution resumes at the correct place after the subroutine returns. Stalling the PC leaves us the simple choice of pushing the current PC in the EX phase, where the stack most likely will reside, and thus not cluttering up the pipeline control with complicated conditions of when and how to push what.

Having investigated how a jump will affect the program flow, we are now ready to implement the PC FSM. But there is still one more option to consider. Any FSM can be implemented as either a Mealy or Moore machine. From a hardware system point of view, Moore machines usually are considered safer, but in our case we want to be able to react to input as soon as possible, and for that reason we have to choose a Mealy machine. If you do not know the difference between a Moore and Mealy machine (see Chapter 10), refer to any text in basic switching theory.

The functionality of our PC controller can be derived from the previous investigation. The PC FSM needs to be able to count to three to keep track of how many nop:s to insert. It

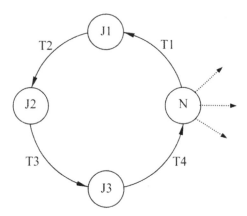

FIGURE 14.20

Jump and call part of the PC FSM.

Table 14.8   Output from PC FSM Depicted in Figure 14.20.

Transition	Condition	Output
T1	jump or call	stall PC
T2	true	insert nop, stall PC
T3	jump taken	insert nop, set PC to target address
	jump not taken	insert nop, stall PC
T4	true	insert nop

must also control the stalling and setting of the PC register. The resulting FSM is shown in Figure 14.20 and Table 14.18. The normal operation of the PC FSM is to stay in state **N**, which is just to increase PC. If a jump or call is executed, transition **T1** is taken, and then the PC FSM goes through the jump states. The PC FSM will do the same thing for a call, but in addition to a call the PC will be pushed to a stack in the EX phase. Table 14.7 also has been augmented with the states of the PC FSM. The gray dashed arrows in Figure 14.20 show that the PC FSM has to deal with more transitions than just jumps and calls (e.g., repeat or interrupts). But this is outside the scope of this example.

14.4 INSTRUCTION DECODER

The instruction decoder decodes the fetched instruction into control signals. The control signals are divided into control signals toward the datapath (ALU, MAC, and register file), control signals inside the control path, and control signals and addresses toward the memory subsystem. Peripheral modules usually are controlled by configuration registers rather than from control signals from the control path.

From the instruction decoding point of view, a processor can be a "data-stationary architecture" or "time-stationary architecture". There is an assumption that the data is available in time-stationary architecture whenever an instruction is fetched and decoded. It is up-to the compiler and instruction set simulator to give an off-line guarantee: that the data is not dependent during the runtime. Most DSP, especially VLIW DSP, is built based on time-stationary architecture.

In a data-stationary architecture, an instruction is dispatched and decoded either the execution data is available or not. The execution will be interlocked by the arriving data. The execution can be issued as soon as the data is available in a data-stationary architecture. Superscalar is a typical data-stationary architecture. In following section, we discuss the data stationary architecture by default.

14.4.1 Control Signal Decoding

There are mainly two kinds of instruction decoders: the ROM-based instruction decoder and the instruction decoder consisting of logic circuits.

A ROM-based instruction decoder is usually used in CISC architectures. Control signals are stored as microcode in a ROM. The fetched instruction is decoded to find addresses for the further fetch of microcode from the microcode memory. There could be multiple microcode memories—for example, one microcode memory storing the execution control signals and another microcode memory storing operands addressing signals.

Numerous advantages could be recognized from the architecture using microcode. First, if the microcode memory is SRAM, the microcode can be modified so that the function definition of an assembly instruction could be modified even after the chip fabrication by the end users, which brings more flexibility. Another benefit could be the orthogonal features among the subsets of microcode if subsets of microcode are allocated in several microcode memories. By fetching microcode from several memories in parallel instead of from one memory, more possible combinations give more flexibility. For example, by separating the microcode for the datapath control and for the memories, the datapath operation and the memory access can be orthogonal. The most important advantage is the possibility of fully utilizing the hardware resource in the datapath. Another benefit is that bug fixing of the microcode is possible.

The disadvantages are the longer pipeline, the extra memory (silicon area) cost and extra power consumption. The long pipeline depth includes reading assembly code, decoding the assembly code into microcode address, and loading microcode from microcode memory in a sequential order. If the microcode-controlled machine is a slave processor and the subroutines in the slave processor are short, the disadvantage can be negligible. A good example is a 2D vector processor for streaming data processing. The 2D slave processor executes simple and extensive parallel tasks, leaving complex flow control tasks to the master processor. The size of the microcode in the slave processor could be small. Meanwhile, the flexibility of the 2D slave processor can be achieved from flexible microcode. A microcode-based instruction decoder is described in Figure 14.21. In this decoder, the fetched instruction is not for the final execution; instead it will be used for the further fetch of the microcode.

Figure 14.22 illustrates an instruction decoder based on a decoding logic circuit. It decodes machine codes directly into control signals. The decoding logic usually is based on AND-OR logic. Each AND-OR circuit supplies one control signal; a control signal can be decoded from several instructions (OR function), and a subfield of an instruction comes to one AND gate.

According to the design of the processor pipeline, some control signals need to be registered and some control signals will be used directly as combinational signals. The processor pipeline architecture can be simplified and can be more efficient when using unpipelined control signals. The unexpected transmission delay on control signals can be eliminated if it is clocked locally where it is used. Compared to the decoder of a microcode machine, the advantages of the logic circuit-based instruction decoder are simple, with low silicon cost. However, the disadvantages are the limited flexibility and longer debugging time.

A control signal might be driven by many instructions and one instruction will drive many control signals. A control signal toward a bus will drive every bit on

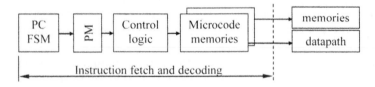

FIGURE 14.21

Microcode-based instruction decoding.

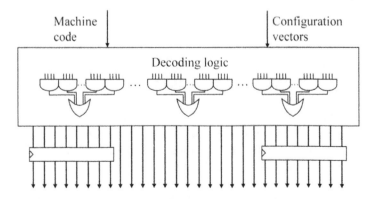

FIGURE 14.22

Instruction decoder using custom logic.

the bus, which means that the fan-out of signals of the instruction decoder could be very large. One functional control signal may drive hundreds of input pins. In Chapter 11, a typical fan-out of a control signal was analyzed by an example. A total of 512 fan-outs were identified. To manage huge fan-out on a physical level, image signals can be used to share the load. By using image signals, huge fan-out can be distributed physically to several physical drivers. For example, an output signal from one decoding logic block can be split and assigned to multiple modules (if the synthesizer cannot do it):

```
control_signal_100 = decoding_logic (condition1 ...
                     condition n);
control_signal_100_ALU = control_signal_100; // split it to ALU
control_signal_100_MAC = control_signal_100; // split it to MAC
control_signal_100_MEM = control_signal_100; // to memories
```

14.4.2 Decoding Order

The instruction decoder must be implemented following a predefined decoding order. High prioritized conditions mask low prioritized conditions. Most processors have relatively similar instruction decoding hierarchies; carefully planning the decoding hierarchy can avoid design bugs in the decoding logic. The decoding hierarchy should be specified before the instruction decoding. To implement a decoder, the RTL coding must be in a particular way. For example, if the hierarchical order is specified, the IF-THEN-ELSE coding style should be used instead of using CASE-based coding style.

```
if(reset) begin
   // React on reset
end else if(debugging_mode) begin
   // Decoding and supporting the debugging mode
end else if(exception) begin
   // Decoding for exception handling
end else if(interrupt) begin
   // Decoding for interrupt
end else begin
   // Decoding the fetched instruction
end
```

14.4.3 Decoding for Exception, Interrupt, Jump, and Conditional Execution

When an exception happens, the instructions in all hardware pipeline stages can be flushed. It is not necessary to keep the current execution of the damaged execution because the execution of the current task is completely wrong.

When an interrupt is taken, the PC and the current flags must be stacked. At the same time, the interrupt service entry is loaded to the PC. There will be no pipeline stall, and the pipeline flush will not be executed. All instructions in the pipeline, already loaded in the core before the interrupt, should be executed.

A special case occurs when there is a branch or call instruction in the pipeline. If the branch is taken, the current PC may be incorrect and the following instructions will be flushed away. It is important not to push the wrong PC value on the stack in this case. One way to solve this problem is to delay the processing of the interrupt when there is a branch or call instruction somewhere in the pipeline.

There are differences between the return from a function CALL and the return from an interrupt. When executing a return from a function call, only PC will be returned from the stack instead of flags. However, when decoding a return of interrupt, both PC and flags must be returned from the stack. When executing a return from interrupt, an EOI (end of interrupt) signal will be generated.

When a jump is taken, all instructions loaded after the jump instruction usually are flushed (removed from the pipeline) instead of executed. The implementation in the instruction decoder is to generate and use a flush control signal to flush all other control signals except for the control signals that decide the next PC. The flush signal is actually the jump taken signal.

Another way to handle jumps without flushing the pipeline is to identify jumps early in the pipeline and simply to stop fetching instructions. In this case, the time from the jump fetch to the jump execution (PC changed) is wasted time because no useful work is performed. Other solutions include always fetching a certain number of instructions (functionally independent to the jump) after the jump instruction and always executing them (no flush) in the jump delay. In this case, no cycles are wasted if the programmer is able to find useful instructions.

For conditional execution, if the condition is true, the result from running the conditional instruction should be written back; otherwise, the execution will be ignored. When running a conditional execution instruction, the signal "condition is true" should be used as the enable signal to all modules that can accept the result of conditional execution. Typical modules are general register file and data memories.

14.4.4   Issues of Multicycle Execution

There are some instructions that take multiple execution cycles. For example a MAC instruction takes two execution cycles, one for multiplication and one for accumulation. A special decoder handling data-stationary architecture decoding is required to supply control signals for each execution cycle according to the data and hardware resource availability. The decoding principle for the MAC execution, for example, proceeds as follows.

During the first execution cycle,

- The multiplier executes the multiplication.

- The accumulation should be held without doing anything if the previous instruction is a single cycle instruction.

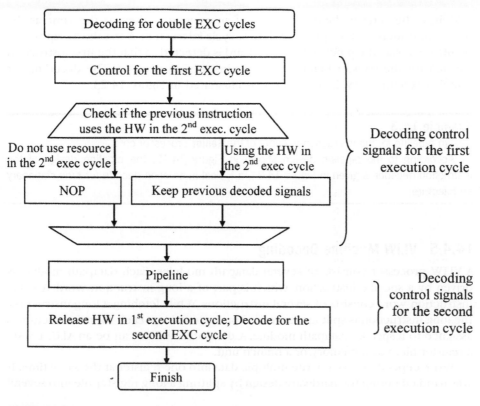

FIGURE 14.23

Decoding for double (execution) cycle instructions.

- If the previous instruction is another MAC, the accumulator should execute the accumulation for the previous instruction.

During the second execution cycle,

- The multiplier should be released to the next instruction.

- The accumulator works for the current MAC instruction, adding the result from the multiplier to the accumulator.

When the instruction following the MAC instruction uses the accumulator register locked by the MAC instruction, a data dependency hazard occurs. There are different ways to handle a hazard. One way is to design a hazard-dependent checker in the instruction set simulator; if there is a data dependent in firmware, a warning or error will be reported. The firmware designer takes care of the hazard by inserting independent instructions. This may not be sufficient because the data dependence may not happen during simulation time. A hardware hazard checker might be useful. The hardware hazard checker locks the destination register. An exception will flush the current running program if an operand address is the same as the destination address of the previous double cycle instruction. A less drastic approach would be to stall the pipeline when a hazard is detected so that the first instruction can release the register before the other instruction needs it. The decoding for double (execution) cycle instructions is illustrated in Figure 14.23.

Example 14.3

Design pipeline for the instruction decoder of the **Senior** processor core.

According to the pipeline design shown in Figure 14.10, the pipeline design of the instruction decoder is given in Figure 14.24, a typical instruction decode for time-stationary architecture.

14.4.5 VLIW Machine Decoding

A VLIW processor consists of several datapath modules. Each datapath module is driven by a specific instruction, which is part of a long instruction word. A VLIW instruction word consists of several instructions. While fetching a long instruction, the long instruction is split into several short instructions. Each short instruction is assigned to a specific datapath module. A datapath module can be an ALU, a MAC, a register file, a data memory, or a branch unit.

Write dependency is to write multiple data into one register at the same time. It was handled during the hardware design by splitting a long register file into several

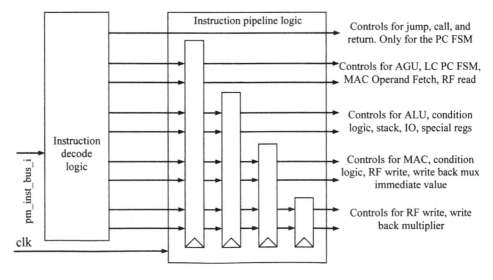

FIGURE 14.24

Pipeline design for time-stationary **Senior** instruction decoder.

cluster register files and assigning each register file to a datapath cluster. Each cluster generates only one result at a time, and a result of a cluster can be written only to its own register file.

Read dependency is to read a value that is not yet valid. It was handled during the code compiling time or code simulation time. The programmer or compiler must identify the read dependency and avoid executing the dependent data access by rescheduling the execution. If the code is written by hand, the data dependencies should be identified by the instruction set simulator.

To conclude, both write and read data dependencies are managed before executing code in hardware. Data-dependent handling and decoding can be eliminated to reduce the hardware costs. The VLIW instruction decoding is therefore relatively simple. There will be several independent instruction decoders, each decoding an instruction, and the decoding for each datapath module is independent.

14.4.6 Decoding for Superscalar

To enhance the performance, a superscalar handles data dependencies by introducing many special hardware units that can identify them. To handle the instruction decoding of the typical data-stationary architecture, the write dependency protection is handled in the reorder buffer, which is the hardware unit releasing results in the correct order to the output ports. The read dependency protection must be handled in both the instruction dispatch unit and the reservation station. All dependencies are handled by checking the instruction fetching order. The order is dynamically changed and managed via instruction renaming. The instruction for renaming is to change an instruction fetching name (PC value) to a name of fetching order. The fetching order will be used to decide the dependencies. Data dependencies can be handled by the following rules:

- When multiple writes occur to the same register, the latest fetched instruction has the right to write.

- When fetching an operand that is the result of an early fetched, yet not executed instruction, the operand fetching operation cannot be executed until the instruction is executed.

This kind of decoding is called interlocked-decoder, the decoder of the dependency-tolerant hardware in the superscalar. It decodes an instruction according to fetching order and execution status of other instructions. If there are dependencies, the decoded instruction will not be executed until the interlock is released. The principle of superscalar was discussed in Chapter 3. The instruction decoding and dependency management hardware in a superscalar can be found in Figure 14.25.

By using reservation stations in a superscalar, decoded control signals, operands, and operand addresses are reserved in one reservation station together with the name of the instruction. Operand dependencies are checked by hardware.

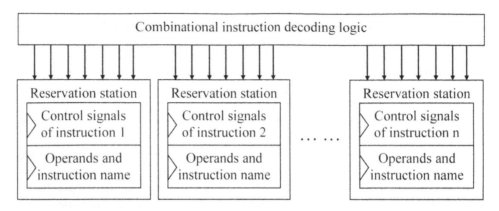

FIGURE 14.25

Instruction decoding for HW with dynamic pipeline.

When there are no data dependencies, operands can be fetched and the instruction can be issued to an execution unit together with the decoded control signals.

14.5 CONCLUSIONS

The organization of the control path and the implementation of the microarchitecture of control path have been discussed. Fundamental knowledge was introduced in the control path organization section. The microarchitecture implementation of a control path includes the design of the PC FSM and the instruction decoder. Loop controller and stack machine were also discussed. The principle of instruction decoder design was introduced. The hardware support for hardware loops is a special functional block of the control path in a DSP processor. The design of the control path in DSP processors is similar to the control path design of other processors.

EXERCISES

14.1 What are the main functions of the control path in a DSP processor?

14.2 Describe functional blocks in Figure 14.12—for example, the PM, the PC FSM, the instruction decoder, the stack, and the loop control.

14.3 Implement the program booting procedure in Figure 14.13 using pseudocode.

14.4   Design and implement a PC FSM using HDL including the circuit schematics, the jump taken decision circuits based on the flags from the ALU including "sign," "zero," and "saturation/carry out." The following cases of Next PC should be included:

a. PC <= PC + 1.

b. PC <= PC.

c. PC <= 0.

d. PC <= Target address; // Jump taken.

e. PC <= Stack pop.

f. PC <= Loop start address.

Jump taken happens when a conditional jump instruction is executed and the jump is taken.

14.5   Design a loop counter including initialization circuits. The output of the loop counter is the loop finish flag = 1 when the loop counter counts down to 1 (next will be 0).

14.6   Design a hardware stack circuit with the depth of 8 storing PC and flags. PC_in[23:0] and flag_in[2:0].

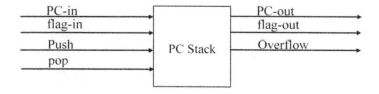

14.7   What is interrupt handling, what is interrupt service, and what is interrupt latency?

14.8   PC jumps to the interrupt service when accepting an interrupt. Why are there no pipeline stall and no pipeline flush? (*Hint:* Compare to branch taken and find the differences.)

14.9   Discuss and explain why (*Hint:* read Chapter 7)

a. When an interrupt is accepted, both the PC and flag values should be pushed onto the stack.

b. When there is a procedure call, only the PC should be pushed onto the stack.

14.10   How do you decode control signals for multicycle instructions?

REFERENCES

[1] Öwall, V. (1994). Synthesis of Controllers from a Range of Controller Architectures. Ph.D. Thesis, Department of Applied Electronics, Lund University, December 1994. CODEN: LUTEDX/(TETE-1010)/1-170(1994).

[2] Tell, E., Olausson, M., Liu, D. (2003). A general DSP processor at the cost of 23K gates and 1/2 a man-year design time. *ICASSP* 2003, Hong Kong.

Design of Memory Subsystems

15

In the first part of this chapter, we will discuss memory subsystem design including addressing circuitry, memory peripherals, and special addressing circuitry for address acceleration. The second part of the chapter gives basic knowledge of memory organization. We start by discussing the bus hierarchy and the memory hierarchy. DMA principles will also be covered within this chapter.

15.1 MEMORY AND PERIPHERALS

A memory subsystem includes physical memories, address generators, circuit around memories, and memory buses. Part of the memory subsystem (address generator) is allocated in the processor core, and the rest is allocated beside the core in the SoC. A memory subsystem is one of the three basic components of a SoC; the other two are the DSP subsystem (including cores and accelerators) and MCU cores. A DMA (direct memory access) controller and a MMU (memory management unit) as hardware peripheral modules functionally are components of the DSP subsystem.

The complete view of a memory subsystem is given in Figure 15.1. Data memories and the address generation will be carefully discussed in this chapter. The memory bus was discussed in Chapter 3. The DMA will be introduced in this chapter from the user's (programmer's) view. The DMA hardware design will be discussed in the next chapter. The SW in memory subsystem is not the focus of the book. However, programming for memory access will be discussed briefly in Chapter 18, and programming for parallel memory access will be discussed in Chapter 20.

15.1.1 Memory Modules

A memory module is a hardware storage component where data can be stored and retrieved to and from a physical position inside the module. Each position is specified by a unique address, and the data access (storing and retrieving) is to point to the corresponding address with associated access control (read or write). When the access control is a read, the data at the pointed position in the memory module

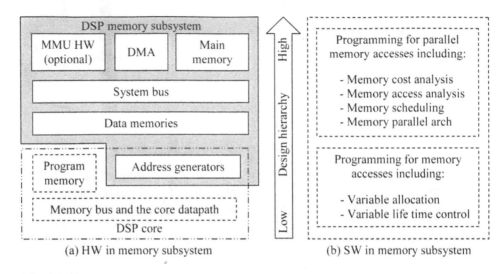

FIGURE 15.1

A complete view of memory subsystem.

will be available on the module output. When the access is a write, the data on the data bus to the memory module will be written into the pointed position in the memory module 0.

From a functional point of view, memories can be divided into read-write memories (RAM) and read-only memories (ROM). From a physical implementation point of view, RAMs can be classified into static memories (SRAM) and dynamic memories (DRAM). A static memory remembers its data as long as the power is supplied and the read operation is nondestructive. A static memory cell (bit) is usually designed with six transistors, which is easy to implement in a standard digital CMOS silicon process. Compared to DRAM, SRAM consume more silicon area. By contrast, the data stored in a dynamic memory cell can only be valid during a period of time. When reading from a dynamic memory, the data in the read storage cell are destroyed. Thus the data has to be restored by a write back operation. Leakage currents will discharge the small capacitors, and the dynamic memory therefore must be refreshed at regular time intervals. The dynamic memory must be fabricated using a special dynamic memory silicon process. Since only one single capacitor, and one transistor are needed for each cell (bit) in the dynamic memory, it consumes much less silicon area than the static memory. However, the silicon process for the dynamic memory is not suitable for implementation of logic circuits. Therefore, most DSP processors use static memories on-chip. Dynamic memories are slower than static memories. The major advantage of the dynamic memory is the higher memory cell density, and it is very often used as large off-chip storage.

If we look at the interface in details, memories could be divided into single-port memories and multiport memories. Single-port memories have one write port and one read port, but read and write cannot be performed at the same time. Multiport memories have more than one read and write port, and multiple read-write access can happen simultaneously. For example, a dual-port memory has two read-write ports; therefore two data words can be accessed simultaneously.

From the synchronization point of view, a memory can be synchronous or asynchronous. A synchronous memory uses the memory clock to control the data access. The address, the data to be stored, and the control signals (enable, write_enable) must be available when the positive edge of the clock arrives. Instead of using a clock, an asynchronous memory uses handshaking to control the memory access.

In this book, we are interested in synchronous single-port static random access memories (SRAM). If it is not mentioned specifically, a block of memory in a DSP processor means a synchronous static random access memory. This concept applies to both data and program memories. A 128-bit memory (128 memory words, one single bit in each word) as a memory component is illustrated in Figure 15.2.

An SRAM circuit consists of memory cells, row decoder, column decoder, and read-write circuitry. A memory cell is essentially a latch that stores a single bit of data (and its inverse, due to the design of the latch). When addressing a memory cell, its row is selected, and all the memory cells in the row are connected to their column line pairs. When performing a read operation, one column line pair is selected, and the sense amplifier of this column samples the stored value and its inverse from the column line pair. Both values are needed since the sense amplifier is a differential

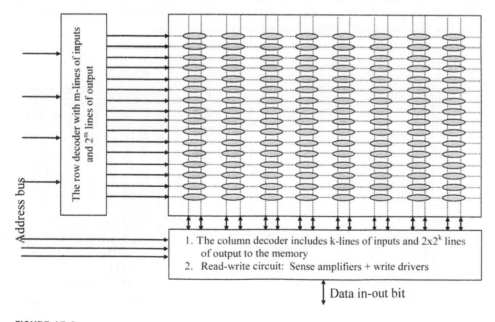

1. The column decoder includes k-lines of inputs and 2×2^k lines of output to the memory
2. Read-write circuit: Sense amplifiers + write drivers

Data in-out bit

FIGURE 15.2

A basic memory schematic (data width = one bit).

amplifier. This also speeds up the read process since the column lines will drive the sense amplifier in the opposite direction. If 1 is stored in the memory cell, the column line will be one and the column-bar line will be zero. The sense amplifier will then sense the value as 1 and send it to the output latch. When performing a write operation, the corresponding column line pairs are connected to the write driver of this bit, and the bit value will be written into the cell because the selected row line connects its cells to the column lines.

If the memory address space is $n = m + k$, the row decoder will convert m bits to 2^m bits row lines; and the column decoder will covert k bits into 2^k pairs of column lines. During one memory access operation, only one of 2^m rows and one of 2^k columns can be selected for one-bit access. The remaining rows and columns will not be connected to the outside. In Figure 15.2, eight memory cells are connected to a row and 16 memory cells are connected to a line-pair of a column. For example, when the first row is selected, all eight memory cells on the first row are connected to their column lines. When the first column is selected, all 16 memory cells on the first column have possibilities to be connected to the outside. However, only the first cell in the first row is connected to the outside via the column lines because only cells on the first row are selected and connected to column lines.

A memory cell is described in Figure 15.3(a). Two inverters are connected as a latch keeping the value stored. When a row is selected, the row line is "high" and every cell on the row is connected to its column lines (column line + and column-bar line −). While reading, the sense amplifier amplifies the output of the cell to the read out circuitry. While writing, the value is written to the column line and column line bar; thus the latch will be updated with the new value.

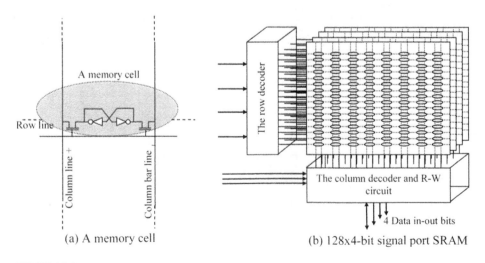

(a) A memory cell   (b) 128x4-bit signal port SRAM

FIGURE 15.3

A memory module.

Finally, a memory of 128 half-byte-words (with 4-bit width) is given in Figure 15.3(b). In Figure 15.3(b), each output of the row decoder drives four row lines over four bit pages. The column decoder connects four bits from four pairs of column lines on four pages to the 4-bit read-write bus.

15.1.2 Memory Peripheral Circuits

Now we define a memory as a black box with fixed functional specification and timing on ports. A basic synchronous SRAM can be specified as in Figure 15.4.

Column and column-bar lines are precharged when the clock is high. When the clock is low, the sense amplifier is connected to a column and column bar line-pair and the data is read out. In most cases, a memory module from a memory synthesizer has both row and column decoders as combinational logic. The write-in port usually accepts data at any time. We therefore need registers around the compiled memory module.

Most memories in a DSP processor are synthesized by a memory compiler from the silicon supplier or supplied by a third-party tool vendor. A memory behavior model described by HDL (VHDL or Verilog) is the input of the memory compiler.

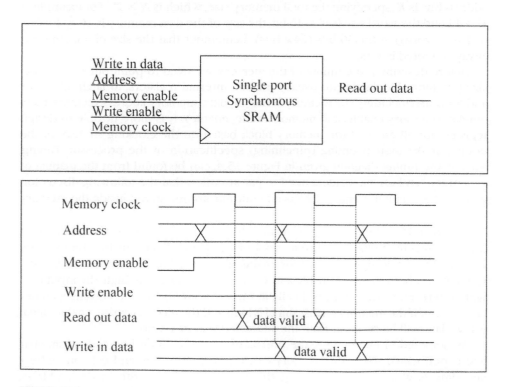

FIGURE 15.4

Black box of a synch single-port SRAM and its timing.

The behavior model gives the functional description of the memory including block size, word size, synchronization, and timing. Detailed physical descriptions are not included in the behavior model. Therefore, more inputs need to be specified during the synthesis of the memory, like row size, column size, speed limit, power constraints, and other specific parameters. After the synthesis, a memory block is generated with its input and output pins specified as in Figure 15.4. The ports are:

- Memory clock: The clock for synch of the memory addressing, read, and write.

- Address: Address bus carrying $K = \log_2 M$ bit address.

- Memory enable: While high, the memory is selected.

- Write_enable: When high, the current memory operation is "write data."

- Write_in_data: The bus carrying N bits data to be written to the memory.

- Read_out_data: The bus carrying N bits data to be read out from the memory.

N is the word width of the memory and M is the memory size. The width of the address bus is K, specifying the total memory size, which is $N \times 2^k$. For example, if $K = 12$ and the word size is $N = 16$-bit, the size of the data memory is 16×2^{12} ($4k$ of 16-bit words) = 65536 bits ($64k$ bits). Remember that the size of a memory is always counted in bits.

Before describing the timing of the memory, we need to point out that synthesized memory blocks do not usually include internal registers, to latch all inputs and outputs, (including the memory write_in_data, memory address, memory read-out-data, memory enable, and memory write_enable). It is more flexible to design registers for all ports of the memory block based on the clocking strategy of the design, or the system timing (pipelining) specification of the processor. Timing, such as the timing diagram given in Figure 15.4, can be found from the manual of a memory synthesizer. Readers need to pay attention that the following discussion may not be useful if memory inputs and outputs are registered inside the memory block.

In Figure 15.4, the memory clock (note that the memory clock may be different from the machine clock or system clock) is used for synchronizing the memory access. The enable signal is used for enabling or selecting the memory to be activated. When both memory enable and write_enable signals are high, the input data will be written to the addressed cells in the memory during the current memory clock cycle. If the memory enable signal is high and the memory write_enable signal is low, data will be read out from the addressed memory cells.

As depicted in Figure 15.3, the precharge of the column line and column-bar line is performed when the memory clock is high. The row and column address decoding is also performed during this time. Therefore, write_in_data, address, memory enable, and write_enable signals should be available and stable at the positive clock edge. On the negative clock edge, if the write_enable is low, the sense amplifier will read out data on the selected column and column-bar lines.

In a DSP processor, the memory clock is often the same as the machine clock. According to silicon scaling rules, the logic speed increases faster than memory speed. Thus the critical path in a DSP processor has shifted from the MAC to the memory. Considering this, it is natural to use multiple clocks in a high-performance DSP processor—for example, a machine clock for the computing logics, and a memory clock for the memory access. Implementing a DSP hardware using a double-clock system is relatively advanced and will be discussed in later chapters in this book.

In general, a synthesized memory block does not include any peripheral logic required by the DSP processor. The DSP processor designers use the synthesized memory block as a basic component and design peripheral circuits to fulfill the functional and timing specification. By checking the timing in Figure 15.4, it can easily be found that we will have timing problems while using the hardware multiplexing technique on the memory peripheral circuits. For example, if exactly the same clock is used for computing logic and memory access, it will not be allowed to have any delay from the logic circuits or interconnections between the memory peripheral registers and the memory input ports. Registers include the address register, the write_in_data register, and the enable signal registers. In the following cases, memory timing problems and possible solutions will be discussed.

Actually, memory interface timing often generates critical paths, and it is not enough to predict the critical paths based only on the computing logics. Experienced designers have to pay attention to both logic circuits and memory peripherals at the very beginning when planning the architecture of a DSP core.

Timing Problem Case 1

Data processing is between the data_out port of the memory and the register receiving the data.

Problem Description

Data processing might be required before registering the data from the memory read_out_data port as illustrated in Figure 15.5. For example, a predecoding circuit might be needed after the program memory, or a data selection multiplexer selecting data from multiple memory pages.

Solution

If the input data to the register is available far before the clock edge, we may not need to do anything. If the result from the speed checking indicates a timing problem, the design must be improved. One solution is to modify the duty of the memory clock to make it unbalanced. In this case, the first clock phase (clock-high time) will be less than 50% of the whole clock cycle. Thus, the read_out_data can be available in time, as can be seen in the third block of Figure 15.5. The drawback is that the design of the clock becomes complicated.

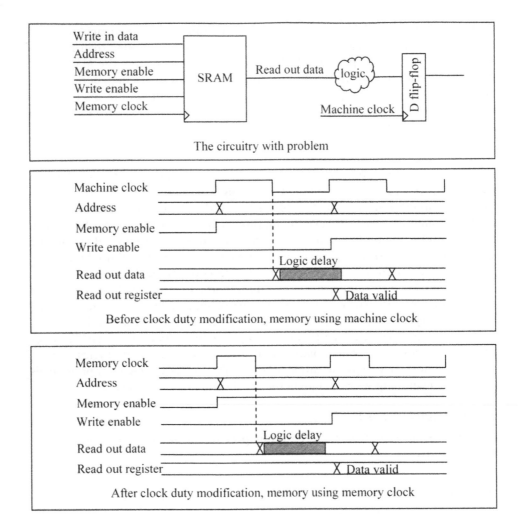

FIGURE 15.5

Modifying the duty cycle of the clock to improve memory readout time.

Problems Introduced by the Solution

If the first clock phase is too short, there might not be enough time for the column line to be precharged, which means that the memory may not work. We have to check the requirement on memory timing carefully and make trade-offs as early as possible.

Timing Problem Case 2

Logic circuitry between address pointer register and the memory block.

Problem Description

Logic circuitry might be allocated between the address pointer register and the write_in_data port of the memory block as shown in Figure 15.6. For example, a multiplexer might be used to select different address pointers. This kind of logic cannot be avoided when a processor has a comprehensive instruction set. This creates two potential problems—a longer critical path and the more power consumption because of the extra toggling introduced by addresses arriving late.

Solution 1

If the critical path is not through the logic circuitry, we could accept this solution. We have to accept the extra power consumption from the row and column decoder inside the memory. If the memory size is rather large, for example, a 16-bit SRAM with 64 K words, the added power consumption from extra toggling is rather high.

If there is a timing problem, one method is to move the memory clock behind the machine clock by using a special PLL (phase lock loop). In this way, there will be extra time available for the logic circuitry as illustrated in Figure 15.6.

Problems Introduced by the Solution

After phase-shifting Δt, the read_out_data from the memory will also have a delay of Δt. If the Δt is too large, the read_out_data may not be available when the next positive edge of the machine clock (not memory clock) arrives. Designers need to check the memory timing carefully before making the decision of phase shifting.

Another problem is the design cost of a PLL and distributing two clocks instead of one. Timing verification becomes therefore difficult. A trade-off is to modify the address circuitry and to move the logic before the address register, which will be discussed as the next solution.

Solution 2

Another solution is to use an image address pointer register directly connected to the memory block. In Figure 15.7, any of the address pointers could be used as the address source and selected by MUX. In a normal case, a designer may use a MUX to select outputs of address pointer registers. In this way, extra logic circuits are generated in between these address pointer registers and the address port of the memory. By adopting the solution depicted in Figure 15.7, the input signal to address pointer register is selected and registered by the "image address pointer register" close to the memory block. Thus the logic circuits in between the address pointer register and the address port of the memory can be eliminated.

Timing Problem Case 3

Similar to the timing problem case 2, the timing problem case 3 is illustrated in Figure 15.8. It is the logic circuit between the data output register and the memory write-in port.

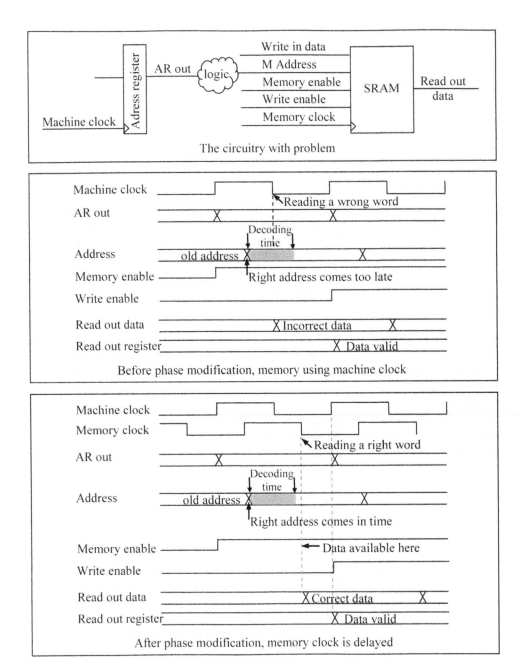

FIGURE 15.6

Shifting the phase of the memory clock.

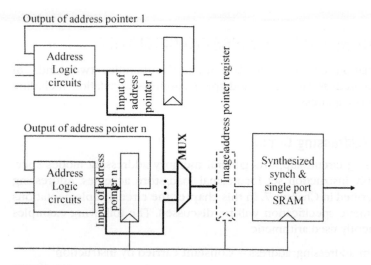

FIGURE 15.7

Image address pointer as a slave register.

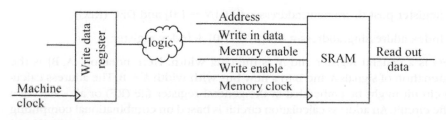

FIGURE 15.8

Design of memory peripheral circuit, case 3.

Problem Description

Data processing logic might exist between the data output register of the computing logics and the write_in_data port of the synthesized memory. For example, a multiplexer might be required to select data from different sources. If possible, this kind of logic should be avoided.

Solution

The solution and possibly introduced problems are similar to those of case 2.

There are other cases. For example, the enable signal could be multiplexed after the register. The solution can be similar to solutions mentioned in Figure 15.7.

15.2 DESIGN OF MEMORY ADDRESSING CIRCUITRY

The address calculation circuitry for data memories handles two kinds of calculations: general address calculation and special address calculation, including modulo and bit-reversal addressing.

15.2.1 General Addressing Circuit

The general addressing circuit supports general memory addressing modes specified by the assembly instruction set for normal load/store accesses. Addressing arithmetic was described in Chapter 7. In this chapter, the circuit implementations of addressing arithmetic specification will be discussed. The following examples give the most frequently used arithmetic:

- Direct constant addressing: address = Constant carried by instruction.

- Register addressing: address = Selected $REG[N - 1:0]$.

- Segment plus offset: address = $REG[N-1:0] + \{(N-M)\text{'b0}, OFFSET[M-1:0]\}$.

- Register post increment: address = $REG[N - 1:0]$ and INC (REG).

- Register post decrement: address = $REG[N - 1:0]$ and DEC (REG).

- Index addressing: address = $REG[N - 1:0]$ + Index register.

REG is a general register and N is the data width. Here notation {A, B} is the concatenation of signals A and B to a new bus with width A + B. The address calculation circuit might be embedded in the general register file (RF) or allocated in a specific circuit. An address calculation circuit is based on combinational computing logic with registered output. The register holding the result of address calculation is called the address pointer or address counter. The output of an address pointer is connected to the memory and back to the address computing logic as feedback. A simplified addressing circuit is given in Figure 15.9. In this figure, the Inputs will be used as the source information from the register file or the instruction decoder. The Initial address is used to initialize the addressing algorithm for iterative computing. The Keeper keeps the old address, and the Addressing feedback is used for iterative computing. The Combinational output is used when an image register is used close to the memory block (mentioned as case 2 in the previous section of this chapter).

The circuit can support both post and pre address computing (increment or decrement). When supplying address computing control signals a cycle before the cycle of using addressing, pre address computing is implemented. When supplying address computing control signals at the same cycle of using addressing, post address computing can be implemented.

Multiple address pointers (registers) might be needed in a DSP processor for two reasons. The first reason is the requirement to support multiple simultaneous

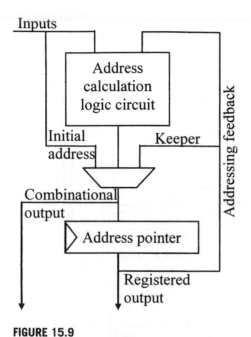

Inputs

Address calculation logic circuit

Initial address

Keeper

Addressing feedback

Combinational output

Address pointer

Registered output

FIGURE 15.9

Simplified addressing circuit.

operands. The second reason is the fact that multiple address pointers can be assigned different values in nested loops so that the address algorithms can be executed with minimized addressing cost overhead.

If we need two operands from two memories simultaneously, two address calculation logic circuits in parallel are required. In principle, each computing path (a datapath cluster) needs two address calculation logic circuits for DSP computing. Moreover, the need of a number of address pointer registers is more than the need of a number of address calculation logic circuits. More address pointer registers are needed to store data addresses of simultaneously running subroutines and nest loops to minimize the cost of swapping pointer values.

Implementation of an address calculation circuit is to map different addressing arithmetic to the address calculation circuits. The circuit in Figure 15.10 gives an example of implementing the address calculation circuit.

In this example, up to seven address pointers can be selected by the instruction set (seven temporary addresses can be stored simultaneously). Two address pointers can be selected in parallel because two identical address calculation logic blocks (I and II) are available in the circuit. Therefore, data and coefficients allocated in two memories can be accessed at the same time (a normal requirement in a DSP processor). Since the address calculation logic I and II are identical, we will describe only one of them.

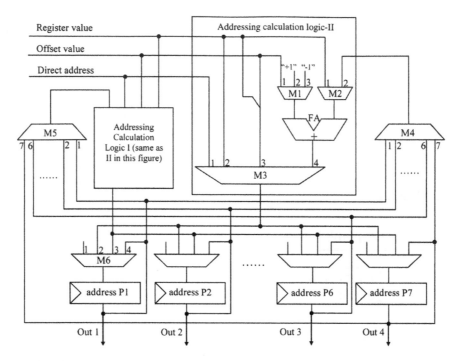

FIGURE 15.10

An example of an addressing circuit.

Address Register

Any address register can be driven by either one of the two address calculation logics blocks. For example, the address point register P1 can be connected to the address calculation logic I when port 3 of M6 is selected by the control signal to multiplexer M6. Similarly, when port 2 of M6 is selected, the address calculation logic II is connected to the address point register P1. If an address point register (e.g., P1) is not selected for addressing any memory, it may hold its value by connecting port 4 to its input (the multiplexer M6). The address register can also be used for purposes other than address calculation; for example, it can be a general-purpose register in a register file. In this case, P1 either keeps its value by receiving the feedback on port 4 or updates its value by connecting the input port 1 of multiplexer M6 to the input bus of registers in Figure 15.10.

Address Calculation Logic

Only one adder is used in address calculation logic for normal address calculation. Thus the hardware cost can be low by hardware multiplexing. Four multiplexers are used to select different addressing modes. In the address calculation logic II, the addressing mode is controlled by the control signals for multiplexing M1, M2,

Table 15.1　Control of Address Calculation Logic.

Addressing mode	M1	M2	M3
Direct constant addressing	X	X	1
Register addressing	X	X	2
Segment concatenate offset	X	X	3
Segment plus offset	1	1	4
Register incremental	2	2	4
Register decrement	3	2	4

and M3. M4 selects one address register as the feedback for iterative addressing $(++/--)$. For example if the address calculation circuit II output is connected to P1 input (by selecting port 2 of M6 to P1), port 1 of M4 should be selected in order to connect P1 output to the address calculation circuit II input via M4. Table 15.1 gives a detailed description of functional multiplexing.

Here, multiplexers M1 and M2 are used to select operands for the full adder in order to perform the address calculation. For example, when M1 selects the offset value, M2 selects data from the segment register, the segment from the register will be added to the offset, and the result will be stored in one of the address registers. The multiplexer M3 is used to select different kinds of addressing modes. The multiplexer M4 is used to select which address register is involved in iterative addressing.

The width of all buses in the address calculation logic is based on the address space 2^A. For example, if the memory address coverage is $2^{16} = 64\,K\,(0-65535)$ the address width will be $A = 16$ bits. The width of segment and offset is based on the design of the instruction set.

In contrast to data calculation, address calculation must be absolutely correct. That is, usually there is no guard, no saturation, and no round operations in address calculation logics. An overflow of address calculation usually introduces an exception. However, in a very special ASIP, the overflow of address calculation might be acceptable to support special mode based addressing algorithms.

15.2.2 Modulo Addressing Circuit

Convolution frequently is used in DSP applications. The details of convolution are defined in the following formula and the flowchart in Figure 15.11.

$$y(n) = \sum_{i=0}^{m-1} x(n-i)c(i) \tag{15.1}$$

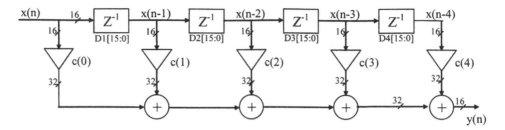

FIGURE 15.11

A detailed view of a 5-tap convolution.

In this equation, $x(n)$ is the vector representing the input data, $c(i)$ is the vector representing the coefficient, and $y(n)$ is the result of the convolution.

A data memory is naturally used as a computing buffer supplying $x(n)$ to $x(n-4)$ to a processor. Although the implementation of both the convolution circuit and the MAC has been discussed, the way that data and coefficients are supplied to the MAC has not been explained. In this section, the detailed discussion is focused on the memory access in order to support convolution.

A part of the memory is defined to be a data FIFO buffer with boundaries TAR = top address of the data FIFO buffer and BAR = bottom address of the data FIFO buffer. A data address pointer DPR is used to point at the position of $x(n)$. Here ACR is the accumulator register. The convolution then is implemented in the following steps:

Initialization
    Define the loop length $m = 5$
    Clear (zero) ACR
    CAR <= starting address of coefficients
    DAR <= starting address of data
    TAR <= top address
    BAR <= bottom address
Write $x(n)$ to the top of the FIFO
Start convolution loop at point $x(n-0)$ let i count from 0 to $m-1=4$
    Load $x(n-i)$ using DPR as the address pointer
    Load $c(i)$ using CPR as the address pointer
    Get the first step result in ACR <= ACR + $x(n-i)*c(i)$
Compare DPR with BAR (check the bottom of the FIFO)
    if equal, DPR <= TAR (jump to the top of the FIFO)
    else, DPR $--$
    CPR ++
Check and finish the iteration loop
    If the loop does not finish ($i < m-1 = 4$) then jump to step 3(a)
    If the loop finishes then write ACR to $y(n)$ after rounding and truncation

The program is presented in Example 15.1:

Example 15.1

```
01 // example 15.1
02 ACR = 0;
03 LCR = m;                              // LCR is loop counter
                                            register;
04 CAR = coefficient_starting_address;
05 DAR = data_starting_address;       // for data memory DM;
06 TAR = top_address;                 // of FIFO in DM;
07 BAR = bottom_address;              // of FIFO in DM;
08 DM(DAR) = input_new_data;          // from in-buffer or
                                         in-port;
09 OPA <= DM(DAR);
10 OPB <= TM(CAR);
11 BFR <= OPA * OPB;
12 ACR <= ACR + BFR;
13 if DAR == BAR then DAR <= TAR
14       else DEC (DAR);
15 INC (CAR);
16 DEC (LCR);
17 if LCR != 0 then jump to 09
18    else Y <= Truncate (round (ACR));
19 end
```

The essential technology used in this program is the software FIFO. Recalling the principle of hardware FIFO, data is shifted one by one into a hardware register chain so that the oldest data is discarded, the next-to-oldest data becomes the oldest data, and so on. In the same way, the arriving data replaces the first data in the FIFO, and the first data is shifted to be the second data in the FIFO.

In software emulated data FIFO, instead of shifting data forward in FIFO, the address pointer is shifted backward; therefore the hardware FIFO is emulated using memory. The software emulated data FIFO buffer has its boundaries marked by its top and bottom address registers. When the backward shifting of the address pointer reaches the bottom boundary of the FIFO, the top boundary value is assigned to the address pointer. A circulation buffer therefore is built up as the software emulated data FIFO.

An implementation of the software emulated data FIFO can be found from code lines 13 and 14 in Example 15.1. If these two lines of code are to be executed within one clock cycle, implementation using dedicated hardware is needed, which is defined as the hardware acceleration of modulo addressing. Usually, accelerated modulo addressing is coded in most instruction sets of DSP processors. The principle of software emulated data FIFO buffer is illustrated in Figure 15.12. The implementation of accelerated modulo addressing is depicted in Figure 15.13.

In Figure 15.12, a part of the data memory is used to realize a software emulated data FIFO. The bottom and top registers of the FIFO are BAR and TAR. The data address pointer is pointing to the position of the first data in the FIFO. The example starts at state 0, where the new data is not yet in the FIFO. At state 1, the FIFO accepts new data, and the data pointer DAP shifts one step backward. The arriving new data word overwrites the oldest data $x(n-4)$. At state 2, another new data

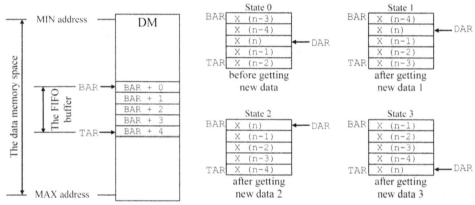

Example: The procedure of a FIFO getting a new data sample

FIGURE 15.12

Modulo addressing using software emulated data FIFO buffer.

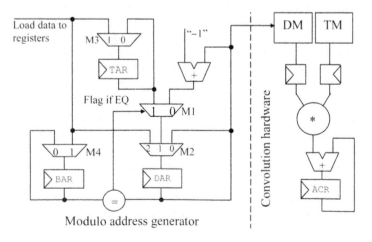

FIGURE 15.13

Circuit for modulo addressing.

arrives and the data pointer equals to the bottom address register BAR after shifting a step backward. The interesting thing is to accept yet another data at state 3. When DAR = BAR, DAT will not shift backward; instead, the top register value is assigned to DAR, and the latest sample is written into the memory cell with address of TAR instead of BAR-1, because BAR-1 is out of the range of the software emulated data FIFO.

The implementation of the addressing acceleration for the FIFO emulation in a data memory is called modulo addressing; it is depicted in Figure 15.13.

In Figure 15.13, the modulo addressing circuit based on the decrement step size of 1 can be used for accelerating the data memory addressing for convolutions (remember, modulo addressing acceleration is designed only for data access instead of coefficient access). While running the initialization program for convolution, the initial value of data_starting_address is loaded to DAR by assigning value 2 to the control signal of M2 (this can be done by one assembly instruction). Then, the value 0 is assigned to the control signal of M2 (default) so that the DAR value will be kept. The initial value of top_address is loaded to TAR by assigning 1 to the control signal of M3 (one assembly instruction); and the initial value of bottom_address is loaded to BAR by assigning the value 1 to the control signal of M4 (one assembly instruction). When the loop is executed, control signals for M3 and M4 will be set to 0 in order to hold the value stored in these registers. By assigning 1 to the control signal of M2, either incremental addressing or modulo addressing is allowed. In case DAR is not equal to BAR, DAR will decrease one step each clock cycle. When DAR is equal to BAR, the comparator will generate a control signal to copy the value in TAR to DAR in the next cycle.

15.3 BUSES

A bus can be a register bus, a memory bus, an I/O bus, and a SoC bus. In this section, we will discuss only the bus inside a DSP core such as a register bus and a memory bus. High-level buses will be discussed in Chapters 16 and 20.

A bus is a shared communication link consisting of a set of wires connecting multiple subsystems. Bus protocols such as physical and link protocols are also part of the bus. The flexibility of the bus allows new devices to be added and removed transparently if the same bus standard is used.

In modern SoC designs, system performance bottlenecks usually show up in the bus system. The maximum I/O throughput of the whole system is limited by the communication throughput of the bus system, which is largely determined by the length of the bus, the number of devices attached to the bus, and the latency of the bus.

Buses usually are defined and classified according to a certain layered hierarchy. For example, the bus inside a processor core is the bus at the lowest level. It connects the processing and the storage units. Therefore it is also called the processor-memory bus (PMB) (see Figure 15.14). The PMB is determined during the processor specification phase, and it will not be scalable after the RTL design. PMB is controlled

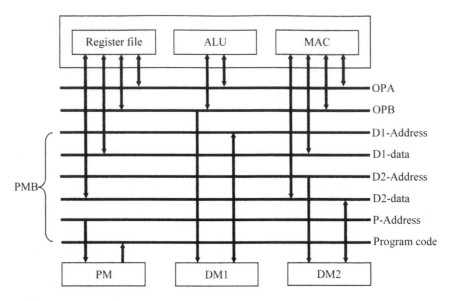

FIGURE 15.14

Processor memory bus in a DSP.

by instructions. For any instruction sequence that can be executed, there may not be any contentions (access conflicts) on the bus. Such structural hazards must be considered and managed during the architecture design phase. Runtime arbitration therefore is not needed. The PMB protocol is send-and-forget, meaning that the reliability of the data on the bus is guaranteed during the physical design of the chip. Error detection or error correction on the PMB is not needed because it has no extension to the external ports of the chip. The requirement on PMB is to maximize the memory-to-processor bandwidth. PMB can be a group of parallel connections to supply multiple data simultaneously. PMB is a bus inside a processor that does not follow any industrial standard or interwork protocol.

15.4 MEMORY HIERARCHY

15.4.1 Problems

The memory subsystem in a DSP processor includes program memory, data memories, addressing circuits, on-chip memory bus, and off-chip memory interface. The so-called memory performance gap is the increased differences between the speed of the computing logic and that of the memories. The speed of logic circuits has been increasing faster than the speed of memories following the scaling of the silicon technology. Thus the memory subsystem in a DSP processor has become the major performance bottleneck.

The memory size required by high-end systems always exceeds the capacity of the on-chip memory subsystem. The cost of putting the same amount of on-chip memory into an ASIP chip is several times higher compared to off-chip memory. One reason is that the silicon technology for off-chip memory is simpler than the one required by logic circuit. Another reason is that the design costs of an off-chip memory is shared by large volume and is negligible compared to the cost of designing an on-chip memory for an ASIP. A trade-off for a DSP system is therefore to minimize the on-chip memory costs and to try to use as much off-chip memory as possible.

To summarize, the challenges and problems for the ASIP designer are:

1. Most DSP algorithms can be executed in parallel, which requires simultaneous memory accesses (multiple read and write to the memories). For example, the LMS (least mean square) algorithm requires three memory accesses simultaneously: read of data memory, read of coefficient memory, and write of coefficient memory.

2. Recall that the memory performance gap, the speed of a CMOS logic circuit, in general is faster than the speed of CMOS SRAM. Therefore, the clock for computing logic in a high-performance DSP processor could be twice or even three times higher than the memory clock. This will increase the design complexity.

3. Moving more DSP applications on chip requires more on-chip memories. However, the size of on-chip memory is limited by the chip size, or in another word by the silicon cost. Issues related to the use of off-chip memory are therefore essential. The design and use of off-chip memory must be specified during the design phase of the DSP core. However, if the use of off-chip memory is specified in too much detail and too early, the system flexibility will be low.

4. As the off-chip memory is large and slow, it is important to maintain high bandwidth and low latency communication between the on-chip memories and the main memory outside the chip. The DMA availability (high bandwidth and low latency) is especially critical when a single DMA channel is used by multiple hardware devices.

15.4.2 Memory Hierarchy of DSP Processors

Generally, both the program memories and the data memories in a simple DSP processor are single-port SRAM. The gate count of a single MAC-based DSP processor is much less than $100\,k$, meaning that the silicon area consumed by logic circuits is less than $1\,\text{mm}^2$ when the silicon feature size is less than $0.18\,\mu\text{m}$. The memory subsystem of a DSP processor consumes the major part of the silicon area. For example, if we use a 32b $\times$ 32 K SRAM as the program memory and two 16b $\times$ 64 K

SRAM as the data memories, the silicon area consumed by the on-chip memory subsystem is at least 10 times more than the area consumed by logic circuits.

In many SoC applications, a DSP processor is only a part of the hardware on the chip. The silicon area for a DSP processor is determined by the system-level specification. We have to be very careful and try to minimize the DSP memory costs on the chip. The main challenge for a system-on-chip design is the communication between on-chip and off-chip memories. Supporting complicated DMA transactions with high bandwidth and low latency is essential.

Many applications require very large memory space. For example, a video encoder may need both reference frames and the current frame in the computing buffer. The size of a frame with VGA size is $640 * 480 * 1.5 * 8$ bits $= 3.68$ megabits. Including two reference frames and the current frame, the computing buffer will be much larger than 11 megabits.

Obviously, the memory costs are too high in this case, and we have to use off-chip memories. What we can do is to partition the memory into two parts: the on-chip memory and the off-chip memory. We have to store most programs and data into the off-chip memory and keep most frequently used and currently running programs and data in memories on the chip. This is called the hierarchical design of the memory subsystem of a DSP processor. The design of the memory subsystem is the major challenge of the current research of an embedded system on a chip. A memory hierarchy of a DSP memory subsystem can be illustrated in Figure 15.15.

A scratchpad memory is a high-speed on-chip data memory used for temporary storage of calculations (inputs, computing buffer, and results). The scratch-pad memory depicted in Figure 15.15 is a SRAM for data buffering, which is interacting directly with the running program. Cache memories in this book are defined as memory

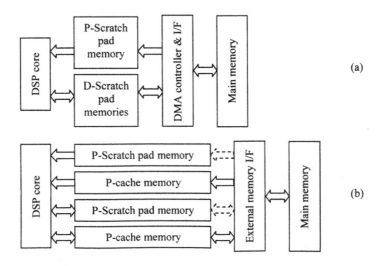

FIGURE 15.15

Memory hierarchy of a DSP processor.

buffers that dynamically allocate and store programs and data. This means that the contents in both program and data cache memories are updated automatically.

Usually only scratch-pad memories are used in most ASIP DSPs, because they are designed for a small number of specific applications and the programmers are relatively sure about what to run on the DSP processor. This is in contrast to a general-purpose DSP processor (which is designed for general-purpose applications). A cache is useful in order to automatically allocate and update data and programs on the chip. Since ASIP DSP is the focus of this book, DSP processors using cache will not be discussed here.

Scratchpad memory-based DSPs give less time uncertainty when scheduling a task. Cache memory-based DSPs are more area efficient because fewer memory cells are required on-chip. To reduce the time uncertainty of task scheduling, both scratchpad memories and cache memories can be found in some DSP processors. The scratchpad memory stores the most frequently used and relatively stable data or programs such as the most used subroutines and the most used coefficients and parameters.

In Figure 15.15(a), the memory subsystem of the DSP core includes a program memory and a group of data memories. Both the program memory and the data memories are connected to the main memory via DMA. The program memory only loads the program code from the main memory, whereas the data memories can either read data from the main memory or write data to the main memory. This architecture is suitable for applications based on static scheduling, which means that the timing of the system is fixed during the product design phase.

In Figure 15.15(b), the memory subsystem of the DSP processor uses both cache and scratchpad memories. The cache is used as the level one buffer for the dynamic data and the scratchpad memory is also used as the level one buffer but used only for storing static or stable data with a longer variable life time. Data in a scratchpad memory can be updated while preparing for a task. It is seldom updated during the runtime and the scratchpad memory is seldom connected to main memory while running an application. Cache memories are connected to main memory whenever a cache mismatch happens. This memory subsystem is more flexible and can be used for both static and dynamic scheduling. However, extra cycle cost introduced by cache mismatch impairs the computing performance. It is not a trivial task to make decisions on the memory hierarchy as well as the size of memories at different hierarchical levels. Benchmarking and modeling on data transactions may help the designers to find a good solution.

15.5 DMA

15.5.1 DMA Concepts

DMA stands for Direct Memory Access, and it is a way to exchange data blocks between memories directly without using the processor for data access. A master device (a DSP processor core in this book) can run a DMA transaction by configuring and issuing a DMA task. The configuration subroutine is a part of the application program. A configured DMA transaction is an asynchronous task running in the background in parallel while running the DSP application. Normally, the transfer of a data blocks takes a long time, but this is hidden because of the parallel execution. The DMA unit is a peripheral hardware module to the DSP core, supporting data transfer between memories.

In this section, DMA principles will be introduced first. Methods of using DMA in DSP applications will be discussed. DMA hardware design details will be covered in the next chapter. In this chapter, the DMA hardware will be introduced very briefly, just for an understanding of DMA handling. A DMA controller (or DMA handler) is a hardware peripheral module. It consists of two FSM and DMA datapath resources; the first FSM is for DMA task configuration, and the second FSM is for running DMA transactions.

Datapath resources in a DMA controller consist of several source ports, several destination ports, one FIFO buffer, and at least two address generators. Datapath resources can be offered for DMA transactions. An offer is a DMA channel, and there are several channels designed in a DMA controller. A channel can be configured for a DMA transaction. When the DMA hardware resource is available, a configured DMA channel will be issued and a DMA task takes place. A DMA channel consists of:

- A time period with fixed priority.
- A source port and an address generator for the source port.
- A destination port and an address generator for the destination port.
- A FIFO buffer.
- A configurable connection network to connect the channel.

If a DMA controller can offer eight channels, it means that hardware resources in a DMA controller are designed and can be organized to offer up to eight DMA configurations. A channel is configured by a DMA task for a data transaction. After a configuration, a DMA channel is in-use. The task in an in-use channel can be issued by activating the channel. The channel then uses the hardware and transfer data within the time slot assigned for the channel. When not within that time slot, hardware assigned to the channel can be used by other channels. After the data transaction, the task is removed and the channel is available for further DMA transactions.

In this chapter, a DMA task is a data transaction between two memory modules, or between one memory module and a peripheral port, or between two peripheral ports. A task is prepared by a master device, a DSP core. The DSP core handles a DMA transaction by running three subroutines. The first subroutine is to prepare for a DMA transaction by forming a DMA transaction (a DMA task) table. The second subroutine is to configure the DMA task by sending the transaction (task) table to the DMA controller. The third subroutine (optional) is to start (issuing) the DMA transaction task in the DMA controller. The third subroutine can be eliminated if the configuration and task issuing is merged into the second subroutine.

Similar to the handling of other peripheral tasks, the procedure of a DMA transaction between the DSP processor and the main memory is as follows:

1. The DSP processor plans a DMA transaction task and makes a DMA task table.

2. The DSP processor sends the DMA task table to the DMA controller to configure a DMA channel.

3. The DSP processor disconnects one memory block from the processor and gives it to the DMA controller.

4. The DSP processor asks the DMA controller for the DMA service (issuing the task).

5. The DMA controller accepts the task and connects the processor memory to a port in the DMA controller.

6. The DSP processor runs other tasks simultaneously as long as the DMA locked memory is not used by the processor.

7. The DMA controller assigns one address generator to the port.

8. At the same time, the DMA controller connects the main memory (or another memory or an I/O) to another port so that the hardware connection for a DMA transaction is established through two ports.

9. The DMA controller assigns another address generator to another port.

10. A FIFO buffer is prepared for the channel.

11. A channel is therefore connected for the DMA transaction.

12. The DMA controller transfers data via the channel until the transfer is finished.

13. The DMA controller sends an interrupt to the task order indicating "task is finished."

14. The memory block (received data from DMA controller) is disconnected with the DMA controller and connected to the DSP processor. The processor uses the new data from the DMA transaction.

According to this procedure, the DMA execution order is scheduled in Figure 15.16. The majority DMA tasks are to prepare data blocks or programs to be used by the DSP processor by loading data from the main memory to on-chip memories. Output data blocks can also be sent via DMA. DMA is also useful for passing data between processors, hardware modules, and peripheral ports. A microcontroller (MCU) interface can be a peripheral port of DSP so that the MCU and the DSP can exchange data packages via DMA.

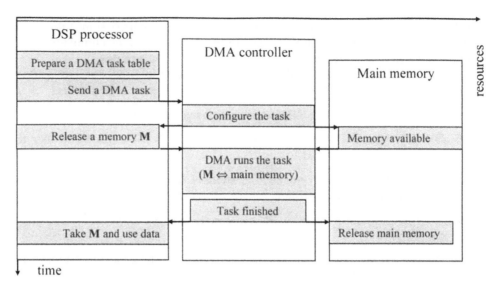

FIGURE 15.16

Scheduling DMA as the interface between DSP and the main memory.

DMA supporting for cache will not be discussed in details in this book because the focus of the book is ASIP DSPs. The DMA controller can be designed to handle transactions between the caches and the main memory with the highest priority. The reason to give the highest priority to a cache is that the processor stalls during a cache miss. If a program cache or a data cache exists in a DSP memory subsystem, the DMA channel with the highest priority will be the DMA cache channel for data transactions handling cache misses between the cache and the main memory.

A simplified DMA controller is shown in Figure 15.17. When a DMA transaction is requested, in most cases, the DSP core requests a configuration of a transaction. It will be handled immediately when the hardware resource is available (a channel is free and can be configured). When the hardware resource is not available (all channels are configured), the DMA request (a DMA task number) will be stored in a task queue in the arbiter waiting for processing.

The arbiter is a circuit managing two cases: The first case is to arbiter which request (task) to serve first when there are two or more requests arriving at the same time to the DMA controller. The second case is to decide which configured task uses the DMA hardware first when there are multiple configured tasks waiting in the queue. When a request is granted, the arbiter will issue the configured task to the DMA controller.

Because the DMA controller must handle two memories and transactions between them, two address generators are required. One address generator generates the address to the source memory, and the other generates the address to the destination memory.

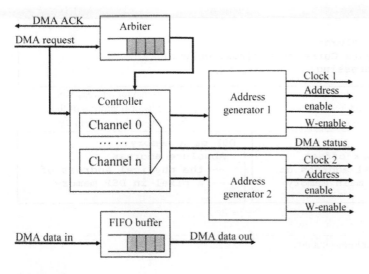

FIGURE 15.17

A simple DMA hardware block diagram.

A data FIFO buffer is used to synchronize the transfer between the two memories. For example, the clock rate and the data width might be different between these two memories. If the DMA controller serves multiple data transactions simultaneously, the FIFO buffer is also used for hardware multiplexing.

15.5.2 Configuring a Program for a DMA Task

Configuring a DMA task begins with specifying addresses. The address of a data word in a DSP subsystem can be in two forms: the behavior address and the physical address. The behavior address is the variable name used in the behavior code. In a video codec, for example, the behavior address of a pixel denotes the position in a frame and the name of the frame. The physical address is the translation of the behavior address to the address of the assembly code. There are at least two physical addresses: the address of a data component in the main memory and the address in the on-chip processor data memory when using the data.

The top block in Figure 15.18 is part of the code in a video processing behavior model. Variables are named as CurrentFrame (row, column). In the assembly code, the behavior address is translated to two physical addresses: the physical address in the main memory and the physical address in the on-chip data memory. In this example, a pixel data is stored in both the main memory and the scratchpad memory of the DSP processor. One physical address is the absolute position of each pixel in the main memory; another physical address of the same pixel is the physical position of the temporary storage in the scratchpad memory of the DSP processor.

```
For row=0 to N {
          For column=0 to M {
          Compute CurrentFrame(row, column)}
}- behavior addressing
```

Behavior address V.S.
physical address of the
main memory

Behavior address V.S.
physical address of the
on-chip data memory

```
Main memory:
Block name: block offset
  -- the physical address of
  -- a pixel in main memory
```

```
DSP data memory:
  Absolute address
  -- the physical address of
  -- a pixel in DSP memory
```

FIGURE 15.18

Relations between two address spaces.

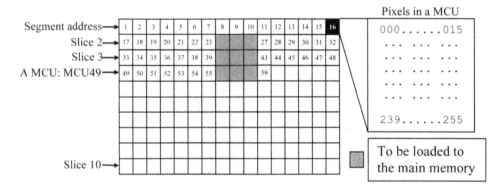

FIGURE 15.19

Nine slices in a frame of video picture.

Memory addresses in a DSP subsystem are always physical addresses. It means that physical addresses of each data in the main memory and in the scratchpad memory must be calculated when preparing for a DMA task. Whenever preparing for a DMA transaction, both the physical source addresses (the data from) and the physical destination addresses (the data to be moved to) must be prepared.

Two advanced issues need to be discussed in this section. A DMA linking table is very useful when transferring multiple noncontiguous data blocks using one DMA transaction. Let us look into the following example. Data pixels in a video frame stored in the main memory are arranged slice by slice as shown in Figure 15.19.

Figure 15.19 shows a picture frame containing nine slices, with each slice containing 16 MCUs (microcode units). Each MCU contains $16 \times 16 = 256$ pixels indexed from 0 to 255. Suppose we need nine marked MCU data blocks (MCU 24, 25, 26, 40, 41, 42, 56, 57, and 58) and want to transfer all these blocks by one DMA transaction; then the DMA controller must fetch nine data blocks from the following addresses:

1. The first part of the data block (768 pixels, including MCU 24, 25, and 26):

   a. Starts from segment address $+ 23 * 256 = 5888$.

   b. Finishes at segment address $+ 26 * 256 - 1 = 6655$.

2. The second part of the data block (768 pixels, including MCU 40, 41, and 42):

   a. Starts from Segment address $+ 39 * 256 = 9984$.

   b. Finishes at Segment address $+ 42 * 256 - 1 = 10751$.

3. The third part of the data block (768 pixels, including MCU 56, 57, and 58):

   a. Starts from Segment address $+ 55 * 256 = 14080$.

   b. Finishes at Segment address $+ 58 * 256 - 1 = 14847$.

If we configure the DMA transactions based on the block-concatenated transfer mode in Figure 15.20, we can merge three data blocks into one DMA transaction using the configuration for these three frames as illustrated in Figure 15.20. Suppose the segment address is 81920 and the linking table of the DMA task table has the following configuration:

1. The start address of the first data block is $81920 + 5888 = 87808$.

2. The start address of the second data block is $81920 + 9984 = 91904$.

3. The start address of the third data block is $81920 + 14080 = 96000$.

In Figure 15.20, a DMA transaction consists of three data blocks. The DMA transaction does not finish when the transfer of the first block is finished because the link value is not zero. Instead, the DMA controller immediately links the current transaction to the next transaction pointed to by the link number. Link $= 2$ means the following data transaction of the DMA is data block 2 with 768 samples and starts at the address of 91904. Link $= 3$ means the following data transaction of the DMA is data block 3 with 768 samples and starts at the address of 14080. Finally, Link $= 0$ means that the DMA is finished and there is no following data block to

DMA block number = 1	DMA block number = 2	DMA block number = 3
Start address = 87808	Start address = 91904	Start address = 14080
Block length = 768	Block length = 768	Block length = 768
... ...	... ...	... ...
Link = 2	Link = 3	Link = 0 (finished)

FIGURE 15.20

Configuration of channel-triggered transfer mode.

transfer. When all three blocks are transferred, the link value is set to zero (link = 0) and the DMA controller terminates the transaction.

A DMA transaction is specified using a linking table. Several data blocks in a Linking table can be transferred. A DMA task table may include the following information:

- The task name or the task number.

- The start address of the source memory block.

- The start address of the destination memory block.

- The size of the data to be transferred.

- The number of data blocks to be transferred (to transfer multiple data blocks).

- Read or write control, data rate, clock rate, and so on.

A linking table-based DMA transaction can be configured using the DMA task table in Table 15.2.

Table 15.2 DMA Task Table with a Linking Table for Transaction of Multidata Blocks.

Spec	Content	Notes
1	Task name	Take 9 MCU (9 × 256 = 2304 pixels) from main memory
2	Mode and priority	High priority
3	Source device	Main off-chip memory
4	Destination device	On-chip data memory DM1, start address: 1024
5	Number of blocks = 3	Take the same example in Figure 15.19
6	Block 1 source address	87808
7	Block 1 destination address	On-chip data memory SEG + 0
8	Block 1 length	768
9	Block 2 source address	91904
10	Block 2 destination address	On-chip data memory SEG + 767
11	Block 2 length	768
12	Block 3 source address	96000
13	Block 3 destination address	On-chip data memory SEG + 1535
14	Block 3 length	768

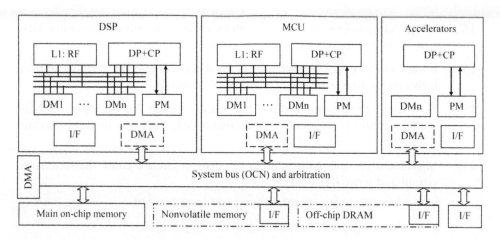

FIGURE 15.21

Memory hierarchy in a SoC system.

15.5.3 A SoC View

There might be more than one DMA controller sharing the system bus and the main memory. When the main memory is shared by two DMA controllers (one for DSP and one for MCU), the right to use the main memory must be granted by an arbiter of the system bus (see Figure 15.21). In this case, the extra latency induced by the system bus arbiter on the system level has to be taken into account.

15.6 CONCLUSIONS

Design of memory subsystem hardware includes design of memory peripheral, memory addressing, memory hierarchy, memory bus, and DMA controller. Basic concepts of designing memory peripherals and addressing have been discussed in this chapter. Advanced design for memory hierarchy also has been addressed without detailed discussion; refer to references [2], [3], and [4] for further information. Principles of DMA transactions have been introduced in this chapter. Detailed DMA hardware design will be given in the next chapter.

EXERCISES

15.1 Why is memory the performance bottleneck in a DSP subsystem?

15.2 Describe timing of memory access (read/write) for a single-port SRAM.

15.3 If logic circuits exist between the memory and the address generator, how can we maximize the memory access speed?

15.4 If logic circuits exist between the general register and the memory in-port, how can we maximize the memory access speed?

15.5 How can we use one address calculation circuit for multiple memory address pointers (registers)? Give an example using a schematic circuit and RTL code.

15.6 Implement the following address calculation circuit. Draw the schematic circuit and write RTL code:

a. Address register = constant from instruction word.

b. Address register = Address register + step register.

c. Address register = Address register − step register.

d. Address register = General-purpose register.

e. Address register = Segment register + General-purpose register GRn.

15.7 Implement a modulo addressing circuit and write RTL code. The FIFO top is FIFO_TOP and the bottom is FIFO_BTM. The step size is simply "positive one."

15.8 Implement a modulo addressing circuit and write RTL code. The FIFO top is FIFO_TOP, and the bottom is FIFO_BTM. The step size can be $-8 < \text{STEP} < 7$.

15.9 Implement a bit-reversal addressing circuit, draw the schematic circuit, and write RTL code.

15.10 Which of the following statements are correct? Physical memory in a DSP chip could be:

a. SRAM.

b. Typically DRA.

c. Used for either program storage or data storage.

15.11 Which of the following statements are true? In address calculation circuits:

a. Round operation on output is not needed in general.

b. Saturation operation on output shall always be conducted.

c. Modulo addressing can be associated with post-increment/decrement.

d. Index addressing is based on the computing of two register values.

REFERENCES

[1] Catthoor, F. et al. (1998). *Custom Memory Management Methodology*. Kluwer Academic Publishers.

[2] Panda, P. R. et al. (1999). *Memory Issues in Embedded System-on-Chip*. Kluwer Academic Publishers.

[3] Furber, S. (2000). *ARM System-on-Chip Architecture, 2nd edition*. Addison Wesley.

[4] Gatherer, A., Luslander, E., Eds. (2002). *The Application of Programmable DSPs in Mobile Communications*. John Wiley & Sons Ltd.

[5] Haraszti, T. P. (2000). *CMOS Memory Circuits*. KAP.

[6] Rabaey, J. (1996). *M. Digital Integrated Circuits, A design perspective*. Prentice Hall.

DSP Core Peripherals

16

Peripherals or peripheral modules are circuit modules around a processor core. Peripherals include on-chip peripheral devices (circuit modules) and interfaces to off-chip devices. In this chapter, we will discuss the design of on-chip peripheral modules, including the timer, interrupt handler, and DMA controller. We will also discuss chip interfaces. In Chapter 19, the integration of DSP cores and peripheral modules in a SoC system will be discussed further.

Requirements and interfaces of an interrupt controller and DMA controller of a DSP subsystem in SoC are different from those of a general-purpose computing system. These differences will also be discussed in this chapter.

16.1 PERIPHERALS

A peripheral module is a piece of custom-designed hardware or an available hardware module that performs auxiliary actions in the system—for example, input/output, backing store, handling communications to others, and handling other external events. Peripheral modules can offload simple and relatively fixed functions from the processor core.

A DSP peripheral module is an I/O device of the DSP core. It is located near the processor core within the processor chip. In the past, when a DSP appeared as a chip, there were always peripherals included on the chip. A DSP processor can be sold today as an IP-core without peripherals, and the customers can add their own peripherals before the chip is manufactured.

A DSP core platform consists of a DSP core and its peripheral modules. Further, by adding memories, a DSP subsystem consists of DSP cores, core peripherals, and on-chip memories. Design of peripherals is becoming increasingly independent of the design of the processor core. Peripheral design therefore is essential knowledge for DSP HW design.

IP-core design became popular because a SoC can be so complicated that the design of all components in a SoC project became impossible. Buying third-party IP is the way to speed up design and relax the design complexity. Design of a processor

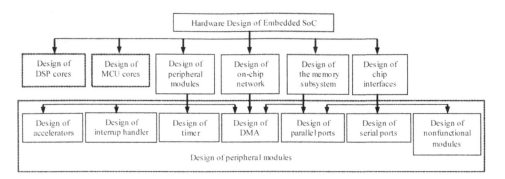

FIGURE 16.1

The hardware design partitioning of an embedded SoC.

core is therefore separated from chip designs and peripheral design is also separated into two families: design of on-chip peripheral modules and design of port drivers. By observing Figure 16.1, we find that most of the hardware design activities for an embedded system are actually designs of peripheral modules.

From a functional point of view, a peripheral device can be a functional module or a nonfunctional module. A functional peripheral module is visible by programmers. Functional peripheral modules can be on chip modules—for example, an accelerator, a timer, an interrupt handler, or a DMA module. Functional modules can also be serial or parallel port drivers. A nonfunctional peripheral device may be invisible to programmers, such as the JTAG1149.1 finite state machine for controls of chip tests.

In general, a peripheral module is configured by its processor core. After configuration, it runs as an autonomous device controlled by its local FSM (finite state machine). The program flow and the peripheral task run independently and are synchronized by handshaking or interrupts between the core and the peripheral module. This kind of working model is called an asynchronous working model from a processor designer's view.

Peripheral functions are regular, and stable, and can be independent of the code execution in the processor core. The main purpose of having several peripheral devices around a processor core is to divide functions from the core and run them in parallel. By moving stable functions into a peripheral device, the processor core can use most of the runtime and the core computing resources for flexible tasks. For example, the data collection from an input port can be offloaded from the processor core and be executed by the data port, and data can be collected, grouped, and buffered in the local data buffer. As soon as the FSM of a peripheral module is configured, the peripheral hardware executes its predefined functions independent of the code execution in the core. When a peripheral module wants support from its processor core, an interrupt will be generated by the FSM of the peripheral module asking for the service by the processor core.

16.2 DESIGN A PERIPHERAL MODULE

There are two cases of how a processor core can cooperate with a peripheral module. In the first case, the processor core initializes a task and sends it to the peripheral device. The peripheral device takes the task, runs it, and gives the result to the processor core. In Figure 16.2, the processor initializes a task and gives the task to the peripheral module by configuring the module, and then triggers the task to run in the peripheral module. The peripheral module runs the task under the control of its FSM.

The second case occurs when a peripheral module initializes a task. The task could be initialized by an external event. The peripheral module executes the task itself. When finishing the task, the FSM of the peripheral module asks the processor core to fetch the result stored in a place the processor can reach. In Figure 16.3, the second case is initialized by an external event.

There are two ways a processor core communicates with peripheral modules. The first way is to let the processor poll a peripheral; that is, it can repeatedly ask the peripheral if it is finished, for example, by checking certain flags in the peripheral's status register. Polling can be quite inefficient since the processor often cannot do other useful things while polling, but it can be useful for low-speed peripherals that can be polled at regular intervals such as keyboard controllers and temperature sensors. The other way is to give the peripheral the ability to signal an interrupt to the processor. When there is an interrupt from a peripheral module, the processor will temporarily suspend the executing of its main program and offer service for the interrupt. The advantage of interrupts over polling is that the processor can easily be programmed to do other tasks while waiting for

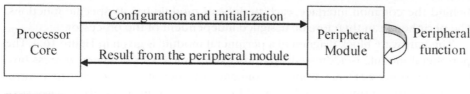

FIGURE 16.2

A processor core configures a task on a peripheral module.

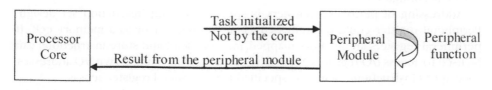

FIGURE 16.3

A peripheral initializes a task by itself or by an external event.

the peripheral to finish. However, care must be taken so that the processor is not overwhelmed by interrupts from different peripherals. There must be enough time for the processor to execute the main program.

If the peripheral has data available for the processor, for example, the result of a calculation, this data must be transferred to the processor or a memory in some way. Again this can be implemented in two ways. The first way is to access I/O by running an I/O access program, and the second way is called DMA (direct memory access) service. By using the first way, the code running in a processor can read the result one word at a time from a data register in the peripheral. This is usually an efficient way to transfer data when the data packet is very small. When a data packet is large, DMA is preferred.

16.2.1 Design of a Common Interface in Peripheral Modules

To design peripheral functions independent of the design of the DSP core, there should be a common interface between the processor core and the peripheral module. At the processor core side, the common interface can be accessed fully by running programs. At the peripheral device side, the design of the custom function modules follows the protocol of the common interface. Custom functions behind the common interface can therefore be connected and accessed by the processor core.

A peripheral module therefore should be divided into two parts: the communication part and the function part. The communication part works as the common interface between the peripheral module and the processor core. The common interface should be the same for all peripheral modules sharing the same interwork protocol between the DSP core and all peripheral modules. The common interface should be specified during the DSP core design. The common interface connects and isolates the DSP core and the function of the peripheral module. Behind the common interface, each peripheral module has its specific functions. The peripheral can therefore be designed independent of the processor.

The hardware block diagram of a peripheral module is given in Figure 16.4. The peripheral module is addressed by a processor core via a peripheral address bus. When the processor core wants to configure the peripheral module, the control register of the peripheral module is addressed. The control and data registers of the peripheral module will be addressed to accept the configuration vectors. The configuration vectors are then sent to the peripheral module, and the peripheral device is configured.

Addressing of peripheral registers is specified during instruction set design. Peripheral register could be addressed as an I/O register or as a memory cell. If peripheral registers are memory mapped, memory load and store instructions can be used to access I/O registers. If peripheral registers have their own I/O addresses, special I/O instructions should be specified for peripheral register access.

A peripheral module could start running the function immediately after the configuration or wait to be activated by the processor core. When the execution in the peripheral module is finished, the peripheral module either sends an interrupt

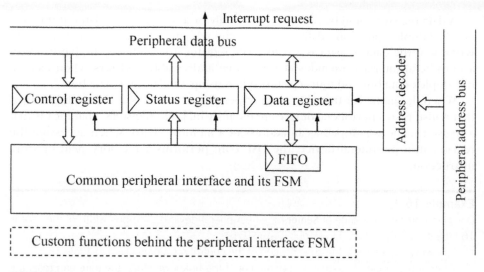

FIGURE 16.4

Peripheral module HW block diagram.

request to the processor core or sets the finish flag and waits for further processing from the processor core.

There are typically at least three addressable registers in a peripheral port: the control register, the status register, and the data register. The control register is used for configuration. The DSP core configures a peripheral module by addressing the control register and sending a control vector to it. Functions to be executed in the peripheral device will then be configured. Both the peripheral interface and the peripheral functions can be configured. The interface configuration includes the setting of the data rate, the synchronization mode, and the way of data transmission. The peripheral function configuration gives the detail specification of the execution. For example, if the peripheral device is a timer behind the interface, the function of the timer can be configured such as the time to count, the way to interrupt when time-out, and so on.

The status register gives the status report to the center processor. Status is from both the peripheral common interface and the function logic behind the common interface. Status of the common interface includes normal/abnormal, working/sleeping, finished (available)/working (occupied), and idling/waiting for service, among others. The specific status of the function behind the common interface could be important information necessary to expose to the running program. For example, if the peripheral module is a DMA controller, the status of the DMA function behind the common interface could be the status of the current user (channel) of the DMA and the priority of the running channel. The status register is usually a read-only register; hence the processor cannot write values to a status register of a peripheral module.

A data register could be read-only, write-only, or a read/write register. If the port is a half-duplex port, one data register is enough, and it should be readable and writable if the data register is a read/write register. To conclude, the hardware of a peripheral interface includes at least three addressable registers when using a half-duplex data bus and four registers when using a full-duplex data bus.

When designing for the addressing of peripheral registers, it is usually assumed that registers in a peripheral device are memory-mapped registers. That means that a register name is mapped as an address of a memory word. When accessing the register in a peripheral device, the DSP core performs a memory read or store instruction.

Example 16.1

Design a common interface of **Senior** for all peripheral devices. The bus width of the core is 16 bits; the width of the address bus is also 16 bits.

Following Figure 16.4, the specification of the control register is given in Table 16.1.

When the peripheral module is abnormal or at the unknown state, the core can reset the module using bit 0 in the control register. The module can be set to sleep mode when it is not used. The data rate setting gives flexibility when the machine clock is different from the memory clock. Setting for synchronization is needed for off-chip communications or if the bus is shared. In this case, synchronization is the enable signal. Some cores may use an 8-bit bus, and some may use a 16-bit or a 32-bit bus. To let the peripheral module adapt to several cores, the result

Table 16.1 An Example of a Common Control Register Specification.	
Bits	**Specification**
[0] = 1	Reset the module, flush the current task
[1]	Shut down the module when [1] = 1 (control for low power)
[2]	Data rate: when 0, peripheral using machine clock; when 1, using local clock
[3]	Synchronization: When 1, special strobe pin is used to synch the data from peripheral module
[5:4]	Result format: 00 : 16b mode; 01 : Byte mode; 10 : 32-bit Big-endian; 11 : 32-bit Little-endian
[7:6]	Report result: when 00 using flag, no interrupt; when 01 normal interrupt; when 10, nonmaskable interrupt; when 11, exception and reset the system
[8]	Reset (clean) the interrupt request
[15:9]	Reserved for applications behind the peripheral common interface

data type might need to be configured. The Little-endian data format is {second-word[15:0], first-word[15:0]}, and the Big-endian data format is {first-word[15:0], second-word[15:0]}. (First and second denote the order of data transfer.) The way to report the finish of a task in a peripheral module can also be configured. If the result from the peripheral module is important, an interrupt might be necessary. If the result is only a reference for the application program, a flag reporting the availability of the result might be enough. Finally, the interrupt can be reset by the running program in the core after the interrupt service.

The status register is specified using Table 16.2. The Mirrors of the interrupt will be used if several peripherals share an interrupt pin/level or if we want to see if a peripheral has generated an interrupt even though interrupts are disabled. By reading the status register, the processor core can decide what to do with the module. Status bits cannot be changed by the processor core.

Addressing and accessing peripheral registers can be done in the same way as memory access of the parameter memory. For example, all port registers in a peripheral module can be addressed and accessed as a data word in a data memory. For example, registers in a peripheral module can be addressed according to the address definition in Table 16.3.

Table 16.2 Peripheral Status Register Specification.

Bits	Specification
[0]	Idle or busy: Idle = 0 and busy = 1
[1]	Acknowledgment: When 1, accept
[2]	Report result: Running task is finished when result flag = 1
[3]	The mirrors of the interrupt pin = 1 give an interrupt indicating the finish of the task
[4]	Abnormal = 1 when there is an exception
[15:5]	Status of the application behind the common interface of the peripheral

Table 16.3 Addressing Common Registers of a Peripheral Module.

Address code	Specification
16'bXXXXXXXXXXXXXX00	Status register
16'bXXXXXXXXXXXXXX01	Control register
16'bXXXXXXXXXXXXXX10	Out port data to DSP core
16'bXXXXXXXXXXXXXX11	In port data from DSP core

By specifying the first 14 address bits, a peripheral module will be activated. The last two address bits will be used further to address one of the four common registers in the peripheral module.

16.2.2 **Protocol Design of Peripheral Modules**

Functions executed by a peripheral module in a DSP chip are controlled by the local FSM of the peripheral module. The functionality of a peripheral FSM is predefined. The function of a FSM of a peripheral module can be configurable. By writing a control vector to control registers of a peripheral device, the DSP core can start, stop, and configure functions of the peripheral device. By reading the status register, the DSP core can monitor the status of the peripheral device.

The FSM and the peripheral interface in Figure 16.4 must be designed following the interwork model between the processor core and all peripheral modules. This core-peripheral interwork model is the protocol between processor core and the peripheral modules. The protocol can be divided further into the normal protocol (the connection and configuration protocol, the task control protocol, and the task terminating protocol) and the exception handling protocol.

Connection and configuration processes can be executed using send-and-forget or using a handshaking protocol. Send-and-forget protocol is used under two conditions. The first is guaranteed data transmission quality when both the core and the peripheral module are on one chip and in one clock domain and the data transmission is reliable (with the guarantee of static timing analysis during the IC implementation). The second is guaranteed data reception. The data will be received, buffered, and handled, in any circumstances. Otherwise, the handshaking protocol must be used when the data reception cannot be guaranteed.

If the processor knows that the peripheral module is a slave device and it is idling and waiting to be assigned a task by the processor, the processor can simply send configuration to the peripheral module. This is called send-and-forget. However, in most cases, the processor does not know the current status of the peripheral module, and necessary handshaking is required.

The handshaking protocol usually is described using handshaking of two FSMs on both the processor side and on the peripheral module side. In Figure 16.5, the handshaking of two FSMs in the processor core and the peripheral module is given under two cases: the peripheral module is available, and the peripheral module is not available during the first try.

The design of a common interface for a peripheral module was discussed in this section. In the following sections, the design of custom functions for typical peripheral modules will be discussed based on examples. The first example will be the design of an interrupt handler.

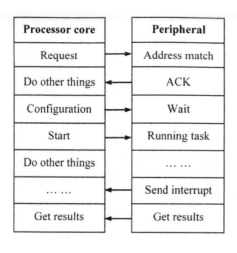

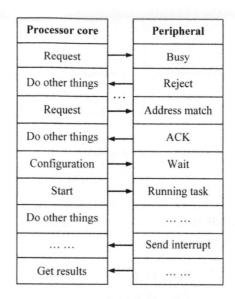

Peripheral is available   Peripheral was not available during the first try

FIGURE 16.5

Peripheral protocol communication examples.

16.3 INTERRUPT HANDLER

16.3.1 Interrupt Basics

An introduction to interrupt was given in Chapter 14; you should reread the section on interrupt handling in that chapter before reading this section. In this section, the design of an interrupt controller will be discussed. An interrupt process, from request to the end of interrupt service, is depicted in Figure 16.6.

16.3.2 Interrupt Sources

A hardware interrupt source is an on-chip or off-chip hardware module requiring service from the DSP core. A list of some interrupt sources is given in Table 16.4.

Principally, there are two kinds of interrupt sources, monofunction and multifunction. A monofunction source requires only one interrupt service. For example, a service to a timer interrupt can flush the current execution and go to the top program entry of the application. The interrupt request of a monofunction source can be a single pin. However, a multi-interrupt source may require several interrupt services if the interrupt source is programmable. For example, when an interrupt is required by the MCU, several services might be requested. In this case, both

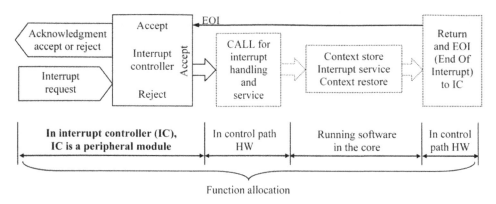

FIGURE 16.6

Interrupt handling process.

Table 16.4 Typical Interrupt Sources in an Embedded System.		
Interrupt source	**To DSP core**	**To MCU core**
Timer or watchdog	Yes	Yes
Power off signal	Yes	Yes
Come to debug mode	Yes	Yes
MCU core HW in general	Yes	No
DSP core HW in general	No	Yes
Data memory address pointer overflow	Yes	Not so often
DMA error	Yes	Yes
DMA complete	Yes	Yes
On chip system bus	Yes	Yes
Accelerator error	Not so often	Not so often
Accelerator complete	Yes	Yes
ADC port	Yes	Sometimes
Input ports or buffers	Yes	Yes
General parallel I/O ports	Yes	Yes
General serial I/O ports	Yes	Yes
Sense key-presses	No	Yes
Memory card plug in or removed	No	Sometimes
Device plug in or removed	No	Sometimes

the interrupt request pin and a register containing the interrupt service entry are needed.

It can be observed that all interrupt sources to the DSP core are on-chip components (even including the power off signal and the signal of debug mode). The DSP core usually is not used to monitor and control off-chip devices such as the keyboard, the mouse, or the memory card. On-chip components (interrupt sources) are fixed during the system design of the SoC.

The knowledge of interrupt sources of DSP core in a SoC is very important. It gives opportunities to dramatically simplify the design of interrupt controller (IC) for DSP core in SoC.

16.3.3 Interrupt Requests

An interrupt request is from an interrupt request pin of an interrupt source. To an off-chip interrupt controller, request pins can simply be connected together and the wired-or-logic behind the collection of interrupt pins can sense an interrupt request.

An interrupt request can be kept until it is served. This kind of interrupt is a level-triggered interrupt. Another type of interrupt is signaled by a level transition on the interrupt line, either a falling edge (1 to 0) or (usually) a rising edge (0 to 1). This kind of interrupt is an edge-triggered interrupt.

A level-triggered interrupt keeps the request pin high until it is served. The service will not be missed. The level-triggered interrupt can prevent the off-chip glitch interference. It is good for an interrupt line shared by multiple interrupt pins. However, a level-triggered interrupt may block other interrupts if the DSP core does not know how to serve it. Although this can be a problem in general-purpose computers, it is not a big problem in an embedded system because all interrupts must be debugged during the design phase.

An edge-triggered interrupt does not block a service of a low-priority device because it can be postponed arbitrarily, and interrupts will continue to be received from the high-priority devices that are being serviced. However, interrupt pulses from different devices may merge if they occur close in time.

An interrupt controller for a DSP core can use either level-triggered interrupts or edge-triggered interrupts. If the interrupt controller is an IP (intellactual property) device, both level-triggered interrupts and edge-triggered interrupts should be accepted by the interrupt controller.

Collect and Latch Interrupt Requests

An interrupt is identified, stored/latched, processed, and terminated as depicted in Figure 16.7. Both level-triggered and edge-triggered interrupts are handled.

A level-triggered interrupt will be handled as soon as the interrupt is requested. If the interrupt is edge-triggered, the interrupt request will be latched (stored) first and processed on the latched request. The interrupt handling (decision of accept or reject) will be discussed later. After the accept/reject decision is made,

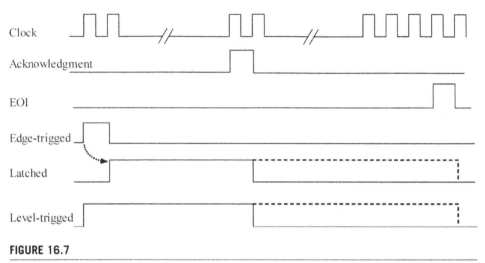

FIGURE 16.7

Collect, store, and terminate an interrupt request.

an acknowledgment will be generated and sent to the interrupt source. In some interrupt controllers, the interrupt request will be reset (the solid line). In some other interrupt controllers, an interrupt request will be reset by EOI (dash line). EOI is given by the DSP core when a return from interrupt instruction is executed.

16.3.4 Interrupt Handling Process

The definition of interrupt handling could be different from different companies or books. In this book, an interrupt handling is the interrupt request handling, instead of interrupt service. An interrupt handling performs the decision of acceptance (accept or reject), context switching, and pointing to the entry of the interrupt service. Interrupt service is the subroutine with the function that is requested by the interrupt source and will be served by the processor core. In this book, interrupt handling is all functions from interrupt request to the calling of the interrupt service executed either in the interrupt controller or in the processor core. Interrupt handling in this book does not include interrupt service.

The process includes two steps. Step one handles simultaneous requests. If multiple interrupt requests interrupt services in the same clock cycle, the interrupt handling process will authorize interrupt to the request with the highest priority. The decision to accept or reject an interrupt is taken in the second step. The priority of the interrupt request and the priority of the current running task are compared. If the current running task holds the higher priority, the interrupt request will be rejected. Otherwise, the interrupt will be accepted. The result of an interrupt handling is the acknowledgment (accept or reject) of the interrupt. If the interrupt is accepted, the interrupt request and service entry will be passed to the DSP core.

An interrupt handling process can be conducted in software or in hardware, or a combination of both.

Priority and Maskable Interrupts

According to the importance of an interrupt, each interrupt has an interrupt priority. Interrupt priority is the definition of the relative importance of an interrupt request and service. An interrupt holding a higher priority will be accepted and served when multiple interrupts are asking for interrupt services at the same time. An interrupt with higher priority can interrupt a running task with lower priority. Notice that we mentioned two cases: the priority used for interrupt arbitration of multiple requests and the arbitration between the interrupt requests and the running task.

Each task, both running task and the task for interrupt service, has its own priority. The task with higher priority has the right to use the DSP core. If the running task holds the higher priority, the interrupt request will be rejected. If the task of interrupt service has higher priority, the running task will be interrupted, giving hardware resources to the interrupt service.

The priority of each task is configured by the DSP core during the initialization of the core or during the starting phase of an application. Each task therefore should have a priority register. By configuring the interrupt controller, a DSP processor gives different priorities to each interrupt source.

To mask an interrupt is to configure the interrupt handler to ignore the request of the interrupt. If an interrupt is masked, its interrupt request will be blocked, and the interrupt requested will not be reacted upon until the mask is removed. It should be possible to mask all normal interrupt requests to minimize the interrupt-induced disturbance while running important tasks.

Nonmaskable Interrupts

Some important interrupts should not be maskable. For example, interrupts from certain timers must be handled whenever they come. Therefore, some interrupts are nonmaskable and must be handled and accepted by the interrupt handler at any time. A nonmaskable interrupt (NMI) generally signals a major or even catastrophic system failure, so it must be served and cannot be masked. It is important to specify which interrupts are maskable and which are not when designing an interrupt handler. The list of NMIs in a DSP subsystem can include:

- Exception from datapath: divide by zero.

- Exception from datapath: Overflow when saturation is off.

- Watchdog timer.

- Loss of power (battery level is critically low).

- Enter to debug mode.

- Data memory address pointer overflow.

- PC overflow (jump to an address that physically does not exist).

- Output buffer overflow.

Configuring an Interrupt Controller

An interrupt controller (IC) is a peripheral hardware module handling interrupt requests or support interrupt handling by the processor core. In most DSP processors, handling of interrupts are conducted by the interrupt controller to minimize the interrupt latency. In GPP, interrupt handling could be conducted in software to keep the maximum flexibility. Here, interrupt latency is the measure of a period of time from the interrupt request until the interrupt service subroutine starts.

An interrupt controller is configured by its DSP core during the initial phase. Here the initial phase could be the power on procedure or the initialization phase of an application. Parameters to be configured are priority of all normal interrupts during the initial phase. While running an application, some interrupts are masked when executing higher priorities subroutines. For example, perhaps some iterative algorithms should not be interrupted because context switching for these iterative subroutines is too complicated to handle. After executing these iterative subroutines, interrupts can be handled.

Handling Interrupt Requests

If the priority control and accept/reject decision is given by the interrupt controller, an interrupt signal and the program memory address to the relevant service entry will be sent to the DSP core from the interrupt controller. If priority control and accept/reject decision is given by interrupt handling software in the DSP core, the interrupt request and interrupt identification will be sent to the DSP core and the DSP core will handle the priority control and accept/reject decision.

More flexibility can be achieved when executing the priority control and accept/reject decision by DSP software, but will result in longer execution time. In embedded systems, interrupt sources of a DSP core are usually on-chip and the requirement on flexibility is low. Interrupt handling latency is critical for real-time applications. Here the interrupt handling latency (or interrupt latency) is the time from the request of an interrupt to the time of starting the interrupt service subroutine (the function of the interrupt).

While handling interrupts in interrupt controller hardware, latency can be dramatically minimized. Therefore, for ASIP or a DSP core, interrupt handling usually is executed in interrupt controller hardware, and sufficient flexibility should be kept. The interrupt handling procedure in the interrupt controller hardware can be illustrated using the following simplified Verilog code:

LISTING 16.1 A simplified interrupt handler with fixed priorities.

```
module interrupthandler(
input wire [2:0] currentlevel,    // The cores current int level
input wire [6:0] interrupts,      // 7 interrupt pins, bit 6 has
                                  // the highest priority.
input wire [6:0] interrupt_mask,  // Mask for the interrupts
```

```
output reg core_interrupt,        // Interrupt req pin to the core
output reg [2:0] core_interrupt_id // Signals interrupt id to core
);
reg [2:0] interrupt_priority; // 0 Means no interrupt
reg [7:0] masked_ints;

always @* begin
            // Default assignments
core_interrupt = 0;
        core_interrupt_id = 0;

masked_ints = interrupts & interrupt_mask;
casex(masked_ints) // A priority decoder
7'b1xxxxxx: interrupt_priority = 3'd7;
7'b01xxxxx: interrupt_priority = 3'd6;
7'b001xxxx: interrupt_priority = 3'd5;
7'b0001xxx: interrupt_priority = 3'd4;
7'b00001xx: interrupt_priority = 3'd3;
7'b000001x: interrupt_priority = 3'd2;
7'b0000001: interrupt_priority = 3'd1;
default:    interrupt_priority = 3'd0;
endcase

        // This module will continuously send an interrupt
        // request to the core until the core signals that the
        // interrupt has been granted by raising currentlevel.
if(currentlevel < interrupt_priority) begin
core_interrupt = 1;
core_interrupt_id = interrupt_priority;
end
end
endmodule
```

16.3.5 A Case Study

Requirements

This case study is to design an interrupt controller (IC) of **Senior** DSP core for real-time application. The DSP core can handle multiple threads. Each thread will be initialized by an interrupt from an input port (e.g., a data packet with high priority arrived). There will be four input ports: IP1, IP2, IP3, and IP4. There will be two timers, T1 and T2—one is the watchdog of the main program, and the other is the watchdog of an interrupt service. (A watchdog is also called a watchdog timer. It is a peripheral hardware that triggers a system interrupt or reset if a running task cannot finish in time. Timers will be discussed later in this chapter.) If there are multiple interrupts served simultaneously, **Senior** will assign the watchdog to the one where the execution time is uncertain.

Interrupt Sources and Requests

To handle an interrupt, each interrupt source must be coded or numbered. This means that an interrupt has a name and a binary code of its name. Some interrupt services can be predesigned, and the service entry will be in the interrupt source table. Service entries will be fetched from the interrupt source or from the shared mail box, a shared memory that can be accessed by both the processor and the interrupt source. Table 16.5 gives an example of coding and priority of interrupts (NMI stands for nonmaskable interrupt).

An interrupt is initialized by an IRQ, including maskable interrupt request and NMI. The information carried by an IRQ includes: (1) the source ID of the IRQ, (2) the priority of the IRQ, and (3) the entry of the service routine for the IRQ.

Table 16.5   Senior Interrupt Sources.

Interrupt source	Source ID	Priority	Service entry
Timer T1	11111	NMI 5	Hardware BIOS code
Timer T2	11110	NMI 4	Hardware BIOS code
DSP datapath exception	11101	NMI 3	Hardware BIOS code
Memory address overflow	11100	NMI 2	Hardware BIOS code
Enter to debug mode	11010	NMI 1	Hardware BIOS code
MCU	01111	15	From system bus
DMA error	01110	14	Configured by user
DMA finish	01101	13	Configured by user
Accelerator 1	01100	12	Configured by user
Accelerator 2	01011	11	Configured by user
Input port 4	01010	10	Configured by user
Input port 3	01001	9	Configured by user
Input port 2	01000	8	Configured by user
Input port 1	00111	7	Configured by user
Cascade	00110	6	Configured by user
Reserved	00000-00101	0-5	Configured by user

Masks and priorities of maskable interrupts can be configured by the **Senior** core. Interrupt from MCU is a vector (or interprocessor) interrupt, and the interrupt service entry is taken from the MCU via the system bus or DSP-MCU communication buffer. The source ID can be the pointer to identify the service subroutine entry of a normal interrupt. Reserved interrupt sources can be either normal interrupts or interprocessor interrupts.

Interrupt cascade is an interrupt source for extension. The cascade pin can connect to another interrupt controller handling low-priority interrupts. It is very useful for a MCU. For DSP applications, it is seldom used.

Interwork Protocol between the IC and the DSP Core

The communication between interrupt controller (IC) and DSP core should be specified during the design of DSP core. IC can therefore be used for other processor cores. As mentioned before in this section and in Chapter 14, the message from IC to the DSP core is "an interrupt with its ID is accepted." The DSP core will call the interrupt service. When the interrupt service is finishing, the last instruction of the interrupt service is "to return from interrupt." This instruction returns PC and flags and sends EOI to the instruction controller.

Example 16.2

List all steps of a vector (interprocessor) interrupt being prepared, requested, and being executed.

The list is given in Table 16.6.

A shared buffer should be designed for communications between programmable devices. Interrupt information, such as the interrupt service entry, should be stored in the shared buffer before requesting the interrupt.

The principle of a shared buffer is depicted in Figure 16.8. A shared buffer can be an I/O device between two processors. Each processor can write messages to another processor via an I/O port write operation. Because the reading buffer and writing buffer are separate hardware units, the interwork protocol can be simple.

The second kind of interrupt sources generates a fast interrupt. Before introducing fast interrupts, let us recall the definition of the hardware multiplexing rate (HWMR) given in Chapter 10, which is the clock rate over the sampling rate. The rate HWMR defines the complexity of software or application algorithms by the number of operations per data sample or per data packet. When the rate is very high, for example, 40,000 clock cycles per sample for voice signal processing, there is no problem to accept the time costs for context switching (push and pop stack) when serving an interrupt. This is because the cost of context saving (50~250 cycles in a typical DSP core) is very low compared to the available clock cycles for processing a voice packet.

However, vector interrupts are not suitable when the HWMR is low. One example is time scheduling for a radio baseband DSP processor. The deadline is the time

Table 16.6 Steps of a Vector Interrupt.

Step	Function	Interrupt source	Interrupt controller	DSP hardware
1	Prepare for the service entry	Send service entry to shared buffer		
2	Request interrupt	Send IRQ		
3	Arbitration step 1		Get the IRQ with the highest priority	
4	Arbitration step 2		Accept or reject, ACK	
If the interrupt is accepted, then:				
5	React to the ACK	Release the shared buffer		Get ID and fetch service entry
6	Context saving			CALL context save
7	Interrupt service			CALL interrupt service
8	Return of service			Return from interrupt, EOI
9	Terminate the IRQ		Terminate IRQ	
10	Context resume			CALL context resume

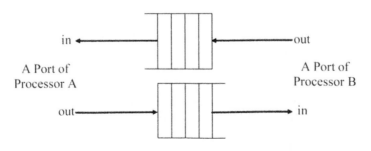

FIGURE 16.8

Shared buffer for interprocessor communication.

to receive the next OFDM (Orthogonal Frequency Division Multiplexing) symbol. All signal processing for the current symbol must be finished within the time of one symbol because the symbols come immediately one after another. Suppose the time of one symbol of IEEE802.11a is 4 microseconds. If the clock frequency is 125 MHz, we have 500 cycles to process all tasks for the received symbol. When one interrupt occurs, the context switch consumes about 50 cycles. In this example, 50 cycles is 10% of the total available time! This is too much and it will not be acceptable. Context switches therefore have to be eliminated for applications with low HWMR.

Fast interrupts therefore are introduced for applications with low HWMR. The service routine of a fast interrupt must be so simple that full-context saving of the interrupted program is not necessary. When serving a fast interrupt, the fast interrupt can use registers that are not used by the main application code; that is, a few registers may have to be reserved for the fast interrupt, and the application code must never write data to them. A minimal context needs to be saved for the interrupted program so that the interrupted program can continue after the fast interrupt service. The minimal context is the PC value and the current flag values. For processors without hardware for context switching, it is possible to emulate this behavior for trivial problems by reserving a number of registers for an important interrupt handler. If the interrupt handler needs to do only very little work (sampling a signal for example), it might need to use only one to three registers. If these registers are not used outside of the interrupt handler, the context switch can happen very quickly since typically only the return address and status register would have to be saved/restored. In general, this is not an option for more demanding real-time applications though.

Example 16.3

List all steps of a fast interrupt being prepared, requested, and being executed.

The list is given in Table 16.7.

HW Interrupt Module—The Interrupt Controller

Following all previous discussions, the interrupt controller for the **Senior** DSP core can be designed as depicted in Figure 16.9.

Registers in the interrupt controller in Figure 16.9 are used to store configuration vectors for configuring the handling of each interrupt request. All registers can be configured via the general I/O port, which was discussed in the first section in this chapter. Registers for each interrupt source are specified in Table 16.8 including:

- A 5-bit ID register for each interrupt source (read only by the core).
- A 2-bit flag register for each interrupt source (read only by the core).
- A 4-bit priority register for each interrupt source (writeable by the core).

Table 16.7 Steps of a Fast Interrupt.

Step	Function	Interrupt source	Interrupt controller	DSP firmware
1	Request interrupt	Send IRQ		
2	Arbitration step 1		Get the IRQ with the highest priority	
3	Arbitration step 2		Accept or reject, ACK	
If the interrupt is accepted, then:				
4	React to the ACK			Get ID and fetch service entry
5	Interrupt service			CALL interrupt service
6	Return service			Return from interrupt
7	Terminate the IRQ		Terminate IRQ	HW or FW EOI

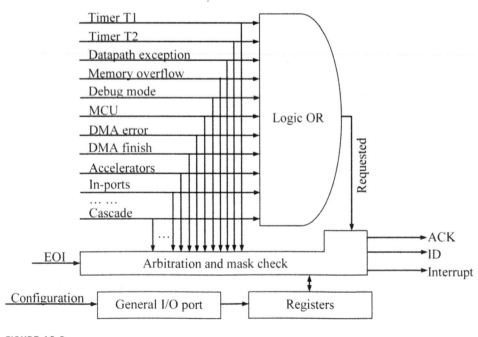

FIGURE 16.9

Block diagram of the Interrupt Controller for **Senior**.

Table 16.8   Registers in the Interrupt Controller in Figure 16.9.

ID register	Flag register	Priority register	Mask register
Pointer to each interrupt source	[6:5] read only	[4:1] writable	[0] writable

- A 1-bit mask register for each interrupt source (writeable by the core).

- The priority register of the current running (main) task (writeable by the core)?

Each interrupt source has a 7-bit register addressed by its ID. The priority register and the mask register can be configured by the processor core, and the flag register is read only. One bit of the flag register is "interrupt requested", and one bit is "interrupt handled." When an interrupt request comes, the interrupt requested flag is set, and when the interrupt is accepted and is under service, the interrupt handled flag is set. When EOI comes, both the interrupt request and interrupt handled flags are cleared. When the mask register is one, the interrupt request will be rejected.

The interrupt controller starts handling interrupts by checking if there is a request. When one or more interrupts occur, the interrupt controller checks the priority and process on the interrupt request with the highest priority.

The interrupt request with the highest priority will be handled further if the interrupt source is not masked. If the current running task has higher priority, the interrupt request will be rejected. If the processor is running another interrupt service with higher priority, the interrupt request will be rejected.

When an interrupt is acknowledged (either accepted or rejected), the acknowledge signal is sent to the interrupt source together with the ID code.

Finally, the interrupt controller is connected to the processor core via a general I/O port. The interrupt controller has two modes: the configuration mode and the handling mode. In configuration mode, the general I/O port passes the interrupt setting to registers in the interrupt controller. After configuration, the module will be switched to the handling mode as the default mode.

Interrupt Latency

As mentioned, interrupt latency is the time from interrupt request to the time starting the interrupt service. In Table 16.9, interrupt latencies are estimated as the design information of quantitative approach.

16.4 TIMERS

A timer, or a timer counter or a watchdog, is a counter that is decremented (or incremented, depending on the implementation) automatically, either at a fixed or configurable rate or when a certain event occurs. When the timer has reached zero,

Table 16.9 Interrupt Latencies.

Task allocation	Request handling		Context switching		Total latency
	General	Senior	General	Senior	Senior
SW arbitration with context saving	3X number of interrupt sources	~ 48	2X number of registers	~ 96	~ 144 clock cycles
HW arbitration with context switching	2 clock cycles	2	2X number of registers	~ 96	~ 98 clock cycles
HW arbitration without context switching	2 clock cycles	2	no	no	2–3 clock cycles

it usually triggers an interrupt and optionally reloads the initial starting value to start another countdown.

In a DSP system, a timer can be used by a program to check a runtime limit of a program. In a real-time system, especially for streaming signal processing, a program must be finished within a limited period of time. Before running the program, a timer is started and keeps counting down at a steady pace. After running the program, the timer is reset by the program because the interrupt from this timer will not be necessary. If the program cannot finish in time because of some unexpected reasons such as the program being stuck waiting for I/O, stuck in a loop or a trap, the timer keeps counting until it reaches zero and then sends out an interrupt request so that the program can be stopped or restarted with other data.

A timer can count based on three events: the machine clock, real-time clock ticks (second, minute, hour, tick counter), or some external event (event counter). The machine clock stands for the clock signal of the synchronous logic circuit of the processor. The counting of external events could be mandatory for some DSP applications such as DSP processors for automatic control.

The counting range is up to the application. For example, 1/10 second can be the period of computing a picture frame for surveillance applications, 20 milliseconds can be the period of a voice frame, and 2 microseconds can be the period to receive a short symbol of wireless connection of 802.11a/g. Suppose the clock speed of a DSP processor is 200 MHz; then the counting of 1/10 second is equivalent to 20 million machine clock cycles, and the counting of 2 microseconds is 400 machine clock cycles. To make the counter efficient, a counter can be configured to count based on one or multiple machine clock cycles—for example, counting can be configured based on one machine clock cycle or every 1024 machine clock cycles. This is accomplished by putting a prescaler unit between the machine clock and the clock input (event input) of the timer.

Example 16.4

Design a timer to supply counting of the machine clock cycle, the clock tick of each millisecond, and external events. The streaming signal packet arrival time could be in every 100 milliseconds. The bus width is 16 bits, and the machine clock frequency is 250 MHz.

To configure for the down-counting on a 16-bit bus, the maximum number is $2^{16-1} = 65535$. Counting up to 100 milliseconds requires 25 million clock cycles when the machine clock is 250 MHz. $25,000,000/65535 = 381.5$. By utilizing a prescaler and counting down once every 1024 clock cycles instead, a 16-bit counter is enough since $25,000,000/1024 = 24414$, which is less than 65535. Following the design of the control register of the common interface in Example 16.1, the control register of the timer is shown in Table 16.10.

Four bits are used for the timer custom configuration. Bits [8:7] are used to select what to count, counting on machine clock, or on clock tick, or on external event. Bits [10:9] are used to configure the counting speed, meaning the trigger arrival time. The counting could be based on each trigger, every 256th trigger, every 512th trigger, or every 1024th trigger. The trigger stands for a machine clock, or every 1 millisecond of clock tick, or every external event. Figure 16.10 shows the timer circuit block diagram. The control register gives control of event

Table 16.10 Timer Control Register Specification.

Bits	Specification
[0] = 1	Reset the module, flush the current task, and clear the interrupt request
[1]	Shut down the module when 1
[2]	Data rate: when 0, machine clock speed, when 1 memory clock speed
[3]	Synch: when 1 using a special strobe signal, when 0, no strobe (see Table 16.1)
[5:4]	(These bits are not used)
[6]	Report result: when 1 using interrupt; when 0 using flag
[8:7]	Counting on: machine clock [8:7] = 01; clock tick [8:7] = 10; external event [8:7] = 11
[10:9]	Down-count once on [10:9]: each = 00; every 256 = 01; every 512 = 10; every 1024 = 11 (Each means, e.g., each machine clock or each clock tick, or each external event. Every 256 means down-count once every 256 clock cycles/256 ticks/256 events)
[15:11]	Reserved

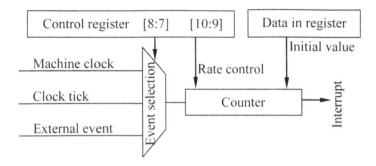

FIGURE 16.10

Block diagram of the designed timer.

selection and rate control. The data in the register gives the initial value of the counter. When countdown reaches zero, an interrupt is generated.

The output from the timer is an interrupt request signal. The type of the interrupt should be an NMI (no-maskable interrupt) if a time-out means that a fatal error has occurred. Many systems have a watchdog timer that is connected to system reset, which must be periodically restarted by the application, typically once for each iteration of the infinite loop. For most timers, the time-out merely requests a service, and the timer interrupt priority depends on the priority of the service.

The timer configuration can be performed by writing a 16-bit vector to the control register of the common peripheral interface in Figure 16.4, and the timer status can be observed from the status register of the common peripheral interface in Figure 16.4. The initial value of a timer can be supplied through the data register in Figure 16.4.

16.5 DIRECT MEMORY ACCESS (DMA)

16.5.1 DMA Basics

DMA of a Processor Core

Direct memory access (DMA) concepts were introduced in Chapter 15. DMA accesses processor memories by DMA hardware (DMA controller) instead of by running a program using the processor hardware resources. By accessing memories using a DMA controller, the processor can do other operations while the transfer is in progress. There are many kinds of DMA controllers for different applications. In this book, the focus is only the DMA controller for real-time streaming signal processing without data cache supporting (program cache might be supported). It handles DMA transactions between memories and peripheral ports.

In most cases, the sizes of on-chip memories are limited by the silicon technology and cost. There are two cases that make DMA essential for streaming signal processing. The first case is when an application requires more memory space than the size of the on-chip memories. That case also includes the support of multithread using on-chip memories with limited sizes. The second case is when transferring data between on-chip memories for streaming signal processing distributed in parallel hardware. By using DMA transactions, simple load/store operations can be off-loaded and more computing capacity of a DSP processor can be used for computing.

DMA hardware is called a DMA controller, DMA handler, or DMA module. By definition it is a peripheral module of a processor core for direct memory access. When volume data needs to be transferred between memories and peripheral ports, the core or the cache controller prepares a DMA request and sends the request to the DMA controller. The DMA controller prepares and transfers data, and the processor core can run other programs. After the data transfer, the DMA controller sends an interrupt to the processor core so that the transferred data or new available memory spaces can be used in the processor. In a similar way, any other device can also ask for DMA services. By using the DMA controller, memory access cost can be significantly reduced because the DMA transaction as shown in Figure 16.11 is running in parallel and behind the running task requesting the DMA transaction.

The Scope of This Section

In Chapter 15, the memory subsystem and DMA concepts were introduced. In this section, the DMA controller hardware in the DSP subsystem will be discussed further in detail. DMA programming will be discussed in Chapter 18. The DMA controller in a DSP subsystem can be identified in Figure 16.12. In Chapter 19, we will finally discuss the system bus and on-chip connection network and integrate the DMA controller into a DSP subsystem.

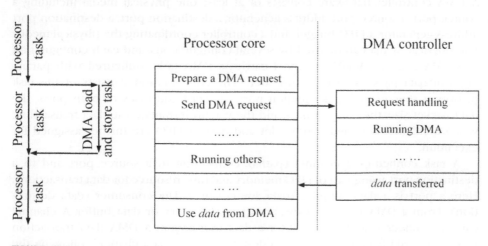

FIGURE 16.11

Using DMA to save processor runtime.

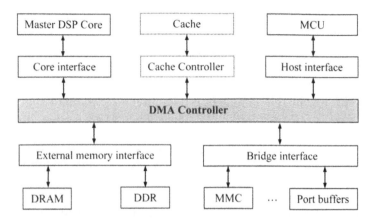

FIGURE 16.12

System view from the DMA controller.

From the DMA controller's point of view, all modules around the DMA controller are DMA clients. The DMA controller handles only block data transactions controlled by the master DSP core. Data formats must follow the interface requirements of the DMA controller. Therefore, all modules are connected to the DMA controller via interfaces. Interfaces adapt the data format between the DMA controller and clients. Discussion of the interface of DDR, DRAM, MMC (a flash memory card standard from Siemens) and other components will not be covered in this book.

DMA Concept

A DMA controller hardware consists of at least one physical media including a source port, a source port address generator, a destination port, a destination port address generator, a FIFO bugger, and a controller coordinating the physical media. The physical media can be used by several configurations, and each configuration is a DMA channel. A DMA channel carries a DMA task configured with parameters of data packets to be transmitted. The configuration consists of the task priority, a source port, a destination port, the start addresses of both ports, the data packet size, the data features, and the way to terminate the data transaction. When a channel is issued, a time slot and part of FIFO are further assigned for execution.

A task configures a channel (paths), connecting to a source port and to a destination port, using part of FIFO memory and time resource for data transactions. Here a port is a data supplier (data source) or a data consumer (data destination). From a DMA point of view, a port is a memory or data buffer. A channel can be configured and activated for a data transaction. A DMA data transaction is to move data from source port to destination port via a channel following the configuration of a DMA task. The simple view of a DMA controller is given in Figure 16.13.

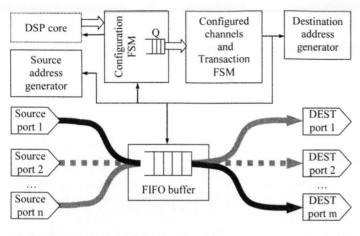

FIGURE 16.13

DMA ports and channels.

A DMA controller can only see ports instead of functions behind a port. DSP core executes a DMA task by configuring a channel with two ports, address information, and data features. At the same time, the priority also is assigned to the channel according to the priority of the transaction and the data behavior. According to the priority, time (full time or time shared) slots also are assigned to the channel. Because the DMA controller must handle two memories and transactions between them, two address generators are required by a channel. One address generator generates the address to the source memory, and the other generates the address to the destination memory. A data FIFO buffer is used to synchronize the transfer between the two memories. Different clock rates and data widths can be matched by separating load and store of FIFO data buffering. If the DMA controller serves multiple data transactions simultaneously, the FIFO buffer also is used for hardware multiplexing.

16.5.2 Design a Simple DMA
A DMA Source, Destination, and Channel

From a DMA controller point of view, a source port is a single port memory with its starting address and the size of the memory.

Example 16.5

Specify DMA ports to link memories of Senior DSP: DM0 as port 1, DM1 as port 2, PM as port 3, and the off-chip main memory as port 4.

Senior specification can be found in the appendix. The main memory is specified by a project using **Senior**. From the data sheet, we specify port 1 and port 2 as the description of data memories DM0 and DM1 (see Table 16.11).

Table 16.11 Port 1 and Port 2 Specifications are the Same.

Specification of port 1 and port 2	Data of port 1 and port 2
Data bus width	1 word = 16 bits
Data rate	Machine clock rate
Data type	Simple, no protocol, no error detection
Address bus width	16b or addressing space is 64k words
Clock	Supplied by DMA controller < 200 MHz
Synchronization	DMA sending strobe signal

Table 16.12 Port 3 Specification.

Specification of port 3	Data of port 3
Data bus width	1 word = 32 bits
Data rate	Machine clock rate
Data type	Simple, no protocol, no error detection
Address bus width	15b or addressing space is 32k words
Clock	Supplied by DMA controller < 200 MHz
Synchronization	DMA sending strobe signal

The program memory can be specified as port 3 (see Table 16.12).

The main memory can be specified according to the project specification (see Table 16.13).

In the specification tables in Example 16.5, clocks are supplied by the DMA controller. A channel is the media for data transaction. A transaction should be specified by configuring a channel. It includes the configuration to the source, the configuration to the destination, and the configuration of the transaction. Basic channel configurations for a data transaction include:

- Task priority.
- Data size: the length of the data block.
- Data from: the name of the source port.
- Data to: the name of the destination port.
- The physical start address of the source port.

Table 16.13 Port 4 Specification.

Specification of port 4	Data of port 4
Data bus width	1 word = 64b
Data rate	Memory clock rate
Data type	Parity check and sending error interrupt to reset the transaction if parity check is wrong
Address bus width	26 addressing bits; or addressing space is 64M word
Clock	Supplied by DMA controller < 60 MHz
Synchronization	DMA sending strobe signal

- The physical start address of the destination port.

- The endian behavior of the source port: Big or Little-endian.

- The endian behavior of the destination port: Big or Little-endian?

Except for the configurations by running SW, physical specifications are predefined yet important to know by both DMA designers and DMA users:

- The maximum source clock rate.

- The maximum destination clock rate.

- Data width of the source port: 8 bits, 16 bits, 32 bits, or 64 bits.

- Data width of the destination port: 8 bits, 16 bits, 32 bits, or 64 bits.

- Data protocol of the source port: error check or correction.

- Data protocol of the destination port: error check or correction?

Specific hardware resources might be required by different source and destination hardware. For example, a parity checker will be required if the data is from off-chip memory. If the data widths of source and destination ports are different, a data buffer will be needed and two clock sources are needed to synchronize two address generators. The data rate of a DMA transaction follows the port with lower speed.

Example 16.6

Specify a channel for a data transaction between DM0 and the off-chip main memory.

A FIFO buffer is needed for data transactions between ports with different data rates, see Table 16.14.

Table 16.14 Specification for a DMA Transaction.

Specification of the transaction	Configuration of the transaction
Task priority	Highest
Data rate	400/100 MHz
Data block length	2 blocks (one block contains 1024 bytes)
Source port	DM0
Source port data width	16b
Source port start address	$0A00
Source port data protocol	No
Endian behavior of source port	No (only 16 bits, no data concatenation)
Destination port	Main memory
Destination port data width	64 bits
Destination port start address	$3000FFF
Destination port data protocol	Parity check (64 + 1 bits)
Endian behavior of destination port	A main memory word = {W1, W2, W3, W4}

DMA Data Transaction Circuit

A DMA data transaction circuit is the DMA datapath, getting data from the source port using the source address generator, and storing data to the destination port using the destination address generator. To handle a DMA transaction, the following hardware modules should be prepared:

- A transaction FSM with a down counter handling data transaction.

- A source address generator and its clock source.

- Source data decoder decoding arriving data.

- A FIFO data buffer storing decoded data.

- Destination data encoder, encoding data and sending to the destination port.

- A destination address generator and its clocks source?

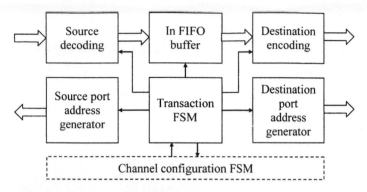

FIGURE 16.14

DMA transaction module.

The circuit for a DMA transaction is prepared, connected, and shown in Figure 16.14.

When the DMA transaction controller is available (no running transaction), a configured channel can be issued from the Channel configuration FSM module to the Transaction FSM module. When the transaction is started, the transaction FSM starts down-counting. The source port address counter generates the address to the source port. The source port supplies data to the source decoding circuit. The source decoding circuit decodes the input data and adapts data format to the FIFO buffer. The FIFO buffer sends data to the destination encoding circuit. The data format finally is adapted to the destination port and sent together with the address generated by the destination port address generator. The transaction FSM in Figure 16.14 handles the transaction until the data block is transmitted. The channel configuration circuit finally generates an interrupt to the master DSP core.

The FIFO speed adapts to the port with lower speed and lower data bandwidth. For example, when the source port speed is lower and the data width is larger, one source word will be divided into several words to the destination port. Data format conversion should be performed during the FIFO access.

DMA Configuration Circuit

To prepare a DMA task, a source port, a destination port, and a DMA channel must be connected and configured. A channel is selected and connected for the DMA task. A DMA task must be named with an identification, a DMA ID. The ID is the name of the DMA task being assigned on a channel, existing in the DMA controller and in the master core until the task is finished. A DMA task can be configured by selecting a DMA channel and writing the DMA task to configuration registers of the channel.

Table 16.15 Specification of a DMA Task To Be Assigned on a DMA Channel.

Specification of the task	Register value assignment
DMA task ID	R11[15:12]
Task priority	R11[11:10]
Source port name	R11[9:7]
Source port data width	R11[6:5]
Source port data width	R11[4:2]
Destination port name	R11[1:0]
Source port start address (24b,)	{R14[7:0], R12[15:0]}
Destination port start address (24b)	{R14[15:8], R13[15:0]}
Data block length (up to 64k)	R15[15:0]
Source port data protocol, with parity check = 1	R16[15]
Destination port data protocol, with parity check = 1	R16[14]
Endian behavior of the source port	R16[13:12]
Endian behavior of the destination port	R16[11:10]
Reserved	R16[9:0]

Example 16.7

Specify configuration registers for a DMA channel.

All configurations can be assigned to several 16-bit registers, see Table 16.15.

R11 stands for register 1 of channel 1. Suppose that there are eight channels. $8 \times 6 \times 16 = 768$ bits are required for DMA configurations of all channels.

Together with the peripheral common interface of the DMA controller, the configuration FSM of the DMA controller handles communications between the processor and the DMA controller. Communications include the following:

1. The processor core requests a channel configuration.

2. The DMA controller accepts or rejects the request for the configuration.

3. The processor reads the status of the DMA controller.

4. The processor writes a task configuration to configuration registers of a channel.

5. The processor activates a DMA transaction.

6. The processor resets (cancels) or suspends a configured channel.

7. The DMA controller sends an interrupt of transaction complete.

8. The DMA controller sends an interrupt of transaction error.

All functions can be interfaced by a DMA control register and DMA status register.

Example 16.8

Specify the control and status registers of a simple DMA controller.

The control register is specified in Tables 16.16 and 16.17.

The status registers are specified in Table 16.17 according to Table 16.2.

Programming the Simple DMA Controller

The program code follows what we discussed in this section:

```
int init_dma(void)
{
    *DMA_config |= 0x200; // Request channel config
    if(*DMA_status & 0x1) {
        return 0; // Busy, try again later
    }
    // Activate a channel
    *DMA_config &= 0x07ff; // Clean bits 15-11 in config reg
    *DMA_config |= 0x5800; // Activate channel 5
    return 1; // Success
}
void DMA_transaction_complete_handler(void)
{
    // Do something here such as:
    // * Wake up a task waiting on DMA
    // * If we have caches we might have to synchronize it
    //   with the DMA
    // * Activate another DMA request
    // * ...
}
void DMA_transaction_error_handler(void)
{
    // Do something here such as:
    // * Ignore it (if the DMA isn't critical)
    // * Retry transaction
    // * Notify operating system
    // * Notify user
    // * Reboot system
    // * ...
}
```

Table 16.16 Control Register Specification of a Simple DMA Controller.

Bits	Specification
[0] = 1	Reset DMA, flush the current task
[1]	Shut down DMA when [1] = 1 (control for low power)
[2]	Data rate: always using DMA clock [2] = 1
[3]	Synch: When 1, a strobe pin is used to synch the data from the main memory
[5:4], [9]	Not used
[7:6]	Report result: when 00 using flag, no interrupt; when 01, normal interrupt; when 10, nonmaskable interrupt; when 11, exception and reset the system
[8]	Reset (clean) the interrupt request
[10]	Activate a task (channel) [10] = 1
[14:11]	DMA task ID
[15]	When [15] = 1, ask for a channel configuration

Table 16.17 Status Register Specification of a Simple DMA Controller.

Bits	Specification
[0]	Idling or busy: Idling = 0 and busy = 1
[1]	A channel can be configured [1] = 1, No channel is available when [1] = 0
[2]	Report result: Running task is finished when result flag = 1
[3]	If the interrupt pin = 1, it requires an interrupt to handle the result
[4]	abnormal = 1 (error interrupt) when there is an exception
[15:5]	Reserved

16.5.3 Advanced DMA Controller

Advanced Features Are Needed

The discussion on the simple DMA controller in the previous section exposed enough fundamental knowledge of a DMA controller. For some ASIP with simple applications, the simple DMA controller can be good enough. However, many advanced real-time applications require heavy and complicated data accesses. There are communication extensive applications such as DMA for large FFTs or DMA for video signal processing with large frame size. These applications need advanced DMA features, which were not discussed. These features are:

- Arbitration of configuration request: If more modules around the DMA controller want to ask for DMA service, multiple requests on channel configurations might come at the same time. The DMA will accept the module with the highest priority for a configuration.

- Arbitration of channel activation: If multiple configured channels are activated by multiple modules at the same time, only the one with the highest priority can be activated.

- Higher priority interrupt: If a higher priority channel asks for a transaction, the transaction with a lower priority will be interrupted and will give the DMA hardware resource to the channel with higher priority.

- Variable data packet size: Normally, the size of a transaction is specified as the number of data blocks. A data block has a normal size of 1 kbytes. However, a special DMA might transfer a very short data packet, only a few bytes. If the smallest packet size is 1 kbyte, there will be unnecessary transaction cost. Flexible data packet size will be useful.

- Merge multiple transactions using linking table: This is especially important for video CODEC and other two-dimensional array signal processing. Several distributed data blocks are required by one DMA transaction. The processor will consume too much time for DMA configurations if each data block requires one DMA transaction. A better solution is to link all these data blocks into one DMA transaction by a linking table. Multiple data blocks will then be concatenated as one large data block of a DMA transaction.

- Simultaneous multichannel transactions: Sometimes, a task cannot start until the complete data packet is transferred (tail triggered). Sometimes, applications can start while receiving or transmitting data. In this case, simultaneous multichannel transaction will be useful. Multiple channels can be activated and running on Time Division Multiplexing (TDM) mode.

DMA Arbitration Hardware

The arbitration can be managed by fixed connection or by configured hardware. The easiest way is to handle a DMA request by a fixed hardware connection called the Daisy chain described in Figure 16.15.

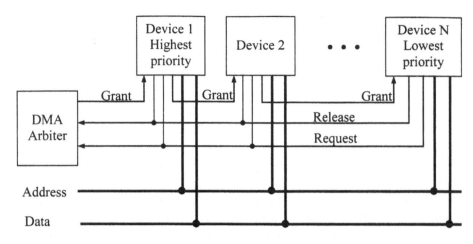

FIGURE 16.15

Daisy chain bus architecture.

In this case, the hardware Daisy chain gives the DMA request arbitration according to the position of the physical connections. Any DMA request can be received by the arbiter immediately. The grant acknowledgment comes to the hardware closest to the arbiter with the highest priority. If the hardware is the request maker, the grant will be blocked. Otherwise, the grant acknowledgment will be passed to the next hardware sequentially. If there are two devices on the bus requesting DMA service at the same time, the device closer to the arbiter gets the right, and the DMA request from the device further away from the arbiter will be blocked and rejected. Following the principle of hardware arbitration, the device that holds the highest priority will be connected closest to the DMA arbiter.

The flexibility of the arbitration based on a fixed connection is not high, and the priority cannot be changed after the hardware design. While running different applications, the DSP processor might want to change priorities of the DMA tasks according to the application. A flexible DMA arbitration can be conducted based on the system connection using the normal bus in Figure 16.16.

In this figure, the DMA request and grants of each device are connected to the DMA arbiter individually. The DMA arbiter therefore can handle DMA requests based on the configuration of the current running application in the DSP processor. Using the connection topology in Figure 16.16, the arbitration can be conducted following the flowchart in Figure 16.17.

Higher Priority Interrupt

In advanced DMA controllers, when a high-prioritized DMA task is requested by a master, an immediate DMA service should be given by the DMA controller. The currently running low-prioritized DMA transaction will be interrupted, and the high-prioritized DMA request will be served. After the interrupt service, the interrupted DMA transaction will continue. Be very careful—a DMA interrupt is not an interrupt

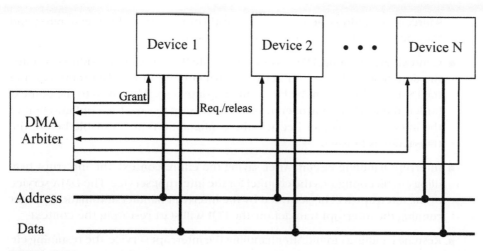

FIGURE 16.16

A normal bus system.

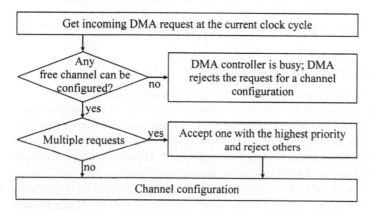

FIGURE 16.17

Arbitration for a DMA channel configuration.

toward the processor core. A DMA interrupt is just the interrupt of a DMA service in the DMA controller, and the core processor actually does not realize the DMA interrupt.

There should be enough hardware support for the interrupt of a DMA transaction. Hardware for interrupt handling includes:

- Interrupt request handling circuit: Similar to the normal interrupt controller, the circuit checks if there is an interrupt request. If there is a request, the priorities of the interrupt and the current running task (transaction) will be compared. If the priority of the interrupt is low, the interrupt will be rejected; otherwise, the interrupt will be accepted. When an interrupt is accepted, the source port address generator, the destination port address generator, and the counter of the data transaction will be stopped.

- Context saving circuit: DMA contexts include the source port address counter, the destination address counter, the down counter of the data transaction in the FSM, and the data in the FIFO buffer. Context shall be saved to the context buffer in the DMA controller. It is not necessary to save the configuration of the interrupted channel, because the configurations are available before the transaction is finished.

- Interrupt handling circuit: After saving the entire context, the interrupt handling circuit configures the channel for the interrupt service. The DMA service of the interrupt will be activated by the interrupt handler automatically. After running the interrupt transaction, the EOI will start restoring the context.

- Resume handling circuit: After handling the interrupt service, the resuming circuit will conduct three tasks: (1) pass the configuration of the interrupted channel to the transaction control circuit; (2) restore contexts to the transaction control circuit; (3) activate and continue the interrupted transaction.

Interrupts will seldom happen, and one group of context registers should be enough.

Merge Multiple Transactions Using Linking Table

Many applications use off-chip memory as the main computing buffer. Sizes of computing buffers might be up to several megawords. The size of an on-chip data memory may not be more than several kilo words. It is therefore essential to use the on-chip memory in an efficient way. The DMA controller shall support a selective load of exactly wanted data for algorithms to be executed. However, data organized in the main memory does not follow a specific algorithm.

Linking tables for advanced DMA transactions is a technique especially important for video CODEC and other two-dimensional array signal processing. Multiple and distributed noncontiguous data blocks are required by an application. The processor and DMA cost will be too high if each DMA transaction handles only one data block. A better solution is to link all these data blocks into one DMA transaction by a linking table. Linking table-based DMA transactions can load several noncontiguous data blocks one after another in the main memory, and merge them into a large data block in the on-chip data memory. Multiple data blocks will then be merged as one large data block of a DMA transaction.

The complexity of a DMA task varies. For example, it is simple to pass a picture frame from a memory in one computing engine to the memory in the next computing engine by a DMA transaction. It is just a simple block transfer. However, to efficiently prepare data for an advanced motion estimation algorithm of video frame encoding, DMA transaction could be complicated. Data from different reference frames will be used at the same time. Data blocks at different locations in the main memory can be required simultaneously by an algorithm. Multiple data

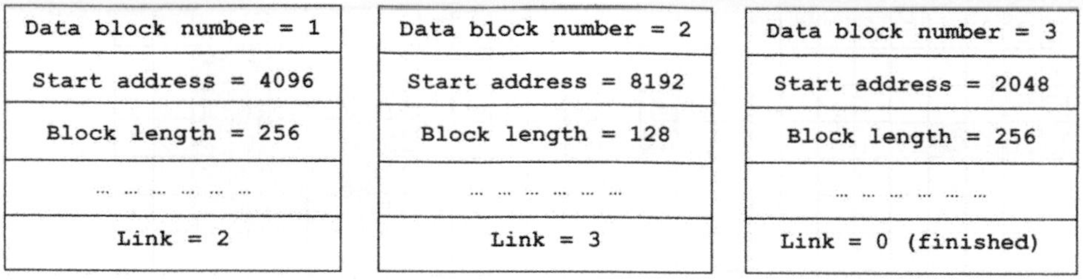

Data block number = 1	Data block number = 2	Data block number = 3
Start address = 4096	Start address = 8192	Start address = 2048
Block length = 256	Block length = 128	Block length = 256
... ...	... ...	... ...
Link = 2	Link = 3	Link = 0 (finished)

FIGURE 16.18

An example of a DMA linking table.

blocks can be loaded sequentially by one DMA transaction if a linking table technique is used. A linking table concatenates several data blocks into a DMA transaction. It describes the relation between data blocks to be transmitted by a DMA transaction. An example of a linking table is given in Figure 16.18.

The first data block starts at the physical address 4096, and the length of the first data block is 256 words; when the first data block is loaded, the loading of the second data block (with data block number = 2) is immediately followed. The starting address of the second data block is 8192, and the length is 128 words. After loading the second data block, the loading of the third data block is activated immediately at the beginning point of the data block number = 3 by link = 3. After loading 256 words from the starting point of 2048, the DMA transaction is finished because link = 0 is reached at the end of the configuration. Using linking table, three DMA transactions for three noncontiguous data blocks are merged to one DMA transaction.

Let us continue the discussion of addressing acceleration for 2D data for parallel data processing in Chapter 7. When estimating the motion of a data unit in a picture frame, the data units close to the region in several reference frames will be required.

Example 16.9

Prepare a DMA transaction for motion estimation of the marked data unit in the current frame in Figure 16.19.

In the example in Figure 16.19, the current picture frame is divided into $8 \times 7 = 56$ data units; each unit contains $16 \times 16 = 256$ pixels, and one picture frame contains 14336 pixels. Running the motion estimation algorithm finds two correspondent data units, the data unit in the reference frame equivalent to the data unit 28 in the current frame, in two reference frames. The motion vectors will thus be generated to indicate the relative move of that data block in the current frame based on reference frames.

Suppose the segment address of the current picture frame in the main memory is 32768, the segment address of the reference frame 1 is 47204, and the segment address of the reference frame 2 is 61540. To prepare for the motion estimation of the twenty-eighth data unit in the current frame, the following 11 data blocks (marked data units) in Figure 16.19 should be prepared in Table 16.18.

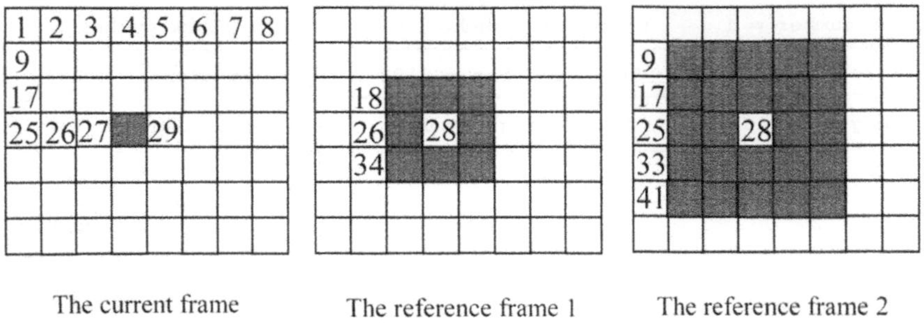

The current frame   The reference frame 1   The reference frame 2

FIGURE 16.19

An example of a DMA transaction containing multiple data blocks.

Table 16.18 Preparing a DMA Transaction for Motion Estimation.

Block No.	Block description	Segment address	Start offset address	Length of data packet
1	1 data unit of the current frame	32768	27*256 = 6912	256*1 = 256
2	3 data units 19, 20, 21, in reference flame 1	47204	18*256 = 4608	256*3 = 768
3	1 data unit 27 in reference frame 1	47204	26*256 = 6656	256*1 = 256
4	1 data unit 29 in reference flame 1	47204	28*256 = 7168	256*1 = 256
5	3 data units 35, 36, 37 in reference flame 1	47204	34*256 = 8704	256*3 = 768
6	5 data units, 10–14, in reference frame 2	61540	9*256 = 2304	256*5 = 1280
7	5 data units, 18–22, in reference frame 2	61540	17*256 = 4352	256*5 = 1280
8	2 data units, 26, 27, in reference frame 2	61540	25*256 = 6400	256*2 = 512
9	2 data units, 29, 30, in reference frame 2	61540	28*256 = 7168	256*2 = 512
10	5 data units, 34–38, in reference frame 2	61540	33*256 = 8448	256*5 = 1280
11	5 data units, 42–46, in reference frame 2	61540	41*256 = 10496	256*5 = 1280

In Example 16.9, the segment address is the starting address of a frame. The starting offset address of a data unit is the unit number times the unit size, 256. By modifying the configuration table of a DMA transaction in Example 16.9, Table 16.14 can be modified to be Table 16.19. The DMA transaction of Example 16.9 can be configured.

The DMA transaction specified in Example 16.9 can be further illustrated:

1. The transaction loads data blocks from the main memory (source port).

2. The data comes to data memory DM0 (destination port) in the DSP processor.

3. The transaction will link, load, and merge 11 data blocks.

4. The data block 1 from source port starts at 39680 and the packet length is 256.

5. The data block 2 from source port starts at 51812 and the packet length is 768.

6. The data block 3 from source port starts at 53860 and the packet length is 256.

7. The data block 4 from source port starts at 54372 and the packet length is 256.

8. The data block 5 from source port starts at 55908 and the packet length is 768.

9. The data block 6 from source port starts at 63844 and the packet length is 1280.

10. The data block 7 from source port starts at 65892 and the packet length is 1280.

11. The data block 8 from source port starts at 67490 and the packet length is 512.

12. The data block 9 from source port starts at 68708 and the packet length is 512.

13. The data block 10 from source port starts at 69988 and the packet length is 1280.

14. The data block 11 from source port starts at 72036 and the packet length is 1280.

When loading data from the main memory, the DMA controller links and merges 11 data blocks, and send the merged data block to DM0 in the processor.

Table 16.19 DMA Channel Specification Using a Linking Table.

Specification of channel 1	Register
DMA task ID	The identification of the transaction
Task priority	The priority of the transaction
Data rate	The data rate of the transaction
Number of linking blocks	= 10 in this example (10 links after the first block)
Source port	The main memory
Source port data width	16 (to simplify the example)
Destination port	DM0 of the DSP core
Destination port data width	16 (to simplify the example)
Destination port start address	From this point, all transactions will be merged
Link 1 source port start address	6912 + 32768 = 39680
Link 1 block length	256
Link 2 source port start address	4608 + 47204 = 51812
Link 2 block length	768
Link 3 source port start address	6656 + 47204 = 53860
Link 3 block length	256
Link 4 source port start address	7168 + 47204 = 54372
Link 4 block length	256
Link 5 source port start address	8704 + 47204 = 55908
Link 5 block length	768
Link 6 source port start address	2304 + 61540 = 63844
Link 6 block length	1280
Link 7 source port start address	4352 + 61540 = 65892
Link 7 block length	1280
Link 8 source port start address	6400 + 61540 = 67490
Link 8 block length	512
Link 9 source port start address	7168 + 61540 = 68708
Link 9 block length	512
Link 10 source port start address	8448 + 61540 = 69988
Link 10 block length	1280
Link 11 source port start address	10496 + 61540 = 72036
Link 11 block length	1280

Handling Multiple Channels

Simultaneously handling multiple DMA payload transactions is not usually the case in DSP applications, because most DSP tasks cannot be executed until a complete data set is available in the on-chip data memory (DMA tail triggered task). However this is necessary for DMA controllers of other ASIPs, such as a network processor, NPU. In this case, the DMA controller schedules multiple channels for multitransaction tasks in a time division interleaved mode.

16.5.4 DMA Benchmarking

DMA benchmarking can be general performance benchmarking of data transactions or benchmarking of specific applications. Parameters to be benchmarked include bandwidth, task setup latency (the time cost of a channel configuration), and latency of the data transaction (the time cost of a data transaction). The definition of data latency is different for different applications. Normal data latency is the time from fetching a word from the source port to the time the destination received the word including the FIFO buffering time. For most applications, data cannot be used until a complete data packet is transferred. In this case, the special data latency is the previously mentioned normal data latency plus the size of the data packet.

Benchmarking a good DMA controller shows about ten cycles of setup latency and a few cycles (less than 10) of normal data latency during data transaction. The benchmark of bandwidth can become worse when the average data size of a data transaction is small.

The DMA benchmarking for a specific application can be much worse because of the irregularity of transaction data.

16.6 SERIAL PORTS

A serial port transmits a single bit at a time, which gives the serial port a low pin cost. When the requirements on bandwidth are low, a serial port is preferred. A serial port can be a chip peripheral device or a device for intercore (module) communication on a chip.

16.6.1 Bit Synchronization

To receive data, a receiving port must synchronize its sampling clock to receive data bits. On bit-level synchronization, a serial port can be synchronous or asynchronous. A port is a synchronous port if the data transmitted and received are based on the transmitting clock, the reference clock. It means that the sending part and the receiving part use the same clock. In this case, the clock could be transferred using a specific physical clock line or carried within the transferred data. An asynchronous port receives data without clock synchronization. The reception is based on integration or accumulation of multiple samples.

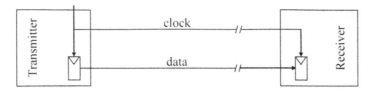

FIGURE 16.20

Synch a bit using carried clock.

Bit Synchronization Using Carried Clock Pins

The easiest way to detect and receive bits transmitted using a serial port is to send the clock signal along with the data. The clock is a physical pin transmitted from the data source, and data detection is conducted using the transmitted clock on the receiver part. The principle is given in the block diagram shown in Figure 16.20.

The transmitting and receiving ports use the same clock, provided by the transmitter. The advantage is simplicity, and the drawbacks are the extra pin cost and extra power consumption of the clock transmission. The physical interface of the clock and the data connections should be carefully designed when the circuit speed is high or the distance is long. The wire lengths of clock and data need to be balanced with the same physical delay so that data on the receiver side can be reliably sampled when the transmitted clock arrives.

Bit Synchronization Using Recovered Clock

Clock and strobe pins can be removed for synchronous data transmission over a long distance. In this case, a data transparent protocol is required, and the clock can be coded into data on the transmitter side and recovered from the receiver side.

The clock can be coded into the preamble of a data packet. In this case, the efficiency of data transmission is $E = L_{data}/(L_{data} + L_{preamble})$. Here the L_{data} is the payload length of a data packet, and the $L_{preamble}$ is the length of the preamble. The efficiency is high if L_{data} is very long and $L_{preamble}$ is very short. However, if the L_{data} is too long, the accumulated frequency error will be significant. A digital phase-locked loop (DPLL) is required to recover the clock carried by the preamble. The implementation of the digital phase-locked loop can be found in most ASIC books. In Figure 16.21, the principle of clock recovery is depicted. The preamble with predefined length and code of 1010 1010 1010 1010 is used to carry the clock. The receiver can recover the clock from it.

Instead of using the preamble to carry a clock, a simpler solution is to carry the synchronization clock using every data bit based on redundant coding. For example, DBPL (differential biphase level) is one kind of coding of clock into each bit transmitted. DBPL coding is given in Figure 16.22.

The coding and decoding principle of DBPL is to change the signal level once if the source code is 0 and change it twice if the source code is 1.

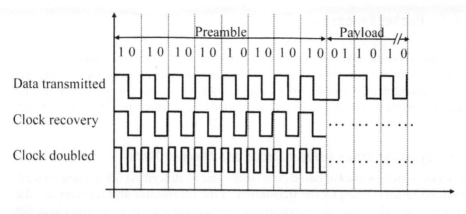

FIGURE 16.21

Clock recovery from data packet preamble.

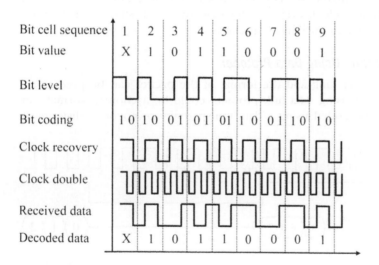

FIGURE 16.22

DBPL coding.

Asynchronous Detection

Asynchronous detection uses a clock that is independent of the transmitter clock. The detection or reception therefore is performed using a local clock at the receiver end (instead of transmitted), and the clock rate is much higher than the data rate. The reception is therefore a kind of integration of the oversampling on the received data. For example, if the received data rate is 1 Mb/s, the oversampling rate might be 16 MHz or more.

16.6.2 Packet Synchronization

Packet synchronization is to find the start time for a received data packet. It can be performed using a strobe, a masked clock, or it can be data based on strobe protocols. A strobe is a physical pin enabling and disabling the clock of receiving data. A masked clock is enabled when data is valid and disabled when data is invalid from the sending part. A masked clock is strobed by the strobe pin from the sending part.

Packet Synchronization Using a Strobe or a Masked Clock

The packet synchronization based on a strobe can be described using Figure 16.23. The strobe signal is given by the transmitter. If the transmitter and receiver use the same clock and the data rate is low or the connection wire is short, the clock pin can be eliminated, and using a strobe is a good choice.

To save a pin, the strobe signal can be merged into the clock, and the transferred clock becomes a masked clock (see Figure 16.24).

When the data rate is high or the connection wire is long, the delay on the serial port can be more than a clock cycle. In this case, the clock shall be transferred together with data to match the delay of the data.

Packet Synchronization Using Data Protocol

If a data packet is rather long and the format of the packet can be predefined, packet synchronization using data link protocol could be a choice. Normally in this case, data must be coded using a data transparency protocol to distinguish

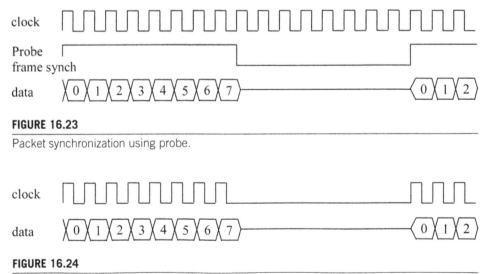

FIGURE 16.23

Packet synchronization using probe.

FIGURE 16.24

Packet synchronization using masked clock.

```
1111 1111 1010 1010 1 control packet
```

```
1111 1111 1010 1010 0 data packet
```

FIGURE 16.25

Preamble for packet synchronization.

the difference between a data packet, control signals, and the preamble. Packets synchronized by data are suitable for data transmission between two devices with a long distance in between.

Example 16.10

Design a preamble for transferring both a control packet and data packet. Here, the control packet carries control information of the serial port. It means that the control packet will be consumed by the serial port.

The protocol can be designed in Figure 16.25.

In this example, 1111 1111 1010 1010 is occupied and used as the preamble so that all data transferred must be precoded to avoid the pattern. For example, one 0 can be inserted after every 8 bits to avoid the pattern of 1111 1111 1. After receiving the preamble 1111 1111 1010 1010, if the bit following the preamble is 1, the following data will be a control packet; otherwise, the following data will be a data packet. 1111 1111 1 can be used to synchronize the data packet, and 1010-1010 following 1111 1111 1 can be used to recover the clock.

16.6.3 Arbitration

Arbitration is not required if two serial ports are full-duplex point-to-point connectors. Arbitration is mandatory when multiple masters can access the same serial bus. Arbitration gives permission to one of the masters to use the port when multiple requests or transactions arrive at the same time. Arbitration can be performed by the application program or in hardware.

The principle of arbitration for transaction requests in a serial port is similar to the arbitration of DMA described in the previous section. In the operating system or application program, the priority of using a serial port is predefined, and the arbitration should be based on the predefined priority. When a port is receiving two requests at the same time, the port FSM will recognize and keep the request with higher priority. The request with lower priority will be rejected. Rejection could be an acknowledgment of NO or a no-reply. If a master device cannot get an acknowledgment of the request in time, the master device can skip the current request and prepare for a new request.

16.6.4 Control of a Serial Port

A serial port is controlled by a FSM. The FSM manages requests of transactions, coding of transmission, synchronization and decoding of reception, and error control. The FSM can be shared with the FSM of the common peripheral interface in Figure 16.4, or it can be independent. A serial port can be activated by the master, the DSP core through addressing its common peripheral registers (master mode). A serial port can also be activated by other port devices connected to the serial bus (slave mode). The master can be a transmitter or a receiver. The slave can also be a transmitter or a receiver.

The task of a transaction will be initialized by the master sending a control packet to the serial port. The acknowledgment is required to report both the acceptance of the request and the finish of the data packet transaction. The serial port will be configured using the configuration vector carried by a control packet. When a data transaction is finished, the serial port changes its status in the status register in the common peripheral interface or sends an interrupt. When a serial port is locked or in an abnormal situation, an exception interrupt is requested and the serial port will be reset.

16.7 PARALLEL PORTS

Parallel ports are similar to serial ports. The main difference is its performance. Parallel ports can supply much higher performance but consumes more I/O pins. A parallel port may carry an address bus so that a parallel port can access memory data directly. A typical parallel port is the off-chip memory interface.

16.8 CONCLUSIONS

Peripheral design is an important and large part of a DSP design project. In this chapter, design of peripheral modules was discussed. In Chapter 19, system integration using peripheral modules will be discussed.

The common interface in a peripheral module is designed together with the processor core. It offers the connection between the processor core and the function module of the peripheral. A task to be executed in a peripheral module can be configured and issued by the processor core. Execution of the task is controlled by the local finite state machine of the peripheral. A peripheral module can offload work from the processor core.

Peripheral modules can be a general IO or a generic peripheral module of a DSP core. Interrupt handler, timer, and DMA controller are three general peripheral modules of a DSP core. An interrupt controller handles asynchronous communications between the processor core and peripheral hardware modules. It also offloads interrupt handling from the processor core. A DMA controller offloads memory transactions from the processor core. A timer counts and monitors the code execution time and interrupts the execution when the execution time is longer than the specified limit.

EXERCISES

16.1 Find an example to support Figure 16.2.

16.2 Find an example to support Figure 16.3.

16.3 How many addressable registers can be found in a peripheral module—for example, the general module in Figure 16.4?

16.4 How many steps are needed to handle an interrupt? Give descriptions of each step. Allocate each step to certain hardware (DSP core, interrupt controller, interrupt source).

16.5 What is HWMR, and how do you use it to decide the use of fast interrupt or normal interrupt?

16.6 What are the main differences between fast interrupt and normal interrupt?

16.7 What is NMI? What interrupts should be NMI?

16.8 How do we specify and calculate interrupt latency?

16.9 List all action steps in order in the DSP core, the timer, and the interrupt controller while a timer sends a NMI.

16.10 What stands for a channel when running a DMA transaction?

16.11 What is a Daisy chain bus, and why is DMA arbitration not flexible enough when using a Daisy chain for the arbitration?

16.12 Discuss the following figure and describe functions of each component.

16.13 Design all connections between a DMA module and a DSP memory subsystem and the DSP core, using a general I/O port described in this chapter as the control and configuration interface, using one general memory peripheral circuit discussed in the previous chapter as a data port. The data port is multiplexed by two data memories and one program memory. Two bits of the control register in the general I/O port can be used for the control of the multiplexing.

16.14 How do you design bit synchronization based on data patterns for a synchronous serial port using only one data pin without clock and strobe pins?

REFERENCES

[1] Dave Comisky, D., Sanjive Agarwala, S., Charles Fuoco, C. (2000). A scalable high-performance DMA architecture for DSP applications. *Proceedings of 2000 International Conference on Computer Design*, 414–419.

[2] PrimeCell mDMA Controller (PL230) Technical Reference Manual, http://www.arm.com/.

[3] http://www.ti.com/ TMS320C55x DSP Peripherals Reference Guide Literature Number: SPRU317B, May, 2001.

Design for DSP Functional Acceleration

17

The design of a DSP accelerator can be divided into the design of the interface between an accelerator and the processor core, and the design of accelerator functionalities. The most important issue is to design a scalable processor that allows accelerators to be attached. Therefore, in this chapter, the focus will be on the design for the scalability and extendibility of the processor core, the design of the interface, and the design of the accelerator behavior model. Functional design of an accelerator can be found in any DSP ASIC book [1].

17.1 FUNCTIONAL ACCELERATION

An accelerator is a hardware module attached to a processor core, which enhances the performance or functionality by executing certain functions in the accelerator instead of the processor core. An accelerator could be a simple hardware driven by port configuration instructions, or a slave processor containing an instruction subset. The following is the list of cases to accelerate:

- The processor core cannot offer enough computing power to certain extensive computing functions. The accelerator will take over these extensive computing functions such as FFT or motion estimation for video encoding.

- Certain functions were missing during the instruction set design. For example an integer division instruction might be needed when the processor is used to handle a certain function; however, it wasn't included in the original instruction set design of the processor core.

- For ultra-low-power consumption: Offload heavy computing from the processor core, so that the supply voltage can be very low because the system clock rate can be low. The power consumption can be reduced dramatically.

Functional acceleration is popular in ASIPs because an ASIP is designed for a specific class of systems. In particular, an accelerator is beneficial to the system if the product volume is high, and the function to be accelerated is known and fixed. Accelerators are dedicated for specific functions so that accelerated solution usually has lower silicon cost and lower power consumption.

597

Table 17.1   Advantages and Disadvantages of Functional Acceleration.

Advantages	Disadvantages
Higher performance	
Lower power consumption	Less flexible and longer verification time
Less silicon costs	

The first question is whether to accelerate or not. Acceleration has both advantages and disadvantages, see Table 17.1. Reduced flexibility is one of the major disadvantages. In cases where the flexibility is essential, functional acceleration may not be favorable. For most DSP applications, adding accelerators is feasible and suitable because DSP functions are relatively known, stable, and computationally extensive.

Most DSP kernel algorithms are based on vector computing, which is relatively regular and stable. Another kind of popular DSP algorithm is for coding and decoding (CODEC) for data compression. A CODEC is a finite state machine (FSM), and many codec FSMs are relatively regular. Regularity is the condition of acceleration, and high computing demand is the main reason of acceleration. In general, the functions to be accelerated are, for example:

- Heavily iterative and computationally extensive functions.

- Functions that are most frequently used by an application.

- Accelerated functions to dramatically reduce the code size.

Besides three kinds of listed functions, functions can be accelerated only when the functional specifications are stable, the flexibility requirements are known and manageable, and the acceleration is really required.

17.1.1 Loosely Connected Accelerator

Accelerators can be divided into loosely connected accelerators and tightly connected accelerators. A loosely connected accelerator is a functional module behind the common interface of a peripheral module. A peripheral module was described in Figure 16.4 in Chapter 16. The communication between the DSP core and the accelerator can be conducted by reading and writing the peripheral register in the common interface of the peripheral module.

The general peripheral module was discussed in Chapter 16. A peripheral module consists of a common interface FSM and custom function modules behind the common interface of the peripheral module. In this case, the custom function module is the accelerator. I/O instructions, read port or write port, can be used to start, stop, configure, and monitor an accelerator. The control register of the peripheral module can be used for accelerator configuration. The status register of the peripheral module can be used for monitoring the accelerator. The data register of the peripheral module can be used to access data to and from an accelerator.

A loosely connected accelerator can be added easily to a processor core without changing the instruction set, the tool chain, and the core RTL code because I/O port in/out instructions can support all communications between the accelerator and the DSP core. However, the data communication bandwidth is low, and communication latency is long between the processor core and the accelerator. The code cost of the interwork between the core and a loosely connected accelerator could be high.

17.1.2 Tightly Connected Accelerator

Designing and adding a loosely connected accelerator is easy, and there is nothing special to discuss. However, a loosely connected accelerator cannot have much interwork with the processor core. For example, operands cannot be carried and delivered to the accelerator in parallel by the dedicated accelerator instruction. Proposed core execution mode cannot be associated. If much interwork is required, a tightly connected accelerator will be required instead. In the rest of the chapter, designing and integrating tightly connected accelerators will be discussed.

Accelerations can be classified as the instruction level-acceleration, the architecture-level acceleration, and the module-level acceleration. The solution choice depends on system requirements and the trade-off of flexibility, performance, and costs.

While designing a general-purpose DSP processor, when the application is not known, to keep enough flexibility instruction-level acceleration is avoided. A VLIW architecture is a solution to enhance the performance in such situations. Silicon efficiency is usually measured using performance over silicon cost. From this point of view, VLIW is more flexible and less efficient compared to an accelerator solution.

The decision to add an acceleration module is based on the understanding of applications via source code profiling under system requirements. In a DSP processor, the accelerated instruction could replace a complete subroutine or part of it. By packing more subroutines into accelerated instructions, enhanced performance for certain applications can be achieved. Figure 17.1 depicts the method of design for instruction-level acceleration.

Recall the common feature of a DSP instruction set. The typical assembly instruction set of a DSP processor consists of a RISC subset and a CISC subset. The CISC instructions are designed for functional acceleration. Two kinds of CISC instructions can be found in an ASIP DSP processor: the instructions that are designed during the design of the processor core and the instructions added after the design. Instruction acceleration in Table 17.2 was discussed previously in Chapter 7. Added instructions after the processor core design will be the focus of this chapter.

The accelerator design for an available DSP processor is to enhance its function coverage and performance by adding hardware besides the processor core

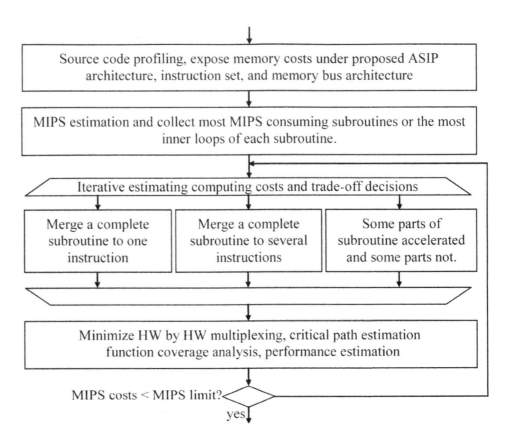

FIGURE 17.1

Profiling and decision of function acceleration.

Table 17.2 Acceleration Mode Trade-offs.			
	Instruction-level accelerations	**Architecture-level accelerations**	**Module-level acceleration**
Decision time	During instruction set design	During instruction set design	During or after instruction set design
Performance enhancement	Certain percentage, not very high	Normally around factor 1.5 to four	Could be very high; also for function extensions
Examples	MAC, auto correlation, look-up tables, bit manipulation	DUAL MAC, VLIW	Motion estimator, FFT module, DCT module, Huffman accelerator, Viterbi accelerator, etc.

without touching the available design of the processor core. Accelerator design is a special technique or method used in industries when the requirements on flexibility are predictable and the complexities of the accelerators are acceptable. When the requirements on flexibility are unknown or when the complexity of the tasks is too high, more flexible solutions such as multiprocessor solutions may have to be used. Since accelerators are dedicatedly designed for one or a group of applications, the hardware cost of an accelerator is lower compared to the cost of more flexible solutions.

From another point of view, acceleration is a specific custom design. Excessive instruction-level acceleration makes the instruction set less flexible. Since the accelerated instructions are not regular, the control and decoding of these instructions are usually complicated.

17.2 ACCELERATOR SPECIFICATION

17.2.1 Principle

The function of an accelerator can be partitioned to one instruction or to multiple instructions. More flexibility can be achieved if the functions in an accelerator are partitioned to several instructions. At the same time, more instructions will bring higher complexity and more control overhead. In this section, the discussion will be around the partition and allocation of functions either to the processor assembly code or to the hardware accelerator.

Theoretically, perfect partition does not generate extra communications. Tasks running in an accelerator should not have any communications with the processor core until the tasks in the accelerator are finished. In special cases, if the communications between the core and the accelerator are not avoidable, the data dependencies should be schedulable; that is, the accelerator should not require exact clock cycle timing of all data transfers but should be able to wait for data.

17.2.2 An Accelerator with One Single Instruction

In many cases, accelerating a simple algorithm is needed. A typical example is an integer divider, "one over x" machine, or a square root machine. In these cases, only one accelerated instruction is needed by a single instruction accelerator. The accelerator instruction can carry information such as the control configurations, operand addresses, or immediate data.

Real-time algorithms with fixed cycle cost can easily be mapped to this kind of accelerator. Hardware cost and cycle cost always have to be traded off according to the system requirements. The silicon cost is higher if the computing is more in parallel. The cycle cost is higher while reducing silicon costs by eliminating parallelization. Usually, an accelerator design is to reach the minimum silicon cost under the fixed cycle cost.

Example 17.1

Taylor series machine.

$$Y = a_0 + a_1 x + a_2 x^2 + a_3 x^3 + a_4 x^4 + a_5 x^5$$

$$= (a_0 + a_1 x) + ((a_2 + a_3 x) + (a_4 + a_5 x)x^2)x^2$$

Usually, the approximation from the first five order computing gives sufficient precision for most embedded applications.

To simplify the hardware and minimize the silicon cost, the accelerator can be implemented based on $(a + bx)$ unit and x^2 unit. The execution cycle cost can be four clock cycles.

The implementation will be an instruction to carry:

- An operand, x.
- The name of the algorithm, for example, $\sin(x)$, or $1/x$, or $x^{1/2}$.

All coefficients of certain Taylor series are stored in ROM, and the name of the algorithm will be the pointer of a group of coefficients, a_0 to a_5.

For example, $\sin(x)$ can be executed while loading coefficients a_{01} to a_{51}, $1/x$ can be executed by loading coefficient a_{02} to a_{52}, and $x^{1/2}$ can be executed by loading coefficients a_{03} to a_{53}.

17.2.3 An Accelerator with Multiple Instructions

In this case, the functions mapped to the accelerator are implemented into multi-instructions. A multi-instruction accelerator is not a slave processor. Each instruction is issued by the master processor, and the accelerator does not fetch instructions. An important feature of a multi-instruction accelerator is its simple instruction flow. Instructions are issued and decoded by the master processing, and the data dependencies between the processor core and the accelerator can be managed reliably offline before the execution.

The following reasons support the partition of an accelerator function to several instructions. One reason is the hardware multiplexing of the accelerator for several applications. For example, a complex-valued MAC accelerator can be used for complex-valued data manipulation such as convolution, convolution with conjugate, auto-correlation, and accumulation.

The use of multiple instructions can simplify the design of accelerator control FSM. For example, to simplify the FSM in an FFT accelerator, each layer FFT can be executed by one instruction. N point FFT requires $\log_2 N$ FFT instructions. In this case, the FSM of the accelerator is simple because it controls the execution of only one loop for one-layer FFT. The FSM for this FFT accelerator consists of only one simple counter and three address generators. The processor core is usually advanced enough to configure the instruction for each FFT layer so that the computing and addressing reconfigurations of each FFT layer are handled by the processor using different instructions for different processing layers.

Example 17.2

Specify an accelerator executing *N* point DIT butterfly-based radix-2 FFT (see the algorithm description in Chapter 7).

The following steps should be performed:

Function partition: To simplify the FSM in the accelerator, only one layer of FFT will be executed in the accelerator.

Design of instructions: To run one layer FFT, all configurations of the execution should be performed. Configurations include:

Bit-reversal addressing (data addressing mode of the first layer).
The data memory addressing mode (from layer 2 to the last layer).
The coefficient memory addressing mode.
The result memory addressing mode for layer 1.
(The data and result memory addressing are different in layer 1).

Datapath was designed in Chapter 13. Refer to Figure 13.17 and corresponding text.

FSM is based on a counter counting up to $N/2+$ PEC. Here, PEC is the total cycle costs of prolog and epilog.

The pseudo assembly code to run a 64-point FFT (32 butterflies) is:

```
FFT (B=32, layer=1, Bit_reversal=1, W=1, D_step=1);
    The core can execute another 39 independent instructions
FFT (B=32, layer=2, Bit_reversal=0, W=2, D_step=2);
    The core can execute another 39 independent instructions
FFT (B=32, layer=3, Bit_reversal=0, W=4, D_step=4);
    The core can execute another 39 independent instructions
FFT (B=32, layer=4, Bit_reversal=0, W=8, D_step=8);
    The core can execute another 39 independent instructions
FFT (B=32, layer=5, Bit_reversal=0, W=16, D_step=16);
    The core can execute another 39 independent instructions
FFT (B=32, layer=6, Bit_reversal=0, W=32, D_step=32).
```

B is the number of butterflies, *W* is the way to fetch coefficient, and D_step is the step size of incremental addressing of data fetching and result store. The total FFT runtime will be $40 \times 6 = 240$ clock cycles. While running the FFT, another $39 \times 5 = 195$ instructions can be executed in the core in parallel. The overhead count of the 64-point is $240 - 195 = 45$ clock cycles.

17.2.4 An Accelerator as a Slave Processor

A master processor can be attached with several slave processors. The master processor core initializes a task by sending a task name to a slave processor core. Since the slave processor is programmable, tasks can be accelerated with more flexibility.

However, having more flexibility also means inducing more overhead. The silicon cost will be significantly higher when an accelerator is programmable. Since a slave processor runs its own programs (in other words, it is loosely connected to the master processor), the control of the master over slave processors is relatively weaker and the execution time of the slave processor might be uncertain. The communications between master and slaves is usually more complicated than accelerators discussed before. Data dependency and interlock are problems.

A programmable slave processor can take over more controls and other complicated tasks from the master processor so that the computing capacity of the master processor can be used for other things. For example, by using a programmable FFT machine all FFT instructions in Example 17.3 can be moved to the slave machine. The master processor can run other programs in parallel. A typical architecture of master-slave multiprocessor is the Sony Toshiba IBM CELL [3]. Multiprocessors will be discussed in more detail in Chapter 20.

17.3 SCALABLE PROCESSOR AND ACCELERATOR INTERFACE

17.3.1 Configurability and Extendibility

Scaling can be either upward or downward. When talking about the scaling of CMOS silicon process, scaling means scaling down. When mentioning the scaling of a hardware design, it usually means scaling up or in other words extending the design.

Scalability of a system is therefore the capability to change the size (scale) of the hardware or software system by adding or removing modules with limited design cost. Scalability of a system should be a fundamental feature of a processor core when the core is used as an IP (intellectual property).

The scalability of a processor could also mean the scalable data precision, which is the configurability of the native data width of the processor. However, in this book, the scalability means the capability to add/remove hardware or to add new instructions to a processor core. For example, the number of execution unit (e.g., MAC) can be scaled up from one unit to many.

In this chapter, scalability means the capability to add hardware accelerators and accelerator instructions to a designed processor core during the system design phase. In this situation, the processor and the instruction set was designed and verified. While adding an accelerator, none of the instruction set, the core hardware, the assembler, and assembly instruction set simulator will be touched or changed.

Scaling of a design is based on an available hardware platform. It usually means that hardware is attached as an accelerator to the processor core as the available hardware platform. New instructions will be introduced to invoke the functions of the newly added hardware. The scalability of the hardware platform shall be extended further to the scalability of the assembly instruction set, the scalability of the hardware, and the scalability of the assembly programmer's design tools. In the following sections, three different scalabilities will be discussed in details.

To plug an accelerator into a design, the interface of the design should act as a bridge between the platform and the plug-in component. Therefore, a scalable protocol should be specified during the design of the platform (the processor core), so that the accelerator can easily be plugged in by following the protocol. The scalable protocol will regulate the coding of the assembly instruction set and the hardware interface.

Extendible Instruction Set

An extendible assembly instruction set and extendible interface are essential features of a DSP core as an extendible platform. Platform-based designs are popular currently (2006) [6]. It provides a platform (a configurable and half-done system design) to speed up the SoC design and verification containing sophisticated silicon IP from different sources.

An extendible instruction set has two basic features: the extra coding space for accelerator instructions and the shared code following the so-called codecoding protocol (to be discussed) between the core and accelerators so that the shared code can be decoded by both the processor core and the accelerator. There the processor can supply supports to the accelerator following the decoding of the shared code. The coding space is well prepared when designing the processor core as the platform. Dedicated instruction type (code multiplexing) should be specified for the accelerator instructions of an assembly instruction set with orthogonal features. The codecoding protocol exposes ways to decode part of the accelerator instruction by the processor core and another part by the accelerator. To design a processor core scalable for adding accelerators, the following definitions should be proposed and used for the design of both the processor core and accelerators.

Codecoding Protocol

Definition I: MUX Code

The MUX code defines code types and multiplexes the following code for different types of functions. For example, an instruction set is divided into four types: instructions for move operations, instructions for arithmetic computing, instructions for program flow control, and instructions for functional acceleration. The MUX code can be specified as shown in Table 17.3.

For example, if the length of the instruction binary code is 32 bits, two bits are used for MUX code and 30 bits can be used for function coding of each subset of the instruction set.

Definition II: Shared Code

Part of the binary code of an accelerator instruction, the shared code will be decoded by both the processor core and the accelerator. On one hand, the processor core decodes the shared code and knows part of the functions supplied by the accelerator. Required supporting to the accelerator by the processor can be

Table 17.3 An Example of MUX Code.

MUX code	Instruction types	Specification: the scope of an instruction subset
00	Move instructions	Memory load and store, register move, load immediate
01	Arithmetic instructions	Single step or iterative ALU and MAC instructions
10	Program flow control	Branch, call/return, NOP, repeat (loop), sleep, etc.
11	Accelerate instructions	Added accelerator instructions to an available instruction set for function acceleration

coded and supplied after decoding. For example, the processor core should supply two operands on OPA and OPB buses to the accelerator. Shared code will be discussed in detail in the following examples.

Definition III: Local Code

This is part of the binary code of an accelerator instruction. The processor core passes the local code to the accelerator instead of decoding the code by itself. The accelerator decodes the local code and executes it accordingly. Local codes have two advantages:

1. Different functions can be carried by different accelerators based on the same platform (the core hardware and the Toolchain for assembly programming). The platform can support different system designs by adding different accelerators.

2. A knowledge isolation exists between the system designer and the core IP suppliers. The core IP supplier supplies only the platform without touching the confidential custom functions. The second advantage becomes more important when IP-based design is well accepted.

The codecoding protocol is the agreement on decoding of the shared code between the designers of processor core and the designers of accelerators.

Coding for Shared Code

Shared machine code carries information decodable by both the processor core and the accelerator. For example, shared machine code carries at least the following information:

- Accelerator pointer (AP): It stores the name (address) of an accelerator. M bits will be used to identify N accelerators where $M = \log_2 N$.

- Operands (OP): If there are operands from the general register file of the core, OP indicates the way to supply operands from the general register of the core.

- Execution mode (EM): An accelerator may contain several execution modes. This code defines and reminds the core of the current execution mode.

- Status Configuration (SC): Configure the way to report the status of the accelerator—for example, the way to send a notification when the result is ready.

The code efficiency will be high if operands can be supplied while executing an accelerator instruction. OP code can be used to synchronize the supply of operands from the core and the consumption of operands by the accelerator. For example:

- OP = 00: No operands from the core.

- OP = 01: Operand A is available on OPA bus.

- OP = 10: Operand B is available on OPB bus.

- OP = 11: Both operands A and B are available on OPA and OPB buses.

The following example sends operands to an integer divider, which is an accelerator. The OP code is 11. When the processor decodes the instruction, the processor sends the dividend onto the OPA bus and the divisor onto the OPB bus. When the accelerator decodes the instruction, it fetches the dividend from the OPA bus and the divisor from OPB bus immediately during the execution cycle according to the current decoding instruction.

Execution mode (EM) is useful when the application is complicated and dynamic task scheduling is required. When the processor decodes EM, the processor core can plan a way to fetch results according to the execution mode. EM could be:

- EM = 00 Not specified.

- EM = 01 The instruction consumes one execution cycle.

- EM = 10 The accelerator runs in parallel with the processor core.

- EM = 11 The execution in the accelerator blocks the processor core.

When EM = 01 or 10, there will be no data dependency between the processor core and the accelerator. The processor can run programs in parallel with the accelerator. When EM = 11, the data dependency between the processor core and the accelerator is unknown; the execution in the accelerator has to block the execution in the core. When running a fully predictable program based on static scheduling, EM coding is not necessary.

Status configuration (SC) code is used to regulate the way status information is fetched from an accelerator. For example, it could be a flag of result ready. The flag will be read by the processor core when the core checks the result. This happens when the priority of the result is low. However, when the priority of the result is high, an interrupt might be required when the result is available. By configuring the way to expose status, the processor core can define the interwork between the core and the accelerator.

MUX	AP	OP	EM	SC		OPA	OPB
11	00	11	00	00	xxx xxxxxxxxxx	aaaaa	bbbbb

FIGURE 17.2

Coding of interdivider instruction for **Senior**.

Example 17.3

Add an integer divider to the **Senior** processor (the appendix provides more information about **Senior**). Design the division instruction for the accelerator integer divider.

The accelerator name is 00 for this design. The processor supplies the dividend to OPA bus and the divisor to OPB bus. Because the cycle cost of the integer division can be fixed, the code of execution mode is not needed. A result ready flag is enough, and interrupt is not necessary. The coding of the instruction is shown in Figure 17.2.

When instruction type MUX code is 11, the instruction will be an added instruction for functional accelerations. AP points to the integer division accelerator with the name of 00. EM and SC codes are not needed because the cycle cost of the execution in the accelerator is fixed. When OP code is 11, both operand A and operand B are supplied from the register file to the register bus OPA and OPB.

The shared code will be decoded by both the processor core and the accelerator. The decoding of an accelerator instruction on both sides is shown in Table 17.4.

By decoding the shared code following Table 17.4 in the processor core and in the accelerator, the processor core and the accelerator are synchronized.

17.3.2 Extendible Hardware Interface

The extendible hardware interface is the interface between the core and accelerators offering connections for tightly connected accelerators. The design of the core will not be modified while connecting an accelerator to the core via the extendible hardware interface.

A processor with scalability feature means that accelerators can be added to the processor core without changing the design of the core hardware. Enormous design and verification cost induced by changing the RTL code can be avoided. Modern SoC design is based on available IP (Intellectual Property). An IP core might be designed by other vendors, and there might be no right for the integrator to access the RTL code. Therefore, in case a processor core is provided as an IP, it must have open interfaces to accept accelerators.

Core Hardware Support for Accelerators

To add acceleration functions without changing the current design, the processor core must be designed to give enough supply to accelerators. Let us observe and find which parts inside a processor core will be involved when an accelerator is attached.

Table 17.4   Decoding an Accelerator Instruction.

Code	Decoding in the core	Decoding in the accelerator
Type/MUX	Identifies an accelerator instruction	The accelerator cannot see the MUX code
AP	The core does not care about the AP code	The accelerator pointed to is enabled by AP
OP	The core supplies no, one, or two operands on operand buses	The enabled accelerator accepts no, one, or two operands according to OP
EM	Tells the core to continue or wait	No special operation
SC	The core configures a way to report results by the accelerator	The way to report the result is configured
OPA	Associated with OP code, the core prepares data on the data bus OPA	The accelerator is told to fetch or not fetch operand A on OPA bus
OPB	Associated with OP code, the core prepares data on the data bus OPB	The accelerator is told to fetch or not fetch operand B on OPB bus

To decode an accelerator instruction, the core knows:

- It is an accelerator instruction.
- An accelerator is selected by the AP code.
- How to supply operands from the register file.
- What the cycle cost of the accelerator instruction will be.
- What the accelerator execution mode will be when EM is used.
- How to get the result from the accelerator.

The processor core therefore:

- Sends the shared code and the local code, as part of the accelerator instruction will be sent from the processor core to the accelerator.
- If there are operands, connects the register buses to the accelerator and sends operands on the register buses.
- Sends other configurations to the accelerator.

When sending data to an accelerator, the address and data buses from the core to accelerators can be a broadcasting bus. The pointed accelerator can be enabled by the matching of address. However, it is not very convenient to use a shared bus to collect the data and status from all accelerators. This is due to the physical limitation of logic gates. The reason is that though the inputs of logic gates can simply be connected together, the outputs from different accelerators cannot simply be connected except for the case using the tri-state outputs. Tri-state output is usually slow, and it is not widely used in modern digital VLSI systems. The processor core thus has to collect outputs from N accelerators using N ports.

In Figure 17.3, the scalability is limited by the number of ports accepting results from the accelerators. The coding size of accelerator names also is limited by the size of AP in Table 17.4. It is actually impossible to connect an unlimited number of accelerators to a core due to the limitation of code size.

Finally, to allow accelerators to be attached, a processor core must be designed to process all scalable features discussed in this subsection. The instruction decoder must support the decoding of the MUX code and shared code. The processor core must react to the decoding of an accelerated instruction. To ease the accelerator design, the way the processor decodes the accelerator instruction should be documented as the reference document to the accelerator designers.

Linking Protocol

Linking protocol in this section is the predefined way of communication between the processor core and accelerators. It can be divided into control protocols and data protocols. The control protocol includes the set up and monitor of a connection. The data protocol regulates data formats during the transmission.

A nondeadlock connection procedure is the principle behind the control protocol, which includes the request of a communication, acknowledgment, arbitration, confirmation, priority setting, interrupt handling, constraints, reacting on higher priority interrupts, monitoring the link, and the way to terminate the link (including watchdog).

Any asynchronous communication protocols can be used as the template to design the link control protocol between the processor core and accelerators. Link control protocols must be designed during the hardware design of the processor. In most cases, accelerators are slave modules of the processor core. An accelerator handles only one or a limited number of functions. The link control

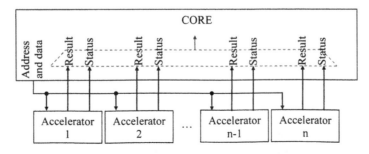

FIGURE 17.3

Core-accelerator hardware interface.

protocol between the processor core and accelerators can therefore be simple. Control linking protocols were discussed in Chapter 16.

Example 17.4

List the contents of a link protocol between a processor core and nonprogrammable accelerators.

The list should include complete control and acknowledgment between the core and an accelerator, including both handling the normal tasks and exceptions.

- The way the processor configures the priority of an accelerator.

- The way the processor checks if the accelerator is busy.

- The way the accelerator acknowledges the status, for example: (1) Available, (2) Busy, yet can be interrupted, (3) Busy and cannot be interrupted.

- The way the processor executes accelerator instructions.

- How the accelerator reports the result, requests an interrupt, and reports an exception.

- How the processor gets the results and releases the accelerator.

- How an interrupt handler handles interrupts while an accelerator is running.

- How an accelerator reacts on an interrupt request/exception from either the interrupt handler or the processor core.

- How to set a timer to monitor the runtime for an accelerator.

A data link protocol regulates the data type and format on the data bus during transmission—for example, the data widths, endianness, and the data packet format. When data and control signals are merged and transmitted along the same bus, data transparency protocol (such as HDLC [4]) is required to distinguish the control vector from data. Since all accelerators are on-chip, data and control can be well separated and data link protocol is not used much between the processor core and accelerators.

17.3.3 Extendible Programmer Tools

Accelerator instructions should be added to the C-compiler, the assembler, the assembly instruction set simulator, and the linker [7–11]. Programming tools were discussed in Chapter 8. The cost of designing a toolchain for assembly programming is high; especially after adding accelerator instructions to a released assembler and ISS (assembly instruction set simulator), the verification costs might be very high.

In most cases, a scalable processor is an IP from an external IP provider. The source code of the assembly instruction set simulator, the assembler, and other assembly programming tools may not be available or may not be modifiable. Therefore, it is preferable for the assembly programming toolchain to remain untouched when adding a new accelerator instruction.

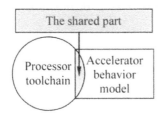

The shared part includes:
1. Shared codes
   a. Instruction word
2. Shared devices
   a. I/O buses
   b. General registers
   c. OPA and OPB busses

FIGURE 17.4

The common part of the toolchain and the accelerator.

To avoid changing the toolchain while adding accelerated instructions, protocols are needed to regulate the design of the toolchain with scalability so that extra instructions can be added later. The same protocol can also be used as regulations to design the behavior models of accelerators so that accelerator instructions can be plugged into the processor toolchain.

To design a protocol that interfaces two systems, the common part—the interfaces and shared resources in the toolchain code—has to be recognized first, see Figure 17.4.

Extendible ISS

An extendible ISS is an assembly instruction set simulator, which can accept new accelerator instructions without changing the source code of the ISS. In this case, the behavior model (the instruction behavior code) of an accelerator must be compiled independently as an included (objective) file. It will be included into the assembly ISS during linking.

If the object file of the assembly ISS is not available (as usual), the accelerator model must be compiled to API format, be included in an API library, and be called and used while running the ISS. An Application Programming Interface (API), a source code interface, specifies details of how two independent computer programs can interact. By using it, an operating system can provide services (another executable program code) requested by a running program. The running program is either the assembler or the ISS. "Another executable program code" is both the behavior model (to be used by ISS via API) and the binary assembly coding of an accelerator instruction (to be used by assembler via API). It is compiled following API format and stored in an API code library that the assembler and the ISS can reach.

While running the assembler, the assembler checks if the assembly instruction is specified in the assembly instruction set manual or not. If it is not specified by the assembly instruction set manual, a call of an API library will be performed. The binary coding for the accelerator instruction will be called, and the binary code of the accelerator instruction will be supplied by the called API code.

While running the simulator, the simulator checks each instruction fetched. If the fetched instruction is an accelerator instruction, it will be partly decoded and partly executed. The accelerator instruction will induce a call of the accelerator behavior model in the API library via an API call. The share code and local code of the behavior model (the function description code) of the accelerator will be passed to the called API function as the instruction to execute by the accelerator.

By checking Figure 17.5, the following variables should be shared by the ISS, assembler, and the API library code:

- The register file output variables: share the selected register values between the core ISS and the accelerator.

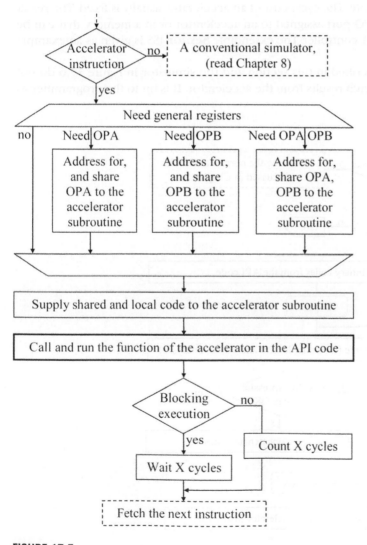

FIGURE 17.5

Extendible assembly instruction set simulator.

■ The cycle cost counter: share the cycle cost counter between the core ISS and the accelerator.

■ The instruction word: share the shared code between the core ISS and the accelerator. The accelerator local code also is shared because the ISS passes the local code to the accelerator, though the ISS does nothing with it.

The accelerator instruction will be executed by both the ISS and the behavior model of the accelerator. The protocol described in this section does not include the way to use the results from the accelerator. The result could be a large data array, or a single sample. The cycle cost of an accelerator usually is fixed. The result data can be on the I/O port assigned to an accelerator or in a memory that can be accessed by the DMA controller. The extendible **Senior** ISS is given as an example in Figure 17.5.

The protocol described in this section and the simulator in Figure 17.6 do not include the way to fetch results from the accelerator. It is up to the programmer to

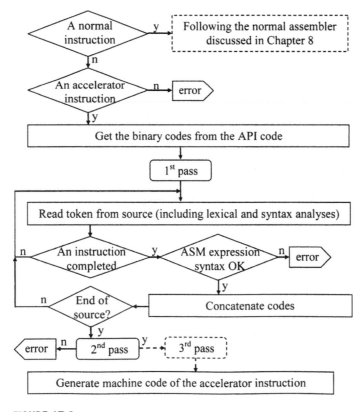

FIGURE 17.6

Part of assembler for accelerator instructions.

decide the way to use the accelerator result, which could be a large data array or a single sample. The cycle cost of an accelerator usually is fixed. Interrupt might be required to report the result ready if the runtime on either side (the processor or the accelerator) is not predictable. The result data can be on an I/O port assigned to an accelerator or in a memory that the DMA handler can access. The accelerator behavior model is compiled and stored as an executable API routine in the API library. When the accelerator is needed by a design, the API code is called by the simulator.

Extendible Assembler

An extendible assembler is the assembler of an extendible core, which can accept and assemble new accelerator instructions without changing the source code of the assembler. Handling code extension by an extendible assembler is easy. The assembler simply checks if the coming instruction is an accelerator instruction or not. If it is true, the accelerator API code in the API library binary will be called. Grammar checking for accelerator instructions should be implemented in the assembler following the protocol discussed earlier. To simplify the relation between the core and accelerators, an accelerator should not use the data memory of the processor core and should not involve any branch operation. Therefore, code and data memory allocation can be skipped during the assembling and linking of the accelerator instruction. An example of an extendible assembler is given in Figure 17.6.

Figure 17.6 is part of an assembler exposing only the binary code generation for accelerator instructions. If the first token (instruction name mnemonic) is an accelerator instruction (cannot be recognized as a normal instruction), the assembler will check if an included file contains the coding of the accelerator instruction (can be recognized as an accelerator instruction). Binary code versus tokens will be matched and fetched from the included file until all tokens of the assembly instruction are fetched. After checking the completeness of the generated binary code, the assembler skips the third pass and generates binary code of the accelerated instruction. The third pass is not required if an accelerator does not have branch and data memory access.

Coding of Accelerator Behavior Model

Coding of the accelerator behavior model should follow the requirements extracted from Figure 17.5 and Figure 17.6. An example of accelerator behavior model is given in Figure 17.7.

Global variables will be used by both the processor simulator and the processor assembler. An accelerator may contain one or more instructions. Each instruction can be coded as a function call. In each function, the code will be divided into two parts. One part is the binary code definition, which assigns each mnemonic a unique binary code in the assembler. Another part gives the functional specification. Functions will be executed by the ISS. Global variables, including cycle counter, register buses, shared code, and local code must be consistent.

```
Behavior model of an accelerator AP(accelerator pointer)
Global variables
  register busses, shared code, local code,
  all mnemonics, cycle counter;
Function AP_I1 // instruction I1
  {
    I1 mnemonics binary coding; // to the assembler
    I1 functions and cycle costs; // to the simulator
  }

Function AP_I2 // instruction I2
  {
    I2 mnemonics binary coding; // to the assembler
    I2 functions and cycle costs; // to the simulator
  }
...
```

FIGURE 17.7

Coding of accelerator behavior mode.

Accelerator behavior model will be compiled as API code and stored in an API library where the simulator and assembler can reach. While designing a system with an accelerator, the API code of the accelerator will be included in the design. While assembling and simulating the design, the included API will be called by the simulator and assembler of the core.

17.4 ACCELERATOR DESIGN FLOW

The simulator and assembler picks up API routines according to the hardware configuration. An assembler picks up the API code of the accelerator and coexecutes the binary coding of the accelerator instruction for code translation. The simulator picks up the API code of the accelerator and coexecutes the accelerator function and counts cycles consumed by the accelerator execution (if the DSP core does not run other instructions in parallel and waits for the accelerator execution). Figure 17.8 gives a flowchart as a review of this chapter.

17.5 CONCLUSIONS

When accelerated functions are relatively stable and predictable, it is preferable to use hardware accelerators for functional acceleration. This is the typical case in many applications using DSP ASIP. To design and add accelerators to a legacy processor

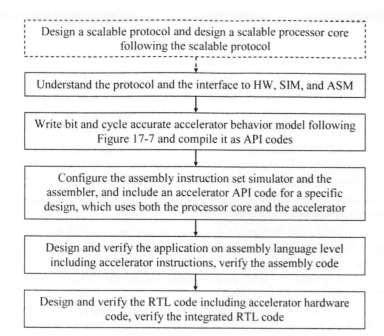

FIGURE 17.8

Accelerator design flow.

core, the core should be prepared to allow new accelerators to be attached, and there should be a protocol that can be followed by the accelerator designers.

The extendible or scalable design of a processor core and accelerators has been discussed in this chapter. The discussion can be divided into two parts: the design of a scalable processor and the design of accelerators following the scalable protocol of the core. Regulations are given through examples on issues such as instruction coding, hardware interfaces, and toolchain extensions.

EXERCISES

17.1 Try to collect all protocols to regulate the instruction set design, the hardware design of the DSP core, and the design of the assembly toolchain for adding acceleration instructions after the design of the DSP core.

17.2 Design a square root accelerator.

17.3 Design a $x^{4/3}$ accelerator (used in MP3 decoding).

a. It may take multiple clock cycles.

b. What are the advantages/disadvantages?

17.4 Design a $y = \sin(x)$ with not more than three execution cycles.

REFERENCES

[1] Wanhammar, L. (1999). *DSP Integrated Circuits.* Academic Press.

[2] Panis, C., Grünbacher, H., Nurmi, J. (2004). A scalable instruction buffer and align unit for xDSPcore. *IEEE Journal of Solid-State Circuits* **39**(7).

[3] Flachs, B., Asano, S., Dhong, S. H., Hofstee, H. P., Gervais, G., Roy, K. et al. (2006). The microarchitecture of the synergistic processor for a cell processor. *IEEE JSSC*, 63–70.

[4] ISO 4335:1987 Information processing systems—Data communication—High-level data link control elements of procedures.

[5] Rowen, C. (2004). *Engineering the Complex SOC.* PTR.

[6] Claasen, T. (2006). An industry perspective on current and future state of the art in system-on-chip (SoC) technology. *Proceedings of the IEEE* **94**(6).

[7] Nie, X., Engel, F., Gazsi, L., Fettweis, G. P. (1999). A new network processor architecture for high-speed communications. *IEEE SIPS'99,* pages 548–557, 20.-22.10.99.

[8] Limberg, T., Winter, M., Bimberg, M., Klemm, R., Matus, E., Ahlendorf, H., Robelly, P., Tavares, M., Fettweis, G. (2008). A fully programmable single-chip 45GOPS SDR baseband for LTE/WiMAX terminals, *ESSCIRC* 2008.

[9] Sucher, R., Niggebaum, R., Fettweis, G., Rom, A. (1998). Carmel – A new high performance DSP core using CLIW. *International Conference on Signal Processing and Techniques (ICSPAT'98)*, pages 499–504, 13.-16.09.98.

[10] http://www.tensilica.com/

[11] http://www.arc.com/

Real-time Fixed-point DSP Firmware

In this chapter, firmware design will be discussed briefly. Discussion throughout the chapter will be divided into the behavior modeling, the HdS (Hardware-dependent Software) design, and the assembly coding. Advanced firmware design techniques can be found in books for professional software engineering, embedded design, and DSP programming [1–3]. This chapter is prepared for novice software (SW) and hardware (HW) designers.

18.1 FIRMWARE (FW)

There is no absolute definition of what should be considered to be firmware (FW) and what should not, but typically, the definition of firmware is fixed software that runs in an embedded product. It is usually stored in a ROM or EEPROM, but not necessarily. Customers typically do not change the software when using the system. However, there are software that cannot be changed by customers, and they are not firmware. For example, customers perhaps do not change the Microsoft Windows software when using a personal computer. Microsoft Office is not in the firmware category because of the implicit scope of FW, the software in embedded systems.

Firmware can be application firmware or system firmware. System firmware is used to manage system hardware and manage other application software. A typical system firmware contains a real-time operating system (RTOS). In this book, the discussion will be focused on the application firmware, the DSP programs. The scope of FW is further implicitly limited to real-time fixed-point DSP FW. Typical DSP FW examples are a voice coder and decoder in a mobile phone.

To implement DSP applications in an embedded product, the total cost of SW design is higher than the total cost of HW design. As a hardware designer, it is essential to minimize the total design cost by supplying good HW. Design of DSP ASIP is HW/SW codesign. Better understanding of SW can enhance cooperative competences for the HW design team. Better understanding of SW design is essential for design of efficient assembly instruction sets.

The knowledge gap is large between the basic understanding of DSP theory and the quality design of DSP firmware. There are three aspects to make a good firmware for an application. The first is the knowledge of the application and the skill of application modeling under hardware constraints. The second is the programming technique on finite precision hardware. The third is the real-time programming technique.[1] The three skills will be discussed in the following sections.

Firmware design flow was briefly introduced in Chapter 4. In this chapter, a relatively detailed discussion will be given based on the introduction given in Chapter 4. Before reading the material in this chapter, reread Chapter 4, and particularly, make sure you understand Figure 4.11.

18.2 APPLICATION MODELING UNDER HW CONSTRAINTS

At the early stage, a design begins with a specification:

1. What is the product (product spec), and what is the project (project plan)?

2. What is the purpose, and what are the objectives?

3. What are the inputs (may not be explicit, designers must collect all of them)?

4. What are the outputs (refinement of the goal and objectives)?

5. What is the functionality (the relations between inputs and outputs)?

6. What are the cost limits (of code size, runtime, data memory cost, design costs, and power consumption)?

7. What are other requirements in general?

8. What are other major constraints (such as compatibility, portability)?

Because the scope of the book is ASIP for DSP, firmware will be discussed only for DSP subsystems. Especially in this chapter, a product stands for DSP firmware running in the ASIP designed by readers.

Firmware design efficiently implements applications on cost-limited hardware with acceptable data quality and runtime. (To simplify the discussion, other requirements such as SW compatibility issues, design costs, etc. will not be considered.) Suppose the input of the firmware design is based on paper documents. Activities included in the early phase of behavior modeling are listed in Figure 18.1.

18.2.1 Understanding Applications

Whether designing a new FW or updating an existing one, the more time spent in understanding the application and planning for the design, the smoother the rest of the development process will go. In the early design phase, sufficient understanding

---

[1]Actually, there is a fourth aspect, which is more important and more challenging. It is the parallel programming technique for predictable DSP applications based on hardware constraints, especially the constraints of the parallel memory subsystem. This aspect will be the focus of Chapter 20.

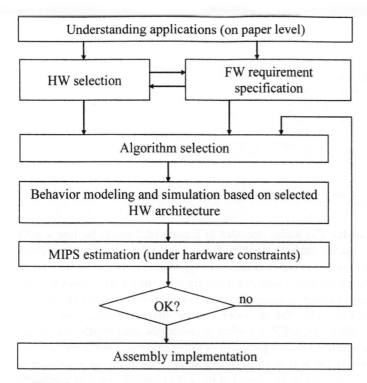

FIGURE 18.1

Activities in early FW design phase.

of applications is essential in order to specify requirements on hardware and on firmware.

An application represents the product and function specification of a DSP subsystem. After market study, understanding an application can be achieved by reading related textbooks (to understand the principle of the application); reading standards (to understand regulations and constraints); and reading product and market documents including product descriptions and product requirements. Of course reference products from competitors also should be investigated.

A DSP subsystem can be broken down into three levels of hierarchies described in Figure 18.2. The product is specified on the first level. A product can be specified as a system for end users—for example, a portable audio player (an MP3 player) or a digital video player that gets video data from either broadcasting channels or from storage component. A product either is used by end users (through human-machine interface) or is part of a higher level system (through the system interface). A product consists of several application components. Each application component is an implementation of a standard or the collection of a class of standards. For example, a portable audio player may consist of an audio decoder and a voice codec (coder and decoder). The audio decoder plays audio for entertainment, and the voice codec

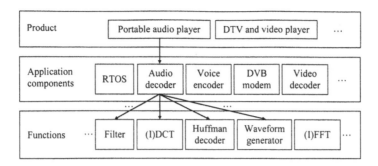

FIGURE 18.2

Understanding applications.

is used as a voice recorder. The audio decoder in Figure 18.2 could be just a MP3 decoder or a decoder supporting both the MP3 format and other audio compression formats such as WMA (Windows Media Audio) or Ogg Vorbis [11].

In this book, the firmware programming starts from an application component. For example, an audio decoder is an application component. To start the implementation of an audio decoder, the decoder can be divided further into function modules such as filters, DCT/IDCT, Huffman decoders, and finally waveform generators. Each function module is a program or subroutine in a software library.

To further demonstrate the way to understand an application, a MP3 decoder is given in Figure 18.3 as an example. The input of the MP3 player is the data packages containing compressed audio data (MP3 format) following the compression standard [4]. The output is the decoded audio streaming data (waveform format) to the DAC (digital-to-analog converter).

The first step in understanding an application is to read related textbooks, for example, a data compression book or audio compression book. The second step is to read standard specifications and divide the application into several function modules. In this example, function modules were partitioned and specified following the encoding format from the standard specification of the audio encoder.

Following the flowchart in Figure 18.3, the first step is to synchronize the packet reading by finding and reading the MP3 header—in other words, finding and reading the synchronization word and other decoding parameters that mark the starting bit of a valid audio frame from the incoming data stream or from the input buffer. The synchronization word is part of a header that contains information to guide decoding, such as the layer number (select and configure a decoder), the sample rate (decide the in-out rate), and the channel configuration (output format). Parameters also include the information to select a Huffman decoding table, the bit rate (frame size), and the scaling for dequantization.

Huffman coded sample values are read and decoded by Huffman decoder; the basic principle was discussed in Chapter 1. After Huffman decoding, quantized samples are recovered. Sample dequantization will be performed. In this step, the bit-stream is dequantized and transferred to audio samples in frequency-domain.

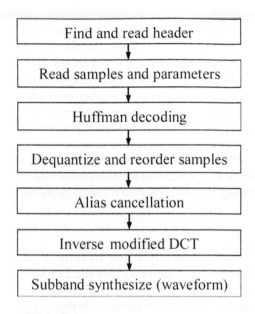

FIGURE 18.3

MP3 decoding flow example.

After being raised to the power of 4/3 during the dequantization process, frequency-domain samples become linear values.

In MP3, the compression is conducted in the frequency-domain after dividing audio signals into several subbands. Before transforming the compressed audio signal in frequency-domain to time-domain, aliasing between neighbor subbands shall be eliminated to compensate for the frequency overlap from the subband filter bank.

By running IMDCT (Inverse Modified Discrete Cosine Transform), each subband will be transformed into the time-domain. Other miscellaneous computing should be given on time-domain samples after IMDCT to compensate the frequency inversion of the subband filter bank. Finally, subband synthesis is conducted by combining time-domain signals from all the 32 subbands covering the whole frequency spectrum.

Results from a real-time system are correct when both functions are executed correctly and the execution is done within a defined time. The first requirement on a DSP subsystem is that the execution time must be less than the time interval (the real-time input arrives). The performance of the firmware therefore should be able to supply the highest computing capacity to decode the compressed audio with the highest sampling rate and the highest bit rate at the real-time. By checking the specification, the requirements on performance should be defined as capable of decoding the compressed MP3 bit stream with the highest bit rate of 320 kb/s and the highest audio sampling rate of 48 kHz.

Requirements include precision on outputs and performance of the decoder. The precision on outputs follows that of reference [5], the compliance requirements. The difference measured using RMS (Root Mean Square) should be less than 3×10^{-5} between a provided reference output file and the outputs of the designed decoder.

Other requirements are specified at the beginning of the project. Requirements are based on product specification. For example, the power consumption usually is not specified as milliwatts; it is specified as battery lifetime instead. The power consumption therefore should be converted from battery life time to milliwatts.

This discussion gives a rough example of how to understand an application, the MP3 decoder. To summarize, understanding of an application includes understanding the function of the application and the basic requirements on implementations.

18.2.2 Understanding Hardware

The behavior model (source code) is the functional description of an application. The behavior model does not contain hardware constraints so that the behavior model can be portable to different hardware platforms. It is not preferred to directly translate the behavior code to assembly language because the distance between the behavior code and assembly code is too long.

To reduce the distance, the behavior code should be modified to adapt to the hardware and the assembly instruction set. To adapt source code to hardware, the target hardware should be analyzed; parameters and features of the hardware should be extracted and added to form the hardware-dependent source code. In particular, when scratchpad memory is used, the understanding of the memory subsystem is essential for further programming.

Hardware understanding for firmware coding can be formulated as two questions: what HW do we need and what HW do we have? The answer to the first question formulates hardware requirements from an application. According to these requirements, a DSP processor is designed or selected to be the hardware platform of the application.

As soon as the hardware is decided, the second question, what HW do we have, should be discussed and answered, and is summarized in Table 18.1, which formulates the hardware features and understanding of hardware for programmers.

The architecture and instruction set of the selected processor should be sufficiently understood in order to select algorithms maximally utilizing the hardware and instruction set features. In Chapters 7 and 8, features of instruction sets and compilers were discussed. Ways to understand an assembly instruction set were discussed in Chapter 7. For example, if the instruction set architecture strongly supports frequency-domain DSP algorithms and transformation, frequency-domain algorithms could be considered.

Most embedded systems require application-specific performance (the hardware performance is sufficient to run an application) instead of the maximum achievable performance (maximum reachable performance of the hardware). The measure of the performance versus supply voltage on the processor can be used to minimize

Table 18.1 Understanding the HW of an ASIP DSP.

Specification	Description
Instruction set	The features of the instruction set, special instructions, and addressing for the application.
Performance	The clock speed and the number of instructions that can be executed in parallel.
Native data width	Precisions of input and output data, datapath, and data stored in data memories.
Internal data width	The data precision inside the datapath for iteration, including guards.
Memory structure	Especially for scratchpad memories, how many memory blocks, constraints to use.
Memory bandwidth	The number of simultaneously accessible memories, the memory speed.
Data memory size	Memory addressing space and sizes of the on-chip program, data memories.
Program memory	The size of the program memory, the longest jumping distance.
I/O bandwidth	The number of I/O ports, the port width, and port access latency.
Peripheral devices	Requirements on peripheral devices, such as timer and interrupt controller.
Toolchain support	The availability and quality of compiler, assembler, linker, simulator, and debugger.
Power consumption	Power consumption of the core logic per MHz, or consumption of the chip including memories, size of the memory, configured size of the design.

the power consumption by voltage scaling. Voltage scaling could be planned while designing the firmware.

Data precision of the hardware can be specified as the native data width of the datapath, the memory data width, and the internal data width for iterative computing. Understanding hardware precision is essential for assembly-level coding and implementation. If the native data precision is not enough, part of the double-precision code should be developed to emulate double-precision computing. That induces significantly longer execution time and code size.

The architecture of the memory subsystem, parallel memory features, memory bandwidth, and addressing models will impact computing overhead. Understanding parallel memory features, addressing, and bus architecture will be essential for early computing time cost estimation. For instance, the execution time cost of an MP3 decoder based on a dedicated instruction set for audio decoding could be much different from the computing time cost of a general DSP processor for MP3 decoding.

For example, suppose that **Senior** (see the appendix) will be used as the hardware platform for audio decoder; an understanding of the processor **Senior** can be summarized in Table 18.2.

18.2.3 Algorithm Selection

DSP functions are implemented using DSP algorithms. Algorithm selection is to select between functionally equivalent algorithms to achieve some kind of optimization. Algorithm selection could also be called SW design optimization on the block/module level. The purpose of algorithm selection can be runtime efficiency, data memory efficiency, program memory (code) efficiency, or adaptation to the hardware architecture. Algorithm selections can be conducted by designers or by CAD tools. To evaluate the quality of algorithms, profiling techniques and profiling tools usually are required.

Table 18.2 An Example: Understanding the **Senior** Processor.

Specification	Description
Instruction set	Single issue, single MAC, conditional execution instructions, repeat instruction, convolution instructions, and rich FSM instructions. Addressing: variable step size and bit-reversal addressing.
Performance	Issue one instruction per clock cycle and highest clock rate (0.12 μ digital CMOS technology) is at least 300 MHz.
Precision	16-bit fixed-point processor. 32 bits + 8 guard bits for iterative computing.
Parallel memories	Two data memory accesses simultaneously; both support FIFO modulo addressing.
Memory bandwidth	Two 16-bit random access data memories, both address spaces are 64 K.
I/O bandwidth	Port addressing space is 64 K, port access latency is one clock cycle.
Peripheral devices	Available interrupt controller, DMA controller, timer, accelerator, I/O instructions.
Toolchain support	Available assembler, linker, simulator, and debugger.

An algorithm selection process starts from an available high-level behavior model. Computing performance demanding functions, precision critical functions, or code cost extensive functions shall be carefully investigated and optimized. Less time should be used to optimize small and miscellaneous algorithms.

Algorithms Following Specifications

It is essential to select algorithms to match the function and requirement specification of the product. Function specification gives input-output relations, and requirement specification gives the quantitative description of the functional specification. For example, a filter algorithm shall offer the amplitude and phase requirements in the pass band and the attenuation requirement in the stop band and transition band. If the phase response of the filter must be linear, IIR filters should be carefully used. To reach the best linear phase response in pass band, Butterworth IIR filters can be used, though the transition band behavior of the filter is not the best. FIR filters give linear phase response and require more cycle costs. The runtime cost of an IIR filter is relatively low, but the filters give poor phase response. Most filter algorithm selection skills are discussed in basic DSP books.

Phase response also is referred to as linearity of the algorithms. Some applications have strong requirements on linearity. For example, linearity is strongly required by radio baseband signal processing for OFDM system (orthogonal frequency division multiplex) with complicated coding such as 64QAM. However, phase response is not important in voice and audio signal processing.

Another example is to select a motion estimation algorithm to reach a certain video compression ratio. For a high-end video camera, compression ratio is important because the minimization of memory cost is essential. For a video camera embedded in a mobile phone, the picture size is small and the power consumption of running the compression algorithm is more important.

Yet another example is to select acoustic echo cancellation algorithms. For high-quality echo cancellation of a large theatre, long echo must be cancelled from the data with high sampling rate. Complicated subband echo cancellation algorithms should be used to minimize the computing cost. For a normal hands-free telephone at home, a low-cost and low-complexity echo canceller is required; the complexity of subband echo cancellation algorithm will be too complicated.

Algorithms Adapt to an Architecture

In addition to the exact functionality and sufficient precision on results, algorithm selection is based on system requirements and hardware. Algorithm selection is to find the best suitable algorithms for the applications following the limitations of the selected hardware. If the hardware computing capacity is much higher beyond the computing cost of an application, algorithm selection may not be essential. One algorithm will be selected from a variety of functionally equivalent algorithms as the trade-off and the best matching of requirements.

There are always many algorithms available from text or reference books and from research publications. However, publications usually do not expose limitations of the algorithms. Different algorithms match different architectures. The implementation of computing-demanding algorithms usually requires specific hardware architectures. To discuss the relation between algorithms and architectures, algorithm-architecture classifications are required. Table 18.3 is designed to help readers to select a suitable algorithm from enormous opportunities. Rows of the

Table 18.3 An Example: Algorithms versus Architectures.

Algorithm features	Descriptions	Suitable HW
Optimized for single-issue hardware	Fine-tune inner-loop algorithms, e.g., Newton-Raphson method.	Single issue core
Optimized for instruction level parallel computing	Explicitly parallel data, task, addressing features, e.g., Taylor Series.	VLIW, SIMD
Suitable for data level partition	A DSP algorithm kernel can be partitioned and executed in parallel.	SIMD, VLIW, multicore
Not suitable for data level partition	A DSP algorithm kernel cannot be partitioned and executed in parallel.	VLIW, superscalar, single issue
Suitable for task level partition	Simple functions, no interfunction dependence.	All streaming architectures
Not suitable for task level partition	Complicated functions cannot be partitioned.	VLIW or single issue or superscalar
Low complexity for parallel computing	With high data and task regularities.	All parallel architectures
High complexity for parallel computing	With low data and task regularities.	VLIW, superscalar
Low vector computing regularity	With low data and addressing regularities.	VLIW, superscalar, single issue
High vector computing regularity	With high data and addressing regularities, e.g., Taylor Series matches very well the architecture and avoid using Newton-Raphson method.	SIMD
Inner loop parallelization hard	Kernel subroutines cannot be further partitioned and executed independently in parallel.	VLIW, superscalar, single issue
Inner loop parallelization easy	Kernel subroutines can be further partitioned and executed independently in parallel.	All parallel architectures, SIMD, multicore

table classify algorithm features. Column 3 of the table suggests the most suitable DSP ASIP architectures.

Most publications contribute algorithms that get minimized vector computing costs regardless of hardware regularities. For a simple processor, such as **Senior** (in the appendix), a superscalar, or a VLIW processor, these algorithms can be used to enhance the performance. These algorithms are usually irregular and with added intratask data dependencies. For example, the Newton-Raphson method is a typical algorithm not so easy to map to a parallel datapath. These algorithms are therefore not suitable for parallel computing machines, especially not for SIMD machines. Research on algorithms also is driven by parallel computing; the foci of the parallel algorithm design are high data and addressing regularities, balanced partition, data and control isolation, and localized data features. For example, the Taylor Series based algorithm is suitable for SIMD and especially suitable for multi-in single-out SIMD datapath, to be discussed in Chapter 20.

Parallel programming will be discussed in detail in Chapter 20. Some basic concepts listed in Table 18.3 will be discussed in this chapter only to help the reade understand the table. Data-level partitioning means that a low-level kernel algorithms (such as the FW implementation of sin, $1/x$, and square root) can be further partitioned and run in parallel because data of the kernel is independent when executing the kernel in parallel. Task-level partition is to partition an application into different task steps and allocate them to different hardware. Execution of these task steps is in a streaming chain (pipelining); the result of early task is the input of the following task. An architecture that supports chains of tasks running in parallel is called streaming architecture.

Some ASIPs have dedicated hardware acceleration for certain algorithms. Special algorithms should thus be selected to adapt to the architecture. Understanding the ASIP instruction set and available accelerators is therefore important for algorithm selection.

Example 18.1

FFT algorithm selection.

Let us look into an example. Consider a special architecture supporting hardware acceleration of radix-2 DIT FFT algorithm. The datapath supports one complex data multiplication and two complex data accumulations in parallel. A DIT FFT butterfly consists of exactly one complex data multiplication and two complex data additions. The benchmarking of the datapath in Figure 13.19 in Chapter 13 thus shows the performance of "one butterfly computation per clock cycle." However, when executing a radix-4 FFT in the advanced datapath for Radix-2 FFT, the achieved performance is very low because the hardware does not accelerate the radix-4 FFT algorithm. Radix-4 FFT is suitable for either of the architecture accelerated radix-4 computing, or for VLIW as well as for superscalar architectures.

Both Radix-2 FFT algorithm and Radix-4 algorithm are mapped and executed on a datapath for Radix-2 FFT given in the last figure in Chapter 13. The total cycle cost in table 18.4 includes both the inner loop cost and the overheads (prolog and epilog). It is obvious that an advanced radix-4 FFT consumes more clock cycles when executing it on a datapath

Table 18.4 Comparing the Cycle Cost of Implementing Two 64-point FFT Algorithms.

Algorithm and costs	Radix-2 FFT with HW acceleration	Radix-4 FFT without HW acceleration
Number of butterflies	$(64/2) \times \log_2 64$ = 192	$(64/4) \times \log_4 64 = 16 \times 3 = 48$
Number of clock cycles for arithmetic operations in a butterfly	1	Containing 8 complex ADD and four complex MUL: $8/2 + 4/1 = 8$
Number of extra clock cycles for memory access in a butterfly	0	Depends on the address generator, it could be at least 4 cycles
Total cycle cost for initializing a layer	1	2
Total cycle cost for terminating a layer	1	2
Prolog cost including bit reversal	0	64 (does not match HW for radix-2)
Total cycle cost	$192 \times 1 + (2 + 1) \times 6 + 0 = 210$	$48 \times 8 + (2 + 2) \times 3 + 64 = 460$

dedicated for radix-2 FFT. However, the number of basic arithmetic operations of Radix-4 FFT is the same number of computing of Radix-2 FFT when counting them based on hardware-independent C-code.

The example shows that the cycle cost could be much higher by executing a low computing cost algorithm if the algorithm does not match the architecture. The conclusion of the example is that algorithm selection must match the hardware of the processor; a very advanced algorithm might not match a parallel architecture.

Reuse of Other Algorithms

The selection of one algorithm also should associate with the selection of other algorithms of the system. The total computing load can be decreased dramatically when an algorithm can start at the partial results of another algorithm (meaning partly reuse of other algorithms). Runtime of the algorithm can be shorter because part of the function of the algorithm is executed already. To utilize the intermediate result from previous tasks, the previous algorithm should be modified and implemented, with the intermediate result exposed to other algorithms.

Minimizing Computing Costs

Computing costs can be minimized by selecting an algorithm to minimize the number of arithmetic computations, to minimize or hide memory access costs (to implicitly run the memory access in parallel with the computing), and to minimize overhead of communications between function modules and program flow controls.

The first possible decision to minimize computing costs is to decide the processing in time-domain or in frequency-domain. Input signals are usually in time-domain. Time-domain signal processing may give low computing latency. If there are only a limited number of simple algorithms with limited numbers of iterations, the signal processing in time-domain is preferred.

Without counting, the cost of transformation, computing cost in frequency-domain is usually much lower than the computing cost in time-domain. A typical example is the implementation of a FIR filter in both time-domain and frequency-domain. The computing cost of a time-domain filter is the convolution with $N \times L$ steps; N is the number of samples in a data packet, and L is the number of taps of the filter. In frequency-domain, the filter is simply the weighting factor of frequency-domain samples. The filter computing cost of N data sample in frequency-domain is N steps of multiplications. Obviously, the time-domain computing cost is L times higher.

However, the overhead from frequency-domain signal processing must be taken into account. The overhead includes the computing cost of the transform from time to frequency-domain and the computing cost of the transform from frequency to time-domain. The decision will be:

```
Comparing computing costs
{
   If FFT + IFFT + F-domain computing < time-domain computing:
   F-domain computing is feasible,
   Else keep computing in time-domain.
}
```

As just mentioned, methods of computing cost minimization can be found from large-volume publications. Normally, program complexity can be higher and regularity can be lower while achieving lower computing cost. When porting the design to a parallel architecture, most of the optimization has to be removed.

Minimizing Code Costs

A regular algorithm with explicit parallel features will decrease the code costs. When code cost is critical in a design, simplicity instead of minimizing computing costs will be the first choice. A simple rule of thumb is: less algorithm optimization means smaller code cost.

If an algorithm is used several times, it should be implemented as a subroutine CALL. Finally, loop unrolling always consumes more code. If a subroutine is not an inner loop, using or not using loop unrolling to an algorithm shall be carefully checked. To minimize code cost, loop unrolling should be eliminated unless the loop unrolling is for the inner loop code in the runtime critical path of the CFG (Control Flow Graph) of the application.

Minimize Data Memory Costs

Frequently used high runtime cost functions usually are implemented as a LUT (look-up table). A LUT is a data array stored in a data memory. The address x is the variable of the function, and the data at address x is the result of the function, $y = f(x)$. LUTs are commonly used in real-time DSP system to minimize the runtime costs of certain arithmetic functions—for example, $1/x$ or $x^{1/2}$. However, a LUT may consume much data memory if the precision on the LUT result is high.

The size of on-chip data memory can be minimized by using off-chip memory as the buffer for not frequently used data. The extra runtime cost of running on-off chip memory swapping should also be taken into account.

Others

Algorithm stability, such as stability of IIR and FIR, are discussed in most DSP textbooks. It is obviously a condition of algorithm selection. Precisions of algorithms can be checked. Design and verification costs of algorithms are also conditions of selections.

Case Study: Square Root Acceleration

Three ways to compute the square root are described here: using LUTs (look-up tables), Newton-Raphson method, and Taylor series. Using a look-up table is a very fast method since the result is found immediately. One potential drawback is the large size of the table. To reduce the size of the table, a sparse table can be used, and the final result can be computed using linear interpolation between neighboring table elements. Other ways to refine the result are possible, such as spline/parabolic interpolation or the Newton-Raphson method, which are described next.

The Newton-Raphson method for computing the square root is based on iterative refinement using the formula $g_{n+1} = (g_n + x/g_n)/2$. The square root $y = x^{1/2} = g_n$ is found when $g_{n+1} \approx g_n$. The accuracy and the cycle cost depend on the number of iterations. It is important to select a good initial guess, g_0, perhaps from a small look-up table, to ensure convergence and to minimize the number of necessary iterations. The method can be used for real-time systems since it is possible to prove that a certain number of iterations will always find an answer with enough accuracy for all inputs. In practice, other methods are sometimes more suitable due to high cycle cost (in this case, each iteration requires a division that is generally a very slow operation). Sometimes, it is possible to reduce the number of cycles by not computing the final result directly. For example, the iteration formula for $z = x^{-1/2}$ does not involve division, and the square root can be computed using $y = x \cdot z = x \cdot x^{-1/2} = x^{1/2}$.

The Taylor series is another way to compute $y = x^{1/2}$ in a fixed number of clock cycles. For a positive fractional data $0 \leq x < 1$, the Taylor series of $(1 - x)^{1/2}$ is approximately:

$$(1 - x)^{1/2} = 1 - \frac{1}{2}x - \frac{1}{8}x^2 - \frac{3}{48}x^3 - \frac{15}{384}x^4 \ldots \qquad (18.1)$$

For example, if $y = 0.15^{1/2}$ should be computed, $(1 - 0.85)^{1/2}$ will be computed. If the requirement on precision is not high, computing $1 - x/2 - (x^2)/8 = 1 - 0.5x$ $(1 - 0.25x)$ will be sufficient. The instructions used in the code include two right shifts, one multiplication, and three subtractions. The cycle cost is fixed. When using the datapath for video signal processing (to be discussed in Chapter 20), an order 4 Taylor series can be executed in one clock cycle, and an 8-order Taylor series can be executed in two clock cycles. Most low-level algorithms (such square root and $1/x$ in normal data range) can be implemented using an 8-order Taylor series when 16-bit data precision is sufficient.

18.2.4 Language Selection

Early High-Level Modeling

Early HW independent functional (behavior) design gives only the modeling and the simulation of the DSP function on the mathematical level. Programming languages or data flow language are popular because DSP functions can be easily and explicitly modeled. The most commonly used languages for design of signal processing algorithms are C, MATLAB [8], or Fortran. Advanced features of modeling mathematical, matrix-based iterative computing is essential for the language.

MATLAB is a popular language used for algorithm development. However, it is not used to generate assembly code for embedded processors. Researchers are trying to use MATLAB as the source code for code generators (MATLAB compiler). There is hope that it can be the source code for assembly code generation in the near future. MATLAB is used to validate algorithms. In most cases, validated algorithms are rewritten as a C program so that the source code can be compiled to the machine language of a processor. Tools compiling MATLAB to fixed-point C are available from CAD providers.

A dataflow language provides possibilities to implement functional block diagrams or flowcharts directly to executable language because the design can be expressed as a set of connected process blocks by the language. When running a program, data flows through the datapath connections. Connections pass intermediate computing results following in-out specification and the predefined execution order. Each block in the system takes in its data in the right order, processes it, and passes it to the next block. A dataflow language thus exactly and explicitly describes functions of DSP subsystems. Simulink [9] is a typical dataflow language.

A dataflow language is perfect for modeling and simulating streaming signal processing algorithms if computing is regular and data moves in a predictable way. Some DSP algorithms (for example, entropy coding-decoding) employ much conditional branching and cannot be easily modeled by dataflow language. Control and interfacing extensive applications do not fit into dataflow languages either. When selecting a dataflow language, one must check if the application involves much bit-level manipulations or control FSM. If a dataflow language has to be used in this case, sufficient configurable libraries for bit-manipulation algorithms should be supported. Otherwise, bit manipulations and other nonsignal-processing portions of the system should be modeled in another language. Possibilities of cosimulation using these two languages must be checked before the language selection.

HdS (Hardware dependent Software) Modeling and Compiling

As soon as the hardware is selected, architecture and constraints should be involved in computing so that the cost and the quality of computing on target hardware can be measured. To adapt architecture and constraints to the functional model, the software becomes hardware-dependent software HdS.

The purpose of early source coding is to validate behavior functions. There are two purposes of coding HdS. One is to adapt the early source coding closer to the target architecture and assembly code and finally compile the HdS to assembly code. Each DSP assembly instruction set has its own special instructions, for example, CISC instructions for function accelerations. Because these instructions may contain several arithmetic operations and advanced memory addressing operations, it cannot be matched to one or a few TAC (three address code) operators of IR (intermediate representation). It is not easy to configure a compiler to match these instructions to the C code. There should be code adaptation during the coding of HdS. When using parallel architecture, source codes should be partitioned and allocated to hardware, and running task in different hardware will generate extra communications, memory accesses, and control overhead. HdS coding for parallel computing will be further discussed in Chapter 20.

Another purpose is to measure cost and quality, including the runtime cost, the memory cost, and the data quality of the software running on target hardware. General computing cost and the cost of vector memory access can be identified from the hardware-independent source code. There are extra costs that cannot be identified from the source code. For example, costs of intertask synchronization and communication cannot be identified until tasks are partitioned and allocated to hardware modules. Architecture-dependent cost cannot be exposed without using HdS of the target hardware.

The early source code holds high precision of the host machine (the machine running the simulation). The impact of finite data length cannot be exposed from running the early source code. The HdS can expose the finite data precision of the hardware by coding bit-accurate codes matching the hardware behavior.

Requirements of a behavior description language for HdS coding is thus close to HW and a short distance from the assembly language. So far, the most frequently used language for HdS coding is C because it is the most used and acceptable high-level language close to machine languages. C was designed as the source code language for system programming (e.g., operating systems). C is perfect as the input code of general-purpose processor compilers. However, in DSP and ASIP processors, unusual data types can be found, and conversion between data types is conducted implicitly in DSP and ASIP hardware. These features are not supported by standard C because it was introduced before the DSP age.

There are two approaches to match C to signal processing applications. The first is to add intrinsic functions that the compiler converts directly into efficient assembly code. It guides compiling to compiler-known code libraries using pragma. Another approach is to extend the C language with specialized data types and operators that closely match the demands of signal processing algorithms and ASIP DSP instruction sets. Different data types and operators should be specified into an extension of the C runtime library. By including and using the library, the modified C can be closer to DSP and ASIP hardware. Both cycle accurate functional model and quality compiler can be exactly implemented.

DSP-C was introduced by ACE (Associated Compiler Experts of Amsterdam) in the later 1990s. Later, other Cs such as SystemC and embedded C also were introduced to adapt more hardware features, including concurrency as parallel executing hardware, reactivity such as waiting and watching, data types, operators, and time. Large DSP suppliers also defined data type and operator extensions that are understood by their compilers.

Typical data-type extensions can be different fixed-point data types to the C language. One is the fract (fractional) type, to define values in the range $(-1.0, +1.0)$. Another is accum (accumulative), a typical custom data type for iterative computing in MAC, where the value range depends on the width of the accumulator. The two new data types can be further extended to short, long, signed, and unsigned types. For some very specific ASIP, special data types can dramatically improve the modeling and compiling. For example, in a programmable radio baseband processor, two special data types are essential. One is complex data and its extension to complex_fract and complex_accum. The complex data format can be either complex integer data $x + jy$ where $[-2^{n-1} \leq x < 2^{n-1})$ and $[-2^{n-1} \leq y < 2^{n-1})$, or complex fractional data $x + jy$ in the range $[-1.0, +1.0) + j[-1.0, +1.0)$.

Operator extensions should also be specified, such as a library of complex data operators, a library of circular buffer with modulo addressing, and a library using X (data) and Y (coefficient) memories simultaneously.

However, these adaptations of C set barriers to port source codes to other hardware and other compilers because the DSP adaptation to C is not yet standardized. This might be good for large DSP semiconductor companies because it blocks customers from running away. However, it is not good for product houses because the DSP code (the design) portability is very low.

Most dataflow languages are used for simulation; however, some can also be used to generate processor code from a dataflow design. One method of code generation is to supply retargetable kernel functional assembly code of each functional block in a code library. The dataflow code (based on dataflow languages) can then be used to generate additional code to handle scheduling, buffering, and the movement of data between the blocks.

18.2.5 Real-time Firmware Implementation

A DSP application can be modeled as a chain of function modules controlled by a top-level FSM (the FSM handling a DSP application). In each function module, the module level function can be further divided into three parts: the vector computing part, the memory access part, and the FSM attached with miscellaneous computing, see Figure 18.4.

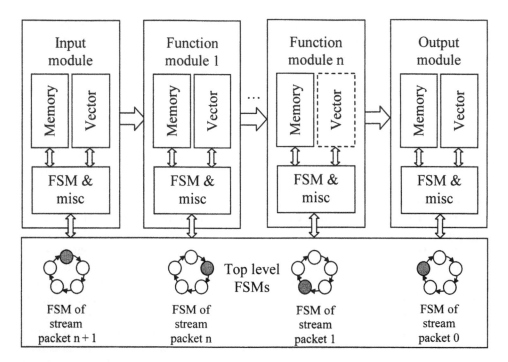

FIGURE 18.4

Modeling of streaming DSP SW.

While running the code, the top-level FSM monitors an input data packet when the input packet is arriving. When the data packet is completely received, the SW FSM sends its data packet to the first signal processing module and controls the module for the signal processing. When the processing in the first module is finished, the data packet is sent to the following module, and the FSM in the following SW module conducts signal processing for the coming data packet from the previous function module. Interlocked with its data packet, this top-level FSM requests each function module in the signal processing chain to provide processing until all signal processing tasks are finished by all modules.

If there is only one task running in one processor, there will be actually only one acting FSM controlling all signal processing modules to conduct all computing tasks during the period of receiving a data packet. If there are multiple processors, there will be multiple-acting FSM; each FSM is bundled to one data packet, and it controls signal processing through the processing chain.

Signal processing for one data stream might require multiple FSMs when a complete streaming processing is divided into several tasks and each task is allocated to different hardware running in parallel during the streaming signal processing. A typical case can be the baseband signal processing for synthetic aperture radar (SAR) shown in Figure 18.5(a), where the period of channel/sample signal processing is usually hundreds of microseconds, and the period of a Doppler signal

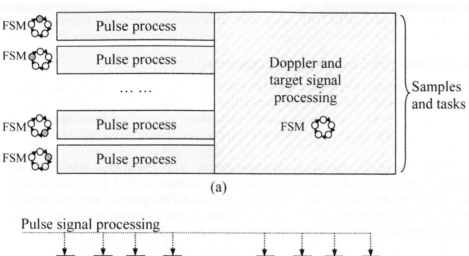

(a)

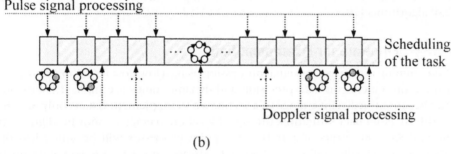

(b)

FIGURE 18.5

SAR task presentation and scheduling.

processing is typically several milliseconds. If a large vector processor is used, channel signal processing will be high prioritized threads (short period time) and Doppler signal processing will be lower prioritized. When a scan is received, the task of channel signal processing will interrupt the Doppler signal processing (see Figure 18.5 (b)). Several FSMs have to be used to handle a single application.

Implementing Vector Processing Subroutines

A vector processor is the device to handle mathematical operations on multiple data elements simultaneously. The vector processing subroutine is the code of regular computing for a data array with regular operations and addressing. The vector processing subroutine can be handled in a single scalar machine without parallel features.

A function module may contain a vector processing algorithm or an FSM-dominated algorithm. In modules containing vector processing algorithms, sublevel tasks can be further assigned to the vector signal processing part, the data access part, and the FSM code. The vector processing subroutines are the implementation of kernel algorithms. The data access part will supply data on demand for processing. The data access part introduces overhead of execution time. The overhead of data access should be minimized. The FSM code inside a function module handles program flow control of the subroutine. Data quality control can also be implemented in the FSM.

Implementing Branch-Extensive Subroutines

A branch-extensive subroutine is part of a code of program flow control or algorithms based on test-and-decisions, for example, Huffman codec [12]. It is good to separate the branch-extensive subroutine from vector processing subroutines and optimize them separately, because the behavior and the challenge of the optimization are different.

Branch-extensive algorithms are always annoying because most DSP processor architectures are suitable for vector computing instead of FSM. FSM is difficult to accelerate or difficult to run in parallel. It is thus preferred to separate branch-extensive algorithms from the hardware running normal vector- and iterative-based DSP algorithms [10].

18.2.6 Firmware for Fixed-point Data

Programming is not easy for signal processing using a low-cost fixed-point processor if the requirements on data precision and dynamic range are high. It is obviously true because the dynamic range of a 16-bit two's complement data is only 32768 or 90 dB. To avoid data saturation, the signal level on average cannot be high so that the real SNR on results of a 16-bit fixed-point processor will be much less than 90 dB. To maximally utilize fixed-point hardware and get sufficient data quality, deep understanding of different ways of data quality control is essential. Quantization noise, dynamic range, and SNR were discussed in Chapter 2.

Quantization errors, rounding, and saturations were discussed in Chapters 2, 12, and 13. Impacts on finite data and hardware precision were discussed in many textbooks based on the assumption that the full hardware data range can be utilized. Many publications addressing the signal round-off/truncation noise were based on the following assumptions that the signal level and the noise level were uncorrelated; noise is white and uniformly distributed, and human perceptual behavior was not taken into account. In reality, it is not sufficient. For applications such as voice, audio, and video, signals and noises are correlated. Relatively strong noise level can be acceptable if the current signal is sufficiently larger than the level of noise or distortions. Other limitations should be taken into account, such as if the real data range of the hardware cannot be fully utilized.

Errors Introduced by Fixed-point DSP

There are two problems in a fixed-point DSP subsystem: the large signal distortion and the quantization errors of small signals.

In an analog system, when the signal value is larger than the representing range, the extra value of the signal will be naturally "clamped," and the signal value will be the largest representable value instead of the real value. Clamping introduces distortions, and the distortion is uncomfortable yet acceptable by users.

In finite-precision digital systems, results larger than the representable numerical range will induce an exception. During an exception handling, the input signal will be scaled down and recomputed until there is no exceptional result. In real-time digital systems, there is no time for exception handling and the overflow will be saturated. Data saturation may trigger a scaling subroutine, scaling down the input signal. The level of coming input signals will be lower, and further saturation can thus be avoided.

To avoid saturation, sufficient headroom should be prepared before the signal processing. The headroom is the distance between the largest possible input signal and the maximum representable value of the hardware system. It means that the dynamic range of a fixed-point DSP processor cannot be fully utilized. In consequence, the real dynamic range of an application implemented on a fixed-point DSP subsystem is always less or much less than the dynamic range of the DSP subsystem hardware.

The quantization error can be accumulated, and a few bits from the LSB side may not be reliable. Usually, the quantization error can be exaggerated by DSP processor suppliers offering processors with high precision. However, by careful firmware design, the propagated quantization error during iterations on a fixed-point DSP processor can be minimized.

Example 18.2
Fractional computations with and without rounding.

In many cases the difference between a fixed-point calculation with rounding and without rounding is not very significant. In other cases the difference can be substantial. The following example shows where the difference is very obvious. In Listing 18.1 we rotate a two-dimensional vector by 0.001 radians for a number of iterations. This is done using both fractional arithmetic and double-precision floating-point arithmetic as a comparison.

LISTING 18.1   Vector rotation using fractional arithmetic.

```c
#include <math.h>
#include <stdio.h>
#include <stdlib.h>
#include <stdint.h>

// Default values
int fracbits = 16;
int doround = 1;

void init_fractional(int bits, int roundmode)
{
        fracbits = bits;
```

```
            doround = roundmode;
}

// Make sure the multiplication result is not too large
void validatemulresult(int64_t r)
{
        if( (r >= (1LL << (2*fracbits -2))) ||
            (llabs(r) > (1LL << (2*fracbits -2)))) {
            fprintf(stderr,
            "Result 0x%llx is out of range\ n", (int64_t)r);
            exit(1);
            }
}

// Fractional multiplication
int64_t frac_mul(int64_t a,int64_t b, int doround)
{
        int64_t r;
        r = a*b;
        validatemulresult(r);

        if(doround) {
                r = r + (1 << (fracbits - 1 -1));
        }
        r = r >> (fracbits - 1);
        return r;
}

// Convert value to fractional datatype
int64_t doubletofrac(double f)
{
        int64_t x = f * (1 << (fracbits - 1));
        return x;
}

// Convert fractional value to double
double i2f(int64_t x)
{
        double f;
        f = x;
        return f / (1 << (fracbits - 1));
}

int main(int argc, char **argv)
```

```
{
        if(argc != 3){
                fprintf(stderr,
                "Usage: trunctest <numbits> <roundmode>\n");
                exit(1);
        }

        init_fractional(atoi(argv[1]),atoi(argv[2]));
        // The rotation matrix is: [c -s]
        //                         [s  c]
        double c = cos(0.001); // A small angle is used to really
        double s = sin(0.001); // show the truncation errors.

        // Fractional version of the above
        int64_t frac_c = doubletofrac(c);
        int64_t frac_s = doubletofrac(s);


        double x=0.99; // (+1.00 is not an allowed value)
        double y=0.0;
        int64_t xf = doubletofrac(0.99);
        int64_t yf = doubletofrac(0.0);

        int i;
        for(i=0; i < 500;i++){
                // Double precision version
                double newx,newy;
                newx = c*x-s*y;
                newy = s*x+c*y;
                x = newx;
                y = newy;

                // Fractional version
                int64_t newxf,newyf;
                newxf = frac_mul(frac_c,xf,doround) -
                        frac_mul(frac_s,yf,doround);
                newyf = frac_mul(frac_s,xf,doround) +
                        frac_mul(frac_c,yf,doround);
                xf = newxf;
                yf = newyf;

                // Print the double precision result
                // and the fractional result. If there are few
                // fracbits this will quickly diverge.
```

```
                    printf("%f %f\n",y,i2f(yf));
        }
      return 0;
}
```

The result is presented in Figure 18.6 for 12 bits and 13 bits using truncation and 12 bits using rounding. (With 13 bits and rounding, the result is almost exactly the same as the 12-bit rounded version.) The reason for the large difference is that the correct result will very slowly approach zero for a number of iterations, whereas it will approach zero more quickly using truncation instead of rounding since truncation will always cause the result to be rounded down here. On the other hand, if we use a 31-bit representation (the largest number of bits that can be handled by the source code listed), the error after 250 iterations is less than 0.001%, and it does not matter very much if we use rounding or truncation. This is also yet another demonstration of why iterative computations should be performed using a MAC unit with higher precision instead of using standard ALU instructions.

The conclusion from Example 18.1 is that if it isn't possible to perform computations inside the MAC it is very important to apply rounding on the result.

Specification of Fixed-point Systems

The specification of dynamic range and computing precision is based on sufficient understanding of applications and the target hardware. The requirements on precision and dynamic range can be extracted from the application. The hardware precision can thus be best utilized if the requirements are carefully specified. An

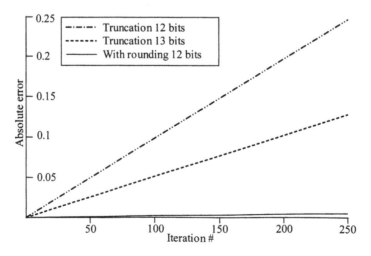

FIGURE 18.6

Iterative computing errors from finite precision systems.

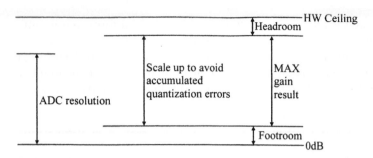

FIGURE 18.7

Specification of data ranges for a fixed-point DSP subsystem.

application on a DSP subsystem can be modeled from ADC to the DSP processor and finally to the result consumer, the DAC. Parameters involved in the specifications are represented in Figure 18.7.

In this figure, the HW ceiling is the hardware resolution. For a 16-bit processor, the HW ceiling is 16 bits or 90 dB for two's complement numerical representation system. The resolution of the ADC will absolutely not be necessary above the HW ceiling. For example, a 14-bit ADC is sufficient to adapt to 16-bit DSP processors. The system will be overspecified while using more than a 14-bit ADC for a 16-bit DSP system. Gain control of analog amplification before ADC can be used to promote the resolution of an ADC input. When the input of ADC is low, the gain control subroutine running in the DSP processor will modify the gain of the analog amplification so that the dynamic range of the ADC can be best utilized without distortion. To increase one bit resolution, the price and power consumption of an ADC will be significantly higher. If a 13-bit ADC, associated with gain control of analog amplification, is sufficient, a 14-bit ADC will not be used.

To avoid overflow during computing, headroom is needed. The headroom in Figure 18.7 is the buffer to avoid overflow. The more bits used for headroom, the more reliable the system will be. At the same time, the fewer bits used for headroom, the larger the system dynamic range. Normally, the design of headroom should be based on the number of guard bits. In a 16-bit DSP processor, there are usually eight guard bits in MAC units and one guard bit in ALU. If iterative computing can be allocated only to MAC, 2-bit headroom plus one guard bit for ALU computing is usually enough. The headroom for computing inside MAC is not necessary if eight guard bits can be utilized efficiently. As more computing is allocated in MAC, the requirement on headroom will be less, and the computing will be more reliable.

The lower precision buffer seldom is discussed, and there is no specific terminology for it. In this book, we simply name it *footroom*, to correspond with the terminology *headroom*. Footroom is the precision buffer, the redundant bits used to avoid the influence of the truncation/rounding error. In a system design, the data within the footroom may not be reliable, and the data above the footroom must be trusted.

Suppose all truncations are executed after rounding; one bit footroom is sufficient when the number of rounding operations is not so high. By experience, "not so high" means that the number of rounding operations is not more than 10 through an execution path from inputs to outputs of an application. To minimize the number of rounding operations, it is preferable to keep computing continually inside a MAC unit.

Example 18.3

Based on a 16-bit fixed-point voice processor, select an ADC, and specify headroom and footroom to best utilize 16-bit resolution.

According to the previous discussion, two headroom bits and one footroom bit are required. To best utilize 16-bit hardware resolution, the ADC shall be 13 bits see Figure 18.8.

Figure 18.8 is a typical case of most voice signal processing. Using only two headroom bits is based on an assumption that occasionally saturation on results is acceptable.

How to Scale Data

Scaling data includes data quality measurements and scaling. After measuring qualities of results and intermediate results, the data quality should be improved by data scaling (see Figure 18.9(b)).

The behavior model of a hearing aid system is given in Figure 18.9(a). Because the implementation of the behavior model is based on double-precision C code, no data quality control is required. While implementing the behavior model on a fixed-point DSP processor, the available dynamic range is much lower and data quality must be enhanced during the execution. The data quality enhancement includes insertion of data quality measurements and insertion of gain control.

A data measurement point is a defined point in a CFG, usually the in-out data of a function, at which the data quality (the maximum data, the minimum data, the average value, the count of saturation) shall be measured. A data measurement point can be inserted at any point in the signal processing chain. However, because some subroutines (a function block such as DCT, FFT, or a filter) might be designed by the third party, measuring at the input or output of a function is better than measuring

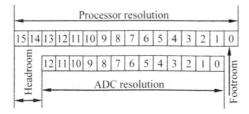

FIGURE 18.8

Graphical representation of Example 18.3.

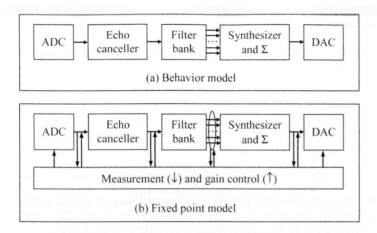

FIGURE 18.9

Implementing a hearing aid to a fixed-point processor.

data inside a function block. Scaling points should also be at the input or output of a function block because of the observability. Inserting a scaling point inside a function subroutine may induce functional errors, so it should be avoided.

It is especially important to insert scaling points before or after transformations, and before or after the data merging (Σ) point because the signal amplitude may dramatically change after transformation or after merging of multiple signals. Gain control of analog amplification for ADC and DAC is an essential way to maximum resolutions of them. In principle, more measurement and scaling points will enhance the system precision and dynamic range. However, inserting each measurement and scaling point will increase the code cost and execution time. The trade-off will end up at a point of sufficient data quality following the system requirement specification.

Except for the number of scaling points to insert, the frequency (how often) to measure and calibrate each scaling point also will be discussed. There are two ways to decide the frequency of scaling: static and dynamic. By supplying a static way, data will be measured periodically with a fixed time interval. Static scaling is easy to design and debug. By using a dynamic way, overflow and underflow flags will be prepared in either HW or in SW. Measurement and scaling subroutines will be enabled when there are overflows or underflows. If the requirements on data quality are not high and low-power consumption is more important, the dynamic way is preferable.

Measuring Data Quality

An intuitive way of measuring data quality actually was illustrated in Examples 18.1 and 18.2. The quality of a result from a fixed-point DSP processor can be specified using the difference between the result from the behavior model with double precision and the result from the fixed-point processor, by running the assembly code of the behavior model. The assumption is that the result from the behavior model is perfect.

The difference can be specified as the RMS difference or the maximum absolute difference. The RMS difference is:

$$D_{RMS} = \sqrt{(R_1 - r_1)^2 + (R_2 - r_2)^2 + \ldots + (R_n - r_n)^2}/N \qquad (18.2)$$

where N is the length of a data packet, R is a result sample of the behavior model with double precision, and r is a result sample from the fixed-point hardware. The maximum absolute difference is:

$$D_{ABSMAX} = MAX\{|R_1 - r_1|, |R_2 - r_2|, |R_3 - r_3|, \ldots, |R_{n-1} - r_{n-1}|, |R_n - r_n|\} \qquad (18.3)$$

D_{RMS} is the most used measure. For example, the difference D_{RMS} of a fixed-point MP3 audio decoder should not be more than 3×10^{-5} according to [5]. The best reachable SNR of the firmware running on the fixed-point HW is specified as:

$$SNR = 20 \log_{10} \left[MAX_{headroom} / D_{RMS} \right] dBV \qquad (18.4)$$

$MAX_{headoom}$ is the maximum signal under the headroom. For example, the $MAX_{headoom}$ in Figure 18.7 is $2^{15-2} = 2^{13} = 8192$, D_{RMS} in Figure 18.7 is 2^1. The SNR in this case is $20 \log_{10}(8192/2) = 72$ dB.

Block Floating-Point

The block floating-point technique uses a fixed-point processor and software emulation to reach or at least approach the quality of a floating-point processor. Block floating-point was introduced in Chapter 2. Part of a DSP software (application) could be implemented based on the block floating-point if the part of computing is precision critical or requires high dynamic range [13]. The way to implement block floating-point firmware is:

- Scan data, analyze data behavior.

- Partition input and intermediate sequential data into groups.

- Assign an exponent to each group.

- Define periods to measure and change exponents of data groups.

- Program subroutines.

- Transfer the block floating-point result data to normal fixed-point result data.

18.3 ASSEMBLY IMPLEMENTATION

18.3.1 General Flow and C-Compiling

As soon as the HdS code is released, it should be translated (compiled) to assembly code. The assembly-level programmer will evaluate the quality (performance and code size) of the compiled code, which will almost certainly not be good enough to meet the constraints of the application. At this point the programmer may spend some time on C-level optimizations, trying to help the compiler do a better job. Or the programmer might jump directly into assembly code, replacing compiled code by subroutines from assembly libraries and directly code part of it. In either case, the programmer uses an iterative process to refine the code (whether in C or assembly), debugging it, and profiling it to identify bottlenecks.

The assembly code programming flow given earlier indicates that good assembly-level coding skill is essential for top-level assembly coding (planning), making assembly code libraries, and fine-tuning compiled assembly codes. That is why readers are required to learn techniques of assembly-level programming. Assembly language programming is discussed in references [1] and [3]. In this book, only a brief introduction will be given.

18.3.2 Plan and Specify for Assembly Coding

From C code to assembly code, physical memory management will be exposed and extra assembly codes will be added. The compiler can do all the work for you. However, the compiler shall be guided to let the code be more structured and compacted, and the data memory is better used.

The use of data memories should be carefully planned according to the data types and behaviors, the data block size, and the lifetime. Before assembly coding, data should be classified into the following groups:

1. **Input data:** A data packet from an input port, from the main memory, or as the result of another task. For sample-based signal processing, an input data buffer is a FIFO queue, and as soon as data samples arrive, data is processed. Using a FIFO queue avoids jitter-induced problems. For packet-based signal processing, an input data packet cannot be processed until the input packet is completely received (tail triggered DSP). The lifetime of input data is specified as the interval between the data arrival time and the time when all input users load the input data into their computing buffer.

2. **Output data:** A data packet to the output port. If the output port is connected directly to a data consumer, for example, a DAC, an output sample can be used immediately by the data consumer. A FIFO buffer usually is used to adapt the rate of data consuming. The lifetime of output data is specified as the interval between the time the data is produced until the time when all users use the output data.

3. **Long lifetime buffer:** A long lifetime computing buffer is a computing buffer used by a function. There are two kinds of computing buffers: long lifetime and short lifetime. The long lifetime computing buffer exists whether the subroutine is under execution or not. The long lifetime computing buffer usually is used for iterative computing—for example, the data buffer of a FIR

filter or the data and result buffer of an IIR filter, the data buffer of an echo canceller, or the data buffer of a statistics algorithm. The lifetime of a static (long lifetime) computing buffer is permanent or as long as the application is active. Data in a long lifetime computing buffer must be kept if the physical memory storing the long lifetime buffer has to be used by others. The buffer must be saved before the physical memory can be used for others. When an algorithm with long lifetime buffer is interrupted or suspended, the buffer must be reserved and it will be restored when the application is active again.

4. **Short lifetime buffer:** A short lifetime computing buffer exists only when running the function or the subroutine. For example, the FFT computing buffer is a typical short lifetime computing buffer. It will not be used after the FFT algorithm is executed. The lifetime of the output buffers of the FFT is decided by the following function consuming the data instead of by the FFT algorithm itself. A short lifetime computing buffer can be overwritten by other functions after using it.

5. **Parameters:** Parameters are not filter coefficients. Parameters usually are loaded to general or special registers before executing a function, whereas coefficients are loaded and consumed directly from data (coefficient) memories during the execution of a function. Function parameters consist of data memory pointers, length of iteration, buffer limiters, counters, and such. Each function has its own parameter buffer. Part of function parameters, such as data pointers, can be changed during or after running the function. The lifetime of parameters is permanent.

6. **Coefficient:** Most coefficients are constants. Adaptive filters consist of variable coefficients. The lifetime of coefficients is permanent.

Example 18.4

Plan of memory usage for an application.

The memory usage, program, and the data allocations should be structurally specified with scalability. The super-mode program and data should be allocated and started at the reset position of PC and address counters. Data with a fixed size and permanent lifetime shall be allocated together. The sizes of computing buffers and stacks which vary from time to time should always be sufficient. An example of code and data allocation is given in Figure 18.10.

18.3.3 Fixed-point Assembly Kernels

The bottom-up approach starts from inner-loop kernels and ends at the top assembly code. An inner-loop kernel is a kernel algorithm at the lowest code level hierarchy in an iterative loop of a function. Saving one clock cycle while running an inner-loop kernel subroutine can result in hundreds or thousands of saved clock cycles at the top level, because the kernel subroutine often is used in the iterative

Program memory		Data memory		Coefficient memory	
0000	Super-mode	0000	Super-mode D	0000	Parameters
0100	Main-APP1	0100	FIFO of FIR1	0100	Coefficient 1
0300	Main-APP2	0200	FIFO of FIR2	0200	Coefficient 2
0500	... ...	0500	... ...	0300	... ...
0F00 – 0FFF	Subroutines	1000 – FFFF	Computing buffer	0400 – 0FFF	Stack

FIGURE 18.10

An example of memory allocation of a DSP subsystem.

part of DSP algorithms. The quality of kernel libraries is thus important, and much depends on the assembly-level design skill. Assembly language design skills can be divided into the skill to design assembly kernel subroutines and the skill to integrate assembly codes. In this section, the skill to design assembly kernel subroutines will be discussed. Scaling and data quality control of fixed-point firmware, as well the best reachable SNR of the firmware running on the fixed-point HW, were discussed earlier.

For signal recovery functions, such as the function to recover radio baseband signal, the data quality is not measured by quantization errors and dynamic ranges; it is measured by acceptable bit error rate instead. For example, if acceptable data packet loss is less than 1% and a packet size is 10^4 bits, the bit error rate should be lower than 10^{-6}. Measurements of bit error rate, quantization error, and dynamic range usually are conducted at the higher level during the integration of the assembly code.

18.3.4 Low Cycle Cost Assembly Coding

As soon as the data quality is acceptable, kernel execution time should be minimized. Most skills used in the compiler optimizer were discussed in Chapter 8, and they are suitable as well for assembly-level programming. To minimize runtime, the first is to find where most runtime is consumed. Usually, if there is much iteration, the runtime minimization should be conducted inside the inner loop, the iterative body. If the kernel is not iteration extensive, the focus of runtime minimization should be on the prologue and epilogue of the module program. Sharing prologue and epilogue is one of the choices.

Runtime can be dramatically decreased by carefully promoting the efficiency of the inner-loop kernel. The runtime overhead of a program is the extra runtime except for the time running kernel algorithms based on available data. The runtime overhead is usually from the memory access and the pipeline stall when jumps are taken. The memory access can be minimized by adapting the following check list:

- Try and keep useful data in the register file as long as possible so that data reloading can be minimized.

- Try to run special instructions with memory accesses in parallel with the computing so that the latency of memory access can be hidden.

- Try to group data and use "load double word" or "load multiwords" instructions (if such instructions are available).

To minimize pipeline stalls of jump taken, the following check list is useful:

- Try to use a REPEAT instruction instead of conditional jumps to carry out iteration.

- If a REPEAT instruction is not available, try to minimize pipeline stalls by arranging the code to minimize the jump taken. This may require full or partial loop unrolling and is usually acceptable if the inner loop is small.

- Try to use conditional execution arithmetic instructions instead of using conditional jumps.

    If a REPEAT instruction is not available, inner loop unrolling is popular to minimize overheads from pipeline stall. The following example does unrolling of a loop accumulating 100 samples.

```
for (int x = 0; x < 100; x++)
{
    ACR = ACR + memory[x];
}
```

The equivalent assembly code is:

```
L1: LD0    R0,(AR0++) // 6 clock cycles are needed for iteration
    ADD    ACR,R0
    DEC    R1          // R1 contains loop counter
    JUMP.NE L1         // jumps (three cycles) if loop is not yet
finished
```

Obviously, $600 - 2 = 598$ clock cycles are required if the jump taken consumes three clock cycles. By using loop unrolling, the cycle costs will be significantly decreased in the following code:

```
for (int x = 0; x < 100; x += 10)
{
    ACR <= ACR + memory(x);
    ACR <= ACR + memory(x+1);
    ACR <= ACR + memory(x+2);
    ACR <= ACR + memory(x+3);
    ACR <= ACR + memory(x+4);
```

```
    ACR <= ACR + memory(x+5);
    ACR <= ACR + memory(x+6);
    ACR <= ACR + memory(x+7);
    ACR <= ACR + memory(x+8);
    ACR <= ACR + memory(x+9);
}
```

The equivalent assembly code is:

```
L1: LD0 R0,(AR0++)
ADD ACR,R0        //step 1
LD0 R0,(AR0++)
ADD ACR,R0        //step 2
LD0 R0,(AR0++)
ADD ACR,R0        //step 3
LD0 R0,(AR0++)
ADD ACR,R0        //step 4
LD0 R0,(AR0++)
ADD ACR,R0        //step 5
LD0 R0,(AR0++)
ADD ACR,R0        //step 6
LD0 R0,(AR0++)
ADD ACR,R0        //step 7
LD0 R0,(AR0++)
ADD ACR,R0        //step 8
LD0 R0,(AR0++)
ADD ACR,R0        //step 9
LD0 R0,(AR0++)
ADD ACR,R0        //step 10
    SUB R1, 1     // loop counter minus 1;
                  // initial value in R1 was 10 instead of 100.
JUMP.NE   L1      // jumps if loop is not yet finished
```

The total cycle costs will be

```
Cycles <= 10×(20+4)-2=238
```

The cycle costs are reduced from 598 to 238, and the improvement factor is more than 2.5. The code size of the inner loop is increased from four instructions to 22 instructions. The increased code size is usually acceptable in most applications.

Fine-tuning on algorithm and arithmetic computing can also minimize the runtime. For example, a choice is to use a look-up table instead of using the Taylor series for special arithmetic functions. Using a look-up table can minimize the runtime cost if the appearance of special arithmetic function is very high.

Other loop transformation techniques can be used. Loop fusion replaces multiple loops with a single one. For example, the original C code is:

```
int i, a[100], b[100];
for (i = 0; i < 100; i++) a[i] = 1;
for (i = 0; i < 100; i++) b[i] = 2;
```

It is equivalent to:

```
int i, a[100], b[100];
for (i = 0; i < 100; i++) {
    a[i] = 1;
    b[i] = 2;
}
```

By merging two loops into one, two significant improvements can be achieved. The jump taken induced overhead is 398 cycles from running each loop of 100 steps (suppose a jump taken consumes three clock cycles). The jump taken induced overhead from the original code is 796 cycles, and the jump taken induced overhead from the equivalent code is only 398 cycles. After merging loops, the prologue and epilogue code will be half compared to that of the original C code. It is the second improvement. (Notice that the loop size in this example is 100, larger than the size of a normal register file.)

If the code size of a function is not large, it might be good to avoid using function calls inside inner loop kernels. A function CALL and RETURN induces two pipeline stalls. Suppose the pipeline stall consumes three extra clock cycles; a CALL-RETURN will consume at least eight extra clock cycles including six stall cycles and the cycle cost of running CALL and RETURN instructions.

18.3.5 Storage Efficient Assembly Kernels

During the code translation from C to assembly, the cost of storage should be minimized. Storage minimization includes reducing the on-chip (processor) data memory cost and reducing the general register costs.

Reduce the On-Chip Data Memories

On-chip (processor) data memory optimizations are:

- To find algorithms requiring less on-chip data memories. For example, some algorithms require half as many coefficients.

- To select algorithms with full memory access predictability. Much data can be stored in the off-chip memory and prefetched when needed.

A program with very large loops using very large memory buffers can be broken into several small loops. By breaking down a large loop body into smaller ones, small vector memories are sufficient to support the current running subroutine. The rest of the vector data can be stored in off-chip memory, allowing the on-chip memory size to be smaller. This programming technique can be found in high-end acoustic echo cancellers. A large acoustic data buffer is divided into several subfrequency bands, and each subband is processed at a time. The remaining subbands can be stored in the off-chip memory so that the size of the on-chip data memory can be smaller.

On algorithm level, loop interchange can be an example to minimize on-chip data memory needs and register lifetime. It is the process of exchanging the order of two loop variables. By changing the order, memory access will be more efficient following the order of vector computing while running the inner loop. Both runtime costs and on-chip memory costs can be less. For example, if the code was originally like this:

```
for(i=0;i<10;i++)   //"i" is the row pointer
  for(j=0;j<20;j++) // "j" is the column pointer
    a[i][j] = i + j;
```

following behavior descriptions, the row pointer is usually the inner pointer and the column pointer is usually the high-level pointer while accessing a data matrix. Obviously, the example code is not the best choice because the step size of the addressing algorithm of the inner loop is not continuing. After the loop interchange, the i loop will be the inner loop, and data will be accessed sequentially and continually in the inner loop. The lifetime of variables loaded from data memories will be much shorter (the shorter the better). The execution of the i loop now requires simple postincremental memory addressing, and the memory access cost can also be minimized. (This can be even more important on processors with caches.)

```
for(j=0;j<20;j++)
    for(i=0;i<10;i++)
      a[i][j] = i + j;
```

Reduce the Register Costs

Ways to reduce register costs is very important. If the application is relatively simple and sufficient registers in a register file can be reserved, interrupt-induced context saving may not be necessary and cycle costs from context saving can be eliminated.

If the number of variables is larger than the register-file size, some variables cannot be stored in the register file. Some variables have to be stored in the data memory, and data swapping between data memory and register file might be required while running a subroutine. Variable lifetime minimization will reduce the data swapping between data memories and registers. Variable lifetime minimization includes:

- Quantitatively analyze the lifetime of all variables; as soon as a variable is no longer used, the register can immediately be used for another variable.

■ Modify or reorder execution so that the variable lifetime can be further minimized.

A simple way of variable lifetime analysis is to expose the lifetime of variables by drawing a lifetime graph (LTG). An example of a LTG is given in Figure 18.11. In Figure 18.11, all variables are listed vertically and register usage at each clock cycle is exposed horizontally. Variable lifetime therefore is explicitly displayed. An example of a complex data MAC operation is emulated using a fixed-point real data DSP processor. The pseudo behavior code of a complex data MAC is:

```
ACR0+jACR1 = ACR0+jACR1 + (a+jb)*(c+jd)
```

ACR0 stores the real part for the complex MAC accumulation, and ACR1 stores the imaginary part result for the complex MAC accumulation. The pseudo assembly code (not including data loading) of MAC computing based on fractional complex data is:

```
s=a*c
t=b*d
u=a*d
v=b*c
x=s-t
y=u+v
ACR0=ACR0+x
ACR1=ACR1+y
```

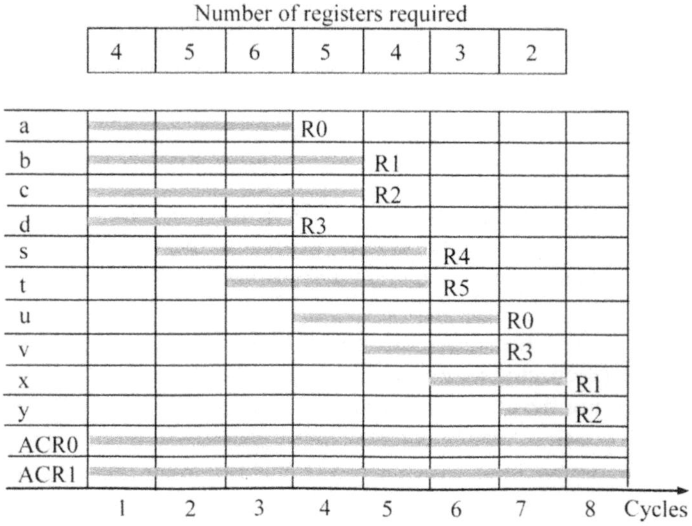

FIGURE 18.11

LTG example.

Counting from a to y, there are 10 variables. ACR stands for accumulator registers and are not counted as variables to be allocated in general registers. The lifetimes of ACR are observed separately. If the design is not carefully conducted, 10 general registers might be required. However, by carefully checking the register lifetime in Figure 18.11, we see that six registers are enough for the algorithm.

Register allocation attempts to keep as many operands as possible in a register file to maximize the code execution speed. One basic method is to draw the lifetime graph (LTG) of variables in a subroutine. As soon as a variable in LTG is no longer used, the lifetime of the variable is over and the register is available to accept a new variable. In Figure 18.11, as soon as a and d are no longer needed, R0 and R3 are assigned to another two variables, u and v. LTG is easy and sufficient for programming of assembly kernels by hand.

For handwritten assembly codes, LTG is an old yet simple and useful way. In compilers, register allocation by graph coloring is the dominant method. After assigning variables to registers using LTG, register usages are specified and the subroutine can be coded. A subroutine code can be divided into three parts: the prologue, the kernel, and the epilogue, see Figure 18.12.

The following code is an example of a FIR filter process frame of 40 data samples. Register assignments are in Table 18.5.

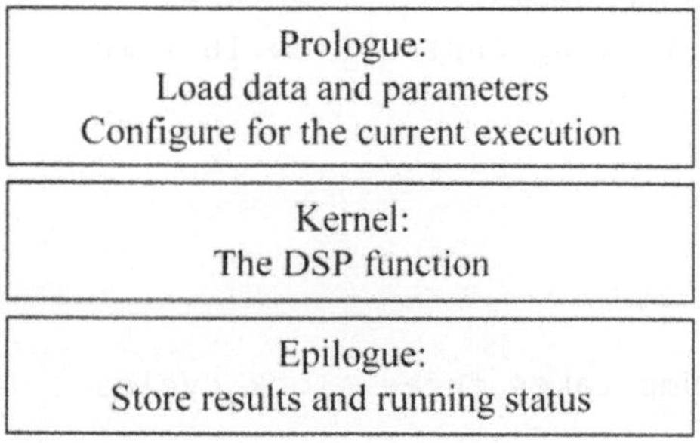

| Prologue: |
| Load data and parameters |
| Configure for the current execution |

| Kernel: |
| The DSP function |

| Epilogue: |
| Store results and running status |

FIGURE 18.12

An assembly subroutine can be divided to the prologue, the kernel and the epilogue.

Table 18.5 Register Assignments for the Implementation of 16 Tap FIR Filter.

Registers	Assignments	Registers	Assignments
R0	Temporally used buffer	R1	Data in buffer pointer
R2	Loop counter	R3	Result Y vector address pointer
R4	Coefficient pointer		

```
; Name: Frame FIR 40 data samples are processed by a 16tap FIR
;------------------ The prolog --------------------
set ar1, SEG_FIR        ; parameter segment address to DM1pointer
ld1 r3, (AR1++)         ; result pointer DM1(DM1pointer++) to R3
ld1 r0, (AR1++)         ; data pointer from DM1(DM1pointer++)
move ar0, r0            ; Load data pointer
ld1 r0, (AR1++)         ; Load FIFO bottom from DM1(DM1pointer++)
move bot0,R0            ; Load FIFO bottom pointer
ld1 r0, (AR1++)         ; Load FIFO top from DM1(DM1pointer++)
move top0, R0           ; Load FIFO top pointer
ld1 r4, (AR1)           ; coefficient pointer from DM1(DM1pointer)
;-------------- The prolog consumes 10 cycles -----------------
;Suppose 40 samples are available at in buffer DM1 pointer by R1
;------------------- The kernel body ------------------------
startfir
set r2,40               ; Outer loop counter
ld1 r0,(r1)             ; load a sample from DM1 to R0
inc r1                  ; in buffer pointer increment
st0 (ar0++%)            ; a sample data from r0 to FIFO buffer
move ar2, r4            ; Load coefficient pointer
clr acr0                ; Clean the accumulator buffer ACR1
repeat endconv,16       ; Repeat following instruction 16 times
convss ap acr0,(ar0++%),(ar2++)
endconv
        move r0,sat rnd acr0
        st0 (r3),r0
        inc r3
        dec r2
        jump.ne startfir          ; Jump takes three clock cycles

; Total clock cycle of the loop is 40*(16+14)+1 = 1200
;-------------------- Post processing ----------------------
set ar1, SEG_FIR -- parameter segment address to DM1pointer
st1 (ar1++), r15 -- Store Y pointer
st1 (ar1++), ar0 -- Store X pointer of the FIFO filter
```

To process a frame of 40 samples using 16-tap FIR, the total cycle cost is $10 + 1200 + 3 = 1213$ clock cycles. But this could be improved if the convss loop was unrolled so the hardware loop could be used for the startfir loop instead.

18.3.6 Function Libraries

A function library (also referred to as an algorithm library) is a collection of software modules for computationally intensive signal processing algorithms. Example algorithms that might be part of a function library include filters, transforms, and coding-decoding modules. A typical list of software modules can be the list of sub-routines of BDTI benchmarking. Function libraries are usually only a small portion of the code in most applications. However, runtime of codes from function libraries can be the majority part of the runtime in an application. This follows the typical code locality rule. Optimization of library code thus becomes essential. Table 18.6 lists typical general-purpose function modules in a general-purpose DSP library.

Except for general function modules given in Table 18.6, special functions could be modules for special processors of special applications. For example, a radio baseband function library of a radio baseband signal processor is shown in Table 18.7.

Use of assembly code library modules as off-the-shelf building blocks are increasingly popular. This is because signal processing applications and DSP processors are more complicated. Signal processing technology is growing rapidly. Most

Table 18.6 Firmware Modules in a General-Purpose DSP Library.

Algorithm kernel	Descriptions or specifications
Block transfer	To transfer a data block from one memory to another memory
N point FFT	N point FFT including all computing, addressing, and memory access
Single FIR	A N-tap FIR filter running one sample
Frame FIR	A N-tap FIR filter running M samples
Complex FIR	One sample complex data processed by a N-tap complex data FIR filter
Simple IIR	A Biquad IIR (2nd order IIR) running one sample
Autocorrelation	N-step autocorrelation (using two FIFO data buffers)
16/16 bits division	A 16-bit positive data divided by a 16-bit positive data
Vector add	$C[i] <= A[i] + B[i]$ where i is from 0 to $N-1$
Vector window	$C[i] <= A[i] * B[i]$ where i is from 0 to $N-1$
Vector Max	$R <= MAX\ A[i]$ where i is from 0 to $N-1$
DCT	8×8 Discrete Cosine Transform

Table 18.7 Modules in a Radio Baseband DSP Library.

Algorithm kernel	Descriptions or specifications
Radix 2 FFT	N points complex data FFT with in-order inputs and outputs
Complex data FIR	Complex data FIR filter or cross-correlation
Autocorrelation	Complex data autocorrelation with two FIFO data buffers
Vector rotation	A vector times a complex constant (immediate data)
Square root	Square root of a complex data
Complex 1/x	Computing 1/x of a of a complex data
Vector plus	A N-sample vector plus another N-sample vector
Conjugate vector plus	A N-sample vector plus another conjugated N-sample vector
Vector product	A N-sample vector times another N-sample vector
Conjugate vector product	A N-sample vector times another conjugated N-sample vector

semiconductor design teams lack special knowledge of applications and DSP software. Pressures of design costs and time-to-market are increasing. These are all reasons why you should accumulate knowledge of assembly-level library design.

For example [6], a processor can support one butterfly operation in one clock cycle if it contains a datapath supporting complex data types. The FFT kernel module can be described or specified using the example in Table 18.8. CC is the cycle cost of the kernel, and it does not include prologue and epilogue cost.

18.3.7 Optimize Control Codes

Usually code optimization is implicitly the inner loop optimization. However, in applications containing nested loops, the execution time cost of control codes within an iteration loop could be high. In this chapter, the control code optimization is to optimize the execution time instead of code size.

The conditional execution is always a feature to be utilized. If conditional execution can be used, do use it.

A branch will lead the program either to the left or to the right. If "goes to left" is dominating, "goes to the left" should be "the jump not taken" in the assembly

Table 18.8 An Example of a Firmware Library Module Description.

Parameters	Specification: *N*-point FFT
N	Size of FFT
L	Number of layers $L = \log_2 N$
Cycle costs	$CC = L(N/2 + 3) + 1$ for example $N = 64$, CC = 211
Input / output	$2 \times 16 \times N$ bits input/output buffer, included in DM cost
DM cost	$64 \times N + 32 \times N/8$ for example 64p FFT, DM cost = 4352 bits
Code cost	The kernel code cost is $2L$ instructions
Prolog code	6 instructions: configuring bus, load parameters, pipelining
Epilog code	Not necessary for FFT

code. The conditional branch should jump to the right instead of to the left. The extra cycle cost of pipeline stall can thus be minimized. A dynamic profiler in the assembly simulator should be used to analyze the application's behavior. The final statistics of each "branch taken" from extensive execution will expose the way to minimize "jump taken."

The low-cost control flow decision is usually the "jump not taken." If the jump is always taken, the program is possibly a loop. In this case, it is very nice to reprogram the code using REPEAT to replace the conditional branches.

18.4 ASSEMBLY-LEVEL INTEGRATION AND RELEASE

In general, assembly-level integration is to write the top-level FSM and link all function modules together by the top-level FSM. It is perfect to compile the top-level code using the C-compiler of the processor if the compiler is available, because the size of the top-level code is large and the runtime cost of the top-level code is very low. The quality of the top-level code impacts code size more than performance.

Top-level codes coordinate function modules under it. The top-level codes also coordinate the operating system in the MCU of the system. Normally, there is no operating system in a DSP subsystem. If the processor is running multiple applications simultaneously, there will be a super-mode code above all top-level codes of all applications. The super-mode is the task threading manager handling interrupts, context saving, and resource sharing.

Extra runtime is added during the integration of a DSP subsystem from:

■ Synchronization of input arrival time.

■ Synchronization of the time to send output.

■ Extra time for top-level memory transactions.

■ Extra time for top-level program flow control.

The extra runtime listed does not include overhead of parallelization. Discussion on overhead from parallelization will be found in Chapter 20. Runtime can be minimized by scheduling execution orders. Extra memory transactions could be eliminated if an output data packet of a previous module could be used directly as the input data packet of the following module [7]. Memory transaction is predictable for most streaming signal processing, and DMA can be used for preloading from the off-chip memory. Inserting a DMA task at the right time is a challenge of top-level scheduling. A transaction of loading data using DMA cannot be issued too early to waste on-chip memory resource, and it cannot be issued too late to introduce extra computing latency.

In principle, the WCET (the worst-case execution time) should be analyzed based on static timing analysis.

In the chapter on profiling, static timing and dynamic timing are both introduced. The principle of static timing on assembly level is the same as the static profiling on C language level. A control flow graph of the assembly code should be extracted from the code to release. On each branch, the runtime can be annotated on all branches of the tree. Runtime finally is accumulated and the longest path exposes the WCET. A longest path might be a true path or a false path. A false path will never be executed by the application. False paths should be ignored during the WCET analysis.

If the program is very simple, the WCET could be derived from running the assembly code on assembly simulator. However, most applications are not that simple and the WCET from running simulation cannot be trusted because it may be the case that the simulator did not execute all possible paths and therefore did not actually find the path of the worst-case.

If multiple tasks (applications) are running on the hardware simultaneously, task threading is required. Saving and restoring context in registers requires at least two operations on each general register. Running multiple tasks requires more data and code memory resources. If the memory is not enough for all applications, on-off chip memory swapping will be conducted. Extra runtime might be required for memory swapping.

Stack overflow should be checked if multiple tasks are running simultaneously associated with many interrupts and subroutine calls.

Finally, a checklist can be used to further promote code quality.

■ Try to use double-precision instructions and keep computing inside MAC.

■ Insert and optimize data measurement and scaling subroutines.

- Use guard and shift together to avoid overflow.

- Truncation and round at the right time.

The last problem in code integration is the reliability issue. Coding integration may induce invisible data dependencies. The first and last instruction of each subroutine shall be normal instructions or the instruction with a single-execution cycle.

18.5 CONCLUSIONS

Quality firmware design is based on rich experiences and deep understanding of the application. Readers should not expect to find a short cut or a formal way. A formal design will never offer a good design. More practice and combination of hardware knowledge and coding experiences is the only way approaching your success.

Firmware design can be divided into three steps: the behavior modeling, the C-code under hardware constraint, or the HdS (hardware-dependent software, bit accurate, or even memory access accurate) design, and the assembly language coding.

REFERENCES

[1] Kuo, S. M., Gao, W-S. (2005). *Digital Signal Processors, Architectures, Implementations, and Applications*. Prentice Hall.

[2] Embree, P. M. (1995). *C Algorithms for Real-Time DSP*. Prentice Hall.

[3] Warren Jr., H. S. (2003). *Hacker's Delight*. Addison-Wesley.

[4] ISO/IEC. (1992). Information Technology—Coding of Moving Pictures and Associated Audio for Digital Storage Media at up to About 1.5 Mbit/s, Part 3: Audio.

[5] ISO/IEC. (1995). Information Technology—Coding of Moving Pictures and Associated Audio for Digital Storage Media at up to about 1.5 Mbit/s, Part 4: Compliance Testing.

[6] Tell, E. (2005). Design of Programmable Baseband Processors. Ph.D. thesis, Linköping Studies in Science and Technology, Dissertation No. 969, Linköping University.

[7] Liu, D., Tell, E. (2005). *Low power Baseband Processors for Communications, Low Power Electronics Design*, C. Piguet (ed.), Chapter 23. CRC Press.

[8] MATLAB www.mathworks.com.

[9] Simulink: http://www.mathworks.com/products/simulink/.

[10] Flordal, O., Wu, D., Liu, D. (2006). Accelerating CABAC Encoding for Multi-standard Media with Configurability. *Proceeding of IEEE IPDPS06 (Reconfigurable Architectures Workshop (RAW))*, Rhodos, Greece.

[11] http://www.vorbis.com.

[12] Salomon, D. (2006). *Data Compression, the Complete Reference, 3rd ed.* Springer.

[13] Kobayashi, S., Fettweis, G. (2000). A Hierarchical Block-Floating-Point Arithmetic. *Journal of VLSI Signal Processing – Systems for Signal, Image, and Video Technology*, 24(1):19–30.

ASIP Integration and Verification

19

Hardware integration and verification can be divided into two parts, which are discussed in this chapter. The first part involves the integration of an ASIP core and the integration of a SoC using ASIP DSP cores as components. Basic verification techniques will be introduced in the second part. In Chapter 16, peripheral modules were discussed. In this chapter, peripheral modules and ASIP cores will be used as components, and interconnects of modules will be introduced.

19.1 INTEGRATION

System integration combines components and subsystems into one system and ensures that the subsystems function together as a system. It is the process of linking together different hardware modules and software applications physically or functionally.

Integration is a main design activity including hardware system integration, software system integration, and final embedded system integration. The word "integration" may implicitly mean the last design steps—but that is not correct. Actually, integration is part of design through all design steps from product concept design down to the design release to the product verification. This chapter will therefore offer concepts and early sign-off for integration in each design step. Table 19.1 gives an example of relations between design, verification, and integration of each design step.

It is not easy to understand details listed in Table 19.1; they will be discussed throughout the chapter. The purpose of showing the table at the beginning of the chapter is to expose the relation between design, integration, and verification. Table 19.1 also demonstrates that the integration and verification activities go through the entire design process. It is only according to the pedagogy that discussions on integrations and verifications are arranged at the end of the book. Known from design experiences, the more detail achieved from early sign-off for integration, the less will be the total design cost by minimizing the cost from design iterations. Detailing early sign-off may induce extra design cost. Trade-offs will be practiced during the real design.

Table 19.1 Relations between Design, Integration, and Verification.

Step	Design Step	Specification and Design	Integration	Verifications and Evaluation
1	Product concept	Implementation proposal	SW and HW concepts	Profits, cost, competitors
2	Project plan	Project specification	Relation between modules	Cost, risk, consequence
3	HW/SW partition	Modeling and profiling	Plan for module interfaces	Behavior model, trust of profile
4	HW system	Function partition, allocation	Interwork modeling and cost analysis	Interconnections and module interfaces
5	HW modules	Microarchitecture design	Early sign off of interfaces, physical parameters	Module cost, Bus model, Physical layer protocols
6	SW modeling	Design of SW modules	Early runtime sign off	Behavior verification
7	SW integration	Application, API, RTOS, and machine interface	SW system integration	System behavior verification
8	HW integration	Interconnect	Circuit and function integration	Compliance test on functions and physical
9	FW design	Bit, memory cost, and cycle accurate system	FW code integration	Application code verification
10	System release		HW/SW integration	Product verification
	...	...	...	...

Hardware integration will be the focus in this chapter. Hardware integration includes the integration of RTL codes and the integration of the hardware behavior models. A hardware behavior model is the assembly instruction set simulator of a processor core and the bit-cycle accurate behavior description of hardware modules (peripheral modules and accelerators) using high-level languages (C/C++ or system C).

Functional communications between modules will be implemented and verified during system integration. Physical issues such as clocking, synchronization, and timing are considered during the hardware integration phase. Integration will be discussed from the bottom to the top in this book, from the integration of a core, up to that of a system-on-a-chip.

19.1.1 HW Integration of an ASIP Core

The HW integration of an ASIP core connects the functional blocks such as the datapath, the control path, and the memory subsystem (address generation unit and memory peripheral) together. In Figure 19.1, the thin lines carry control signals. The thick black lines are memory buses, and the thick gray lines are register buses. Core integration is not just about connecting these lines. The challenges are also from the handling of new emerging critical paths by modifying pipelines and moving registers between modules to balance the path delays among pipelines.

Modifying pipeline should be avoided if it changes the specification of the assembly instruction set. While designing a superscalar, changing the pipeline will not change the assembly instruction set because the control and datapath functions are interlocked by the fetched instruction in a superscalar. Otherwise, changing the pipeline is a large design change that will significantly increase the design cost.

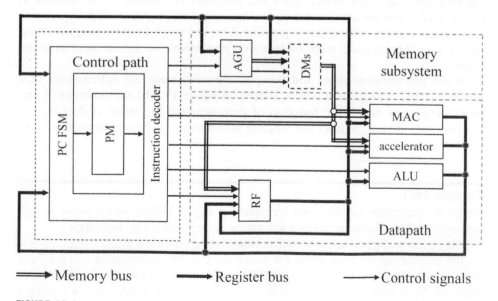

FIGURE 19.1

Integrating a DSP core.

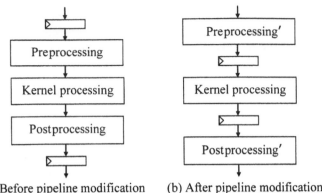

(a) Before pipeline modification   (b) After pipeline modification

FIGURE 19.2

An example of pipeline modification.

Principle of Pipeline Balancing

Moving pipeline registers to balance the path delays is popular because it does not change the pipeline specification. If the critical path is located inside a certain pipeline stage and there is no critical path in the neighboring pipeline stages, it will be useful to move functions from the pipeline stage to neighboring pipeline stages. The critical path will therefore be relaxed. As mentioned in Chapter 10, a module usually is designed by cascading preprocessing, kernel processing, and postprocessing in one pipeline. A typical way is to move preprocessing to the previous pipeline stage or to move postprocessing to the next pipeline stage.

If the preoperation stage and the postoperation stage are submodules of a pipeline stage, moving them will be simple. In Figure 19.2(b), the function of preprocessing' must match the function of preprocessing in Figure 19.2(a) from a system point of view. Neither the pipeline skeleton nor the assembly manual shall be changed.

Example 19.1

Design an accelerator for arithmetic computing in Figure 19.3(a), which works at high frequency (2 GHz). Modify the pipeline to meet the design need of high frequency.

The preprocessing part is identified and marked using a dash line in (a).

The pipeline register is moved to the output point of the preprocessing and the input of the kernel component in Figure 19.3(b). The circuit speed can therefore be significantly improved. Try to code the circuit and check the improvement from (a) to (b) by running a logic synthesizer, such as Synopsys Design Compiler (dc).

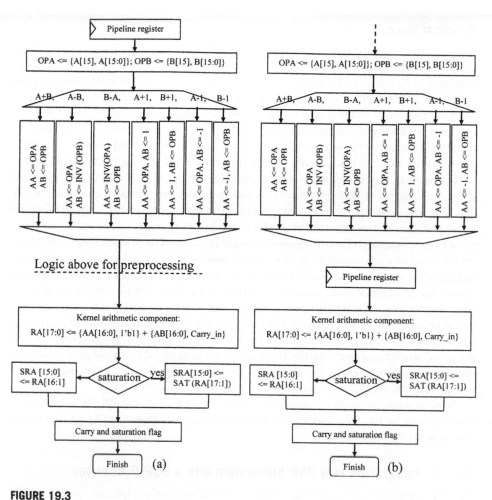

FIGURE 19.3

Another example of pipeline modification.

Redesign for Predictive Computing

If sequential executed logic functions induce critical path, they should be executed in parallel. A typical example is to compute flags in parallel using flag prediction circuits. For example, $A[15:0]$ and $B[15:0]$ are inputs to a multiplier. The predicted sign bit of the output SignMO should be:

$$SignMO = A[15] \,\hat{}\, B[15];$$

It is simple and useful because the sign flag can be available and be used far before the time when the multiplier result is available. It will be especially good if the multiplier is in a critical path.

Practical Issues

The register file supplies operands to many functional modules inducing large fan-out and unexpected delay. To minimize the delay, usually two methods are used. One way is to balance the load of operand buses. There are at least two operand buses, OPA and OPB. From the functional point of view, it is easier to use one operand bus, such as OPA, and to broadcast the output from the register file to all modules, including the datapath, control path, address generators, and data memories. However, this will eventually induce a heavy load on the OPA bus. A reasonable way is to balance the load to OPA and OPB buses during the instruction set design.

The second way is to specify a pipeline stage for operand fetching. Actually, this is the way that most DSP processors are designed. An extra cycle for operand fetch can hide all problems such as heavy load on the register bus or excessive fan-in or fan-out of the control signals. It was discussed in Chapter 11.

When designing the control signals, designers may not have sufficient knowledge on the load of each control signal from the instruction decoder. For example, the length of the connection wire and the fan-out of the control signal will not be known before the integration and placement-routing of the chip. During the integration of the core, further information such as the fan-out will be exposed, which may result in the change of RTL-level design. For example, the flip-flop at the output pin might need to be moved to the input pins in the succeeding modules. This happens when the physical distance between the two modules is long and the critical path is located in the succeeding module. Another case is to move a flip-flop from the input pin of the current module to the output pin of the previous module.

Modification on memory interfaces might be required around data memories, which could be on the addressing circuits and the data buses between memories and general registers.

19.1.2 Integration of a DSP Subsystem and a DSP Processor

A DSP subsystem consists of one or several DSP cores and peripheral modules such as the interrupt controller, the DMA controller, the timer, accelerators, and the main memory interface. A DSP processor stands for a chip including a DSP subsystem surrounded by off-chip peripheral components, such as serial I/O and parallel I/O.

Design for integration includes the integration of the DSP subsystem, the design of linking protocols, the hardware connection protocol, and the design of the physical connections between the core and the peripheral components, and the design for off-chip ports. The linking protocol is the way of communication between the processor core and peripheral components. The hardware connection protocol is the way to connect peripherals to the processor core, including the clock, the data width, the definition of connection pins, and requirements on speed and fan-in and fan-out.

A DSP core (inside dotted line), a DSP subsystem (inside the gray area including the core), and a DSP processor (all included) are drawn in Figure 19.4. The main memory could be on-chip, off-chip, or partly on-chip. Peripherals include generic

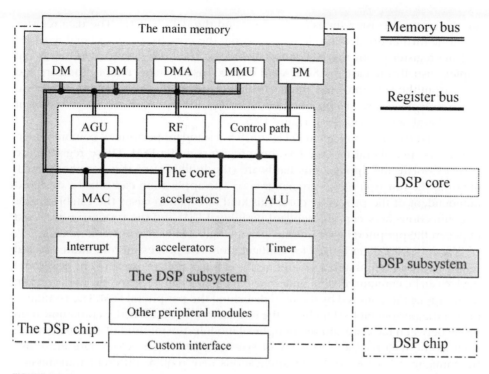

FIGURE 19.4

A simple DSP processor with a single core.

peripherals (DMA, interrupter, and timer) and custom interface (chip serial port, chip parallel port, and other functional ports).

The integration of a DSP subsystem or a DSP chip could be divided into two categories: hardware integration and the integration of a DSP subsystem or chip behavior simulator. The simulator is the assembly instruction set simulator plus behavior models of peripheral modules.

Integration of DSP Subsystem

As specified in previous chapters, a DSP subsystem consists of one or several DSP cores, the memory subsystem (including data, program, and the main memories), the interrupt controller, the DMA controller, the timer, and accelerators. Usually, a SoC solution consists of one or several DSP subsystems, a MCU subsystem, and the chip level memory subsystem.

Peripheral modules can be found in a DSP subsystem. A peripheral module conducts peripheral functions controlled by its local FSM of the common interface. The FSM of the peripheral module handles communications between the processor core and the peripheral module. It is configured by the processor core via reading and writing three registers of the FSM. These three addressable registers usually are the control register, the status register, and the data register. The data register might be further divided into in-register (from the core to the peripheral module) and out-register (from the peripheral module to the core) if the data bus is full duplex (usually the case in ASIP). A full duplex bus separates read to the core and write from the core into two physical media. Application functions are connected to the processor core via the common interface of the peripheral. Specification of a peripheral device is given in Figure 19.5.

By using the structure given in Figure 19.5, a standard peripheral common interface in the peripheral module has three registers and a FSM. These registers are used to isolate the applications hardware (the function) of the peripheral module from the function of the processor core. The application can thus be designed independent of the processor core. This kind of isolation eases the establishment of connections between the standard peripheral FSM and the processor core, and between the peripheral FSM and the custom logic inside a peripheral module.

The control register is used for configuration. The processor core addresses and sends control configuration vectors to the control register so that the peripheral device can be configured. It includes the configuration of FSM and the configuration of the application offered by the module behind the peripheral FSM. The configuration of the peripheral FSM includes the interface configuration, the communication mode between the core and the peripheral module, the method of communication such as the data rate, and the way of synchronization. The clock rate must also be configured because different applications may require different transmission rates.

There are usually available bits in the control register of peripheral modules, which can be used as custom control bits. For example, if the peripheral device is an A/D converter, the configuration bits of the A/D converter can be connected to

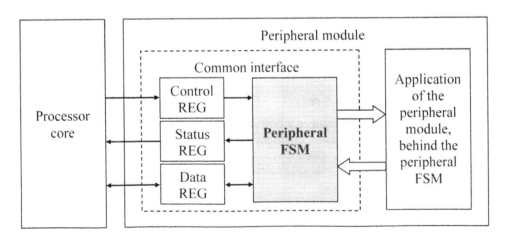

FIGURE 19.5

A peripheral module.

some bits of the control register of the peripheral FSM. For example, the sampling rate could be configured via writing available bits in the control register.

If the configuration of the application function is complicated, the available bits in the control register will not be enough for configuration of application functions behind the peripheral interface. In this case, the data register in Figure 19.5 can be used to pass the configuration vector to the application module behind the peripheral FSM.

The status register in a peripheral module is used to expose the status of the module to the processor core. By reading the status register, the processor core knows the status of the peripheral module. The status of a peripheral module includes the status of the peripheral FSM and the status of the application behind the FSM. The status of the peripheral FSM could be idle, busy, or abnormal. For example, if the device behind the peripheral interface is a DMA controller, the status of the DMA function behind the standard interface could be the current task name, the priority of the task, and its execution status (done, doing, or waiting).

The data register could be a read only, write only, or read/write register. If the port is a half-duplex port, one readable and writeable data register is enough. Otherwise, one read register and one write register are both required.

The hardware integration process of a peripheral module is depicted in Figure 19.6. The integration can be conducted in four steps. There are two main design inputs; one is the core-peripheral interface protocols. The protocols were specified during the core design, including the physical connection protocol and the linking protocol. Both protocols were discussed in Chapter 16. Another input is the function design specification of the application module behind the peripheral interface, at the right part in Figure 19.5, which describes the application carried by the peripheral port.

According to the core-peripheral interface protocol, the core can be connected to the peripheral module. The application module of the peripheral can be connected

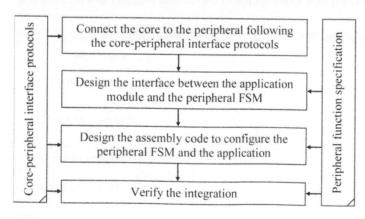

FIGURE 19.6

Design a peripheral device.

to the peripheral FSM according to the function specification of the application. After connecting both the core to the peripheral and the application to the peripheral, assembly codes can be designed according to both inputs. The assembly code includes the handshaking between the core and the peripheral, the configuration for the peripheral module and the application module.

Example 19.2

Example: design configurations for ADC.

The application behind the peripheral FSM is a dual channel 14-bit ADC. Specify and design the assembly code including configuration of the ADC and the way to get sampling data.

The first step of the design is to connect the ADC configuration pins to the control register of the peripheral FSM. The connection can be described using Table 19.2.

The assembly code is:

```
1:    in    r0,ADCStatus    ; load status to R0
2:    add   r0,0            ; test sign bit
3:    jump.ne PASS  ; if sign bit=1, the ADC is used by
                             others.
4:    set   r0,0x7f ; Ctrl[7:0] = 8'b01111111
5:    out   ADCcontrol,r0
6:    ret    ; Return from ADC configuration subroutine
7:    PASS … ; Cannot configure the ADC
```

To test if the ADC is available or not, status [15] is loaded as the sign bit, the second instruction tested as the sign bit is the status. If it is one, the ADC will not be available. If the ADC is

---

Table 19.2   Connecting ADC Control Pins to the Control Register of the Peripheral FSM.

ADC Pin Name	Peripheral Register Bits	Function Specification
Enable	Ctrl [0]	When [0] = 1, ADC is enabled
Channel 0	Ctrl [1]	When [1] = 1, the channel 0 of the ADC is selected
Channel 1	Ctrl [2]	When [2] = 1, the channel 1 of the ADC is selected
Sampling rate	Ctrl [3]	When [3] = 1, full rate, the clock rate is the sampling rate When [3] = 0, half rate, the half clock rate is the sampling rate
Gain control	Ctrl [7:4]	[7:4] gives 16 gain levels to the ADC pre-amplifier

available, the ADC will be enabled. Both channel 0 and 1 are switched on; the gain factor for channel 0 and 1 will be 0111 after running instruction 4.

Processor Simulator Integration

The chip behavior simulator is the processor simulator consisting of the ISS and all behavior models of all peripheral modules. It must exactly represent the hardware function of the chip, including consistencies of the data width (bit accurate), the cycle cost (cycle accurate), and I/O pins visible by programmers (pin accurate). While running any assembly code on both the behavior model and on the RTL code, the result must be exactly the same.

Behavior models and their equivalent hardware functional RTL codes are listed in Figure 19.7. Following modular (reusable) design rules, the behavior model and the RTL code should have the same design hierarchy so that adding or removing any module can be conducted in the same way both on the behavior model and on the RTL code.

At the top level, a DSP subsystem or a DSP processor consists of a core and several peripheral FSM modules. All peripheral FSM should be the same and designed following the core I/O specification. A function module can be found behind each peripheral FSM. These function modules include a DMA handler, an interrupt handler, timers, accelerators, and chip I/O.

As discussed in Chapter 17, tightly coupled accelerators are connected directly to the core instead of via a peripheral FSM. Loosely connected accelerators are connected as a normal I/O device via a peripheral FSM. Each module is usually a function call. Inputs and outputs of a functional module are global variables in the behavior C code. A function carries only its own inputs and outputs as global variables used by the function. Use of other global variables by mistake can therefore be avoided. Writing to one global variable more than once in one clock cycle will be checked as a design bug.

To maintain cycle accuracy, a clock as a global variable is mandatory in a behavior model of a DSP subsystem or a DSP chip. The clock should be the same clock of the processor core. The clock enables a step of hardware operation in each pipeline step of each module. Modules include the assembly instruction set simulator of the core, all peripheral FSM, all function modules behind peripheral FSM, and accelerators directly connected to the core. The clock as the global variable is usually a counter. While the counter value is increased by one, all flip-flop inputs pass their values to the flip-flop outputs, and all hardware in each pipeline step is executed once.

The way of synchronization between behavioral modules works well if all interfaces (inputs and outputs of all modules) are clocked. However, the situation might not be very formal if there is a combinational logic signal connected between behavior modules. In this case, the easy way is to merge these two modules into one larger module, and handle data transfer on the combinational signal according to the function specification. Otherwise, the execution order must follow the system-level flowchart, the CFG (Control Flow Graph) of the application.

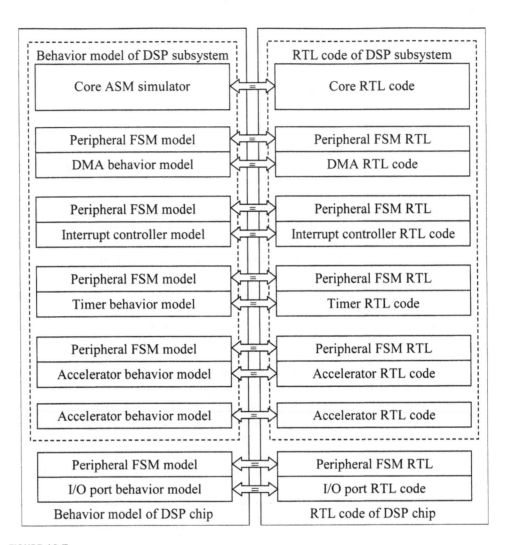

FIGURE 19.7

Implementation of behavioral models and RTL codes.

The synchronization using a clock as a global signal can be described using the following code as an example.

```
/*The top behavior module*/
{
CC=CC+1;/*Cycle counter plus one, to reach another clock
        cycle*/
        /*CC is used in each module*/
  {All_D_flip_Flop_Q = All_D_flip_flop_D;
```

```
    }
...
// Function for the adder hardware as an example
void hw_adder(int CC, int in1, int in2, int *out1){
    static CCold = 0;
    static newout;
    if(CC != CCold){// Don't do anything if the clock hasn't
advanced
        *out1 = newout;
        newout = in1+in2;
        CCold = CC;
        ... ;
        Flip_flop_D = assignment of logic operations;
    }
  }
}
```

19.1.3 HW Integration of a SoC[1]

There is no principal difference between the integration of a DSP chip and the integration of a SoC. The difference is from the scale. When the SoC scale is large enough, the SoC integration should be more formal so that the integration can be managed and the design cost can be minimized.

The definition of SoC varies according to different people. In industry, SoC normally stands for a digital subsystem on a chip containing multiple processor cores with different instruction sets and accelerators. Applications are partitioned, allocated, and executed in multiple cores in the SoC in parallel. These instruction sets may be heterogeneous or homogeneous. In this book, SoC stands for on-chip multiprocessors (MP). In this section, only basic issues on MP will be discussed. Detailed and advanced issues will be addressed in the next chapter.

SoC design and integration can be classified into four kinds of activities:

- **SoC Function specification**. It is the system design including understanding the system, profiling, architecture decision, function partition, IP selection, analysis of dependencies, and scheduling. Experiences and know-how are kept in large companies. Basic concepts can be found in many SoC books.

- **SoC EDA for function specification**. Much research can be found regarding design automation tools for functional specification of SoC. Only a small part of the research can be used, such as scheduler (as of 2007).

---

[1]This section is based on our research on NOC (SoCBUS), based mostly on the research from Daniel Wiklund [1]. The author acknowledges the enormous profiling work conducted by our radio baseband researchers (Eric Tell and Anders Nilsson) and our media DSP researchers (Di Wu, Johan Eilert, Per Karlström, and Andreas Ehliar).

- **SoC integration and verification**. Integration was discussed by many research publications. Integration platforms are so-called NoC (Network on Chip) or OCNs (on-chip networks), which are available integration environments for SoC designers. Verifications will be discussed later in this chapter.

- **SoC EDA for integration and verification**. System-level synthesis and NoC generators can be found. Many verification methodologies are proposed for SoC verifications.

Network Centric Interprocessor (Memory) Communication

Since the late 1990s, researchers have been working on flexible SoC connection networks for multiple simultaneous connections. The purpose is to enhance the bandwidth and flexibility of interprocessor communications and memory as well as I/O accesses.

A SoC as a typical DSP subsystem is given in Figure 19.8. There might be a MCU if the system requires advanced task control. There are at least one or more DSP cores and accelerators. All the DSP cores, accelerators, MCU, main memories, and system I/O are connected to the SoC interconnection network.

One SoC can be much different from another SoC. A classical SoC is given in Figure 19.8. If there is only one DSP core and one MCU in a SoC, a time division bus (e.g., the Daisy chain bus discussed in Chapter 16) can be the interchip connection network. When more cores are integrated in a chip, advanced NoC might be required.

Requirements of NoC from Applications

There are many research activities and publications proposing many kinds of NoC platforms and platform-based design methodologies. Currently (in 2007), there are

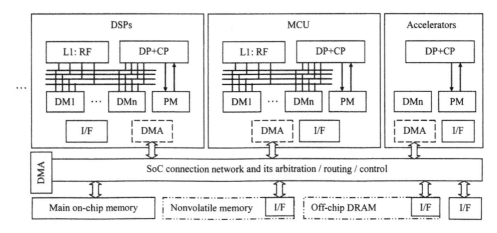

FIGURE 19.8

A typical SoC.

many companies that offer NoC as IP or as a platform. Students might get dizzy from so many things to learn; technology buyers (designers or product managers) will face a difficult situation when so many opportunities are available, especially when each supplier says that his or her offer is the best. This section will help you understand NoC for real-time computing (DSP subsystem).

A general-purpose NoC will never be optimal because different applications have different requirements on the NoC. To understand the requirements, we should start from analyzing communication behaviors of applications. Here the communication behavior is the description of communications between hardware modules while running an application mapped on the specific hardware architecture with multimodules running in parallel. When a function of a SoC is partitioned and allocated, the cost of intermodule communications on system level was not counted. To expose these communications, we need to allocate data (inputs, outputs, parameters, and computing buffers) to physical memories. The time cost of data transfers between modules or between memories can be the sum of:

1. The physical network setup time.

2. The link connection setup time.

3. The data transfer time.

1 and 2 together are the time for preparing a transaction task; 3 is the time for a data transaction. In most DSP subsystems, it is difficult to start computing until a full data packet is received. The memory transfer time will thus be the time for a complete transaction including 1, 2, and 3. The physical network setup time plus the link connection setup time can be from a few clock cycles up to several hundred clock cycles. The memory transfer time varies a lot, according to applications. One typical example is video signal processing. A data packet size is several hundreds of kilobytes or as much as megabytes.

The time cost of data transfer will be the impact of SoC design in two cases:

■ The data transaction cannot be fully predicable and performed in advance.

■ The size of a complete data packet is large.

If memory accesses are fully predictable and there are redundant memory resources available for data loading in parallel with computing, the cost of data transfer will be scheduled in parallel with DSP computing. Input and output data transfer can be predictable, and data transaction time can be hidden by data prefetching. However, the time to transfer intermediate data (swapping computing buffers or context switches) during computing may not be fully predictable, and the time of a data packet transfer will be an impact of SoC design. The impact is measured by the relative communication overhead. For example, if the computing time is 10 ms and the time of all data packet transfers is less than 100 μs, the overhead will be only 1% and it will be trivial.

To design an on-chip connection network, applications can be classified by a HWMR (hardware multiplexing ratio). It is the number of clock cycles used by running an infinite loop of a DSP application once, consuming one input data packet. For example, the infinite loop of the baseband signal processing of an OFDM (orthogonal frequency division multiplexing) symbol for IEEE802.11a/g should be $4\,\mu s$ (the input packet arrival time). If the system clock frequency is 250 MHz, HWMR will be 1000 or there will be 1000 clock cycles for one symbol processing. If the data transaction cost is 10 clock cycles during the period of symbol processing (1000 clock cycles), the overhead can be considered as negligible (1% of the computing cost).

Another example is the MPEG4 video decoder. The period of a video frame decoding is 1/30 second; it is equivalent to HWMR = 10 million clock cycles when the system clock rate is 300 MHz. One percent of 10 million cycles will be 100k clock cycles. In such a case, only if the data transaction cost is not more than 100k clock cycles during this period (10 million cycles), the overhead can be considered as negligible.

Two examples expose absolutely different figures; less than 10 cycles or up to 100 k cycles can be used for data transaction. The conclusion is obviously that different applications have different constraints on transactions. The cycle cost for SoC data transaction can be negligible if it is about 1% of the HWMR.

In a DSP subsystem with several parallel computing devices, up to several kilowords should be transferred between tasks in parallel hardware while processing an OFDM symbol in the period of 1000 clock cycles. Up to several megawords should be transferred between the main memory and the vector processor while processing a HDTV frame (the frame size is 1900×1200) in the period of 10 million clock cycles. *The total cycle cost for data transaction actually is equivalent to the number of clock cycles for signal processing.* To minimize the transaction cost, data transactions must be executed in parallel with signal processing. The 1% overhead will be used mostly for configuring and managing data transaction channels (for example, the setting of a DMA channel)—it is not enough for data transactions.

Master-Connected NoC

If the data transactions can be executed in parallel with computing, the minimization of the cycle cost for data transactions is actually the minimization of the time for preparing a transaction task. Master-connected NoC is popular for a deterministic or predictable system, such as parallel DSP subsystems, because the setup time for a data transaction can be short. The master is usually a master DSP controller running the top-level firmware of an application. The NoC is usually an interconnection circuit controlled or configured by the master. There are two popular master connected NoC circuits: the ring network and the 2D mesh network.

The ring network is popular for parallel DSP processors. A typical example is the STI CELL [2]. The ring network is suitable for architectures sharing one main memory. The network can connect each computing module (core) to the main memory. The use of the main memory is scheduled using a round-robin scheduling algorithm. The algorithm gives rights to only one core on the ring. The core can be connected to the ring, and the memory transaction between the main memory and the core can be conducted.

Figure 19.9 (a) is a simple ring network called a token ring. There is only one token on the ring; the token holder gets rights to communicate with the main memory. The ring in Figure 19.9(a) can be used in advanced ways—it can be divided into pieces, and each piece can be a connection between cores (see Figure 19.9(b)). When more main memories are required, each main memory spins off a subring, and all subrings are connected using a bridge. Each ring carries one main memory; for example, two rings are connected by a ring bridge (see Figure 19.9(c)).

When the system scale is very large, or distributed memories have to be used for distributed computing, the ring network cannot be used. More advanced NoC should be introduced. A simple crossbar (knockout switch) network is given in Figure 19.10. By using the simple knockout switch network, each memory can be

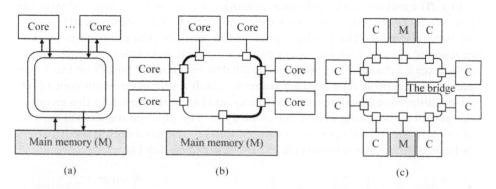

FIGURE 19.9

On-chip ring networks.

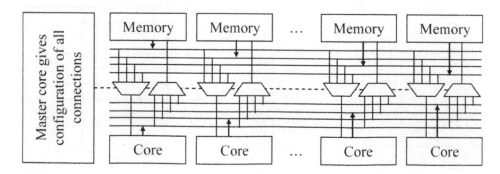

FIGURE 19.10

Simple 2D network (knockout [10] switch).

used by all processor units, meaning all processor units can read and write to all chip-level memories simultaneously in parallel. The configuration of each connection between a core and a memory can be given by the system-level controller, the master core.

The interconnection cost of a simple knockout switch network is large. When the number of memories and the number of computing units are large enough, the interconnection cost and interconnection-induced power consumption will not be acceptable. The 2D mesh network connection network therefore is introduced.

Packet-Connected NoC

The 2D mesh network in Figure 19.11 is the network with routers (circles), cores (C), and memories (M). A connection in the network is connected by a routing packet instead of by a master machine; therefore it is called packet-connected NoC. Packet-connected NoC can be divided further into packet-connected circuit (PCC) and packet network.

In a 2D mesh network, each router connects a core or a memory and switches up to five inputs (north, source, east, west, and local) to five outputs (north, source, east, west, and local). The full-duplex router can pass five data channels in parallel. Because the switch of "return" is not necessary (functional irrelevant), each output multiplexer is actually 4-to-1 switch controlled by two control bits. Four 4-to-1 multiplexers are controlled by a master controller; each router needs eight control bits and six addressing bits (less than 64 routers on chip in this case). In this example, one router can be configured in one clock cycle. The pipeline registers at the output of each direction is optional or can be added or removed during the hardware synthesis time. Similar solutions are available on the market [4].

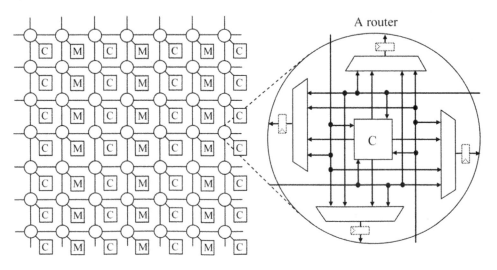

FIGURE 19.11

2D mesh network.

The PCC NoC

The procedure to set up a communication from a core (memory) to a core (memory) in the system in Figure 19.11 should be predefined and programmed as the program in the master core. Off-line scheduling of on-chip communication is mandatory. If applications dynamically request intercore or core-memory communications, the request should be sent to the master processor from the core. The master sets up the communication channel and informs the core to use it. After using it, the slave core will inform the master core to release the connection. It is obviously not convenient, and the setup time will be too long. When dynamic communications between cores or core-memory is the dominant chip level communications, the master-connected NoC is not a solution.

PCC (packet-connected circuit) is a realistic and available solution when dynamic intercore communication is dominating. The SoCBUS is a typical example [1]. The SoCBUS-based NoC is useful for multiple applications and multiple-task parallel systems. Multiple application streams are executed in massive parallel, and the processing of each stream is executed in multicores in pipelined and massive parallel. (Parallelization details will be discussed in Chapter 20.)

Figure 19.12 gives the principle of PCC. Any core can be a master sending a control packet for core-to-core or core-to-memory circuit connection. A control packet should be small enough to minimize the routing time. A control packet consists of the source address and the destination address. If necessary, priority and packet length may also be carried by the control packet.

A control packet comes to the router controller, distributed in each node, first when it moves from one node (router) to another node. The router controller

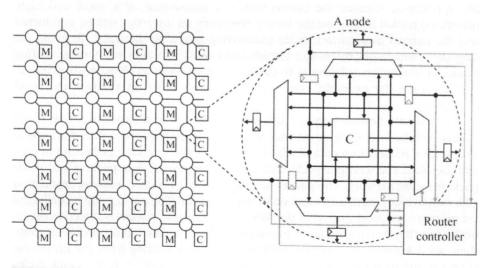

FIGURE 19.12

PCC 2D mesh NoC.

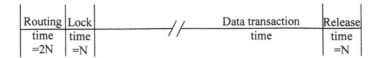

N is the number of nodes from the source to the destination

FIGURE 19.13

Overhead measure of PCC NoC.

decides the circuit connection according to the destination address. The connection can end up to the minimum distance if the network is not busy. Otherwise, an alternative can be routed. As soon as a circuit is connected, the circuit is locked for the data transaction. After the transaction, the reset signal will be sent to the connected circuit and all nodes will be released.

The communication overhead is measured in Figure 19.13. This measure is based on the shortest routing distance from the source to the destination. Suppose the longest distance is $N = 12$—the routing-induced overhead will be $2N + N + N = 4 \times 12 = 48$ clock cycles. It is negligible when data transaction time is about thousands of clock cycles. The advantages of PCC-based SoCBUS are the flexibility and dynamic multimaster routing. The drawback is the relative longer connection setup time.

The Packet NoC

The general drawback of a circuit-based network is the lack of QoS (quality of service). As soon as a circuit is configured, it is locked by an application until the application releases the connection. If a transaction of a small and high-prioritized packet has to use the locked resources, an interrupt will be conducted and the current data packet under transferring might be damaged. To avoid the damage by interrupts, redundant circuits and context storages are needed, costing more. An alternative solution is to design a circuit-switched NoC with extra circuit resources (for example, extra nodes) to supply more alternatives and avoid congestion.

The ultimate flexible way, yet the most inefficient and expensive way, is the packet-switched network or on-chip Internet. A core requests a data transaction and sends a request or a data block to its wrapper, the interface between the core and the NoC. The wrapper packs data into one or several data packets and sends packets to the network. The packet header consists of source address, destination address, payload length, and other control information. The packet will be delivered via routers (switching nodes) of the NoC. In a router, the packet is stored in the in-buffer first. It is then routed to an output port by a packet-switching circuit. Suppose there are 1 k words in a packet including the header; routing from the northwest corner to the southeast corner will need at least $12 \times 1000 = 12000$ clock cycles when there is no special routing algorithm, such as cut-through algorithms. If each node consists of 5 in-buffers and the size of each in-buffer is at least 1 k words, the

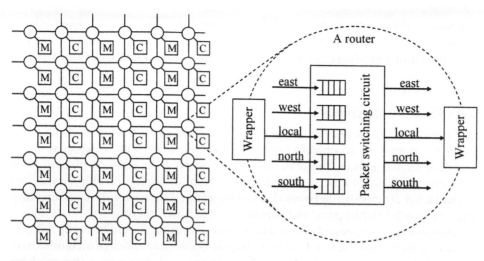

FIGURE 19.14

NoC as an a packet-switched network.

total memory cost of all nodes in the network in Figure 9.14 will be 42 × 5 × 1 = 210 k words or 3.36 megabits, about 8 mm^2 using state-of-the-art technology in 2007.

Actually, the real cycle cost could be much less if the packet can be transmitted as soon as part of the packet is buffered in the FIFO (this kind of data transmission is called cut-through). In this case, the cycle cost can be less. This is actually feasible for on-chip data transmission because data transmission quality can be good enough to eliminate error control algorithms. Nevertheless, the relative cycle cost of packet network is much higher than the cycle cost of a circuit-connected network.

Deadlock and livelock should be discussed when designing an NoC. A deadlock is a situation wherein two or more competing actions are waiting for the other to finish, and thus neither ever does. Deadlock in an NoC system occurs when none of the processes meet the condition to move to another state. A livelock is similar to a deadlock, except that the states of the processes involved in the livelock constantly change with regard to one another, none progressing.

Deadlock or livelock problems can be avoided when using a master-connected 2D circuit because the master can always offer deadlock-free connections. When using packet-connected 2D circuits, the deadlock can also be avoided by carefully designing the connection protocol [1]. When routing on a packet-switched network in Figure 9.14, both deadlock and livelock might happen because a routing could be unpredictable. Deadlock and livelock avoidance is not the focus of this book. Interesting discussions can be found in references [1, 9].

Up to now, after discussions on different NoC for different systems, the conclusion is that there is no general NoC for all applications. Different applications require different NoCs. The structure of the memory subsystem gives strong impact on NoC selection. An application-specific NoC can be specified by its memory subsystem, the parallelization, speed, bandwidth, size (number of nodes), configuration latency, data transaction latency, and silicon cost. A low-cost and low-latency NoC has limited flexibility, and the latency and cost of an ultimate flexible NoC will be high.

SoC Hardware Integration

In Figure 19.15, challenges of SoC hardware integrations are classified. To integrate multiple processors, an on-chip interconnection network will be selected as the platform for integration. The network could possibly be a classical point-to-point network, a traditional bus, or multiple buses. More often, the on-chip connection network is a 2D crossbar network that can be adapted to master-slave architecture, massive parallel architecture, and streaming parallel architecture.

To minimize computing latency, a network connection should be set up within a short time (few clock cycles). Different processor cores and different memories have different data types (different data width, data rate, and data transaction models). Data must be buffered so that different data types can be adapted first to the SoC network format and finally to the data format required by the destination device. Dynamic adaptation of data types and data buffering will eventually increase the connection time and computing latency. One of the design challenges therefore is the trade-off between network-induced latency and flexibility of the connection network.

When the interface is not consistent, wrappers should be inserted in between the network and a core or a memory. The wrapper of the source node converts the data

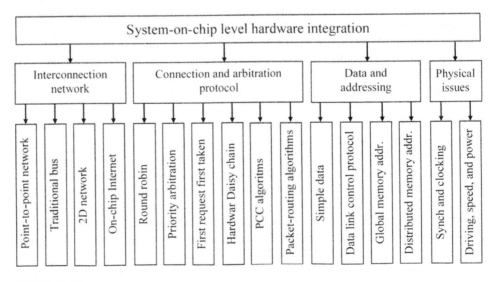

FIGURE 19.15

Specification of chip level HW integration.

type adapting to the NoC data type. The wrapper of the destination node converts the NoC data type to the local data type of the destination node. The wrapper of the destination node also buffers data.

The connection setup in the SoC interconnection network is done by master or packet routing, or arbitration protocols. Each SoC connection network has its own connection and arbitration protocol. For best-effort and ultimate flexible computing systems, packet data transmissions based on routing protocols are probably the right choice because the systems do not require real-time data transmission, and the routing protocol can supply the maximum flexibility. For real-time computing, master-slave architecture and circuit routing are preferred. The master is a processor, which switches the SoC connection network according to scheduled upcoming transactions. The master-slave architecture provides a guarantee to the real-time requirements. For the system with enormous on-off chip swapping, each main memory might need a connection ring attached to all devices, and the configuration of connections to the main memory will be based on either round-robin or priority-based arbitration.

There will be a day in the future when a real-time system becomes significantly more complicated compared to today's solutions. A single master (even it is a superscalar) will not be able to handle all tasks and SoC managements. In this case, distributed control will be a better solution, and a packet-switched interconnection network will be popular for real-time systems with such high complexity [1].

A data packet can be small, containing few words as a control vector. It also could be large, containing the data swapped between computing buffers. If a data packet contains both data and control information, data link protocol might be needed to distinguish the data and the control vectors.

Finally, a plan of physical level integration should be conducted during the early system planning time. Physical implementation is not within the scope of the book.

19.1.4 Integration of SoC Simulator

In this section, a SoC simulator is the integration of all assembly instruction set simulators of all processor cores, the behavior models of all accelerators, the behavior model of the chip-level connection network, the behavior models of all system-level memories, the behavior module of all glue logic modules and wrappers, and all behavior models of peripheral and chip I/O devices.

The functional simulation of an entire DSP subsystem is essential for system designer and verification engineers. If a DSP subsystem is partitioned and allocated to multiple DSP processor cores, cosimulation based on multiple simulators will be necessary. A SoC simulator is a platform that can integrate behavior models of all processor cores and other functional modules. To simulate SoC hardware, a SoC simulator must:

- Integrate all behavior models of all chip-level hardware devices together.

- Have custom interface and match functions between cores and modules.

■ Coordinate runtime and run order according to system HW connections.

■ Simulate all core functions, module functions, and interface functions.

When integrating multiple processors in a SoC, bugs might be introduced by interprocessor communications and resource sharing of multiple processors. These bugs may not be locatable without running all applications including the interprocessor communication and interprocessor resource sharing. The support of cycle-accurate simulation is necessary for the debugging of the interprocessor and interface functions. The simulation of RTL code gives exact cycle count but is too slow.

To reach both cycle and bit accuracy and fast simulation, special C libraries are needed. An available solution is SystemC, which is the library of C++ classes. It supports all hardware features that C cannot support and supply C equivalent run performance. For example, it supports concurrency as parallel executing hardware. It supports clocks, hardware hierarchies such as modules, ports, and signals. It supports reactivity such as waiting and watching. It also supports different DSP data types. The interfaces between SoC level devices can therefore be modeled and simulated. Device simulators based on C can also be integrated and simulated.

19.2 FUNCTIONAL VERIFICATION

Functional verification takes most of the development time and costs. Professional verification technique is essential [6], though it might not be easy for undergraduate students. In this chapter, verification will be introduced only for the design of ASIP. However, the method introduced in this chapter can also be used for hardware functional verification in general. In this section, discussion on functional verification will be limited from instruction set down to RTL code. VLSI verification will not be discussed. The VLSI verification, including verification of the netlist, LVS (layout versus schematic), DRC (Design Rule Check), and post-silicon verification can be found from reuse design manual [5]. Formal verification technique (such as the assertion-based test) will not be discussed in this book. Physical performance verifications (timing closure and power consumption) will not be discussed in this chapter.

The focus of this section is the verification of a DSP core. The SoC verification is out of the book's scope.

19.2.1 The Basics

Verification is a process used to demonstrate the functional correctness of a design (Janick Bergeron [6]).

The design flow discussed in Chapter 3 shows that an engineering design starts from the documentation of its early functional specification. The early documents offer product requirements and descriptions. During system design, system descriptions are implemented into SW or HW. Verification is conducted in parallel with system design. Finally, the final system design will be fabricated.

System design is a kind of top-down approach, and a step of system design is a translation from one level to the level lower. Through each design step or translation, functions must be correct by checking the consistency of functions before and after the translation of the design step. One step of the verification proves the correctness of the step of translation. The complete verification process of a design proves the correctness of all steps translating from the early paper document down to the final design release. For example, when translating an assembly instruction set document (assembly manual) to an assembly language simulator, the correctness of the translation has to be ensured. This means that the execution of each line of assembly codes must give exactly the result as the assembly language specified in the assembly instruction manual.

Following the major design milestones for an application-specific instruction set processor, necessary steps of functional verification are depicted in Figure 19.16. The right part of the figure gives a general view of the processor verification flow from the verification of requirements by document review down to the verification of hardware design, release of the RTL code.

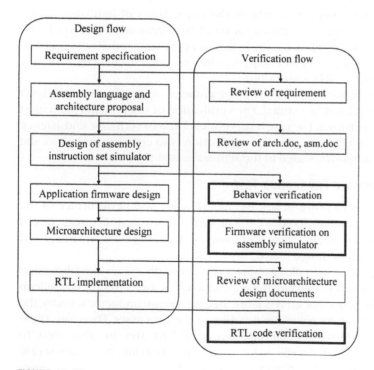

FIGURE 19.16

Design versus verification.

Paper-level design will be reviewed by engineers instead of verified by running tools because a paper document is not executable. By review, we mean compliance check by proofreading. The first heavy verification task is to verify the assembly instruction set simulator, and the second heavy verification task is the RTL code verification.

Behavior, Hardware, and System Verifications

In this section, verification is divided into three parts: behavior verification, hardware verification, and system verification. Behavior verification verifies the processor behavior model, the assembly instruction set simulator. Hardware verification is the compliance test of the hardware design specification. After hardware verification, the system verification confirms the functional correctness of the application firmware. We discuss only the behavior verification and hardware verification in this section.

The verification of the assembly instruction set simulator is to verify the processor's behavior-level design. It verifies the consistence between the behavioral model (the instruction set simulator) and the specification of the instruction set. It is the toughest step because there is no executable functional model above the assembly instruction set simulator.

The hardware verification only confirms the correctness of hardware design following the design specification. Processor hardware verification is to verify the correctness of the RTL code following the behavior of the instruction set simulator. The consistency should be proven by comparing the execution result of the assembly simulator and the result of RTL simulation.

After RTL code verification, the silicon implementation of a processor at gate level and physical layout level will be verified. The verification of silicon implementation is not the focus of this chapter. The gate-level verification is to verify the consistency of RTL code and the net-list. The physical verification is to compare and prove the consistency of the components used in the schematic and the layout.

Verification vs. Hardware Fabrication Test

Verification is a process to prove or test the functionality and performance of the system according to a certain specification. The terminology "test" is usually mixed with "verification" and it might cause confusion. Usually "test" means functional verification in the function design phase.

When testing engineers are talking about a test, it usually means the fabrication test, which is not a test of functionality, but instead, a test of the correctness of silicon fabrication. When the processor chip is fabricated by a semiconductor foundry, the foundry engineers are not aware of the function of the processor. Therefore, there is no way for them to test whether all functions are alright after the fabrication. To test the result and the yield of the silicon fabrication, function independent tests are introduced, such as design for (hardware) testing (DFT).

During the net-list design, the silicon designer implements the test supporting circuit together with the functional circuit. There are two steps to accomplish the test of a digital circuit. Step one is to reconnect all flip-flops into a scan chain when the circuit is switched from working mode to test mode. The second step is to generate a set of test vector patterns to the scan chain, and the test patterns can toggle all logic gates. By scanning-in the test vectors via the test supporting circuit to the circuit under test, a scan-out vector can be generated by both the net-list (the gate-level design) and the fabricated circuit.

After the fabrication of the chip, the same test vector will be supplied to the fabricated test chip and the output from the fabricated chip is collected. In case the collected results from the test chip are different from the results collected during the net-list verification, the fabricated chip is proven with some problems [11].

19.2.2 Verification Process

The overall verification strategy involves two major concerns: the quality and the cost. The quality is the verification coverage, and the cost includes the time cost and EDA cost. Full coverage usually means the design is touched (toggled) 100% during the verification. However, 100% coverage of processor verification does not mean that a processor is 100% verified. Bugs could happen in very special cases. For example, a bug may happen while accepting an interrupt when running a stack pushing subroutine. It is not related to the verification coverage. The conclusion is that a processor is only partly verified while reaching 100% toggling coverage.

The verification quality analysis starts from the reconvergence model (see Figure 19.17). The model gives a conceptual representation of the verification process. A design is defined as a group of transformations from behavior level to the hardware description language level. The reconvergence path in verification is illustrated in Figure 19.17.

It would be perfect if all design steps could be verified based on a reconvergence path. However, human factors cannot be verified completely because the early paper-based specification cannot be executable. It cannot be translated to an executable behavioral description language. The real reconvergence path is more frequently used as depicted in Figure 19.18.

It could be assumed that errors in the paper specifications can be corrected during the design review phase, yet many specification errors cannot be identified during reviews. There are high-level specification languages—UML, MATLAB,

FIGURE 19.17

The concept of a reconvergence path in verification.

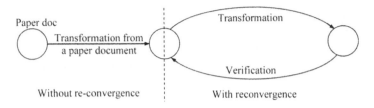

FIGURE 19.18

Real reconvergence path in verification.

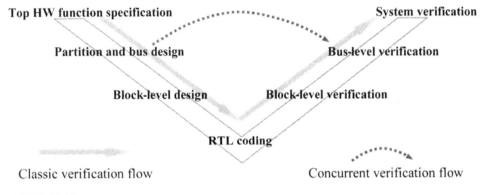

FIGURE 19.19

V-flow and concurrent product design flow.

or Simulink. Unfortunately, executable high-level specifications languages cannot completely replace specifications on paper.

Classical verification is coordinated with its design following the so-called V-flow in Figure 19.19. The classical way usually takes a lot of time for verification because the process is sequential (not concurrent). The concurrent development flow also is depicted in Figure 19.19. Verification time can be decreased dramatically by early verification of high-level functions in parallel with low-level function design. For example, the top-level RTL code can be generated as early as the partition and bus design is done. As soon as the bus design is finished, the bus-level verification can be performed by verifying the bus-based operations. For example, load and store instructions can be verified as soon as the top-level RTL code is ready.

There are usually two opportunities to speed up the verification. One opportunity is from the partition of the design and the verification. If a design can be partitioned and conducted in parallel, the module-level verification can be conducted in parallel. This is important because the iteration time cost at module level is low.

Another opportunity is to plan a concurrent verification flow so that the verification of the early high-level function design can be executed in parallel with the low-level function design. In the example given in Figure 19.19, the top-level partition and bus are specified at the early design phase. The interconnection between modules and buses can be verified by executing move instructions. When the memory and register buses are available, microoperations such as operand fetching and result storing of ALU or MAC instructions can also be verified. The verification work load can therefore be distributed.

19.2.3 Verification Techniques

To carry out high-quality verification, there are several types of classical verification techniques used in industries.

Compliance Testing

Compliance testing verifies that the design complies with the functional specifications. It is the verification of design specifications. Specifications of the instruction set simulator include the assembly instruction set manual and the user manual of the instruction set simulator. The specifications of RTL design include the architecture description and the microarchitecture description. Compliance tests include the compliance test of the instruction set simulator and the compliance test of the RTL code of the core.

The compliance test checks functions and specifications including the specification of each instruction and all possible interinstruction problems. The verification of each instruction is to verify the correctness and completeness of the execution by the instruction set simulator and the RTL code, by:

- Performing the right functions following the instruction set manual.

- Consuming the right number of clock cycles.

- Writing back during correct cycles.

- Checking all memory addressing models of all memory access instructions.

- Checking all register addressing models for all RISC operations.

- Checking all branch instructions, all jumps taken following the right conditions.

- Checking all jump target addresses for branch and call instructions.

- Checking all configurations to every instruction.

Interinstruction problems are interferences to one instruction by executing another instruction. It could appear when the following instruction cannot be executed because of the dependence induced by a previous running instruction. It also could appear during jump taken. For example, pipeline flush of a jump taken might flush an instruction that should not be flushed. Another example might be insufficient flush, where the iterative results of some early execution (executing predecrement addressing, for example) were executed and stored and cannot be flushed. Interinstruction problems may also happen when high-priority interrupt is accepted at the wrong movement. For example, an interrupt is accepted at the time when executing stack operation, loop operation, or jump flushing.

A compliance test shall be executed to verify that:

- The function exists and it works.

- The function is completely verified for all cases.

The function exists when one instruction using the function can be executed. For example, the Add with carry-in can be executed if $R0 = R0 + R0 + 1$ after running ADDC $R0, R0, R0$ when the carry flag is 1 (the first R0 is the destination address, and the second and the third R0 are OPA and OPB). However, this instruction is not yet completely verified until all combinations of inputs are executed. It takes too much time to run all data combinations for an instruction. A realistic and efficient way is to identify and execute corner cases.

Compliance testing is not easy because of two reasons: the difficulty to prove the translation from paper documents and the extremely large number of patterns to test.

Corner Testing

A corner means a part of the design that is irregular or the design that is based on informal methods. A corner test identifies corners and tests them during the compliance test. A corner test is part of the compliance test. Corners can be classified to data/bus corners, addressing corners, and control corners. For example, data corners are induced by the design around MSB and LSB in the datapath, the combining and splitting bus signals, concatenation for special operands, flag computing, and around interfaces between modules. Addressing corners can be identified close to the maximum address, the minimum address, and the boundary of a FIFO buffer. Control corners include all data hazard and control hazard cases, interinstruction problems and possible PC overflow or underflow. A control corner case could also be from the interwork with peripheral components, such as working with the interrupt controller. The corner cases are complex scenarios that are most likely to cause the design to fail.

A corner case test is a part of the compliance test. Corner case testing includes identifying and specifying corners, writing corner test benches, and executing corner test benches. To carry out a compliance test, all corner cases shall be specified and identified, and sufficient test benches should be supplied to expose all possible corner cases to speed up the compliance tests. Some typical corner cases are:

- In the integer datapath:

  a. All data truncation points, data concatenation points.

  b. All guards, rounds, and saturate points.

- All result patterns affecting flag values

  a. Corner cases to set and reset each flag.

   b. Which instructions keep flags and which instructions change flags.

- All cases changing data formats

   a. Change data types between ALU data and long precision in MAC.

   b. Change other data formats (such as the fractional multiplication).

Example 19.3

Identify corner cases and write test benches for the **Senior** instruction ABS.
  ABS is absolute operation on operand A.

```
SET R0,0x8000  ; Decimal -32768
SET R1,0x7FFF  ; Decimal 32767
ABS R15,R0     ; The result will be 0x7FFF (Saturated from -
                 32768)
ABS R15,R1     ; The result will be 0x7FFF (not saturated)
```

Around and in the addressing path and program memory addressing (PC), compliance and corner tests are at least the following:

- All control and status registers are cleaned after the system reset.
- All registers can be accessed according to the hardware specifications.
- Reaction of all specified exceptions induced by

   a. Data dependency.

   b. Computing exception from the datapath.

   c. Overflow and underflow of the hardware stack.

- Corner values for memory and register addressing (e.g., close to MAX/MIN).
- Corner values for destination addressing of branches.
- Starting/stopping points of hardware loop functions.
- FIFO limiters (functions of the top and the bottom registers).

Corner tests and coverage tests are closely related. Tests of memory access coverage and coverage of accessing peripheral registers are part of the corner test.

- Check if memory cells with corner addresses can be accessed.
- Check if port registers with corner addresses can be accessed.
- Check if all accelerated functions can communicate with the core.
- Check the memory bus and memory address space of the design.

There are two sets of corner memory addresses; one is the corner address set of the instruction set simulator, and another is the corner address set of a specific design. (It is the instruction set simulator configured to match a specific hardware implementation.) For example, the data memory address space of **Senior** is 64 k. The data memory configuration of a certain project using **Senior** is only 16 k. If the address toward a data memory is 16384 < address < 65535, it will be correct while running on a project-independent simulator. It will be wrong while running on the simulator for a project with a data memory configuration of 16 k.

Some control corner cases can be exposed only when two events happen sequentially. Verification of interinstruction functions is to check and to find out all possible dependency problems. The first problem is data dependency—the current instruction wants to load an operand that is not yet available and locked by the previous instruction. The second problem is the control dependency—the control (branch) hazard. It happens when an executing program is jumping from one program stream to another. Whether a jump is taken or not will not be identified until the jump instruction is executed. Typical cases are:

- During the execution of two instructions that consume a different number of execution cycles.

- Normally, when a branch is taken, the following fetched instructions shall be flushed. Possible bugs include insufficient flush (some pipelines are not completely flushed), and excessive flush (branch-independent registers and functions are flushed). Registers to be flushed shall be specified and listed for verification.

- The correct clock cycle penalty during a branch taken, a CALL, or a RETURN.

Asynchronous control dependency is a case of the program stream inter-work with interrupts or port accesses. An interrupt may induce bugs in the following combined corner situations:

- An interrupt comes while executing a branch instruction.

- An interrupt comes while cache and main memory exchange data.

- An interrupt comes while executing context switching.

- An interrupt comes while executing stack operations.

Example 19.4

Identify corner cases and write test benches for the **Senior** instructions ADDN, ADDC, ADDS, and ADD.

ADD: Plain addition

```
SET R0,0x7FFF
SET R1,0x0001
SET R2,0x0000
```

```
SET R3,0x8000
SET R4,0x00FF
SET R5,0x7FFE
ADD R15,R0,R1        ; Overflow flag will be 1
ADD R15,R1,R1        ; Overflow flag will be 0
ADD R15,R1,R1        ; Zero flag will be "0"
ADD R15,R2,R2        ; Zero flag will be "1"
ADD R15,R3,R4        ; Sign flag will be "1"
ADD R15,R2,R4        ; Sign flag will be "0".
```

ADDC: Addition with carry-in

```
ADD  R15,R3,R3       ; Carry flag after computing will be "1"
ADDC R15,R2,R2       ; The result will be "1" instead of "0".
ADD  R15,R1,R1       ; Carry flag after computing will be "0"
ADDC R15,R2,R2       ; The result will be "0" instead of "1".
```

ADDS: Addition with saturation

```
ADDS R15,R0,R1       ; The result will be 7FFF instead of 8000.
ADDS R15,R5,R1       ; The result will be 7FFF (not saturated)
```

ADDN: Addition without flag change

```
ADDN R15,R0,R1       ; Flags shouldn't change
```

Random Testing

Before introducing random testing, the concept of RTL versus ISS verification shall be introduced. RTL versus ISS is the verification process used to send the same best bench of assembly code and data to both the instruction set simulator (ISS) and the RTL code. The result from the ISS and the result from the RTL code will be compared. If two results are the same, the behavior model (ISS) is the same as the function of the RTL code. However, it does not mean that the function verified through RTL versus ISS is correct. It is only proved that both the ISS design and the RTL design are the same. Design errors could exist if the specification is wrong or if the ISS and RTL are designed by the same designer. Running real application code is the best way to prove the correctness. Nevertheless, the RTL versus ISS is an efficient way to speed up verification.

Random testing is based on the concept of RTL versus ISS verification. Random testing is the process of generating a random test bench (including the assembly codes and the data), and using the generated test bench as the stimuli for running both the assembly instruction set simulator and the RTL code. If both results are the same, the test is passed. If the result is different, there are design errors either in the ISS or in the RTL code. The corresponding stimuli will be stored in the log file. To speed up verification, a random test can be prepared and executed under an automatic test environment during the night. The log file will be analyzed during the day time.

A random test bench generator supplies both random test data and randomly generated assembly code. It is easy to generate random test data, but random code generation is more complicated. It is necessary to constrain the generated code so that it does not contain code sequences that would cause undefined behavior. Irrelevant code should be removed from the generated random assembly code. (The verification coverage of random testing is therefore not sufficient.) The randomly generated assembly code will be executed by both the simulator and the RTL simulator. The consistency will be proven by comparing results from the execution of ISS and the execution of RTL code. Following is a list of irrelevant codes:

- For an instruction following a multicycle instruction, operands needed by the instruction shall not be locked by the previous multicycle instruction:

  a. MAC instruction locks an ACR one more clock cycle.

  b. A specific vector instruction may lock the vector memory and some iterative computing hardware resources, such as memories and the loop controller.

  c. To check if special instructions such as repeat instruction can be nested.

  d. For some processors, the convolution instruction uses the hardware loop resource and cannot be executed inside the repeat loop.

- To avoid unpredictable problems in the simulation, the following situations must be handled carefully:

  a. Stack instructions and instructions with implicit stack operations.

  b. REPEAT instructions.

  c. The target address of a CALL must not be random data; otherwise, a target address may point to a position that never returns.

Constraints such as data ranges and address spaces should be checked during the random generation of data and addresses. The coverage of random data includes immediate data, memory addresses, target addresses of the program memory, general register values, and special register values.

Random tests can create scenarios that the engineers never anticipated, and often uncover the most obscure bugs in a design. The tricky problem is how to remove irrelevant code and the data that are not suitable as test stimuli. To find the right stimuli set for random testing is an experience-based task. The method to run random tests is:

- Generate a random stimuli set and remove irrelevant code and data.

- Run the random stimuli set on the instruction set simulator first.

- Run the same stimuli set on RTL code.

- Compare the results from the preceding two steps.

- If the result is not the same, data stimuli and code stimuli will be stored for further analysis.

Real Code Testing

By real code we mean both the code and data are based on real applications. It is obvious that an ASIP is designed for a class of applications. Application source code (assembly code in this chapter) should naturally be used to verify the specification of the ASIP. The way to design assembly language programs was discussed in Chapter 18.

19.2.4 Speed-up Verification

Verification of a processor design is enormous work, and the efficiency of verification is essential. To speed up the processor verification, several methods will be discussed.

Assembly Simulator versus RTL Code

This method was briefly mentioned during the discussion of random texting. As soon as an assembly instruction set simulator is available and verified, the simulator can be used as the bit and cycle accurate reference model. All test benches will be executed on the simulator first, and the results of such simulation will be compared to the results from RTL simulation of the same test benches. Figure 19.20 gives the principle to verify the RTL code.

Regression Test

One of the typical problems in verification is that new bugs can be introduced when other bugs are corrected. It is important to continually and repeatedly run all test benches on both the assembly simulator and RTL simulator after each correction. The simulation results should be automatically compared and the difference stored in test log files. This is called regression testing. Usually, a running script for the regression test will be necessary for any processor design project. The script will run all test benches using all corner data patterns. Normally, regression tests are

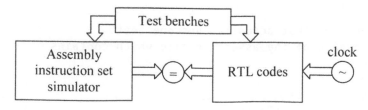

FIGURE 19.20

Using the assembly simulator to speed up RTL code verification.

performed overnight. A normal verification plan based on a regression test for a day could be:

- Check the error reports from the random and iterative verification.

- Find, analyze, and fix bugs in the design. Update a new version of the design.

- Run regression verification to check if there are new errors introduced.

- Check all introduced bugs and debug.

- Prepare enough random tests before going home for the day.

- Run as many iterative and random tests as possible during the night.

The regression test should be done whenever the RTL code is changed.

Example 19.5
Write a LINUX/UNIX script to run a regression test.

This is an example of how an engineer could use a script to verify that the RTL code simulation behaves in the same way as the instruction set simulator.

```
#!/bin/bash
# Small regression testing script that runs the instruction
  level simulator on all assembler files in the tests directory
# and compares the result with the output from the RTL
  testbench.
#
# Files used:
# asm: Assembler
# sim: Instruction level simulator
# senior_tb: Name of RTL testbench
# test.hex: Assembled program. Will be read by the RTL
                testbench
# and the instruction level simulator
# simoutput.hex: Output from the instruction level simulator
# rtloutput.hex: Output from the RTL testbench
# vsim: Modelsim executable

LOGFILE=testlog.txt

# This is the function that actually runs one test.
# $1 contains the name of the assembler file which contains
# the test bench.
  function runtest {
      # Remove old outputs
      rm -f simoutput.hex rtloutput.hex

      # Assemble the file to test
      asm -i $1 -o test.hex || return 1
```

```
    # Simulate using the instruction level simulator
    sim -i test.hex || return 1

    # Start modelsim in batch mode and run the testbench
    # The testbench environment in senior_tb should fill the
      program
    # and data memories of the processor from the file
      test.hex
    vsim -c -do "run -all;quit -f" senior_tb || return 1
    diff -q simoutput.hex rtloutput.hex || return 1

    # If we reach this line, the output from the instruction
    # level simulator and the RTL testbench is identical
    return 0
}

# Empty log file
  cp/dev/null $LOGFILE

# Iterate over all assembler files in the tests directory:
  for i in tests/*.asm
  do
  runtest $i || echo "Test $i failed" >> $LOGFILE
  done
  echo "All tests finished" >> $LOGFILE
```

19.2.5 Simulation or Emulation

Simulation is to run a program that attempts to simulate the abstract model of a particular system on a host machine (e.g., PC). Emulation is to implement the functions of the design to a platform system for fast prototyping, such as a FPGA development board, so the function of the platform system is the same as the design to be emulated. Emulation means fast prototyping of a hardware system (e.g., DSP) onto an FPGA.

At the beginning of the verification, many design errors will be found, and the debugging time should be the main concern. Therefore, during the early verification and debugging phase, the assembly simulator and RTL simulator are used because they offer easy switching between the simulation mode and the debugging mode.

After fixing most design errors, the design becomes relatively stable. It will take a very long time to find the rest of the bugs, so it is necessary to speed up the verification. Emulation can utilize the parallelism in the prototype hardware (e.g., FPGA). The execution time can be reduced dramatically by emulation. To switch from simulation to emulation at the right time can minimize verification time costs. If the change from simulation to emulation is too late, too long a time will be consumed by simulation. However, if switching to emulation is too early, too much

Table 19.3 A Comparison of Simulation and Emulation.

	Time Cost for a Debugging	Execution Performance
Simulation	Low	Low
Emulation	High	High

time might be used for debugging and implementing the hardware on the emulator, see Table 19.3.

19.2.6 Verification of an ASIP

An ASIP can be verified using methods and techniques discussed previously and depicted in Figure 19.21. The verification is partitioned in three stages. Careful verification of the instruction set simulator and RTL codes should be done using methods including compliance, corner, real code, and random tests. The assembly instruction set simulator should be verified first as the reference model. RTL code verification consumes most of the time. Net-list verification is not the focus of this chapter. Relevant methods for net-list verification can be found in reference [5].

19.2.7 Writing Testbench

DUT (Design Under Test) is the top-level hardware module of the hardware design. When verifying an ASIP core, the top-level code of the core is the DUT. The test suit

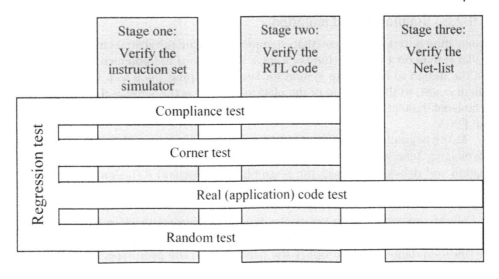

FIGURE 19.21

Verification of an ASIP.

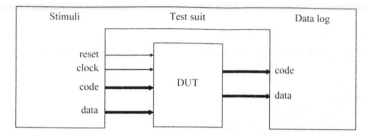

FIGURE 19.22

A test suit and DUT.

```
module clk_gen(output reg clk);        module reset_gen(output reg
                                       reset_n);
    always begin
        clk <= 1;                          initial begin
        #10;                                   reset_n <= 0;
        clk <= 0;                              #150;
        #10;                                   reset_n <= 1;
    end                                    end
endmodule                              endmodule
```

FIGURE 19.23

The fixed part stimuli.

is a HDL design running in parallel with the DUT. Figure 19.22 illustrates the test setup.

While running for verification, the test suit is the top HDL. A test suit consists of three modules: the stimuli, the DUT, and the data log. The stimuli are files including fixed part and variable part. The fixed part stimuli supplies clock and reset signals as shown in Figure 19.23.

The variable part consists of data and program code. Data can be loaded to an in-port by the loading program in Figure 19.24.

Finally, the code for writing code and data into a log file can be implemented by modifying the code in Figure 19.24.

19.3 CONCLUSIONS

Hardware integration includes the integration of the hardware HDL code and the integration of a hardware behavior simulator. A hardware behavior simulator includes all assembly instruction set simulators of all cores and the behavior model of hardware modules. The method to implement the chip-level cycle accurate behavior simulator was introduced. NoC was introduced from a user's view.

```
module io_read
  #(parameter nat_w = 16,
    parameter reg [255*8-1:0] FILENAME = "RTLinput.hex")
(
input wire clk_i,
input wire reset_i,
input wire id_io_rd_en_i,
output wire [nat_w-1:0] io_rd_data_o
);

//local declarations
integer file, r;
reg [nat_w-1:0] in_data;

initial
begin
    file=$fopen(FILENAME, "r");
    // Read the first value from the file immediately
    r=$fscanf(file, "%x", in_data);
end

assign io_rd_data_o=in_data;

//read from file
always @(posedge clk_i)
begin
    if (reset_i==1'b1)
    begin
        if (id_io_rd_en_i==1'b1)
        begin
            if (r==1) r=$fscanf(file, "%x", in_data);
        end
    end
end
endmodule
```

FIGURE 19.24

Loading code and data.

It is a great challenge to verify a processor and to ensure it is bug-free in electronics designs. There is no perfect way to achieve this goal. Especially, there is no way to get a bug-free assembly instruction set simulator in a short time.

EXERCISES

Homework in this chapter is implicitly linked to the appendix of the book, the **Senior** assembly instruction set manual.

19.1 Identify and verify the corner cases of addressing for load/store instructions using **Senior** addressing models INDX, INC, DEC, and OFC.

19.2 Identify and verify all corner cases of **Senior** instruction SUBx.

19.3 Identify and verify all corner cases of **Senior** instructions RR and LR.

19.4 Identify and verify all corner cases of **Senior** instruction PostOP.

19.5 Identify and verify all corner cases of **Senior** instruction CONV.

19.6 Is the coverage of random test high enough? If it is not, what corner cases shall be specified to enhance the verification coverage?

REFERENCES

[1] Wiklund, D., Liu, D. (2003). SoCBUS: Switched network on chip for hard real time embedded systems. Parallel and Distributed Processing Symposium.

[2] Flachs, B., Asano, S., Dhong, S. H., Hofstee, H. P., Gervais, G., Roy Kim et al. (2006). The microarchitecture of the synergistic processor for a cell processor. *IEEE JSSC*, Jan., 63–70.

[3] Nilsson, A., Tell, E., Liu, D. (2008). An $11\,mm^2$, $70\,mW$ fully programmable baseband processor for mobile WiMAX and DVB-T/H in $0.12\,\mu m$ CMOS. Accepted to ISSCC, San Francisco, Feb.

[4] http://www.PicoChip.com.

[5] Keating, M., Bricaud, P. (1998). *Reuse Methodology Manual*. KAP.

[6] Bergeron, J. (2000). *Writing Testbenches Functional Verification of HDL models*. KAP.

[7] Van Campenhout, D., Hayes, J. P., Mudge, T. (2000). Collection and analysis of micro-processor design errors. *IEEE Design and Test of Computer*, October-December.

[8] Smith, M. J. S. S. (1997). *Application-Specific Integrated Circuits (ASIC)*, Chapter 14. Addison-Wesley VLSI Systems Series.

[9] Dally, W. J., Towles, B. (2003). *Principles and Practices of Interconnection Networks*. Morgan Kaufmann.

[10] Tannenbaum, A. S. (1996). *Computer Networks, 3rd edition*. Prentice Hall PRT.

[11] Wang, L. T., Wu, C. W., Wen, X. (2006). VLSI Test Principles and Architectures: Design for Testability (Systems on Silicon), Elsevier.

Parallel Streaming Signal Processing

In this last chapter of the book, advanced streaming signal processing and challenges of parallel DSP signal processing will be discussed together. The discussion is based mostly on my personal view, although other opinions will be discussed briefly as well. This chapter is prepared for researchers and engineers working for predictable parallel streaming signal processing.

20.1 STREAMING DSP

20.1.1 Streaming Signals

Streaming signals or streaming data are digital signals arriving periodically within a regular time interval. The unit of streaming data is usually a data packet instead of a data sample. Streaming signals either are sampled with fixed data rate, transmitted through communication channels, or stored in storage devices, and are consumed with a fixed data rate.

Real-time behavior is the most important feature. Signal processing on the current data packet must be finished before the new data packet arrives. Streaming signal is seldom represented using 64-bit or 32-bit floating-point data. Most streaming DSP applications have custom data types and finite data precision, which is enough to satisfy the SNR requirement. For example, an 8-bit fixed-point format is sufficient for most of video signal processing. Sixteen-bit fixed-point precision is sufficient for voice signal processing.

20.1.2 Parallel Streaming DSP Processors

Parallel (streaming) signal processing was introduced in Chapter 3. Parallel processing partitions a system into several independent tasks and runs them in parallel hardware. The partitioning can be done in two levels: task level partitioning and data level partitioning.

A parallel task is a logically discrete section of computational work that can be partitioned and executed in parallel by multiple processors yielding correct results. A task is typically a program or program-like set. Simply said, a parallel task is a

705

section of program that can be executed independently of others. Task level partitioning therefore is the partition of programs. A task in a CFG (Control Flow Graph) is a part of a flowchart with only in and out, without depending on other running tasks. As soon as the inputs are available, a task can be executed independently.

Data level partitioning partitions data that executes in parallel using the same program. It is required when the set of regular data (array) is large. Including its task level partition and data level partition, a streaming signal processing flow is described using Figure 20.1. In this figure, P is the streaming period or input packet rate, the time to receive each input packet.

In Figure 20.1, tasks are partitioned into both sequential modules and parallel modules. Sequential modules execute different tasks in parallel, and parallel modules execute the same task on different data in parallel. If there is one processing unit executing a task, the processing time for a data packet must be less than P. Task T3 is heavy, and fortunately, the data array can be partitioned further. The data thus is partitioned into four subarrays and processed in four hardware modules using one task program. Task Tn–1 is heavy, and it can neither be executed within the time P nor be partitioned further. In this case, two hardware modules in parallel are required: E at the input of Tn–1 is the even packet and O is the odd packet.

The stream processing latency is the sum of all the processing time from the time "input packet available" to the time "output packet available." There are n tasks connected in serial, and the task Tn–1 uses two P. The total processing time is $P(n + 1)$. The output packet transfer time is much less than P, so that the stream processing latency is almost (yet slightly more than) $P(n + 1)$. If the feedback is required after streaming signal processing, the latency of the pipelined stream processing might be limited.

To decide architecture for streaming signal processing, a deep understanding of streaming task features and parallel programming is essential. By extensively profiling the source code of streaming signal processing, its features can be

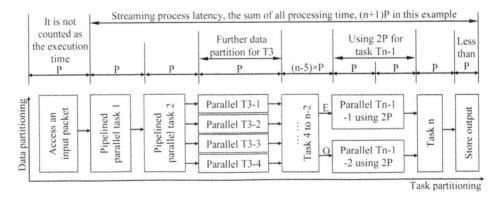

FIGURE 20.1

A general streaming task flow.

summarized. Streaming signal processing on one data packet is largely independent of operations on other data packets, resulting in a large amount of data parallelism and high latency tolerance. During streaming signal processing, data are used only locally, and the possibility of global data reuse is low (except for video CODEC). Memory access of streaming signal processing might be huge yet regular.

20.2 PARALLEL ARCHITECTURE, DIVIDE AND CONQUER

Traditional partitioning for parallel processing gives only the partitioning on the volume of tasks, yet it did not give the best partition on the complexity. Handling of data complexity, handling of memory access complexity, and handling of complexity of the program flow control are not yet considered. In the following text, we try to propose a new architecture and offer the partition of complexity, which is to partition tasks by separating complexity of program flow control, data complexity, and memory access complexity and to handle them separately.

20.2.1 Review of Parallel Architectures

The challenge of architecture selection is to achieve sufficient control capability, sufficient computing and memory access performance, as well as sufficient function coverage under the constraints of silicon cost and power consumption as well as design cost.

Scalable Processor and Accelerators

A scalable processor is a scalable processor architecture platform. The extra design cost is limited and acceptable while adding or removing hardware modules or adding processors to and from the hardware platform.

Let us review what we discussed so far. An accelerator is a nonprogrammable hardware module. If an accelerator is programmable, it will be classified as a slave processor in a heterogeneous multiprocessor system. Accelerators usually work in parallel with its master processor. Dependencies should be predictable and predefined during the hardware design of an accelerator.

If the communication cost between a processor and its accelerator is negligible, an I/O port-driven accelerator will be sufficient. If a significant amount of communication is needed between the processor and the accelerator, a tightly connected accelerator is required. Loosely coupled and tightly coupled accelerators were discussed in Chapter 17.

ILP Processors

Most DSP processors are Instruction Level Parallelism (ILP) processors because there are always multiple microoperations (for example, data access and data processing) running in parallel. In this section, we particularly mean that the ILP is the processor issuing multiple instructions running in parallel under a single program flow. ILP DSP processors were discussed in Chapter 3. SIMD is the simplest ILP processor, used mostly as a slave processor for vector signal processing. VLIW machines are popular today (in 2007). The essential technology behind a VLIW processor is the compiler. If the VLIW compiler technology is not available to the designers, designing VLIW machines should be avoided.

Superscalar processors consume relatively more silicon area and more power. However, the performance-to-silicon ratio of a superscalar processor is usually higher than that of a VLIW machine because the IPC (instructions per clock cycle) measurement on a superscalar machine is usually higher than that on a VLIW machine.

The main drawback of designing a superscalar is its high design cost. However, code portability and lower assembly code design cost are the main advantages, and these advantages will be more important in the future. We can therefore predict that a superscalar plus accelerators or a superscalar plus multiple SIMD datapaths or SIMD machines will be a popular choice of architectures in the future. The first reason is that designers are more aware of the assembly code design cost, and they always try to avoid the excessive cost of programming on multiprocessor. The superscalar hardware design cost will not be high if half designed superscalar platform as silicon IP (intellectual property) is available. The extra silicon cost of superscalar will be less significant compared to the increasing memory cost on chip. In general, drawbacks of superscalar DSP will be trivial and the advantage of superscalar will be more significant. It will be discussed and proven, in later sections, that most DSP applications can be partitioned to the complicated control flow running on superscalar and the predictable vector processing running on SIMD accelerators.

Heterogeneous Multiprocessors

A heterogeneous multiprocessor is a multiprocessor system consisting of more than one assembly instruction set. To get the highest performance over silicon, different tasks should be processed by dedicated processors. For example, an audio processor is used for audio signal processing, a video vector processor is used for video frame processing, and a data compression codec might be used for Huffman coding and decoding. Heterogeneous multiprocessor architecture offers higher performance-to-silicon ratio if the system fulfills the following conditions:

- The system consists of dedicated ASIP cores for different classes of applications.

- Each suitable task is allocated to a dedicated ASIP core.

In an embedded system, the signal processing flow is fixed and applications are stable. Heterogeneous architectures might be the best solutions. However, there are two problems: restricted compatibility toward other applications, and the relatively high design and system integration cost.

Heterogeneous multiprocessor architectures are used widely in systems aimed at fixed applications. For example, functions in a radio baseband processing subsystem

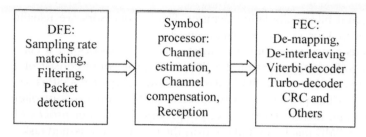

FIGURE 20.2

Multiprocessor for digital TV channel decoding.

in a multistandard digital TV receiver can be relatively fixed during the system design phase. Standards include DVB-C/S/T/H [1] (digital video broadcasting) from Europe and similar standards from Japan (ISDB-T ARIB STD-B31) and China (DMB-T GB 200600-2006). However, all standards consist of three kinds of processors: one is the digital front-end (DFE) for digital filtering and packet detection, one is the symbol processor based on FFT and complex-data convolution, and one is the "soft bit manipulator" for error correction decoding. The heterogeneous architecture is depicted by the block diagram in Figure 20.2 [3].

Obviously, there are three different processors comprising a heterogeneous multiprocessor system. The first core, the DFE, is dedicated for filtering (convolution) and packet detection (auto and cross correlations). The second core, the symbol processor, is used for vector processing based on complex data format and program flow control using integer data format. The third processor, FEC (Forward Error Correction), is used for error detections and error corrections based on short integer and bitwise formats (see Figure 20.2). The heterogeneous multiprocessor in Figure 20.2 is powerful and very suitable for radio baseband receivers. However, it is not suitable for other applications such as video decoders or audio decoders.

A homogeneous multiprocessor offers high flexibilities. All tasks can be allocated to all processor cores in a homogeneous multiprocessor system. The programming and debugging on homogeneous multiprocessors are therefore easy. Assembly instruction sets of cores are different in a heterogeneous multiprocessor system. Each processor core supports only its own instruction set, and programs cannot be moved from one core to another core. A heterogeneous multiprocessor is therefore less flexible. Flexibility and performance over silicon are always trade-offs and cannot be achieved at the same time.

Master-Slave Multiple Processors

Many multicore DSP subsystems consist of master-slave processors. A master processor is the task manager handling task flow control and interprocessor communications. Slave processors handle parallel (vector and iterative) DSP tasks. The master processor could be a microcontroller or could be a single-issue DSP processor with strong FSM handling and bit manipulation features. Slave processors are usually SIMD processors dedicated for a vector signal processing.

Most heterogeneous processors consist of master-slave processors for a class of specific applications. However, most master-slave ASSP processors consist of a master controller and several identical SIMD slave machines. A typical ASSP is STI CELL (Sony-Toshiba-IBM CELL processor) [8, 9]. CELL consists of a microcontroller (PowerPC) and up to eight identical SIMD slave processors. The master processor issues tasks to the slave SIMD machines and controls the execution of issued tasks. Each SIMD machine may have up to eight executions per clock cycle, and eight SIMD machines may offer up to 64 operations per clock cycle. A similar example of master-slave processor is the Imagine processor from Stanford University [11].

Homogeneous Multiprocessors

There are seldom real homogeneous multiprocessors for streaming DSP. Most so-called homogeneous multiprocessors are master-slave processors with one master processor and multiple homogeneous slave processors. This is simply because of the differences between tasks running in the master processor and tasks running in slave processors.

20.2.2 Divide and Conquer

When performance requirements and the complexity of a task (an application or a function) are beyond the limit of a single machine, the task will be partitioned to several processor cores in parallel. Divide and conquer has been the design principle of any advanced or complex system.

Task Partition

If tasks are partitioned and allocated in several computing engines, the execution order of parallel tasks must follow the order of the behavior flowchart. The freedom of task partition and allocation therefore is not high. Execution times of parallel tasks on different processor cores will never be exactly the same because tasks are different, with different programs and different data. Parallelization overheads therefore are usually high.

Coarse granularity partitioning generates less communication overheads but suffers from load imbalances. Fine granularity partitioning can minimize load imbalances but suffers from high communication overheads. The trade-off of partition granularity is actually the essential issue of system design.

According to coarse-grained partitioning, tasks are application components or functions modules such as those depicted in Figure 20.3.

By profiling tasks in Figure 20.3, it exposes the large load (runtime) imbalance. Some processors holding easy tasks will have to wait for processors executing heavy tasks. In such a case, the efficiency of processors with easy tasks will be low. To speed up the processors running heavy tasks, the heavy task will be further

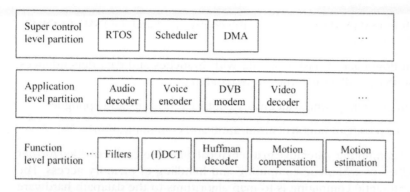

FIGURE 20.3

Coarse-grained partitioning of applications.

partitioned with fine granularity into several sublevel tasks and distributed for parallel processing. The execution of these sublevel tasks should be synchronized, which will induce extra control overhead. For example, if motion estimation is distributed to several slave processors, these processors will keep handshaking to synchronize and share the progress of tasks in each processor.

The main challenge of parallel streaming signal processing is computing efficiency. High-level coarse-grained partitioning will induce a load imbalance problem. Low-level fine-grained partition will induce excessive communication and control overhead. Thus, task partitioning can never be perfect. The balance of granularity is always a challenge of designing a parallel signal processing system. Three dimensions will be discussed: the application, the application-specific parallel architecture, and the granularity of partition. Yet another challenge of parallel computing is the extra computing latency induced by data dependencies. Especially the worst-case is that conditions are unpredictable. It cannot be identified until the condition is used for a decision. The same problem happens also in an ILP level for WCET (worst-case execution time) prediction.

Different Ways to Divide and Conquer

These challenges cannot be handled all together because traditional partitioning is based on task loads instead of on complexities. In this section, we will try to separate control complexity, data complexity, and complexity of parallel memory access (addressing complexity) into three problem-solving domains. The separation is applied to both hardware and software development.

Before discussing the partitioning of complexity, a formalized streaming processing program model should be introduced briefly. A streaming program usually consists of two parts: the parallel computing part and the FSM (finite state machine) part. The parallel computing part handles iterative mathematical functions. The FSM part handles program complexities. Actually, the FSM part can be divided further into two kinds of FSM in a DSP program—the control FSM and the FSM of CODEC. The control FSM handles:

- Program flow control: Task handling with branches and data dependences, HW resource handling, external event handling.

- Miscellaneous computing: data quality control, prologues, and epilogues.

- Control for parallelization: synchronization and communication.

The parallel computing of iterative mathematical functions can be divided further into the handling of arithmetic computing and handling of memory access. The handling of arithmetic computing is to map algorithms to the datapath hardware and maximally utilize the hardware parallel features. The handling of memory access supplies multiple data to the computing units with minimum delay. The ideal case is that as soon as an algorithm is to be executed, sufficient data will be available just in the register file.

20.3 EXPOSE CONTROL COMPLEXITIES

The control of streaming signal processing can be divided into the control of the single processing flow (the conventional control of a DSP application), the control for threading multitasks running in parallel in one processor, and the intertask control between processor cores (communication and synchronization between tasks).

20.3.1 General Control Handling

Except for code of DSP algorithm functions, program flow control programs include:

- Synchronous control functions: Execution of the top-level CFG (running the program according to the control flow graph), input and output handling of each subroutine, prologue and epilogue for each DSP subroutine.

- Asynchronous control functions: Interrupt handling and service, interprocessor synchronization and communication, context switching.

- Hardware management functions: Data memory management, off-chip memory access, I/O port management, and interprocessor management.

- Fixed-point data quality control functions: Decision to measure data quality on each measurement point, gain factor decision on each gain control point.

- Runtime management and reliable computing for both streaming synchronous and asynchronous controls.

These program flow controls are handled using sufficient DSP RISC instructions (moves, ALU instructions, and jumps) and sufficient setting, resetting, and testing of flags and status.

Hardware support of interrupt and exception handling includes at least an interrupt handler, stack memory, and interrupt service routines. Handling includes normal interrupts, exceptions, and fast interrupt (without context switch).

Hardware management includes handling of I/O ports, accelerators, and the peripheral components of the core (interrupt handler, timer, and DMA). The handling includes configuration of I/O and memory access from and to I/O.

Data quality management is essential for firmware running in fixed-point processors. The purpose of the data quality handling is to maximize the signal-to-noise ratios of results from the fixed-point processor. Instructions for conditional execution, mean, MAX, and MIN value of a vector array are important.

Runtime control is part of the program flow control functions based on machine cycle counting. The machine cycle counting is conducted by the running program. When the runtime reaches the deadline, the running program will be interrupted by the timer and the interrupt service will lead the execution to the top of the program, starting to receive and process a new data packet.

Reliable control is to avoid data and control hazards. A typical impact on the reliability can be data hazards accruing when adjacent instructions consume a different number of execution cycles. An operand might not be available when the previous instruction is a multicycle instruction. The compiler must check for this kind of data dependency problems and, for example, by padding with NOP instructions, make sure the result register of a multicycle instruction will never be used by the following instruction until the result is available. In hardware, the result register can also be locked so that its contents cannot be fetched until the result is written.

Another impact on reliability is the data hazard induced by asynchronous events. An interrupt might be accepted right after running a multicycle instruction. If the context saving stores the result register while the result is not yet available from the multicycle instruction, the saved context is wrong. The order of context saving must be modified so that the dependent context will not be saved until the result is ready. For example, contexts in general registers can be saved first, and contexts in accumulator registers can be saved in the last step during context switching.

20.3.2 Exposing Challenges

Separating Control and Signal Processing

The nature of control behavior is based on branching, either two-way "if-then-else" or multiway "case A, B, C...". The common feature of streaming signal processing is iterative vector computing with predictable memory accesses. The exception is the processing of bit-serial CODEC, for example, Huffman coder and decoder, which is a kind of FSM.

Obviously, it may not be a perfect idea to combine the code of control FSM and iterative vector computing into one classic instruction set and to execute them in a machine with conventional architecture. However, industries have been doing it this way since von Neumann's time. The question is: Is it possible to separate control operations from arithmetic computing? The answer is yes, on the condition

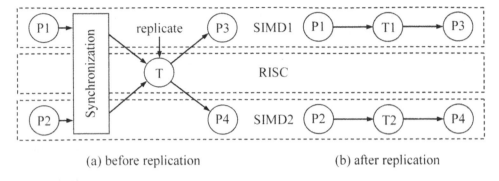

(a) before replication   (b) after replication

FIGURE 20.4

Replication of a common subroutine.

of special applications with predictable features, a class that includes, for example, most streaming signal processing systems.

A simple way is to separate the vector/iterative processing from normal processing by introducing SIMD instructions to a VLIW machine. The machine gives the same performance as the VLIW machine. By using SIMD instruction, the code size can be lower because one short SIMD instruction can be equivalent to several instructions in a long VLIW instruction word [10].

Another advanced way is to separate and accelerate the vector processing to slave SIMD machines from the master machine. The STI CELL architecture mentioned in the previous section is a typical example (see [8, 9]). The RISC machine handles task management, and SIMD machines will handle heavy computing tasks. The SIMD machines in CELL are absolutely separated from the RISC machine. Conditional computing and low-level program flow control have to be handled by the SIMD machine. However, the problem is that the RISC machine will be idle most of the time, and the SIMD machine loses its performance while handling low-level control.

Some condition handling can be conducted in a SIMD machine by the replication of the common subroutines. By replication, the task T in Figure 20.4(a) is modified to be T1 and T2 and allocated in SIMD1 and SIMD2. The execution time of T1 might be larger than the execution time of T; the total execution time is less because of removing the cost of synchronization, see Figure 20.4(b).

Challenges from Separating RISC and SIMD

A SIMD-VLIW mixed machine may not be efficient enough because of the hardware overheads of the VLIW machine. Master-slave machines (STI CELL, for example) may not be efficient enough when handling low-level controls in subroutines in the slave SIMD machines. By checking profile results of streaming DSP applications and by analyzing master-slave multiprocessors, one interesting result is identified. The original cost of miscellaneous programs (organizations and controls of vector computing subroutines running in a RISC machine) between two vector-iterative subroutines (running in a SIMD machine) is actually low. If control function and vector computing are partitioned into master and slave machines, the extra cost of synchronization and communication could actually be the significant part of overheads. Execution of these overheads is conducted by the RISC machine and cannot be in parallel, so extra latency is induced. For example, to handle the conditional branch by the master RISC machine will induce excessive overheads. These overheads could be caused by:

- Sending part of the vector data to the RISC machine for condition tests.

- Handling interprocessor synchronization and communications.

- The vector processor waiting for the decision of the conditional test-branch.

It is obvious that most overheads can be eliminated by merging the execution of RISC computing and SIMD computing into one program flow by running one instruction set. By doing that, each RISC machine can carry several SIMD datapath modules. Each SIMD module executes predictable vector computing in parallel in the same way as running RISC instructions. The RISC machine can thus directly access data used by the SIMD datapath. Overheads can therefore be eliminated.

There are two problems; one is to run unpredictable tasks, another is to run tasks requiring excessive computing power. Obviously, unpredictable computing cannot be handled by the proposed solution. However, runtime of streaming signal processing must be fixed, and unpredictable tasks must be handled within a fix the time slot of streaming signal. In Chapter 18, for example, we discussed that the Taylor series is preferred to the Newton-Raphson algorithm when fixed runtime is required.

When excessive computing power is required, the solution will be the partitioning on coarse granularity level to multiprocessors. For example, excessive computing power is needed for signal processing for synthetic aperture radar or computer tomography. These computing tasks can be partitioned into independent steps, and each step can be allocated to an independent processor without dependency on other tasks. The only two interfaces to the previous and following tasks can be handled by FIFO buffers between processors.

Conclusions are:

- The latency of handling streaming program flow control is the main impact of streaming signal processing because it is difficult to be executed in parallel.

- The latency of handling streaming program flow control shall be minimized by merging control and vector processing into one program flow.

- Only if control and parallel data processing can be separated and the SIMD can be concentrated on vector computing, the RISC controlling SIMD datapath is a preferred solution. Fortunately, this is true for streaming signal processing.

- For streaming signal processing, runtime unpredictable tasks can be avoided.

20.3.3 SIMT Architecture for Low-level Parallel Applications

To reach such a solution, a SIMT (single instruction flow multiple task) machine has been investigated and will be introduced in this chapter. SIMT is a RISC machine carrying several SIMD datapath modules. The RISC machine handles all control functions and issues vector computing tasks to SIMD datapath modules. All program flow controls and vector processing are executed under one program flow so that the latency or inter-processor overheads can be avoided. A SIMD datapath module is a programmable accelerator. By merging the SIMD instructions into the RISC instruction set, a SIMT machine tightly couples miscellaneous functions with vector computing so that a SIMT minimizes control and parallel overheads. Another benefit is the direct access of vector data by the RISC machine because vector and RISC computing both are allocated in the RISC instruction domain with the same addressing space. Monitoring vector data and data quality control can be conducted by the RISC machine. A SIMT machine is simply illustrated in Figure 20.5. All RISC and SIMD datapath modules share vector memories by NoC, the on-chip connection network. The NoC can be a crossbar (simple 2D network) as discussed in Chapter 19.

Both the RISC datapath and the SIMD datapath modules are controlled by the RISC control path. The control path executes the program as a simple program flow. Each vector memory consists of its own addressing module (AGU), and an addressing algorithm can be configured before running a task by the RISC control path. Before running a task, the on-chip connection network is configured, and memories as well as addressing modules will be connected to the related SIMD datapath modules. A task can thus be issued.

While executing a task in one or more slave SIMD datapath modules, other SIMD modules and the RISC datapath are available for executing other tasks. The coding and executing SIMT program is illustrated in Figure 20.6.

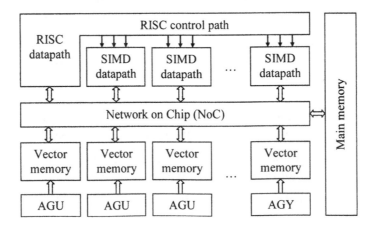

FIGURE 20.5

Concept of SIMT machine.

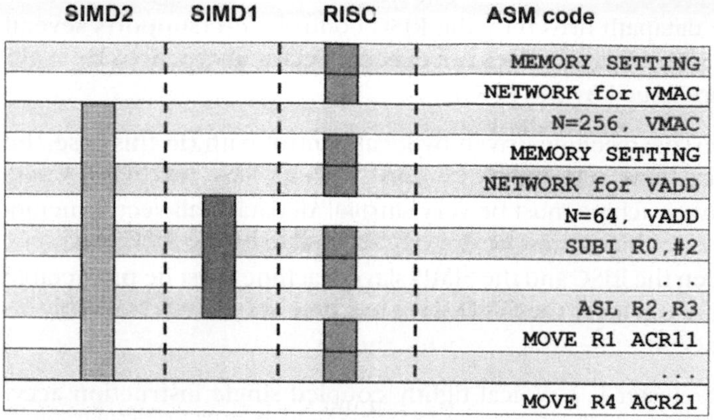

SIMD2	SIMD1	RISC	ASM code
			MEMORY SETTING
			NETWORK for VMAC
			N=256, VMAC
			MEMORY SETTING
			NETWORK for VADD
			N=64, VADD
			SUBI R0,#2
			...
			ASL R2,R3
			MOVE R1 ACR11
			...
			MOVE R4 ACR21

FIGURE 20.6

Coding and code execution on a SIMT machine.

By observing the example given in Figure 20.6, the code and its execution expose the advantages of SIMT. On the top level, the program running in the RISC configures the AGU and network connections for the vector instruction VMAC. VMAC then is issued and executed in the SIMD2 cluster. While executing VMAC in SIMD2, other instructions can be executed in both the RISC and the SIMD1 cluster. The RISC can thus be used to configure another AGU and network connections for VADD to be executed in the SIMD1 cluster. As soon as the VADD instruction is configured and prepared in the SIMD1 cluster, VADD is executed, which will consume 64 clock cycles. Meanwhile, the RISC datapath can be used to run other instructions until VADD is finished after 64 clock cycles. The result in ACR11 in the SIMD1 cluster is moved to the RISC datapath for further processing. Finally, the VMAC is finished after 256 cycles in the SIMD2 cluster, and the result of VMAC is moved from ACR21 in the SIMD2 cluster to the RISC datapath.

In a real-time streaming signal processing system, the cycle cost of vector computing is always fixed or can be predictable when unpredictability is eliminated (for example, avoid using Newton-Raphson algorithms, avoid unpredictability of data accesses). During the firmware debugging phase, the scheduling of tasks in the RISC, the SIMD1 cluster, and SIMD2 cluster are done, and thus hardware resources can be best utilized. There are exceptions of runtime predictability, for example, when iterative optimization algorithms are used. In this case, either the algorithm can be reformed, or a timer can be used to stop the iteration to catch up on the real-time requirements.

After merging a RISC processor and a SIMD machine into a SIMT machine, the SIMD machine became a datapath cluster instead of a SIMD processor. As a hardware module, a SIMD datapath cluster in Figure 20.5 can be designed into three styles:

- A configurable datapath module (supports only one instruction, the instruction is configurable).

- A programmable datapath driven by the RISC control path (supports several instructions, the SIMD datapath does not execute vector instructions by itself, all instructions are issued by the RISC).

- A programmable slave machine driven by local control path (in this case, the SIMD is a slave machine very tightly coupled with its RISC machine). Code executed in a slave machine must be very simple. All data in all vector memories in SIMD slave machines must be directly accessible by the RISC machine. Interwork between the RISC and the SIMD slave machine must be predictable and simple. On the top level, the SIMD slave machine should not be visible to the system engineers.

The SIMD datapath cluster is a typical tightly coupled single instruction accelerator discussed in Chapter 17. The datapath module offers a single function with certain configurability. For example, it can be a convolution machine with configurable iteration size.

The second style SIMD datapath is programmable and can execute multiple vector instructions. Instead of a slave processor, there is no control path in the SIMD module. All SIMD instructions are decided and issued by the RISC control path. SIMD instructions are issued and decoded by the RISC control path, and the decoded control signals are supplied to the SIMD datapath directly. There is no extra parallelization overhead. Table 20.1 gives instructions for a programmable SIMD datapath driven by the RISC control path. This example is a SIMD datapath for digital radio baseband algorithms based on complex-valued data computing for radio communications (LeoCore of Coresonic).

Table 20.1 SIMD Instructions of Complex Data Computing.

Instruction	Cycle costs	Parameter 1	Parameter 2
Complex data convolution	$N + 3$	Conjugate or not	
Complex data autocorrelation	$N + 3$	Conjugate or not	
Complex data Vector-vector product	$N + 2$	Conjugate or not	
Complex data Vector-constant product	$N + 2$	Conjugate or not	
Complex data Vector-vector add	$N + 1$	Conjugate or not	
Complex data MAX-value search	$N + 1$	No	
One layer DIT based radix-2 FFT	$N/2 + 4$	Step size	Bit reversal addressing for the first or the last layer

A SIMD datapath cluster could become a programmable slave machine, and it is still very tightly coupled with its master, the RISC machine, and shares the same memory addressing space. In this case, a short program can be executed in the slave vector machine while the RISC part can execute other tasks. To the RISC machine, the short program running in the slave SIMD machine must be independent and with fixed execution time. Because the cycle cost of each program executed in the SIMD slave machine is fixed and known by the RISC and also because vector memories in the slave SIMD machine are reachable by the RISC, there will be no extra communication between the master and the slave.

The slave machine usually executes an inner-loop task using only a few instructions. Since the number of instructions is very small, the size of the vector program will be very small and the code size will not be a design challenge. There are opportunities to fully utilize the datapath by directly using control signals as the microcode. This is actually one of the main advantages of using a SIMD slave machine; the code length of vector instructions can be independent of the code length of the RISC machine.

Remember what we discussed in Chapter 7. A very low percentage of the datapath functions can be exposed and driven by an instruction set. It was discussed that only about 20 of 2000 combinations available in an ALU datapath are used by the RISC arithmetic instructions in Chapter 7. To expose all datapath functions, control signals must be used directly as binary code without code compression. It is impossible because the code will be too long when the processor is a VLIW. However, when using the SIMT concept, separating the SIMD code from RISC code, the direct use of microcode in SIMD datapath becomes possible.

In Figure 20.7, the main program in the RISC (master) core is implemented using compact binary-coded instructions. Special vector instructions will be allocated to

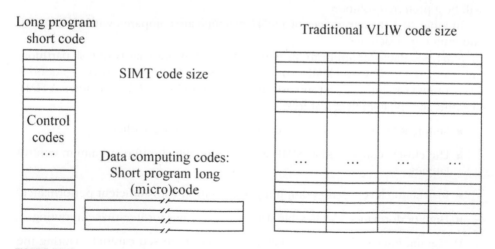

FIGURE 20.7

Using SIMT to reduce the program footprint.

the vector SIMD cores. Because the code size (the number of instructions in a subroutine) is very small (only a few lines), the code width will not have impact on the silicon cost. Ultra-long code can be used to expose datapath features. The extreme case is to directly use microcode (control signals) as the vector machine code (for example, Carmel processor of Infineon technologies). The datapath can be ultimately used. Although the microcode memory is very wide, the code cost of a few lines of instructions will not be significant. In contrast, to reach the ultimate flexibility, a VLIW machine mixes up the control code and data processing code. Therefore, the code efficiency of VLIW is relatively lower, and only a small percentage of datapath functions can be driven by the instructions with compressed binary code.

Scheduling the co-execution of RISC instructions and multi-cycle SIMT instructions is a challenge of using the SIMT machine. There are two ways of scheduling: static scheduling and dynamic scheduling. The static scheduling is based on off-line scheduler and the verification of the scheduling by running programs on ISS (assembly instruction set simulator). The static scheduling of using SIMD instructions is easy and suitable for explicitly predictable applications, such as streaming signal processing for radio baseband.

Dynamic scheduling can be implemented as a scheduling program running above the application program and supported by interlocking hardware in the RISC control path. While dynamically calling a SIMD subroutine, the runtime of the subroutine is stored in a timer. The name of the hardware module and the result address are locked by the timer while the timer is down-counting. The acting timer blocks all executions of instructions using the locked hardware module and the access of the results. When dependent instructions are decoded in the RISC, the RISC will be locked in the "wait mode" until the timer stops counting and raises its flag. Obviously, the performance of a simple RISC might be reduced, and a superscalar will be a preferred solution.

In Table 20.2, the advantages of a SIMT machine are compared with SIMD, VLIW, and multicore machines.

SIMT is suitable for applications with relatively low granularity of task partitioning compared to RISC and SIMD. For ultra-large scale tasks, a reasonable solution is to partition tasks into parallel subtasks running on multiple SIMT machines. A SIMT machine has the following features:

- Simple RISC handling control complexities discussed earlier.

- The closely coupled RISC SIMD architecture, which offers minimum control latency.

- Instruction level vector computing acceleration with sufficient performance.

- High performance and flexible data manipulations for iterative computing.

Designing interrupts for SIMT may need to be discussed carefully. During the interrupt handling process, the context of data registers in the RISC datapath, data

Table 20.2   Advantages of SIMD Architecture.

Specification	SIMT	SIMD	VLIW	Traditional multicore
Handling controls	Good enough	Poor	Very good	Very good
Vector computing	Very good	Very good	Acceptable	Very good
Datapath utilization (datapath flexibility)	Best	Only a few percentage	Only a few percentage	Only a few percentage
Code size[1]	Low	Low	High	Very high
Silicon costs	Low	Low	High	Very high

[1] For code size and silicon cost, "the lower the better."

registers in SIMD datapath (accumulator registers), and address registers in AGU should be reserved if the task of the interrupt service is unknown. The cycle cost of context saving is high if all context saving is conducted by the RISC part. Different solutions might be needed. For example, two groups of accumulator registers and addressing registers can be designed for a SIMD datapath and an AGU. One set of registers is for the main task, and another set of registers can be used for the interrupt service.

Actually in most cases, interrupt services to a DSP subsystem can be well known during the firmware design, and only a limited number of contexts need to be saved.

20.3.4 Design of Multicore DSP Subsystems
Multicore DSP Subsystems

Principally, the execution time of control handling (in the RISC datapath) is much less than the execution time of vector computing (in a SIMD datapath cluster). The RISC datapath in a SIMT machine usually is not fully used if there are only two to four SIMD datapath modules (handling two to four vector instructions in parallel). To enhance performance, more SIMD datapath clusters can be added. By running a single program flow in a SIMT, most problems induced by multicore can be avoided.

Except for adding a SIMD datapath cluster, specific slave FSM cores can also be added to SIMT architecture to offload CODEC FSM functions from the RISC datapath. For example, a Huffman coder and decoder can be offloaded from the RISC datapath by a slave machine. The reason is that a CODEC program is usually self-contained and independent of the RISC datapath. By offloading them to a small FSM machine, the computing load in the RISC datapath can be further reduced.

Nevertheless, there are extreme cases in which multicore DSP subsystems must be used. In this case, a SIMT is no longer sufficient. In communication

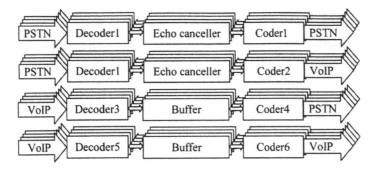

FIGURE 20.8

A simple view of a datapath in a VoIP gateway.

infrastructures, multiple streams of DSP functions are running in parallel to support multichannels or multiusers. For example, the function of a data processor in (the datapath) of a voice over IP (VoIP) gateway is shown in Figure 20.8.

A datapath of a VoIP (Voice over IP) gateway is used to translate voice compression formats between different telephone lines from different phone subsystems and to cancel the line echo induced by analog telephone lines. A datapath chip shall handle as many voice channels as possible in a DSP chip. In this figure, PSTN is a telephone line from a normal telephone network called the Public Switched Telephone Network. The coder-decoders are voice CODEC compressing and decompressing voice from one format to another. The echo canceller cancels the echo induced by the PSTN line imbalance. In a modern gateway datapath, thousands of voice channels need to be handled. Obviously, multiple streaming channels and multiple processors are needed.

Homogeneous-Heterogeneous AIS

Many system designers believe that principally there is not much to do in parallel hardware design. The main challenges in a parallel system with multiple cores are the efficiency of parallel programming and reliability issues. Actually, this is not true. The silicon efficiency, the power consumption, and the design cost of multicore solutions are always challenges.

Heterogeneous and homogeneous multicore structures were discussed in Chapter 3. A heterogeneous structure consists of different cores dedicated to targeting different applications. Compared to homogeneous multicore solutions, high performance and low silicon cost can be achieved by dedicated selection of heterogeneous processor cores with different application-specific instruction sets and custom architectures. By using heterogeneous processor cores, different cores can accelerate different functions. However, the cost of integration and verification of heterogeneous processor cores can be very high or even unaffordable by small and medium-sized companies. The code of one processor is usually not portable to another processor. The scalability of the design is low.

Most currently (2007), successful SoC platforms are based on bottom-up design methodology or based on the integration of heterogeneous instruction set processors. It is believed that the bottom-up methodology can achieve both high performance and silicon efficiency.

In the DSP domain, a multicore system with a homogeneous instruction set is not efficient. This is simply because the applications are heterogeneous, and a homogeneous instruction set can never be suitable for heterogeneous applications. To adapt the instruction set to all applications, the instruction set must be general, meaning not efficient. However, the advantages of a multicore system using a homogeneous instruction set are also significant: the low system integration cost, the portability of firmware, and the low verification cost. The firmware can be executed in any core, and the multicore integration is based only on one instruction set.

Obviously, the ideal case would be to design a multicore architecture with the same flexibility as a homogeneous instruction set system and with the same silicon efficiency as the heterogeneous instruction set system. This is unrealistic.

A compromise solution can be provided based on our comprehensive analysis of streaming DSP applications. The solution is to introduce a compact homogeneous instruction subset as the kernel of all processors. The kernel instruction subset can be used for general scalable RISC computing and program flow controls. The kernel instruction subset can be implemented as the RISC processor core of the proposed SIMT architecture. Custom functions can be implemented and accelerated by the heterogeneous instruction subset, which is the extension of the homogeneous instruction subset (for example, the SIMD parts of the proposed SIMT architecture). Heterogeneous instructions are tightly connected accelerator instructions, and they will eventually be integrated into the homogeneous instruction subset.

Each core in the multicore solution presented in this chapter thus consists of two subsets of instructions: the general subset and the domain-specific subset. From an application SW designer's point of view, all cores will be heterogeneous, and they accelerate domain-specific functions and achieve high performance with low silicon cost. From a HW integration and system designer's point of view, all cores look like homogeneous cores because the instruction subset for general control functions and intercore control functions are the same in all cores. Eventually, heterogeneous instructions will not be visible to system designers during the system integration. The scalable core with homogeneous instruction subset is illustrated in Figure 20.9.

The concept of a homogeneous instruction set supporting heterogeneous computing is illustrated in Figure 20.9. The system design and integration is based on the homogeneous instruction subset. The applications engineers will use both the homogeneous instruction subset and heterogeneous instruction subsets for a class of applications.

An exciting by-product of the solution illustrated in Figure 20.9 is the low design cost. There are usually two embedded firmware design flows in all embedded systems: the MCU firmware design flow and the DSP firmware design flow. By using the

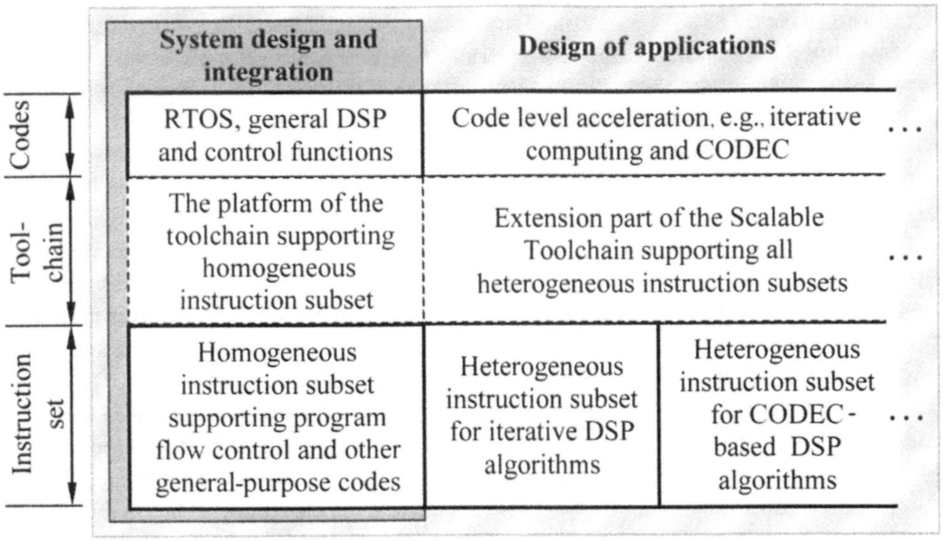

	System design and integration	Design of applications	
Codes	RTOS, general DSP and control functions	Code level acceleration, e.g., iterative computing and CODEC	...
Tool-chain	The platform of the toolchain supporting homogeneous instruction subset	Extension part of the Scalable Toolchain supporting all heterogeneous instruction subsets	...
Instruction set	Homogeneous instruction subset supporting program flow control and other general-purpose codes	Heterogeneous instruction subset for iterative DSP algorithms	Heterogeneous instruction subset for CODEC-based DSP algorithms ...

FIGURE 20.9

System view on homogeneous-heterogeneous cores.

solution given in Figure 20.9, two design flows are combined into one design flow based on the homogeneous instruction subset. A comprehensive MCU can be built based on the homogeneous instruction subset by adding sufficient instructions for string manipulation and peripheral modules.

Because of using the homogeneous instruction set as the kernel instruction set of all processors, the MCU and the DSP as well as all vector processors will be based on the same assembly instruction set kernel and use only one toolchain kernel; thus only one design flow is used. The firmware design cost can therefore be much lower.

Some heterogeneous instructions consume multiple execution cycles and are called multi-cycle instructions. Scheduling the execution of multi-cycle heterogeneous instructions could be a challenge. The same ways of scheduling used for SIMT were discussed previously in this chapter.

Multicore (SoC) Integration

In a multicore SoC, the homogeneous instruction subset is the same for all master cores. This general instruction subset is responsible for flow control and intercore communications. Program code based on the general instruction subset must be portable among master processors. Intercore communications must be conducted by codes of a general (homogeneous) instruction subset. Intercore communications include packing and unpacking data packages, setting up communications, and monitoring and terminating interprocessor communications. By using this general

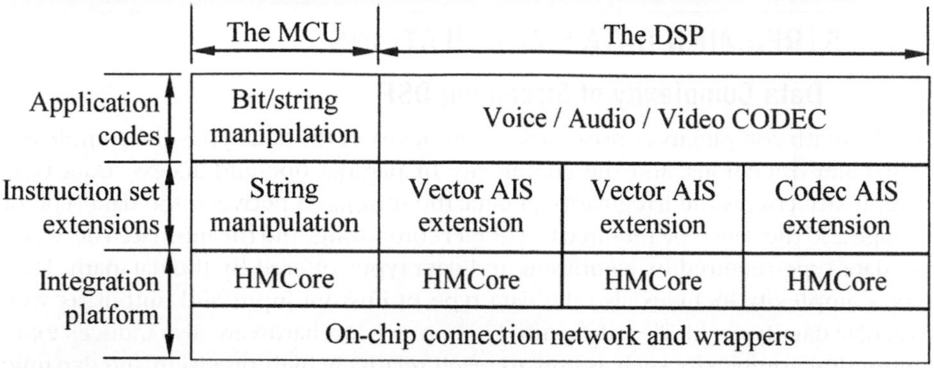

The MCU	The DSP		

	The MCU	The DSP		
Application codes	Bit/string manipulation	Voice / Audio / Video CODEC		
Instruction set extensions	String manipulation	Vector AIS extension	Vector AIS extension	Codec AIS extension
Integration platform	HMCore	HMCore	HMCore	HMCore
	On-chip connection network and wrappers			

FIGURE 20.10

Multicore integration.

instruction subset for interworking, the SoC integration of heterogeneous cores can be as easy as the SoC integration using homogeneous cores.

In Figure 20.10, "HMcore" stands for the processor core with the homogeneous instruction subset. Together with the on-chip connection network and wrappers (introduced in Chapter 19), the integration platform exposes homogeneous instructions to the system designers and integrators. The core with homogeneous instruction subset could also be used as a MCU, the microcontroller. In a MCU, the bit and string manipulation is one of the tasks for human–machine interface functions. Bit and string manipulation instructions can be a subset of heterogeneous extensions to the instruction set kernel. In this case, the MCU of the system is a HMcore with the homogeneous instruction subset and string manipulation extension.

Recall the SIMT concept introduced in the previous section—a SIMT consists of a HMcore (RISC datapath) and its heterogeneous instruction subset (SIMD datapath). The coding of an accelerated heterogeneous instruction follows the technique of tightly coupled acceleration illustrated in Chapter 17. It supplies heterogeneous features, yet these will not be exposed during the multicore integration phase. To system designers, this part is executed as accelerator functions or black box functions.

During the SoC integration, all assembly instruction set simulators of these cores must be integrated together. Since all simulators are based on the homogeneous instruction subset, integration of all simulators can be easy.

The **Senior** assembly instruction set (see the appendix) is designed as a typical homogeneous subset. Design for instruction extension and toolchain adaptation was carefully discussed in Chapter 17. So far, the conclusion of the section is to use the SIMT concept and a homogeneous instruction subset to simplify the control complexity. The data complexity can be managed by using heterogeneous instruction subsets.

20.4 STREAMING DATA MANIPULATIONS

20.4.1 Data Complexity of Streaming DSP

The datapath complexity consists of the complexity of the data types, the complexity of parallel operations, and the complexity of parallel operand access. Data type complexity covers the irregularity of data, the mismatch between the data type of storage and the data type required by parallel computing, and the mismatch between the data type required by algorithms and data types offered by the datapath. Data type complexity includes also the data type of custom input and output, as well irregular data types inside a datapath. Finite precision hardware also induces extra computing complexity such as how to reach relatively high precision and dynamic range at different computing stages using low-cost hardware.

The complexity of parallel operations and the complexity of parallel operand access will be discussed in this section.

20.4.2 Data Complexity: Case 1—Video

Concepts

Image and video compression is usually based on 2D signal processing techniques. Both lossy and lossless compression techniques are used for image and video compression. Both image and video compression techniques are illustrated by Figure 20.11 [12].

The shaded subset (with gray background) in Figure 20.11 is a simplified image compression flow, and the complete flow in Figure 20.11 is the video compression

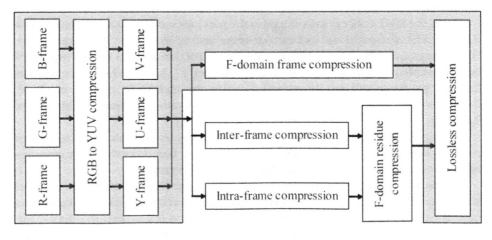

FIGURE 20.11

Image and video compression.

flow, including compression using inter- and intraframe references. The task flow of compression is from the left to the right, and the task flow of decompression is from the right to the left.

The compressed information can be recovered completely without distortion by using lossless compression. Lossless compression recognizes and codes redundancies from the source code. For example, in run-length encoding, in order to compress a series of 1 (or a series of 0), the length of the 1 string will replace the 1 string. Another example is to represent frequently appearing combinations with shorter code patterns (Huffman encoding). Lossless compression and decompression are implemented as CODEC FSM.

Lossy compression causes a loss of information that is negligible by human beings but contributes to a larger compression ratio. The first step of lossy compression for image and video compression is the color transform. Because human sensitivity to chrominance (color) is lower than the sensitivity to luminance (brightness), three original color components (R = red, G = green, and B = blue) can be translated to another three color components (Y = luminance, U and V = chrominance), where U and V frames can be decimated to one-quarter of the original size. Therefore, three full frames of RGB data can be compressed to $1 + 1/4 + 1/4 = 1.5$ frames of YUV data.

Frequency-domain compression is executed after RGB to YUV conversion in image compression. DCT (discrete cosine transform) is used to transfer the image from time-domain to frequency-domain. Since human beings are less sensitive to high spatial frequency details, the information changing fast (with high frequency) can be quantized with lower resolution. After the quantization, long chains of 0 and 1 will appear and will be compressed further by lossless compression. The classic JPEG (Joint Picture Expert Group), a compression standard for still images, can reach 20:1 compression ratio on average [13].

Video compression is the extension of image compression. Video compression is conducted by exploiting two types of redundancies: spatial and temporal. The compression of the first video frame is essentially the same as the compression of an independent image frame. After that, the first compressed frame can be used as the reference frame for further interframe compression. Interframe compression codes and transmits or stores the difference between frames. Several following frames can be compressed further because the differences between the reference frame and the current frames are usually small. Intraframe compression codes and transmits or stores the difference between neighboring pixels. Pixel values are correlated with their neighboring pixels. If part one of a frame is the same as part two in the same frame, part one can be replaced by part two and only one part needs to be stored or transferred.

The classic MPEG2 [4] (Moving Picture Expert Group) video compression standard can reach 50:1 compression ratio on average, and the advanced video codec such as H.264/AVC [5] can reach compression ratios of more than 100:1. Frequently used algorithms in video and image compression algorithms are listed in Table 20.3.

Table 20.3 Most Used 2D Algorithms for Video and Image Compression.

Algorithms	Mathematical presentation	Parallel features	Cost factor
Motion estimation	$\Sigma\lvert a_i - b_i \rvert$; a_i and b_i are positive values	Maximum parallel computing	~ 10
Discrete cosine transform	$F(u) = \frac{2c(u)}{N} \sum\limits_{m=0}^{N-1} f(m) \cos\left(\frac{(2m+1)u\pi}{2N}\right)$	Up to 4 butterflies in parallel	~ 3
Discrete wavelet transform	$\Sigma a(x(n) + x(-n))$	2N way parallel	High
Interpolation	Round $((a + bx_1 + cx_2 + dx_3 + ex_4) \gg N)$	5(6) way accumulative	~ 2
Deblocking	$y = ax_1 + bx_2 + cx_3 + dx_4 + ex_5$	5(6) way accumulative	~ 2
Transfer color space	New color $= ax + by + cz$	3 way accumulative	~ 3

Case Study on Video/Image Datapath

The common datapath features can be exposed after analysis of source codes of audio, video, and image CODEC. The purpose is to find opportunities for hardware reuse, to design for hardware multiplexing and merging algorithms for audio, video, and image signal processing into one parallel datapath. The datapath can be configured for processing high-definition audio signals at 32-bit resolution with certain parallel features; normal audio signal processing and high-end video/image signal processing at 16-bit resolution with certain parallel features; and low-cost video/image signal processing at 8-bit resolution with highly paralleled features.

Typical vector computing can be divided into two groups: multi-in and multi-out computing using massive parallel datapath for data parallel operation, shown in Figure 20.12(a); and multi-in single-out computing using accumulative parallel datapath, shown in Figure 20.12(b), for parallel reduction.

PU in Figure 20.12 is a processing unit; it can be a MAC unit or an ALU. The first datapath in Figure 20.12(a) is used mostly for transformation and vector computing. The second datapath in Figure 20.12(b) is used mostly for accumulative iterations. By merging two types of datapath into one circuit, we get Figure 20.13.

Figure 20.13 shows a typical four-way 16-bit vector datapath for video and audio signal processing. Eight inputs came from a register file. There are four multipliers and four full adders using the eight inputs. The pp1 to pp8 are eight preprocessing units. P-out1 to P-out4 in Figure 20.13 is the same as the P-out1 to P-out4 in Figure 20.12. By adding two 64-bit full adders and a 72-bit full adder to the multi-in multi-out parallel datapath, an accumulative datapath is formed. T-out is the output of the accumulative datapath. The 64-bit and 72-bit long adders consist of two

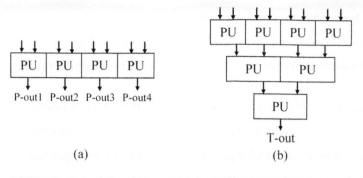

(a)   (b)

FIGURE 20.12

Two types of datapath.

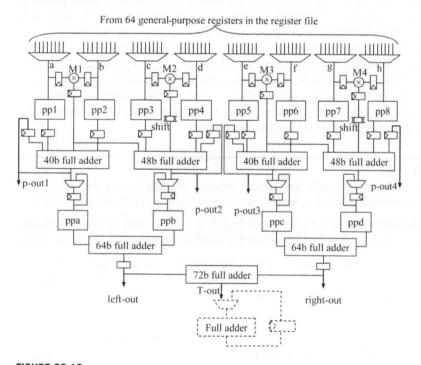

FIGURE 20.13

Datapath for video signal processing.

32-bit and 36-bit concatenated full adders, respectively. Concatenated full adders are linked via the carry bit. The small modules, pp1-pp8 and ppa-ppd, are modules for operand preprocessing. Operand preprocessing could be used to formulate inputs for subword computing (see Example 20.1).

The datapath can be configured into six modes:

- Eight-way 8-bit subword accumulative parallel mode (part of T-out).

- Eight-way 8-bit subword massive parallel mode (8 × 8-bit outputs, P-out1 to P-out4).

- Four-way 16-bit accumulative parallel mode (32-bit output, part of T-out).

- Four-way 16-bit massive parallel mode (4 × 16-bit outputs, P-out1 to P-out4).

- Two-way 32-bit double-word massive parallel mode (left-out and right-out).

- One-way 32-bit double-word accumulative parallel mode (T-out).

The first configuration is for an eight-way motion estimator with 8-bit precision. Eight 16-bit inputs will be split into 16 8-bit inputs when passing through eight pp units. Sixteen 8-bit inputs will be used for the motion estimation of eight pixels in a picture that is carried out within one clock cycle. In this case, the datapath will be an accumulative hardware generating output at the triangle port T-out.

When eight 16-bit inputs come to four MAC, four outputs, P-out1, P-out2, P-out3, and P-out4 are outputs of the massive parallel datapath. The outputs can come from the multiplier, the long arithmetic unit, or the MAC. When further merging P-out1, P-out2, P-out3, and P-out4, the output T-out could be the result of a filter for deblocking, interpolation, and wavelet filters. The T-out can also be the output of a 32b × 32b multiplication for audio signal processing.

Example 20.1

Mapping and configuring an eight-way parallel datapath for motion estimation using 8-bit subword and conditional computing technique on the four-way datapath in Figure 20.13.

Step 1: The algorithm is $\Sigma|a_i - b_i|$, where i is from 0 to 7. From each 16-bit operand port, two 8-bit data can be supplied, so that 8 of 16-bit inputs can carry 16 of 8-bit inputs.

Step 2: Insert two 0 bits between two 8-bits such that each 16-bit operand becomes an 18-bit operand in the concatenated format $\{a_{i+1}[7:0], \text{"00"}, a_i[7:0]\}$. Therefore, a 40-bit or a 48-bit full adder can be used as four of 10-bit full adder.

Step 3: Use the [9:0] of 40-bit full adder to get
$\{\text{"00"}, a_i[7:0]\} - \{\text{"00"}, b_i[7:0]\}$ and send result to out [7:0] if positive
Use the [19:10] of 40-bit full adder to get
$\{\text{"00"}, b_i[7:0]\} - \{\text{"00"}, a_i[7:0]\}$ and send result to out [7:0] if positive
Use the [29:20] of 40-bit full adder to get
$\{\text{"00"}, a_{i+1}[7:0]\} - \{\text{"00"}, b_{i+1}[7:0]\}$ and send result to out [27:20] if positive
Use the [39:30] of 40-bit full adder to get
$\{\text{"00"}, b_{i+1}[7:0]\} - \{\text{"00"}, a_{i+1}[7:0]\}$ and send result to out [37:30] if positive

In this way, the absolute operation of $|a - b|$ is executed by selecting a positive result of $(a - b)$ or $(b - a)$ as the output of $|a - b|$. At the result port of a 40-bit full adder, there will be two results of $S_{i+1} = |a_{i+1}[7:0]\} - b_{i+1}[7:0]|$ and $S_i = |a_i[7:0]\} - b_i[7:0]|$. Therefore, eight outputs of S_{i+7}, S_{i+6}, S_{i+5}, S_{i+4}, S_{i+3}, S_{i+2}, S_{i+1}, and S_i can be achieved at the same time from four 40-bit/48-bit full adders.

Step 4: Split the right-side 64-bit full adder into two 32-bit adders; four inputs, S_{i+3}, S_{i+2}, S_{i+1}, and S_i can be calculated $S_{32} = S_{i+3} + S_{i+2}$ and $S_{10} = S_{i+1} + S_i$. Using the same way, $S_{76} = S_{i+7} + S_{i+6}$ and $S_{54} = S_{i+5} + S_{i+4}$ can be achieved from the left 64-bit full adder.

Step 5: In the same way, the 72-bit full adder can be split into two 36-bit hardware adders. When working for motion estimation algorithm, the carry bit linking two 36-bit adders should be disconnected to separate two adders. The right 36-bit adder gets $S_{high} = S_{76} + S_{54}$ and $S_{low} = S_{32} + S_{10}$. The left 36-bit adder gets the final result $S = S_{high} + S_{low}$.

Example 20.1 demonstrated a method of splitting one computing device for multiple computing simultaneously. Speculative conditional computing was used to carry out the absolute arithmetic. Branches induced by arithmetic computing can therefore be avoided. If low-level branch operations within an inner loop can be managed inside the parallel computing datapath, the control complexity will be much lower. If the condition control of conditional execution cannot be managed inside the parallel computing datapath, it has to be managed by the master core, and partition with fine granularity will not be avoided. Extra synchronizations and communications will induce significant overhead.

To minimize the overhead of fine-grained communications between the master core and the parallel computing datapath, the basic way is to design the vector processor handling conditional executions. This is efficient enough because data dependency involved in vector computing is relatively simple and localized. A good example of conditional execution was given in Example 20.1.

Example 20.2 demonstrates using a parallel datapath for computing with higher precision.

Example 20.2

To emulate multiplication $V[31:0] \times U[31:0]$ using the datapath in Figure 20.13 based on four 17-bits- multipliers and accumulators.

The principle of the example was described in Chapter 7.

Step 1: Assign operands to multipliers in Figure 20.13:

```
as = {V[31], V[31:16]};
bs = {U[31], U[31:16]};
au = {1'b0, V[15:0]};
bu = {1'b0, U[15:0]};
```

Step 2:
```
ppa <= MSS <= as * bs;
ppb <= MUS <= au * bs;
ppc <= MSU <= as * bu;
ppd <= MUU <= au * bu;
```

Step 3: `leftout  <= ppa + ppb>>16;`
        `rightout <= ppc + ppd>>16.`

Step 4: `Tout <= leftout + rightout >> 16;`

The emulation algorithm behind Example 20.2 was discussed in Chapters 7 and 13.

The accumulative circuit is represented using dashed line in Figure 20.13. The accumulative circuit is very useful when running high-order filters and high-order Taylor series. For example, 8-order Taylor series can be executed using two instructions. The first instruction stores the temporal result in the accumulator register and the final result is the sum of the data in the accumulated register and the data at the output port of T-out.

20.4.3 Data Complexity: Case 2—Radio Baseband

Another typical example is the datapath for programmable radio baseband signal processing. A classic radio baseband modem is a nonprogrammable ASIC module. Multiple ASIC modems have to be integrated into one single chip if multiple radio standards are needed in a mobile terminal. For example, GSM/GEPRS, WCDMA, WLAN (IEEE802.11g/b), WiMAX, GPS, and DVB may need to be integrated into a modern cellular phone. The multi-ASIC solution consumes significantly more silicon area. A preferred solution is to accommodate all standards by having one programmable baseband ASIP [2].

Following the design methodology discussed in Chapter 4, careful profiling of IEEE802.11a/g was conducted in Chapter 6. Profiling scores of other standards are also available at [14]. Streaming signal processing for radio baseband can be classified into five categories: classical analog radio (FM, VHF-TV), single carrier digital radio (GSM, Bluetooth), OFDM digital radio, CDMA digital radio, and MIMO enhancement for OFDM or CDMA. The computing requirements of these five groups of radio baseband systems are given in Figure 20.14.

To further analysis, all radio streaming signal processing can be divided into three classes of processing tasks: digital filters with low complexity and ultra high computing demand; the symbol processing with relative high complexity and relative high computing demand; and the bit-level manipulation with simple parallel processing and parallel memory access features. To summarize, a formalized task graph of a multimode digital radio baseband ASIP is depicted in Figure 20.15.

Figure 20.14 gives both the computing cost estimation and the list of essential algorithms used for radio baseband signal processing. To group these algorithms, a formal task flow in Figure 20.15 shows the streaming flow mapping to the architecture of the baseband ASIP. The horizontal streams are:

a) From left to right: Receiving and channel estimation.

b) From right to left Transmission streaming.

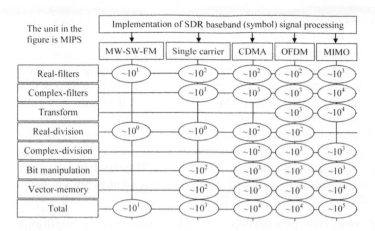

FIGURE 20.14

Computing cost estimation of radio baseband signal processing.

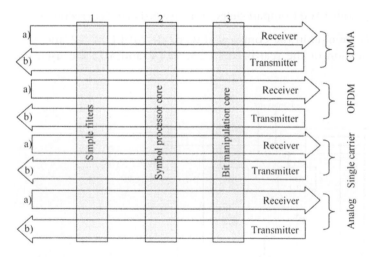

FIGURE 20.15

Task flows in a programmable radio baseband ASIP.

The vertical gray bars represent three heterogeneous processors for:

1. Simple filtering and correlations: The digital front end processor.

2. Complex symbol processing: The symbol and transformation processor.

3. Massive parallel subword signal processing: Bit manipulation processor.

The challenge is to design a symbol processor to reach both high performance and high flexibility. The function coverage of the symbol processor core in Figure 20.15 includes filters of real data, filters of complex data, complex transformation, interpolation based on complex data, and vector arithmetic operations. The program flow control of baseband signal processing can be managed using RISC datapath. Vector and iterative computing can be managed using SIMD datapath handing complex data formats. It is proven that the SIMT concept also is suitable for radio baseband signal processing. One of its parallel datapaths for radio baseband in the SIMT radio baseband signal processor is depicted in Figure 20.16.

Operands A (real AR and imaginary AI) and B (real BR and imaginary BI) are data inputs of FFT. Operand C (real CR and imaginary CI) represents coefficients. Operands A and C are also inputs of complex convolutions. The baseband datapath in Figure 20.16 should be configured to four datapath modes following the listed algorithms in Figure 20.14:

1. Complex MAC mode (complex data filter and correlation, vector computing).

2. Complex FFT (Radix-2 DIT mode FFT transform).

3. Four-way real MAC mode (Normal Filters in parallel).

4. Eight-way ACS mode (Add-Compare-Select for forward error correction).

When the datapath is running in the complex MAC mode, MUL1, MUL2, and AD1 are used for multiplication of the real part; MUL3, MUL4, and AD2 are used for

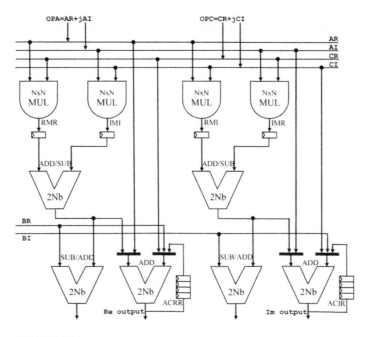

FIGURE 20.16

Modified complex MAC supporting FFT.

multiplication of the imaginary part. AD4 and AD6 are used for accumulation of the real part and the imaginary part, and the output of the complex MAC is O1R + jO1I.

DIT, Decimation-in-time butterfly algorithm, was discussed in Chapters 1 and 7. It is implemented in Figure 20.16 as:

```
(O1R + jO1I) = (BR + jBI) + (CR + jCI)*(AR + jAI)
(O2R + jO2I) = (BR + jBI) - (CR + jCI)*(AR + jAI).
```

Here O1R is the real part of the output-1; O1I is the imaginary part of outpt-1. The O2R is the real part of the output-2; O2I is the imaginary part of outpt-2.

When sampling and receiving analog radio signal, heavy decimation and band pass filtering should be executed. The datapath should be used as a four-way parallel MAC for real data computing. The four-way MAC outputs will come from O2R, O1R, O2I, and O1I.

20.5 NOC FOR PARALLEL MEMORY ACCESS

20.5.1 Design Methods

Computing in the vector datapath discussed in the previous section was based on an assumption of enough vector data being available in a vector register file. This is actually very difficult so that in most cases the parallel datapath is not efficiently used due to the memory access latency. To supply data in parallel with minimum latency, both connection networks and addressing for memory access will be designed.

Data is allocated in the main memory or other on-chip vector memories. Data should be supplied in parallel for parallel computing. However, the off-chip memory can supply only one addressable data word during one memory access. Sometimes, data can be accessed in parallel from the on-chip memory to the register file. Sometimes the memory access cannot be in parallel because of memory access mismatch or address conflict. In this section, memory access challenge in detail will be discussed at the starting point depicted in Figure 20.17.

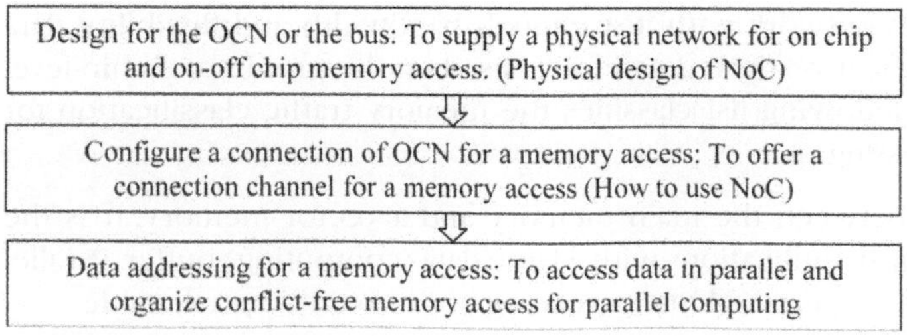

Design for the OCN or the bus: To supply a physical network for on chip and on-off chip memory access. (Physical design of NoC)

⇩

Configure a connection of OCN for a memory access: To offer a connection channel for a memory access (How to use NoC)

⇩

Data addressing for a memory access: To access data in parallel and organize conflict-free memory access for parallel computing

FIGURE 20.17

Three steps for the design of parallel memory access.

To partition the complexity of parallel memory access, design for parallel memory access should be divided into three steps. The first step in Figure 20.17 is to design a network on chip (NoC) according to the parallel architecture and traffic analysis between processors and memories. The second step is to design ways to configure the NoC and use it for each data transaction. The first two steps offer a physical connection channel between the data source and the destination. Finally, the third step is to design for conflict-free parallel addressing for an algorithm to be executed. Here conflict-free stands for the memory access of all data words in parallel from different physical memory blocks. To reach conflict-free access, the number of physical memory blocks must be equal or more than the number of parallel memory accesses.

To transfer data for parallel computing, a physical medium must be designed first, and the medium must be configured for each data (packet) transaction. To access data, two connection channels should be supplied for memory access. One is the NoC (the on-chip connection network discussed in Chapter 19) between memories and processor cores on the chip. Another is the network inside a core with parallel datapath, including the interface to the main memory or NoC and the memory interface to the datapath (the register file), to be discussed in this chapter.

A configured NoC supplies only connection channels; it does not offer the method of parallel memory access to adapt to the execution of algorithms in parallel. While running an algorithm in a parallel datapath, specific data sets and memory access patterns are required. Different algorithms require different data sets and memory access patterns.

The challenge of the design of no-conflict parallel memory accesses is to find data in the main off-chip memory and to allocate and supply the fetched data to the on-chip memories for parallel algorithms to be executed. It is based on an assumption that the connection network is available.

20.5.2 Analyses of Parallel Memory Access for NoC Design

Several NoC topologies were introduced in Chapter 19. In this chapter, the way to select one of an NoC topology will be discussed based on the analyses of parallel memory accesses. Design of NoC is based on analysis (the profiling) of chip-level communications between selected cores and memories. To design a low-cost and low-latency connection network with just enough bandwidth and flexibility, data traffic models need to be investigated carefully based on the profiling of chip-level communications. The following list classifies the memory traffic classification for streaming signal processing:

- Memory access between the main memory and a vector memory: It is the main traffic for most applications with a large data (computing) buffer. Parallel memory access is required. Memory access patterns can be predictable.

- Memory access between the main memory and data memories in RISC DSP: Miscellaneous data transmissions are required. Parallel-memory access usually

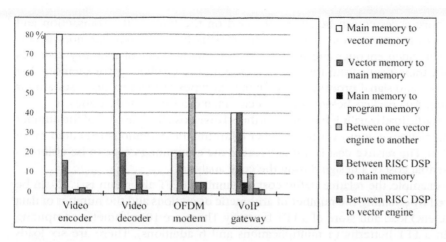

FIGURE 20.18

Relative costs of memory accesses in different applications.[1]

is not required. Memory access patterns may not be predictable. This kind of memory access should thus be as little as possible.

- Data vector from one vector engine directly to another vector engine: It is the typical delivery of streaming data. The output of the previous computing engine is the input of the following computing engine. The access is predictable. The latency should be minimized.

- Short data packets delivery between the RISC DSP and vector cores: A short data packet usually carries control messages. A shortcut channel (point-to-point link) might be preferred.

- Load program-to-program memory from the main memory: It is predictable memory access with regular and continuous addressing patterns. However, the access usually requires very low access latency.

- Load data to the main memory or the vector memory from IN-ports.

- Send data from the main memory or the vector memory to NoC-ports.

Memory accesses from industrial design activities were analyzed (profiled) and summarized, and can be found in Figure 20.18. In this figure, a vector memory is a group of on-chip memory blocks. Each memory block can be accessed once in a clock cycle. With multiple memory blocks, a vector memory can supply parallel memory accesses.

---

[1] The data access profiling of video encoder and decoder is based on the assumptions that the size of the on-chip memories are much smaller than a frame size 1920 × 1080. The OFDM modem is based on IEEE802.11a/g transceiver, and the voice over IP gateway is based on the assumption that the on-chip data memory size is much less than the requirements from multi-channel echo cancellers.

Memory accesses can be analyzed based on the type of the transaction and the load of the transaction. Different applications have different memory access types. For video encoder and decoder, the dominant types of memory accesses are DMA transactions from the main memory to the on-chip vector memory. For radio baseband signal processing, the dominant types of memory accesses are for streaming data packets between local vector memories of vector engines.

The traffic load usually is measured as the relative cost. It is the ratio of the number of memory transactions over the number of arithmetic operations executed by an algorithm. The relative cost is useful because the computing cost can be visible as the essential cost of a design during the code analysis.

For example, the relative traffic cost of running a FFT algorithm is 1. It can be observed by counting the number of arithmetic operations and the number of data (coefficient) load and store of a FFT butterfly. There are 10 arithmetic computing steps in a FFT butterfly (4 multiplications and 6 additions). There are six loads (2 complex input data and 1 complex coefficient data) and four stores (2 complex data). However, the relative traffic cost is not high while running convolutions. There are $2/K$ data transactions (one input and one output) where K is the number of iterations of the inner loop. If the order of the filter is high, the memory access load will be low.

The relative traffic cost also depends on the register file size. If the register file is large enough, the memory access load will be decreased significantly. For example, if the data packet size of a FFT is less than one-third of the register file size, all data and coefficients can be buffered in the register file during FFT computing. The memory cost will be only $3N$, the cost of the load in data (N) and coefficient (N) as well the store of the result (N), where N is the FFT size. The total computing cost of a N point FFT is $(N/2) \times 10\log_2 N$. The relative traffic cost is decreased from 1 (the cost of memory access is equal to the cost of computing) to $3N/[(N/2) \times 10\log_2 N] = [3/(5\log_2 N)]$. (The memory access cost is much less than the cost of computing.)

Obviously, different applications require different connection networks. According to the analyses of the type of the transaction and the load of the transaction, physical circuits of NoC can be selected.

For video applications, the NoC supplies sufficient bandwidth and minimum latency between the main memory and the vector memories in several vector computing engines. Therefore, a ring network with round-robin-based scheduling is preferred. Video processors for ultra-large resolutions may require multiple main memory channels. A double-ring NoC with double main memories might be required. For radio baseband signal processing, the main algorithm might be a large-size FFT. Multiple parallel connections between multiple vector memories and multiple parallel computing engines are required. In this case, a two-dimensional connection network (2D mesh) is preferred. There is no general on-chip connection network for embedded computing. NoC for embedded systems is always a custom- and experience-based design.

20.6 PARALLEL MEMORY ARCHITECTURE

Both the parallel memory architectures and methods of parallel memory addressing will be discussed in this section.

20.6.1 Requirements for Parallel Algorithms

As soon as the connection network is configured and available, the memory access will be prepared for the parallel algorithms to be executed. The memory access is required by the algorithm to be executed. The faster the access, the lower the computing latency induced by memory access. The main latency usually is induced by the main memory access.

Only one word in one main memory can be accessed in one clock cycle, not in parallel. The main memory access channel is usually the bottleneck. A question might pop up: why not simply design a wide memory and access data in parallel? The answer is, except for an extremely wide memory, a normal off-chip memory cannot sufficiently support parallel computing because of the following reasons:

1. Parallel data needed by an algorithm might be scattered around everywhere at different locations in the main memory. It is always a challenge to collect volume data from different locations in the main memory in a very short time. Principally, it is difficult to collect data allocated at different locations in a huge off-chip memory because a DMA control packer (see Chapter 16) carries only the start address and the length of a data block.

2. There exist multiple parallel algorithms in one application. Some require memory access in one way and some in other ways. For example, one algorithm may require parallel memory access of each row of a matrix, another algorithm requires parallel memory access of each column, yet a third algorithm requires diagonal parallel memory access for matrix computing.

For video and image signal processing, different parallel algorithms require different ways of parallel addressing. For example, some video applications require 8×8 pixel values to be accessed in parallel in a row, and some in a column. Typical ways of parallel memory accesses can be found in Figure 20.19. A vector register file supplies multiple data in parallel because there is no limit on the number of output ports from a register file. A vector register file can thus support all addressing requirements in Figure 20.19 if the size of the register file is large enough.

For radio baseband signal processing, multiple data should be accessed in parallel to speed up computing and minimize computing latency. For example, memory access for a radix-4 butterfly of a FFT algorithm should be in parallel. In each clock cycle, eight complex input data and four complex coefficient data should be loaded, and eight complex output data should be stored for computing a radix-4 butterfly.

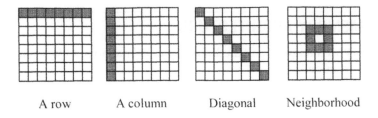

A row   A column   Diagonal   Neighborhood

FIGURE 20.19

Simultaneous addressing and access of multiple data.

If the size N of the FFT is large, it will be difficult to load eight data at the relative positions at $0, N/4 - 1, N/2 - 1, 3N/4 - 1$; and $1, N/4, N/2, 3N/4$ in parallel.

One arithmetic operation usually requires two operands. An eight-way datapath needs 16 operands simultaneously, eight data and eight coefficients. This will never be a problem if the size of a register file is infinite, such that all data can be pre-allocated in the register file. However, in reality, a real register file can never be large enough due to the limitation of silicon cost. The register file therefore has to be used in an extremely efficient way. For example, if only a part of a DU (data unit of a picture frame, a DU usually contains 8×8 pixels) will be used by the algorithm to be executed, it is sufficient to load only the part of DU needed by the algorithm to the register file, instead of loading a complete DU to the register file.

Rather often, distant data units scattered around in the main memory will be loaded sequentially by an algorithm. A typical case is to load several data units from three picture frames (the current frame and two reference frames). The distance between data units might be up to millions of pixels (addressing distance). Questions are:

- How to pick up different pieces of data from different places in the main memory and link them into one data transaction?

- How to support conflict-free access while executing an algorithm and using data from the vector memory? Or how to distribute/allocate data to a vector memory while receiving data from a DMA transaction?

- How to keep the register file most efficiently utilized?

In the following text, different solutions will be discussed.

20.6.2 Cache

Since direct main memory access is slow and the access latency is high, a cache as a main memory buffer often is used to store and supply most used data to general-purpose processors with short access time.

A typical and simple (fully associative) cache is shown in Figure 20.20 and Table 20.4.

Table 20.4   Understanding Cache Parameters.

Parameters	Cache	Main memory
Memory	SRAM with tag, matching circuit, and status	SRAM or DRAM or flash memory
Size	Few kilowords	Giga words
Addressing space	For all data in the main memory	For all data in the main memory
Access time (hit)	Usually 1–4 clock cycles	——
Access time (miss)	10–100 clock cycles	10–100 clock cycles
Miss rate	1% to 20% Miss rate is high for streaming DSP	——

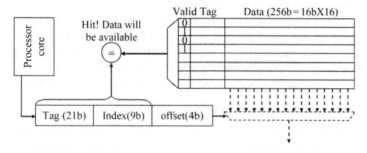

FIGURE 20.20

Schematic of a cache.

The example in Figure 20.20 contains 512 cache lines, and each cache line consists of 256 bits or 16 words. The cache size is 8192 words or 131072 bits. The main memory address space is 32 bits, which can be addressed up to 4 G words. The processor core provides a 32-bit address to the cache. The cache matches the 28 MSB (most significant bits) address bits with the tag. If it matches, a cache line with 16 data words will be exposed to the processor core. The four lower LSB (least significant bits) address bits are used as the offset to select the word in a cache line. The processor core can take one or to up to 16 words from the cache line to the register file according to the architecture of the processor. If data words required by addressing are allocated with continuous addresses, up to 16 words can be supplied by a cache line for parallel computing. Unfortunately, data words required by many parallel algorithms are not allocated adjacently.

When loading one word from a cache, the word can be loaded to the register file from the cache when it is available in the cache (match). One to four clock cycles are used for the loading. If the word does not exist in the cache, a cache miss is reported to the main memory and a cache line containing the data will be loaded from the main memory. Up to a hundred clock cycles might be required to load a cache line from the main memory.

For vector signal processing, multiple data must be supplied to the parallel data-path simultaneously. For example, an eight-way datapath requires 16 data words in each clock cycle. Therefore it is preferred that the cache example in Figure 20.20 supplies 16 words in parallel for an eight-way datapath. The requirement is tough for a cache because the cache architecture originally was not designed for parallel access. The requirement is actually tougher because reusability of video data is not high and most data words are possibly used only once. In other words, only a small part in a cache line will be used by the current parallel operations; the majority part will not be used.

For example, to run the deblocking algorithm for H.264 video decoding [5], the pixel values close to the boundaries of neighboring macro-blocks will be the operands and the pixel values at the central part of the macro-block will not be used. The data addresses required for deblocking operations actually do not continue in the main memory. A cache line contains both the boundary data and inner data of a macro-block. As a result, less than 50% of the data words in a cache line will be used. That is why most cache for video signal processing is actually the cache with access prediction table (in PC processors or in general-purpose DSP processors) or even the scratchpad memory (in custom ASIC solutions) designed to support predictable addressing.

One way to reach high efficiency of memory access for the deblocking algorithm is to preorganize data in the main memory in a special way for the deblocking algorithm. The inner data words of each macro-block are removed. The data read in each cache line will be completely used while running the deblocking algorithm. It is not realistic because the data stored in the main memory will be required and used by other algorithms as well.

A cache is designed for memory access with strong temporal locality. However, streaming data and stream signal processing have very limited or, sometimes, even no temporal locality because the video data reuse rate in cache is very low. The cache miss rate is therefore very high if there is no predictable cache fetching.

Usually for video and image signal processing, a cache supports tasks for lossless codec, program flow control, and vector processing. Temporal locality can be identified from memory accessing of program flow control tasks. But the memory accessing of video vector signal processing does not have temporal locality. The requirements on cache by vector processing and flow control as well as the lossless codec are not consistent.

Cache performance varies dramatically as the picture size changes. For example, register-cache-memory architecture for video and image processing of QCIF (Quarter Common Intermediate Format with 176×144 pixels in a frame) is completely different from the register-cache-memory architecture for video and image processing based on the picture size of 1920 $\times$ 1080—the larger the picture size, the more the cache miss.

Cache performance also varies dramatically with the changes of the vector datapath. Comparing the cache behavior of a two-way datapath (the width of a datapath, two datapath modules or clusters working in parallel) with that of a four-way datapath, the cache efficiency is much different, and the cache miss rate of a four-way datapath is much higher than that of a two-way datapath.

Research shows that multimedia applications get better performance when the cache block size is large enough and the cache line is long enough. For example, a long cache line of 128 bytes is proposed by research. Video cache is not as silicon efficient as other caches. For image and video streaming signal processing, especially when the picture size is large, the cache miss rate can be up to 20%, and about half the CPU runtime is wasted because of memory stall.

To conclude, classical cache is not suitable for parallel video signal processing or other streaming signal processing when the sizes of data packets are large and multiple data must be accessed simultaneously. To improve the cache behavior, predictable fetch is required. But this means that the advantages of cache have disappeared. Remember, the meaning of cache in French is to hide (to hide the complexity of memory access). Predictable fetch means to expose! If predictable fetch is required, scratchpad memory will be better because of its better silicon efficiency. Still, most researches in multimedia memory subsystem designs are based on an available processor with predefined cache architecture. While designing an ASIP for video signal processing, a cache-based architecture is not preferred.

20.6.3 Ultra-large Register File

Obviously, a normal cache is not a suitable solution for parallel streaming signal processing. If cache is not used, the first impact will be the complexity of programming. Cache provides a good way to hide the complexity of memory access. If there is no cache, the complexity will be exposed directly to programmers. Therefore, current problems are:

- What kind of memory architecture can replace cache architecture and efficiently support parallel streaming signal processing?

- How do you program on it?

One solution is to use an ultra-large register file. A register file can supply many operands simultaneously. Suppose that a register file is sufficiently large regardless of the silicon cost, the number of input ports, and the number of output ports. One vector instruction can therefore load multiple operands from the register file and write multiple data to the register file. The parallel access of one row, one column, a diagonal, and neighbor values of a pixel will be easy [11].

For example, the Imagine processor of Stanford University has a 256-Kbyte stream register file (SRF). In each ALU, there are distributed register files. These distributed register files can efficiently support multiple functional units running in parallel. Imagine uses a Stream Register File (SRF) instead of a cache. The SRF allows explicitly loading and storing of multiple data words simultaneously. The datapath in the Imagine processor can therefore be very well utilized.

The register file size of Storm-1 processor from Stream Processors Inc. is 256k Bytes or 2 megabits [31]. There is a $128 \times 128b$ register file in each (of eight) SIMD vector processor of STI (Sony, Toshiba, and IBM) CELL [8, 9]. It supplies multiple data words to the configurable SIMD datapath. The register file is relatively large, though it is not yet large enough for all applications. CELL has a 256 kB on-chip scratchpad memory in each SIMD slave machine. There is no cache in CELL, and the complexity of memory access cannot be hidden. The memory access predictability and DMA operations are thus essential for parallel computing [15, 16].

It is a challenge to handle write dependency while writing multiple data to a register file. The cluster register file architecture introduced in Chapter 3 should be used. A datapath and its local register file consist of a cluster. Each datapath only writes to its own register file and reads data from all registers in all clusters.

The advantages of using an ultra-large register file are the high computing performance and the low cost of parallel programming. The drawback is the high silicon cost and high-power consumption. Usually ultra-large register file architectures are avoided in high-volume products for low-power applications.

Ultra-large register files cannot manage the low locality feature of media data and other streaming data. For example, for motion estimation of a large picture frame, the amount of pixels involved could be up to megabytes. An even worse example is the H.264 motion compensation algorithm for HDTV. The addressing distance could be several million pixels between the pixels in the current frame and the pixels in the reference frame. The conclusion is that the size of the register file will seldom be enough for parallel video signal processing. When the addressing space of data to be processed is larger than the size of the register file, parallel memory access techniques are required.

20.7 P3RMA FOR STREAMING DSP PROCESSORS

P3RMA, Programmable Parallel Memory Architecture for Predictable Random Memory Access, is a scratchpad memory-based solution supplying parallel data for parallel datapath clusters. P3RMA includes hardware architecture, methodologies, and tools for embedded parallel programming [20, 21, 32].

As mentioned previously, a parallel datapath can be easily designed and proposed. However, a parallel datapath can never be efficient because the memory behind the datapath can seldom supply the right data at the right time. Instead of the datapath, the memory bandwidth and latency problem has been the major bottleneck of streaming signal processing in modern embedded systems. This is the main challenge for applications such as media (video, image) signal processing and radio baseband signal processing. This is also the main challenge for other streaming signal processing such as the baseband signal processing of advanced digital radar, real-time image processing of computer tomography, geography signal processing, and accelerations of scientific computing.

20.7.1 Parallel Vector (Scratchpad) Memories

It was discussed in the previous section that cache is not the preferred solution for vector datapath. A popular way is to use an ultra-large register file instead of a cache. If the register file is large enough and carries the complete data set for vector computing, the datapath can be used efficiently. However, for low-silicon cost and low-power applications, the ultra-large register file architecture is not feasible. Yet another hardware architecture, parallel vector (scratchpad) memory (PVM) architecture, can be used. Except for the parallel datapath, the architecture also consists of:

- A small or medium-sized multicluster register file to accept and supply multiple data in parallel.

- A wide bus and multiple read-write-ports between the parallel scratchpad vector memory and the vector register file.

- A scratchpad memory with multiple memory blocks connected in parallel.

- A main memory, preferably with simultaneously accessible multiple physical memory blocks.

- A flexible permutation hardware supporting parallel addressing of all scratchpad memory blocks.

- Most importantly, a high, quality programmer's toolchain and programming methodology to utilize the hardware.

The architecture of PVM is depicted in Figure 20.21. Comparing PVM architecture to others, low-silicon cost is the advantage and the design complexity is the drawback. On-chip PVM consists of several parallel (small) physical memory blocks, and each of them can be accessed independently. If there are eight memory blocks, there will be a possibility to access up to eight data words in parallel. In such a way, the accessing of parallel data with minimum latency is achievable.

An example of a scratchpad vector memory is marked with the dashed box in Figure 20.21. The memory consists of eight physical blocks; from each, 16-bit data can be accessed at a time. The data width of the vector memory can be from 16 bits up to 128 bits. A 16-bit data on the input bus at the left side of the permutation hardware can be connected to any data memory block via the permutation hardware. The permutation hardware can connect any input data words to any one of eight memory blocks. The 128-bit output from the vector memory consists of 8×16-bit data to the vector register file. The vector register file can also write 8×16-bit data to the vector memory.

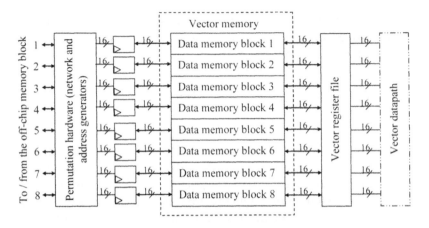

FIGURE 20.21

A 128 b wide 8-way PVM and its surroundings.

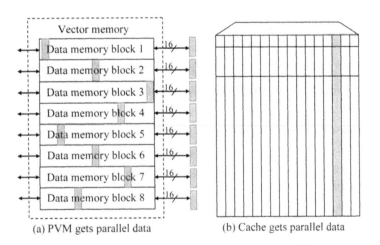

(a) PVM gets parallel data   (b) Cache gets parallel data

FIGURE 20.22

The advantage of a PVM over a cache.

A memory block in the vector memory can be accessed only once during one clock cycle. If eight data are loaded to the register file and all these eight data are unfortunately allocated in one physical memory, they have to be accessed in eight cycles, one value in each cycle. The intuitive conclusion is that data to be accessed in parallel must be allocated in different memory blocks to make the parallel access possible. This also implies that the data to be addressed must be predictable so that the vector data can be statically allocated and scheduled for parallel accesses.

Comparing PVM with a cache, the advantages can be found in Figure 20.22. A cache supplies parallel data from one cache line. If the parallel data words are not organized within a cache line, more clock cycles will be required. The advantage of accessing data from PVM is the possibility of simultaneously accessing multiple data from different places in several memory blocks.

Streaming data and streaming signal processing have enough static features (predictability). In the following sections, the methodology of PVM-based programming will be discussed in detail based on the assumption that the memory access is predictable.

20.7.2 The Memory Subsystem Hardware

The PVM-based memory subsystem for SIMT architecture is given in Figure 20.23.

In Figure 20.23, the RISC DSP core is the DSP task manager. Each dashed box covers a vector engine. The vector engine can be any type of hardware in parallel, a SIMD datapath (in an ILP system), or a slave SIMD processor (in a TLP, Task Level Parallel, system). A DMA linking table (discussed in Chapter 16) is issued by the RISC DSP. The linking table concatenates and merges several data blocks in any location in the main memory into one DMA transaction. The table could be constructed before or during execution time according to the requirements of the algorithm and data allocations in the main memory. The DMA linking table consists of a chain of memory accesses; each memory access is specified with its starting address and its data block length. All data blocks of a DMA transaction will finally be concatenated as a DMA data packet of the DMA transaction (see DMA in Chapter 16).

A permutation table consists of a table of target addresses for all data carried by the DMA transaction from the main memory to the PVM. All data to be used simultaneously must be allocated to different memory blocks in the PVM. A permutation table could be designed before or during runtime. The design of a permutation table is based on the parallel algorithm, the specification of the DMA linking table, the hardware structure of the PVM, and the currently available memory space in PVM. A permutation table is issued to the vector engine by the RISC DSP core before its DMA transaction. During the DMA transaction, the permutation table is synchronized and used by the DMA transaction.

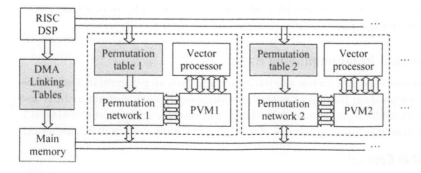

FIGURE 20.23

Scratchpad data memory subsystem.

As mentioned before in this chapter, one RISC DSP core can drive several vector engines. Two vector engines can be found in Figure 20.23. The RISC DSP core must know exactly what is to be executed in each vector engine. The RISC DSP core prepares and assigns tasks to each vector engine. A task to a vector engine consists of five parts:

- The task code (program) entry or the task code (if the task is short enough).
- The DMA linking tables for the main memory access.
- The permutation tables for the vector memory access.
- The trigger to start the task running in the vector engine.
- The execution of DMA transfer in parallel prefetching vector data for the parallel task.

To supply data for a program to be executed in a vector engine, two tables will be prepared, the DMA linking table and the permutation table. The DMA linking table specifies the fetching of n groups of data in the main memory, links them, and sends data to the permutation network of the vector engine. When the RISC DSP core prepares a DMA transaction task, it always makes another table—a table to the permutation hardware in the vector engine. While loading data from the main memory, the permutation engine will distribute this data word to a memory block of the vector memory and store it at a predefined position. The permutation engine is controlled by the vector engine, and writing to the PVM is synchronized by the DMA transaction.

There are two steps of data reorganization during a DMA transaction. The first step happens during the data reading from the main memory configured by the DMA linking table, and the second step happens during the data writing to the vector data memory. In the first step, several data blocks from any place in the main memory with different block length are linked together, comprising a DMA transaction. The data distributed in the main memory will be collected and merged into one DMA transaction. In the second step, each word can be written to any memory block in the vector memory PVM following the permutation table. By implementing the vector memory access into two tables, any kind of predictable memory access in parallel can be manageable.

20.7.3 Parallel Programming by Hand

The next discussion is how to use the hardware and how to program it. The discussion will be focused on how to expose the addressing behavior of algorithms and how to map the addressing to two tables (DMA linking table and permutation table). Finally, the integration of two tables to source and assembly code will be discussed.

The Method in General

The goal of parallel programming is to minimize the memory access time by maximizing the bandwidth of useful memory access. A good program running on a good

1. Formulate the physical behavior model of the PVM
2. Formulate the physical behavior model of the main memory
3. Modeling, partitioning, and mapping algorithms to parallel datapath
4. Modeling the addressing behavior for each parallel algorithm
5. Define the relation between behavior and physical addressing
6. Translate the behavior to physical addresses in the main memory
7. Implement DMA transactions using linking tables
8. Select or design corresponding permutation algorithm for each DMA
9. Implement permutation tables (for conflict-free PVM addresses)
10. Coding for parallel data loading from PVM to the register file

FIGURE 20.24

Programming parallel algorithms based on PVM architecture.

parallel architecture hides most (or even all) of the memory access cycles. To hide as much memory access as possible, DMA should be extensively used and the on-chip PVM should be efficiently utilized. Carefully scheduling the algorithm execution and the data supply will minimize the execution time and maximize the memory efficiency. The method of manual parallel programming for P3RMA based on PVM architecture is listed in the steps shown in Figure 20.24.

The first two steps in the flow depicted in Figure 20.24 give general modeling of a parallel architecture using PVM. The first two steps are independent from applications. Algorithms are analyzed, parallelized, and mapped to the parallel datapath at step 3. The parallel memory accessing thus is exposed according to the execution of algorithms.

A data array as local variables is used by an algorithm allocated to datapath modules. Loading and storing data to and from the data array will be exposed as memory accesses. If the corresponding data is allocated in the main memory, the DMA transaction between the main memory and the local PVM will be identified. If the corresponding data is allocated in another PVM, the DMA transaction between two local PVM in two vector engines will be identified. For example, the original behavior code is:

```
Main /* original */
{
  ...
  CALL Subroutine ALG /* using Array_A directly as global variable
  */
  ...
}
```

In the main code, the Array_A used in ALG is a global variable. The ALG is not allocated to any hardware module. The code of ALG only exposes the behavior function of it. After allocating ALG to a hardware module, ALG will be executed locally in the hardware using the local memory with limited size. Because the data for ALG was in the main memory before executing ALG, DMA subroutines are required to move data from the main memory to the local memory. If the size of Array_A is less than the local memory size, all data can be moved using one DMA transaction. If the Array_A is larger than the local memory size, multiple DMA transactions will be needed while executing ALG in the local hardware. The rewritten parallel behavior code will be:

```
/*
Re-write C model with coarse granularity parallel execution */
Main /* parallel */ {
  ...
  ALG /* Call subroutine ALU */
  ...
}
ALG
{/*Subroutine ALG in HW module*/
  Int Array_A (m, n) /* isolate array A as local variable */
  ...
  DMA_Array_A(start_address_1, length_1) /* Call DMA */
  /* and load data block 1 to Array_A */
  /* executing part of the subroutine */
  ...
  DMA_Array_A(start_address_2, length_2) /* Call DMA */
  /* and load data block 2 to Array_A */
  ...
}
DMA_Array_A (start_address, length)
{
  ...
}
```

By modifying the original main to the parallel main, DMA transactions requested by ALG are identified.

Analysis of Addressing for Parallel Computing

In step 5 of Figure 20.24, the relation between behavior and physical addressing should be further identified and specified according to the parallel execution of the algorithm and the parallel architecture of the PVM. To ease the parallel program coding, addressing of parallel DSP algorithms can be analyzed and classified into several coding templates based on the classification of addressing behaviors:

- Algorithms: Convolutions, or transforms, or FSM for CODEC.

- Computing radix: radix 2 algorithms with relatively simple addressing algorithms or radix 4 algorithms with relatively complicated addressing algorithms.

- Data dimensions: 1D or 2D or more than 2D.

- The size of data to use versus the size of PVM: Data size is larger or smaller than the PVM size.

Algorithms based on convolutions are FIR filters, IIR filters, autocorrelations, cross-correlations, and matrix accumulations. The basic data features of these algorithms include fixed-length FIFO buffer (modulo-based addressing), fixed variable lifetime. Sometimes, the volumes of load and store are imbalanced, and the volume of data loading is larger than the volume of data store.

Transformation algorithms include FFT, DCT, and others. Memory access is always challenging for transforms because the volume of loading and store is huge. The relative memory access could be up to one (discussed earlier in this chapter). From the addressing point of view, executing transformation algorithms produces equal times of data memory reads and memory writes. Modulo and variable stride (step) size addressing will be typical templates.

Matrix manipulation can be classified as a kind of matrix transform algorithm. Typical matrix computing algorithms are matrix OP matrix and matrix inversion. However, the addressing for matrix transform is not as regular as that of FFT.

FSM (branch extensive codes) need to be accelerated using special datapath engines or accelerators [17]. Parallel addressing is not necessary because the relative memory access rate is low.

Data required during the execution of an algorithm includes data in the input buffer, the coefficient set, data in the computing buffer, and the results. In most cases, the variable lifetime of input data is low, and input buffers can be used as computing buffers while running the algorithm. There are mainly two situations while planning for PVM memory access:

- **Situation one:** The size of the PVM is sufficient for memory access while running an algorithm. There is no extra swapping between the main memory and the PVM.

- **Situation two:** The size of the PVM is not sufficient for memory access while running an algorithm. Some data, such as input buffer, coefficient buffer, computing buffer, and the results have to be swapped between the main memory and the PVM. The efficiency of the PVM is critical.

Coding Template for PVM Memory Accesses

A coding template for parallel addressing is the code of an addressing algorithm based on a parallel architecture and available for programmers as reference code. If an addressing pattern can be specified for each algorithm and eventually adapted to the PVM structure, the coding of parallel computing on a SIMT machine using PVM will be relatively easy. The challenges should be (1) to find coding templates, and (2) to develop a way (a toolchain) to use the coding templates.

There are principally two kinds of coding templates: the advanced templates, which support CFM (Conflict-Free Memory) parallel access; and the easy templates,

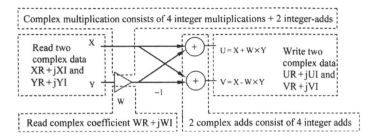

FIGURE 20.25

Radix-2 FFT butterfly.

for memory access without special requirements on conflict-free or other parallel features. Here conflict means two or more data within one physical memory block to be accessed at the same clock cycle [18] and [19]. Because only one data can be accessed from one physical memory in one clock cycle, the conflict happens when multidata from one memory block is required in parallel. Conflict-free parallel access is the ideal case to access N data allocated in N physical memory blocks; all data can be accessed in parallel in one clock cycle.

Requirements for conflict-free access are necessary when the load of parallel memory access is heavy. For example, consider the case while the relative memory access ratio is up to 1. When data cannot be accessed to the register file in time, the datapath will be hungry. A typical case is when running memory access extensive algorithms, for example, FFT (see Figure 20.25). In this case, a butterfly computing consists of 10 basic arithmetic computing and 10 memory accesses (reading 6 integer data and writing 4 integer data). The number of computing operations is equal to the number of memory accesses! In this case, especially when the size of the PVM is less than the size of the FFT data buffer, the requirement for memory bandwidth will be very high. If there is any memory access conflict during FFT computing, the datapath will stall.

Another case that requires conflict-free memory access is when the vector register file is very small. If the register file is large enough, containing all data required by an algorithm to be executed, conflict-free memory access might not be needed. However, when the register file is very small, that is, a register file contains only data for a few execution cycles, the register file will not be able to supply sufficient data to the parallel datapath in time. For example, a register file containing 64 words is too small for an eight-way datapath requiring 16 data words in parallel. In this case, the eight-way PVM has to supply possibly up to eight data in parallel to the register file, and the conflict-free memory access is essential for loading data to the register file and avoiding the datapath becoming hungry.

A coding template example depicted in Figure 20.26 gives CFM for row and column accesses in parallel. In Figure 20.26(a), only rows can be accessed in parallel if there is no permutation while loading data in memory blocks MB0 to MB3. By con-

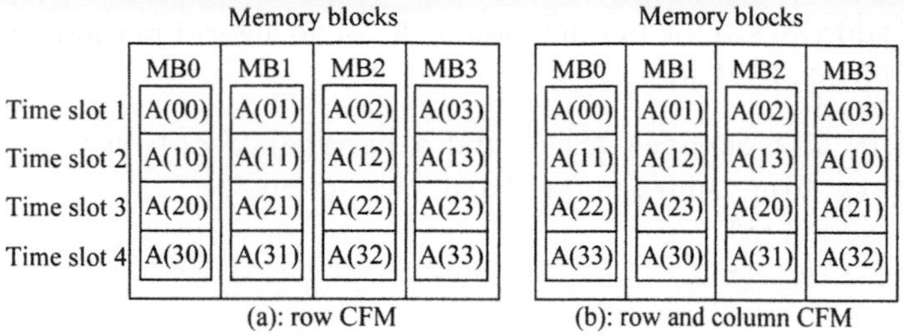

	Memory blocks					Memory blocks			
	MB0	MB1	MB2	MB3		MB0	MB1	MB2	MB3
Time slot 1	A(00)	A(01)	A(02)	A(03)		A(00)	A(01)	A(02)	A(03)
Time slot 2	A(10)	A(11)	A(12)	A(13)		A(11)	A(12)	A(13)	A(10)
Time slot 3	A(20)	A(21)	A(22)	A(23)		A(22)	A(23)	A(20)	A(21)
Time slot 4	A(30)	A(31)	A(32)	A(33)		A(33)	A(30)	A(31)	A(32)
	(a): row CFM					(b): row and column CFM			

FIGURE 20.26

Conflict-free parallel memory access.

ducting permutation while loading data to MB0 to MB3 in a PVM in Figure 20.26(b), both row and column can be accessed in parallel.

A behavioral address coding template (BACT) is specified for modeling a parallel addressing required by an algorithm. The BACT models the address permutation while loading data to memory blocks so that data can be accessed in parallel while running the algorithm. The BACT is coded as a kind of conflict-free memory access by specifying relative positions of each access.

A BACT can be configured to adapt to the target hardware, such as the number of parallel memory blocks and the size of each block. After hardware adaptation, a BACT becomes a coding template of a PVM, a PMCT. Finally, a PMCT can be used for a specific algorithm with specific physical PVM address. An example of a BACT in Figure 20.26(b) is:

```
int array A[N][N]
      /* i denotes a row and j denotes a column */,
  for (i = 0; i < N; i++)
    for (j = 0; j < N; j++)
    {
      Row address = i;
      Column address = (i + j) % N;
    }
```

When it adapts to a PVM with four parallel memory blocks in Figure 20.26, the BACT is converted to PMCT.

```
int array A[N][N] /*i denotes a row and j denotes a column*/,
  For (i = 0; i < 4; i++)
    For (j = 0; j < 4; j++)
    {
      RA = i                  /* the row address */
      CA = (i + j) mod 4 /* the column address */
    }
```

Finally, the physical addresses as the permutation table for an algorithm running on hardware with the 4-block PVM are:

```
int array A[N][N] /*i denotes a row and j denotes a column*/,
int Segment /*starting address to load the array*/
For (i = 0; i < 4; i++)
  For (j = 0; j < 4; j++)
  {
    CA= (i + j) mod 4
    address = segment + 4*I + CA /* the final physical address */
  }
```

20.7.4 Programming Toolchain for P3RMA

Programming tools are essential because manually writing programs for the PVM memory access is always a nontrivial task. Based on discussions of previous sections, the coding template-based programming can be implemented into a toolchain.

Referring to Figure 20.27, the BACT set (a) was the theoretical model of the coding template. BACT is configured to be PMCT (c) based on physical specification of PVM (b) for a physical design of PVM architecture. Six tools (c, e, f, h, i, and j) are needed to guide the parallel programming for memory access based on the PMCT. The relation between these tools for programming based on PVM using P3RMA methodology is illustrated in Figure 20.27.

A coding template library should be prepared, and all the most important BACT should be available there as the architecture-independent model. As soon as the parallel architecture is decided, the PVM specification will be available, and the BACT will be configured to be PMCT. Again referring to Figure 20.27, these PMCT will be the coding templates for parallel programming on the architecture using PVM (see (a), (b), and (c)).

A source code analyzer (vector computing exposer) [20] will be used to expose opportunities of parallel computing. Exposed opportunities will be analyzed further by the Memorizer (memory access exposer), and the memory access flow graph (MFG) will be generated as (g) through step (f) [21].

Matching of MFG (g) with each PMCT (c) will be conducted in the tool Matcher in (h). The matched and selected template from PMCT will be adapted further to the program running in the parallel processor. Two tables will be generated: the DMA linking table for DMA transaction (l) and the PVM address table (k) which will be finally generated from the last two tools (i) and (j).

The code analyzer is the static analyzer in Relief discussed in Chapter 8. The simplified code analyzer is given in Figure 20.28. In this figure, the step for the vector basic block (vbb) identification and vbb cost estimation is important. The memory access patterns of these exposed vector subroutines will be further analyzed by the Memorizer. Other steps were discussed in Chapter 8.

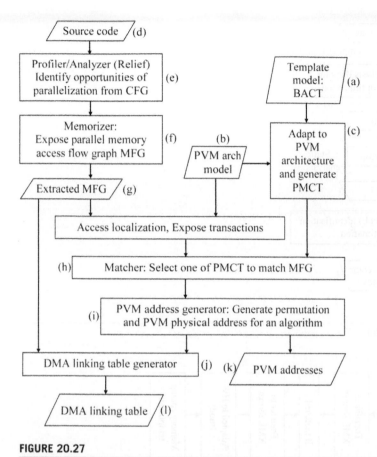

FIGURE 20.27

Tools for P3RMA programming.

Both the code analyzer and the memorizer can be implemented based on GEM (GCC Extension Modules). The memorizer identifies the pointer reference and array references in C. The computing on pointers and array references will be further identified and analyzed. The analyzed pointer computing and array reference computing will be patterns to be matched to the conflict-free parallel addressing templates.

Address coding templates (BACT) will be configured to adapt the parallel hardware and become PMCT. Both the exposed memory access and the PMCT will be represented using XML, eXtensible Markup Language, and coding format for matching. Different matching algorithms can be selected for matching, and the matcher will eventually find a suitable PMCT, see Figure 20.29. Finally, the retargetable physical address of the PMCT will be generated for conflict-free parallel data access of an algorithm.

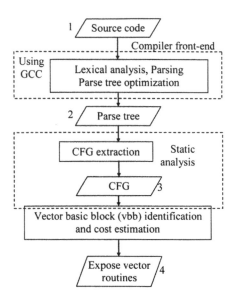

FIGURE 20.28

Simplified code analyzer.

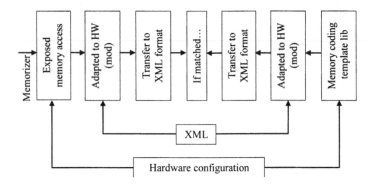

FIGURE 20.29

Principle of the matcher.

Finally, the parallel programming process based on the P3RMA concept is simplified and given in Figure 20.30.

The highlighted box in Figure 20.28 was not discussed in previous text. This step is to localize the memory access to local memories using local variables. After the localization, local memories are loaded to get the data for computing; the DMA access will therefore be exposed.

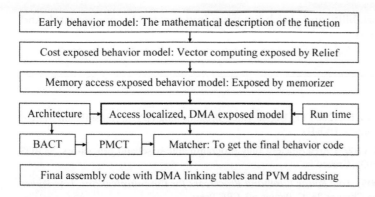

FIGURE 20.30

P3RMA-based programming flow.

20.8 CONCLUSIONS

Parallel digital streaming signal processing was discussed in this chapter. The partition for parallel streaming signal processing can be conducted following either the classical way or the way proposed in this chapter. The challenge of classical partitioning is to minimize the parallelization overheads by trading off the computing load balancing and intertask communication. Classic partitioning for streaming signal processing adds system complexity.

The partitioning method proposed by this chapter is based on the partitioning of complexities. The complexity was divided into three domains: the complexity of control, the data complexity, and the complexity of memory access. To partition on both complexity and tasks, a new architecture was proposed, the SIMT as a scalable processor core. When introducing the SIMT concept to multicore architecture, the concept of heterogeneous multicore architecture based on homogeneous instruction set was introduced. This concept enables easy integration based on the homogeneous instruction set and the high performance from custom heterogeneous multicore architecture.

Data (path) complexity was discussed through cases. Multi-in multi-out and multi-in single-out datapath structures were discussed. A datapath adapts to both structures and can support most low-level DSP algorithms (for example, 1/x, square root). Challenges of data types were also discussed.

Conflict-free parallel memory access is essential for parallel data processing, especially for data access extensive data processing, such as transforms. Complexity of memory access was carefully discussed, and the P3RMA architecture and its programming models were introduced. I believe that the P3RMA architecture will be an efficient solution for future performance demanding low-cost embedded DSP subsystems.

The firmware portability and scalability were investigated by our research and were will be published.

REFERENCES

[1] ETSI EN 302 304 V1.1.1 (2004-11) Digital Video Broadcasting (DVB); Transmission System for Hand-held Terminals (DVB-H).

[2] Nilsson, A., Liu, D. (2007). Area efficient fully programmable baseband processors. SAMOS VII Workshop; SAMOS, Greece.

[3] Liu, D., Tell, E. (2005). *Low Power Baseband Processors for Communications, Low Power Electronics Design*, Chapter 23. C. Piguet, ed. CRC Press.

[4] http://www.mpeg.org.

[5] Telecommunication Standardization Sector of ITU. (2005). Advanced video coding for generic audiovisual services (ITU-T Rec. H.264), March.

[6] IEEE Std 802.11a 1999 Part 11. Wireless LAN Medium Access Control (MAC) and Physical Layer (PHY) specifications.

[7] IEEE Standard for Local and Metropolitan Area Networks, Part 16 (e).

[8] Flachs, B., Asano, S., Dhong, S. H., Hofstee, H. P., Gervais, G., Roy Kim et al. (2006). The microarchitecture of the synergistic processor for a cell processor. *IEEE JSSC*, Jan, 63–70.

[9] O'Brien, K. M., Wu, P., Chen, T., Oden, P. H., Prener, D. A., Shepherd, J. C. et al. (2006). Optimizing compiler for cell processor. *IBM Systems Journal* **45**(1).

[10] http://www.ceva.com.

[11] Khailany B, K., Williams, T., Lin, J., Long, E. P., Rygh, M. Tovey, D. W., and Dally, (2008) W., A Programmable 512GOPS steam processor for signal, image, and video processing, *IEEE J. of Solid-State Circuit*, Vol. **43**.

[12] Salomon, D. (2006). *Data compression, The complete reference, 3rd ed.* Springer.

[13] ISO/IEC IS 10918-1 | ITU-T Recommendation T.81 (JPEG Standard).

[14] http://www.coresonic.com.

[15] Wu, D., Lim, B., Eilert, J., Liu, D. (2007). Parallelization of High-Performance Video Encoding for a Single-Chip Multiprocessor. IEEE ICSPC, UAE.

[16] Wu, D., Li, Y-H., Eilert, J., Liu, D. (2007). Real-Time Space-Time Adaptive Processing on the STI CELL Multiprocessor. European Radar Conference, Munich.

[17] Flodal, O., Wu, D., Liu, D. (2006). Configurable CABAC encoder for multi-standard media compression. *Proc. RAW06*.

[18] Budnik, P., Kuck, D. (1971). The organization and use of parallel memories. *IEEE Transactions on Computers*, December, 1566–1569.

[19] Gössel, M., Rebel, B., Creutzburg, R. (1994). *Memory Architecture and Parallel Access*. Elsevier Science Inc., New York, NY, USA.

[20] Skoglund, B. (2007). Code profiling as a design tool for application specific instruction sets. Master thesis, LiTH-ISY-EX – 07/3987 – SE http://www.da.isy.liu.se/undergrad/exjobb/finished/, Feb. 2007, Linköpng Sweden.

[21] Lundgren, B., Ödlund, A. (2007). Expose of patterns in parallel memory access. Master thesis, LiTH-ISY-EX – 07/4005 – SE, http://www.da.isy.liu.se/undergrad/exjobb/finished/, August, 2007.

[22] Mei, B., Lambrechts, A., Mignolet, J-Y, Verkest, D., Lauwereins, R. (2005). Architecture exploration for a reconfigurable architecture template. *IEEE Design & Test of Computers*, Mar-April, 90–101.

[23] Wiklund, D. Liu, D. (2005). Design, mapping, and simulations of a 3G WCDMA/FDD basestation using network on chip. Proc of the International workshop on SoC for real-time applications, Banff, Canada, July.

[24] Eilert, J., Wu, D., Liu, D. (2007). Efficient complex matrix inversion for MIMO software defined radio. IEEE ISCAS, USA.

[25] Van Berkel, K., Heinle, F., Meuwissen, P. P. E., Moerman, K., Weiss, M. (2004). Vector processing as an enabler for software-defined radio in handsets from 3G+WLAN onwards. 2004 Software Defined Radio Technical Conference, Nov., Scottsdale, Arizona.

[26] Richter, T., Drescher, W., Engel, F., Kobayshi, S., Nikolajevic, V., Weiss, M., Fettweis, G. (2001). A platform-based highly parallel digital signal processor. IEEE 2001 Custom Integrated Circuit Conference, 305–308.

[27] Hinrichs, W., Wittenburg, J. P., Lieske, H., Kloos, H., Ohmacht, M., Pirsch, P. (2000). A 1.3-GOPS parallel DSP for high-performance image-processing applications. *IEEE JSSC* **35**(7). July 2000.

[28] Gatherer, A., Luslander, E. eds. (2002). *The Application of Programmable DSPs in Mobile Communications*. John Wiley & Sons Ltd.

[29] Nurmi, J. (ed.). (2007). *Processor Design, System-On-Chip Computing for ASICs and FPGAs*. Springer.

[30] http://www.fwdconcepts.com.

[31] http://www.streamprocessors.com.

[32] Dai, J. (2008) Automatic parallel memory address generation for parallel DSP computing, Master thesis, LiTH-ISY-EX—07/4065-SE, Linköping University, Sweden, Feb 2008. www.ep.liu.se/index.en.html.

Glossary

3GPP The third-generation partnership project, a group specifies the third-generation mobile communication standards.

90-10% code locality rule (or 10-90% code locality rule) The rule expose that the 10% of instructions take 90% of the runtime and 90% of instructions take 10% of the runtime.

Adaptive filter A filter that adapts its coefficients based on the input signal.

ADC Analog-to-digital converter that changes real-world analog signals into digital code.

Aliasing Refers to distortion effect that causes different continuous signals to become indistinguishable when signal is sampled with too low sampling frequency. It is the effect on a signal when it has been sampled at less than twice its highest frequency.

ALU Arithmetic Logic Unit in a datapath.

Anti-aliasing filter An analog filter that is used prior to sampling signal to limit its bandwidth to less than half the sample rate (generally low pass) to prevent aliasing distortion.

Application programming interface (API) A source code interface, an API specifies details of how two independent computer programs can interact. By using it, an operation system can provide services (another executable program code) requested by running a program.

Application-specific integrated circuit (ASIC) A chip or parts of a chip designed for specific functions or applications.

Application-specific instruction set processor (ASIP) A processor designed for a group of applications with its specific instruction set.

Architecture The software or hardware structure of all or part of a computer system; includes all the detailed components of the system.

ASSP Application-specific standard product. ASSP is an integrated circuit that implements a specific function to a wide market. As opposed to ASICs, which combine a collection of functions and are designed by or for one customer, ASSPs are available as off-the-shelf components. Examples of ASSPs are integrated circuits that perform video and/or audio encoding and/or decoding.

Assembler A software program that creates machine codes from source files that contain assembly language instructions and directives. The assembler substitutes absolute operation codes for symbolic operation codes, and absolute or relocatable addresses for symbolic addresses.

Assembly language A low-level symbolic programming language, closely resembling machine code language and composed of groups of letters, each group representing a single instruction, allowing a computer user to write a program using mnemonics instead of numeric instructions.

Assembly language instructions A low-level symbolic programming language closely resembling machine code language allowing a computer user to write a program using mnemonics instead of numeric instructions.

Attenuation Decrease in magnitude, in particular, reduction of signal strength during transmission.

Autocorrelation The correlation of a signal with a delayed version of itself.

BACT A behavioral address coding template for parallel conflict-free data access.

Band-pass filter A filter that allows only a single range of frequencies to pass through.

Barrel shifter Part of the combinational logic circuit in ALU that allows single cycle shifting and rotating of data words.

BDTI Berkeley Design Technology Incorporation, a company that gives benchmarks of DSP processors on the market; see www.bdti.com.

Benchmark A relative measure of the performance of a processor by running a particular application.

Benchmarking A type of program execution that allows you to track the number of CPU cycles consumed by a specific section of code. It is performance evaluations of different implementations.

Big endian An addressing protocol in which bytes are numbered from left to right within a word. The most significant byte in a word is stored at the memory location with the lowest address. See also *little-endian*.

Biquad (Bi-quad) Typical building block of second-order IIR filters; from the biquadratic equation.

Bit field manipulation Logical, shift, and permutation operations on a bit or group of bits. Used mostly for error correction coding, efficient coding, and framing or deframing a data package.

Bit-reversal addressing Addressing in which bits of an address are reversely shuffled in order to speed up the processing of algorithms, such as Fourier transform algorithms.

Bluetooth A short-range radio link; see www.bluetooth.org.

Boot The process of loading a program into program memory.

Bus A shared electrical connection of multiple hardware modules. Two of the hardware modules use the bus at a time.

Butterfly The smallest constituent part of an FFT, it represents a cross multiplication, incorporating multiplication, sum, and difference operations. The name is derived from the shape of the signal flow diagram.

C A high-level, general-purpose programming language.

Cache A fast memory into which frequently used data or instructions from slower memory are copied for fast access. Fast access is facilitated by the cache's high speed and its on-chip proximity to the CPU.

CISC Complex instruction set computer.

Circular addressing An addressing mode using modulo addressing arithmetic for post-incremental or decremental addressing; also called modular addressing. By using circular addressing, an efficient and low-power FIFO data buffer can be emulated using single port data memory.

Circular buffer Or FIFO buffer, a region of data memory used as a data buffer that appears to wrap around. For example, N-tap FIR requires a circular buffer with size N. Addressing of circular buffers can be accelerated using circular addressing.

Clock cycle The time required for one cycle of the processor's master clock for logic circuit synchronization. It can also be called instruction cycle or machine clock cycle.

Clock tick A second of nature clock. 60 clock ticks consist of a minute. Be aware of the difference between clock tick and clock cycle in this book.

CODEC or Codec Coder-Decoder. Normally means compression and decompression for voice, audio, image, or video signals.

Core Also processor core. The kernel part of a processor including its datapath, the control path, the addressing path, and the memory bus. A processor core does not include memories and peripheral components.

Compiler A translation program that converts a high-level language set of instructions into a target machine's assembly language.

Compress To convert information from one type to another. By eliminating redundancies, compressed information occupies less space and results in greater performance efficiency. For example, video compression with a 100-to-1 ratio is possible with little noticeable change in image quality.

Convolution A time-domain reference for digital filtering that makes extensive use of sum-of-products.

Correlation The comparison of two signals in time, to extract a measure of their similarity.

COTS Commercial off-the-shelf (a component or an available processor on market).

CS Computer science; a class of university education.

CPU Central processing unit; the module of the processor that controls and interprets the machine-language program and its execution.

DAC Digital-to-analog converter.

Datapath A collection of execution units including the lever one data buffer. In a DSP processor, a datapath includes at least a register file, an arithmetic unit, a logic unit, a shift unit, and a MAC.

Data Stationary Coding The instruction (or decoded control signals) travels together with the associated data (operands and intermediate result) in each pipeline stage and controls the sequence of operations on these data in each pipeline stage, e.g. superscalar.

Debugger A software interface used to identify and eliminate mistakes in a program; here, a collection of emulators and GUI. The emulator consists of the instruction set simulator and behavior model of involved peripheral modules.

Design an assembly instruction To specify all microoperations of the instruction, specify mnemonics to expose and select configurable microoperations, and specify the binary code for the instruction.

Die An area of silicon containing an integrated circuit. The popular term for a die is chip.

Digital signal processor A special-purpose microprocessor designed to handle signal-processing applications very efficiently.

Direct memory addressing The direct absolute address is carried in the instruction.

DIT Decimation in time, a way of butterfly-computing for FFT. Another way of butterfly-computing for FFT is DIF, decimation in frequency.

DSP Digital signal processing or digital signal processor.

DSP benchmark A relative measure of the performance of a DSP processor by running a particular optimized assembly subroutine. It is the measure of an instruction set, instead of its compiler.

Direct memory access (DMA) A mechanism by which a device other than the host processor can access a memory in a processor and transfer data without the processor's execution of load and store instructions.

Discrete cosine transform (DCT) A fast Fourier transform used in manipulating compressed still and moving picture data. See also *fast Fourier transform*.

Dynamic random-access memory (DRAM) a kind of random access memory that stores each bit of data in a capacitor within an integrated circuit. Since a capacitor leak charge, the stored information must be refreshed periodically. Because of this refresh requirement, it is a dynamic memory versus static memory. Its advantage over SRAM is its very high density.

DVB Digital Video Broadcasting, a European standard for digital video television.

Dynamic range The ratio of maximum and minimum signal levels.

EE Electrical engineering; a class of education.

Embedded system A system embedded inside another system or a product.

Epilogue Part of the program that terminates the subroutine after running an algorithm.

Exception An unplanned-for event resulting from a software operation, such as division by zero. It gives a condition that is handled outside the normal program flow of a task. An exception gives a hardware interrupt. After the interrupt for exception handling, the processor does not continue running the interrupted program.

Fast Fourier transform (FFT) An efficient method of computing the discrete Fourier transform algorithm, which transforms functions between the time-domain and frequency-domain. The time-to-frequency-domain is called the forward transform, and the frequency-to-time-domain is called the inverse transform.

Fast interrupt An interrupt suitable for hard-real-time SW with limited time budget. A fast interrupt should have very low interrupt latency, and its service should be very simple and very short. The interrupt does not change the data of the current running program so that the context saving can be minimized.

FIFO buffer First-in, first-out buffer. A part of memory in which data is stored and then retrieved in the same order as it was stored.

FIR Finite impulse response.

Finite impulse response (FIR) filter A filter that includes no feedback and is unconditionally stable.

Fixed-point An arithmetic system using fixed number of digits for computing and storage. The inputs and outputs of the system are based on a fixed number of digits.

Fixed-point processor A processor not using floating-point data format for computing and storage.

Floating-point A number scheme that codes a value with a fraction and an exponent and allows a high signal dynamic range.

Floating-point operations per second (FLOPS) A measure of performance for a floating-point processor. It usually is noted as MFLOPS or Million FLOPS.

Floating-point processor A processor not using fixed-point data format for computing and storage.

Frequency The number of cycles per unit of time, denoted by hertz (Hz). One Hz equals one cycle per second.

Full duplex Communications in two directions simultaneously.

GSM Global System for Mobile Communications. The European standard for the second-generation digital cellular service.

Graphical user interface (GUI) A type of user interface between people and computer. It provides windows, menus, dialog boxes, icons, lists, and options that allow you to start a program or perform a task by pointing to a pictorial representation, selecting an item using a mouse, or using a keyboard.

Guard bits Extra bits as sign extension. Before computing, they are the copy of the sign bit of the operand. During computing, they are used to guard the result to prevent overflow.

Half duplex Communications in two directions, but only one at a time.

Hardware component A low-level hardware device with a single simple function. It is also called hardware primitive or RTL component. For example, a hardware component can be a multiplier, full adder, or logic gate.

Hardware loop An instruction set feature of a DSP processor. In a hardware loop, one or multiple instructions repeat under the control of hardware loop control hardware to minimize overheads of iterations.

Hardware module A low-level hardware function block that consists of hardware components and interconnections. Different from a hardware component, a hardware module gives complicated multiple functions. An ALU is a hardware module that gives different arithmetic, logic, and shift operations.

Hardware stack A push-down stack implemented in hardware instead of emulated in SW using data memory. It stores necessary break information, PC and flags, for resuming the interrupted program by a fast interrupt.

HWMR Hardware multiplexing ratio. It is the definition of complexity of SW for streaming signal processing. HWMR could be the machine clock rate over the signal sampling rate. When processing streaming data frames, HWMR could also be the cycle count to process a data packet, for example, a picture frame.

High pass filter A filter that allows high frequencies to pass through.

Host processor A general-purpose processor used for running simulations and debugging the target processor. The target processor is either **Junior** or **Senior** in this book.

ICT Information and Communication Technology, a group of technologies.

Image processing To compress or enhance an image or extract information or features from an image.

Index addressing The address is calculated and stored in a general register as the offset of a data segment.

Indirect addressing The address is referred or carried by a data or address pointer register.

Infinite impulse response (IIR) filter A filter that incorporates data feedback; also called a recursive filter.

Instruction set design Design a complete set of assembly instructions and its memory bus architecture, its function coverage, instruction level accelerations, addressing models, ways to specify mnemonics, and ways for binary coding.

Integrated circuit The formal name for a die, or a chip. Its name resulted from the integration of previously separate transistors, resistors, and capacitors, all on a single chip.

Interlocked pipeline The pipeline is not visible to the programmer or compiler; for example, the execution in a superscalar with OOO (out-of-order) execution. The hardware takes care of the data, control, and resource dependence by hardware dependence control.

Interrupt A signal sent by hardware or software to a processor requesting attention. An interrupt tells the processor to suspend its current operation, saves the current task status, and executes a particular set of instructions.

Interrupt latency The time from an interrupt request until the processor starts the interrupt service.

ITU International Telegraph Union (formerly CCITT) {www.itu.org}.

JPEG Joint Photographic Experts Group. JPEG is also an image compression standard.

JTAG Joint in Test Acting Group.

Junior A simple DSP processor used as an example in this book.

Kernel A small portion of code (assembly program) that forms the heart of an algorithm.

Latency The time delay between an event occurring and the reaction to the event.

Linker A program that combines separate binary object code modules into a single executable object code and resolves cross references (global variables) between modules.

Least mean square (LMS) A typical algorithm to reach the filter coefficient set for adaptive filtering.

Little-endian The definition of ordering of bytes within a multibyte word, least-significant byte first.

Low pass filter A filter that allows low frequencies to pass through.

LSB Least significant bit. The bit having the smallest effect on the value of a binary numeral, usually the rightmost bit.

LTE Long Term Evolution of 3GPP, a fourth-generation mobile communication standard.

Mantissa A component of a floating-point number consisting of a fraction and a sign bit. The mantissa represents a normalized fraction whose binary point is shifted by the exponent.

MCU Microcontroller; a processor or a processor core for embedded applications.

Memory The computer's physical work space that stores programs and data needed to accomplish the tasks executed by the processor.

Memory-based I/O Exchanges data using a shared memory. Data is stored in the shared memory from one instruction domain and is loaded by another instruction domain.

Memorizer A static code analyzer to expose and analysis data access in a C code.

Microarchitecture Architecture of a hardware module and functional description of the hardware module.

Microarchitecture design Designing a hardware module and specifying its function for RTL coding.

Microcode instruction An instruction that directly carries control signals supplying the maximum datapath flexibility with the highest code cost.

Microprocessor An integrated circuit component that can be programmed with stored instructions to perform a wide variety of functions.

Million instructions per second (MIPS) A measure of performance for a fixed-point processor. It refers to the number of instructions performed each second.

MPEG (Motion Picture Experts Group) An industry standard for compressing video and audio. MPEG1, MPEG2, and MPEG4 are included in the standard.

MP3 MPEG 2 audio layer 3. It is an audio coding and decoding standard. It is a most popular audio CODEC. Usually, it is the nickname of handheld MP3 audio player.

MPU Microprocessor unit.

Multiplexing A process of transferring more than one set of signals at a time over a single wire or communications link (also known as muxing).

MAC Multiply and accumulate. It could be an assembly instruction or a hardware module.

Modulo addressing See *circular addressing*.

OEM Original equipment manufacturer; here, customers of semiconductor companies.

OFDM Orthogonal frequency division multiplexing, a digital multi-carrier modulation.

OP Operation.

Open pipeline The pipeline is fully visible to the programmer or compiler. The programmer or compiler is responsible for handling the data, control, and resource dependence.

Overflow A condition in which the result of an arithmetic operation exceeds the capacity of the register that stores the result.

P3RMA Programmable parallel memory architecture for predictable random memory access.

Parallel processing An efficient form of information processing that emphasizes the exploitation of concurrent events in the computing process. Concurrency implies parallelism, simultaneity, and pipelining. Parallel events may occur in multiple resources during the same time instant; pipelined events are attainable in a computer system at various processing levels. Parallel processing demands concurrent execution of many programs in the computer as a cost-effective means to improve system performance.

Pass band The frequency range of a filter through which a signal may pass with little or negligible attenuation.

Peripheral module Peripheral modules are hardware modules around a processor core that execute auxiliary actions in the system; for example, input/output, backing store, handling external events, and handling communications to others. Typical peripheral modules of a DSP processor core include interrupt handler, timer, DMA controller, serial ports, and parallel ports.

PC Program counter. PC is a register that holds the current state of the program. The PC contains the address of the next instruction to be fetched.

PMCT An adaptation of a BACT to hardware architecture of a parallel vector memory.

Processor A processor consists of its processor core, memories, and peripheral modules. See also *microprocessor*.

Process technology The semiconductor manufacturing flow that creates or fabricates an integrated circuit. CMOS process technology (complementary metal oxide on silicon) enables low-power, high-density chips.

Prologue Part of a program preparing for running an algorithm.

PSTN Public switched telephone network.

Reduced instruction set computer (RISC) Computer architecture with reduced chip complexity by using simpler instructions.

Register-based I/O Data is transferred between instruction domains via a register. The register can be pointed (addressed) as a register or as a memory word.

Register file A small area of high-speed memory, located within a processor or electronic device, that is used for temporarily storing data or instructions. Each register is given a name, contains a few bytes of information, and is referenced by programs.

Relief A C source code analyzer for both static and dynamic code profiling.

Resolution Measure of accuracy or dynamic range of an A/D or D/A converter.

RF Here, Register file. Commonly, Radio frequency.

Round-robin scheduling A method for scheduling the execution of a set of tasks of equal priority. The tasks share a processor or a DMA module on a rotating basis. Each of the tasks takes its turn executing for a time interval that is roughly equal for all tasks.

QCIF Quarter common intermediate format,: 176×144 pixels in a picture frame.

Sampling The process of converting continuous signals into discrete values.

Sampling rate The rate at which an analog signal is sampled for conversion to and from the digital domain. The sampling rate is measured as the number of samples per unit of time.

Senior Used as an example in this book, it is a single-issue DSP processor with a comprehensive assembly instruction set.

Serial port A DSP peripheral used for sending and receiving data samples sequentially in order to communicate with serial devices such as A/D and D/A converters and codecs, microprocessors, and DSP.

Simulator A development and verification tool that simulates the behavior of a device.

Stop band The frequency range of a filter through which a signal may NOT pass, and where it experiences sufficient attenuation.

Subroutine A routine that is called from an applications program by means of a standard function call.

Syntax The grammatical and structural rules of a language. All higher-level programming languages possess a formal syntax.

Task A program element as part of job or part of an action.

Time-domain The representation of the amplitude of a signal with respect to time.

Time-stationary architecture There is an assumption that the data is available in time-stationary architecture whenever an instruction is fetched and decoded. It is up-to the compiler and instruction set simulator to give an off-line guarantee: that the data will not be dependent during the runtime.

TSMD Typical single MAC DSP (TSMD) processor. Here, a TSMD is used as the benchmarking reference.

Variable A symbol representing a quantity that can assume any of a set of values.

Virtual memory A program memory uses more memory than a computer actually has available as RAM. This is accomplished by using a swap file on disk to augment RAM. When RAM is not sufficient, part of the program is swapped out to a disk file until it is needed again. The combination of the swap file and available RAM is the virtual memory.

WCDMA (Wideband Code Division Multiplex Access) The third-generation mobile phone proposed by EU.

WiMax Worldwide Interoperability for Microwave Access, a radio communication technology. See IEEE802.16d/e

WLAN (wireless local area network) The IEEE standards 802.11a/b/g wireless data connection network.

Senior Assembly Instruction Set Manual

This appendix describes the **Senior** processor architecture and its assembly language. The assembly language programming tools (assembler, assembly instruction set simulator) and related documents can be found at the web page of this book.

Contents of Appendix:

The appendix can be found at URL:
http://books.elsevier.com/companions/9780123741233

Index

Printed and bound by CPI Group (UK) Ltd, Croydon, CR0 4YY

03/10/2024

01040317-0008